D1028877

CHEMICAL ENGINEERING THERMODYNAMICS

McGraw-Hill Chemical Engineering Series

BUILDING THE LITERATURE OF A PROFESSION

Fifteen prominent chemical engineers first met in New York more than 60 years ago to plan a continuing literature for their rapidly growing profession. From industry came such pioneer practitioners as Leo H. Baekeland, Arthur D. Little, Charles L. Reese, John V. N. Dorr, M. C. Whitaker, and R. S. McBride. From the universities came such eminent educators as William H. Walker, Alfred H. White, D. D. Jackson, J. H. James, Warren K. Lewis, and Harry A. Curtis. H. C. Parmelee, then editor of *Chemical and Metallurgical Engineering*, served as chairman and was joined subsequently by S. D. Kirkpatrick as consulting editor.

After several meetings, this committee submitted its report to the McGraw-Hill Book Company in September 1925. In the report were detailed specifications for a correlated series of more than a dozen texts and reference books which have since become the McGraw-Hill Series in Chemical Engineering and which became the cornerstone of the chemical engineering curriculum.

From this beginning there has evolved a series of texts surpassing by far the scope and longevity envisioned by the founding Editorial Board. The McGraw-Hill Series in Chemical Engineering stands as a unique historical record of the development of chemical engineering education and practice. In the series one finds the milestones of the subject's evolution: industrial chemistry, stoichiometry, unit operations and processes, thermodynamics, kinetics, and transfer operations.

Chemical engineering is a dynamic profession, and its literature continues to evolve. McGraw-Hill and its consulting editors remain committed to a publishing policy that will serve, and indeed lead, the needs of the chemical engineering profession during the years to come.

THE SERIES

Bailey and Ollis: *Biochemical Engineering Fundamentals*
Bennett and Myers: *Momentum, Heat, and Mass Transfer*
Beveridge and Schechter: *Optimization: Theory and Practice*

Carberry: *Chemical and Catalytic Reaction Engineering*
Churchill: *The Interpretation and Use of Rate Data—The Rate Concept*
Clark and Davidson: *Manual for Process Engineering Calculations*
Coughanowr and Koppel: *Process Systems Analysis and Control*
Daubert: *Chemical Engineering Thermodynamics*
Fahien: *Fundamentals of Transport Phenomena*
Finlayson: *Nonlinear Analysis in Chemical Engineering*
Gates, Katzer, and Schuit: *Chemistry of Catalytic Processes*
Holland: *Fundamentals of Multicomponent Distillation*
Holland and Liapis: *Computer Methods for Solving Dynamic Separation Problems*
Johnson: *Automatic Process Control*
Johnstone and Thring: *Pilot Plants, Models, and Scale-Up Methods in Chemical Engineering*
Katz, Cornell, Kobayashi, Poettmann, Vary, Elenbaas, and Weinaug: *Handbook of Natural Gas Engineering*
King: *Separation Processes*
Klinzing: *Gas-Solid Transport*
Knudsen and Katz: *Fluid Dynamics and Heat Transfer*
Luyben: *Process Modeling, Simulation, and Control for Chemical Engineers*
McCabe, Smith, J. C., and Harriott: *Unit Operations of Chemical Engineering*
Mickley, Sherwood, and Reed: *Applied Mathematics in Chemical Engineering*
Nelson: *Petroleum Refinery Engineering*
Perry and Chilton (Editors): *Chemical Engineers' Handbook*
Peters: *Elementary Chemical Engineering*
Peters and Timmerhaus: *Plant Design and Economics for Chemical Engineers*
Probstein and Hicks: *Synthetic Fuels*
Ray: *Advanced Process Control*
Reid, Prausnitz, and Sherwood: *The Properties of Gases and Liquids*
Resnick: *Process Analysis and Design for Chemical Engineers*
Satterfield: *Heterogeneous Catalysis in Practice*
Sherwood, Pigford, and Wilke: *Mass Transfer*
Smith, B. D.: *Design of Equilibrium Stage Processes*
Smith, J. M.: *Chemical Engineering Kinetics*
Smith, J. M., and Van Ness: *Introduction to Chemical Engineering Thermodynamics*
Thompson and Ceckler: *Introduction to Chemical Engineering*
Treybal: *Mass Transfer Operations*
Valle-Riestra: *Project Evolution in the Chemical Process Industries*
Van Ness and Abbott: *Classical Thermodynamics of Nonelectrolyte Solutions: With Applications to Phase Equilibria*
Van Winkle: *Distillation*
Volk: *Applied Statistics for Engineers*
Walas: *Reaction Kinetics for Chemical Engineers*
Wei, Russell, and Swartzlander: *The Structure of the Chemical Processing Industries*
Whitwell and Toner: *Conservation of Mass and Energy*

CHEMICAL ENGINEERING THERMODYNAMICS

Thomas E. Daubert

Professor of Chemical Engineering
The Pennsylvania State University

McGraw-Hill Book Company

New York St. Louis San Francisco Auckland Bogotá Hamburg
Johannesburg London Madrid Mexico Montreal New Delhi
Panama Paris São Paulo Singapore Sydney Tokyo Toronto

This book was set in Times Roman by Science Typographers, Inc.
The editors were Kiran Verma and Madelaine Eichberg;
the production supervisor was Phil Galea.
The drawings were done by ANCO/Boston.
R. R. Donnelley & Sons Company was printer and binder.

CHEMICAL ENGINEERING THERMODYNAMICS

1234567890DOCDOC898765

ISBN 0-07-015413-9

Library of Congress Cataloging in Publication Data

Daubert, T. E. (Thomas E.), date
Chemical engineering thermodynamics.

(McGraw-Hill chemical engineering series)
Includes bibliographies and index.
1. Thermodynamics. I. Title.
TP363.D38 1985 660.2'969 84-21808
ISBN 0-07-015413-9

To Nancy, Steve, Mark, and Linda

CONTENTS

PREFACE

Thermodynamics is the science which deals with the transformation of energy from one form to another and with the laws governing such transformations. Most of the fundamental laws were known a century ago; however, continuous advancements have been made such that today thermodynamic literature is so extensive that no one can claim to be an expert in all phases of the subject. This text attempts to summarize the aspects of thermodynamics necessary for modern chemical engineers to formulate, design, operate, and control chemical processes of the 1980s. No attempt is made to advance new theories, although salient features of current research are summarized. The major purpose of this work is to present thermodynamics in as simple a form as practicable to solve problems of industrial importance. The essential components of thermodynamics, both theory and practice, necessary for practicing chemical engineering are included.

The book is arranged beginning with an introductory chapter which provides necessary background and definitions of common terms and quantities of thermodynamics. The first major chapter treats equations of state, both analytical and corresponding-states approaches, in some detail, emphasizing the methods which are now used by industry rather than the historical perspective offered by most texts. This approach allows immediate use of any equation of state in subsequent chapters on the first and second laws of thermodynamics, phase equilibria, and chemical equilibria. This treatment is unusual in that most texts relegate equations of state to a later chapter following the first- and second-law treatment, requiring that all topics must again be considered since only the ideal-gas law is available for the first pass.

A unique feature of the book is the chapter on the estimation of auxiliary physical properties necessary for thermodynamic calculations. Most equations of state and other thermodynamic methods require critical properties and third parameters. Densities, molecular weights, normal boiling points, and vapor pressures are often also required. Treatment of mixtures is quite necessary in processing calculations. Quite often these data are not readily available, and sometimes experimental values do not exist. Thus, drawing on work as coauthor of the "Technical Data Book—Petroleum Refining" for the petroleum industry through the American Petroleum Institute and the "Data Prediction Manual" for the chemical industry through the AIChE Design Institute for Physical Property Data, the most up-to-date generalized prediction methods for each of these properties has been included in a separate chapter in the text.

The text is arranged in nine chapters. Treatments tend to be brief so that students do not get lost in the prose. Line drawings, diagrams, plots, and examples are included throughout. Exercises are liberally located throughout the text following the appropriate subsections. Such exercises will allow students to more clearly determine if they understand the material and are valuable for both normal classroom instruction or self-study of the material. Lists of problems follow each chapter. References to topics discussed and bibliographies of the most important current literature, reviews, and compendia are included in each chapter. Appendixes listing pure-component physical properties of common compounds, selected thermodynamic properties of model compounds, thermodynamic properties of steam, and conversions among different systems of units are included.

The text is designed for first courses in chemical engineering thermodynamics starting at the late sophomore or junior college level. No previous thermodynamic study is required. Four to five semester credits are required for total coverage. A balanced treatment is attempted between essential thermodynamic principles and the methods actually used in current practice to calculate thermodynamic properties and to use modern equations of state.

Most sections of the book have been tested in the classroom by the author and by some of his colleagues, whose solicited critical comments led to changes in the manuscript. The suggestions of these faculty as well as the tireless work of my secretary, Sue Decker, are gratefully acknowledged.

Thomas E. Daubert

LIST OF SYMBOLS

All symbols used in more than one subsection of the book are listed; in the latter case the variables are defined in the particular section. Nomenclature used conforms to standard symbols of the American Institute of Chemical Engineers wherever possible. Symbols used for more than one variable are easily differentiated by the context of the use.

Capital Letters

A = Helmholtz free energy
A = cross-sectional area
A = constant in van Laar and Margules equations
B = second virial coefficient
B = general property for correction terms of Lydersen corresponding-states method
B = constant in van Laar and Margules equations
C = heat capacity
C = number of components in Gibbs phase rule
C = third virial coefficient
C = interaction coefficient of Chueh and Prausnitz for equations of state
D = deviation term for Lydersen corresponding-states method
E = internal energy
E = electrode potential
F = force
F = value of faraday
F = general function
G = Gibbs free energy
H = enthalpy
H = Henry's law constant
K = equilibrium constant or K value
M = molecular weight
P = pressure

$P°$ = vapor pressure
P = general property
Q = general function $[f(y)]$
Q = heat
Q' = $Q/\dot{m}$
R = general function $[f(x)]$
R = universal gas constant
R = number of independent equilibrium reactions in Gibbs phase rule
S = entropy
S = distance
S = Soave [Eq. (2.22)] or Peng-Robinson [Eq. (2.25)] function of acentric factor
SC = number of special constraints in Gibbs phase rule
T = temperature, absolute
U = alternate symbol for internal energy (used in certain other works)
V = volume; $\overline{V}$, specific volume
V = variance in Gibbs phase rule
W = work
W_s = shaft work
W_s' = $W_s/\dot{m}$
X = molar ratio of two components of mixture
X = general property
Y = molar ratio of two components of mixture
Y = general property
Z = compressibility factor

Lowercase Letters

a = activity
a = van der Waals and Redlich-Kwong type of equations-of-state constants in attractive terms
b = van der Waals and Redlich-Kwong type of equations-of-state constants in pulsive terms
c = constants in Wilson equation
d = symbol for derivative
e = base of natural logarithms = 2.71828
f = fugacity
f = function
g = acceleration of gravity, distance/time^2
h = heat; h_f, frictional heat
k = ratio of constant pressure to constant-volume heat capacity
m = molality of solution
m = mass
$\dot{m}$ = mass flow rate, mass/time
n = number of moles for compound or number of atoms for element
q = heat

r = compression ratio
s = general variable
t = general variable
t = temperature, Celsius or Fahrenheit
u = velocity, distance/time
v = general variable
w = general variable
x = mole fraction in liquid phase
x = general variable
y = mole fraction in vapor phase
y = general variable
z = distance above a reference plane
z = general variable

Greek Letters

α = Reidel's and Soave's function (Chap. 2)
β = coefficient of thermal expansion
γ = activity coefficient
δ = solubility parameter
δ = symbol indicating change or difference
Δ = symbol indicating change or difference
ε = reaction coordinate
η = efficiency
κ = isothermal coefficient of compressibility
λ = latent heat
μ = chemical potential
ν = stoichiometric coefficient
π = pi = 3.14159
Π = symbol indicating product of several variables
ρ = density
Σ = symbol indicating sum of a series of variables
ϕ = fugacity coefficient = f/p
ϕ = volume fraction of component
ϕ = number of phases in Gibbs phase rule
χ = Flory interaction parameter
ω = acentric factor

Special Symbols

ln = natural logarithm with base e
log = logarithm with base 10
∂ = sign for partial derivative

Subscripts

abs = absolute
aq = aqueous solution

b = normal boiling point
c = critical
C = combustion
f = formation
fus = fusion
g, G = gas phase
i = component of mixture
j = component of mixture
l, L = liquid phase
m = mean
M = mixing
P = constant pressure
pc = pseudocritical property
Pr = products
r = reduced
R = reaction
RA = Rackett number
Re = reactants
rel den = relative density
rev = reversible
rxn = reaction
s = system
s = shaft when used with W
s = saturation when used with pressure
S = surroundings
S = constant entropy when used with derivative
t = constant temperature
T = total
T = constant temperature when used with derivative
v = vapor phase
V = constant volume
vap = vaporization
0 = base or reference condition of given thermodynamic property
1, 2, etc. = states

Superscripts

$^{\circ}$ = standard state
or $''$ = modified variable
$^{(0)}$ = simple fluid term for Pitzer corresponding-states methods
$^{(1)}$ = correction term for Pitzer corresponding-states methods
$^{(h)}$ = heavy reference fluid for Lee-Kesler-Pitzer corresponding-states method
$*$ = ideal-gas property
$\bar{}$ = molar quantity; solution property with fugacity
$\tilde{}$ = partial molar quantity

CHAPTER
ONE

PURPOSE, USEFULNESS, AND DEFINITIONS OF THERMODYNAMICS

1.1 BACKGROUND—STATE OF THE ART

Thermodynamics can be divided into several categories—classical, statistical, and irreversible; only the first is discussed in detail in this text. *Classical thermodynamics* begins its analysis with macroscopic properties of matter such as pressure, temperature, and volume and presents its properties on the basis of a given mass or mole of material. Properties can be estimated by using measurable physical properties of the substance. *Statistical thermodynamics* starts on the molecular or microscopic level and attempts to calculate the thermodynamic properties based on observed energy levels, statistical distributions of these levels, and spectroscopic data. Great advances have been made in this field, although the current state of the science is still primarily in the realm of the simple molecule for which necessary data can be obtained to calculate thermodynamic properties of such materials. *Irreversible* or *nonequilibrium thermodynamics* is a relatively new field which attempts to directly (rather than indirectly as does classical thermodynamics) treat systems which are not at equilibrium such as heat, electric, or mass flow systems and systems which are chemically reacting. Applications of these principles are not yet at a stage where they would be useful for practical calculations.

1.2 USEFULNESS OF THERMODYNAMICS TO THE CHEMICAL ENGINEER

Standard chemical engineering practice requires the use of thermodynamics in almost every situation. Consider any chemical process system which normally contains chemical reactors, separation equipment, flow-moving devices, and heat transfer equipment throughout. No single piece of equipment named can be designed or analyzed without the use of thermodynamics principles. The brief listing below gives some examples:

Energy balances on every heat exchanger
Energy balance on an entire process
Pumping- or compressing-power design
Turbine or heat engine analysis
Calculation of pressure loss in pipes and vessels
Calculation of separation in distillation columns, extractors, and absorbers
Calculation of maximum conversion and equilibrium product distribution in reactors
Calculation of heat effects in reactors
Calculation of work necessary to carry out any process

Thus, it is apparent that a thorough knowledge of basic thermodynamics is necessary for success in chemical engineering.

1.3 BASIC QUANTITIES AND UNITS OF THERMODYNAMICS

The basic quantities of importance in thermodynamic calculations are time, length, mass, force, and temperature. Several systems of units have been used, with current usage limited to three—the English system used by chemical engineers for decades, the cgs system used by scientists for centuries, and the Système International (SI) which recently has been adopted by the United States and by the chemical engineering profession through the AIChE and other professional organizations as well as by the U.S. Congress to replace the English system entirely by the late 1980s. This book will use the SI system with proper references to and some examples in English units where appropriate. The basic quantities have units in the two systems as noted below.

	SI	English
Time	second (s)	second (s), hour (h)
Length	meter (m)	foot (ft)
Mass	kilogram (kg)	pound-mass (lbm)
Force	newton (N)	pound-force (lbf)
Temperature	kelvin (K)	°F, °R

An exception to the use of these units is that degree Celsius (°C) may be used for the kelvin in certain circumstances. Certain important derived properties are listed below.

	SI	English
Area	square meter (m^2)	square foot (ft^2)
Density	kilogram per cubic meter (kg/m^3)	pound-mass per cubic foot (lbm/ft^3)
Energy (heat)	joule (J)	British thermal unit (Btu)
Power	watt (W)	Btu per time
Pressure	newton per square meter (N/m^2) or pascal (Pa)	pound-force per square foot (lbf/ft^2)

To gain a better understanding of the magnitude of the various units, the conversions for some common quantities from the English to the SI system are given in Table 1.1. A complete list of conversion factors and units for conversion from common systems to the SI system as compiled by the AIChE (1971) and the American Petroleum Institute (extant 1982) is given in App. D.

Example 1.1 A process is being carried out at 700 K, 2050 kPa total pressure, and consumes 10 J of energy. Your supervisor only understands the English system of units. Convert these values for his/her use.

$$700\ \text{K}\frac{1.8°\text{R}}{\text{K}} - 460°\text{R} = 800°\text{F}$$

$$2050\ \text{kPa} = 2.050 \times 10^6\ \text{N/m}^2$$

$$2.050 \times 10^6 \frac{\text{N}}{\text{m}^2}\ \frac{\text{lbf}}{4.4482\ \text{N}}\ \frac{(0.3048\ \text{m})^2}{\text{ft}^2}\ \frac{1\ \text{ft}^2}{144\ \text{in}^2} = 297\frac{\text{lbf}}{\text{in}^2}$$

$$10^9\ \text{J}\frac{\text{Btu}}{1055.1\ \text{J}} = 0.948 \times 10^6\ \text{Btu}$$

Table 1.1 Common English to SI unit conversions*

Seconds = $3.600(10^3)$ (h)
Meters = 0.3048 (ft)
Kilograms = 0.4536 (lbm)
Newtons = 4.4482 (lbf)
Kelvins = (°F + 459.67)/1.8
Square meters = $9.2903(10^{-2})$ (ft^2)
Kilograms per cubic meter = 1.6018(10) (lbm/ft^3)
Joules = $1.0551(10^3)$ (Btu)
Watts = 0.2931 (Btu/h)
Newtons per square meter = $4.7880(10^{-1})$ (lbf/ft^2)

* To obtain units on the left, multiply the quantity in units on the right by the given factor.

Exercise 1.A Convert the following properties of ethylcyclohexane to the SI system of units.

Normal boiling point = 269.2°F
Vapor pressure at 100°F = 0.483 lb/in^2
Critical volume = 0.064 ft^3/lb
Specific gravity = 0.7922
Liquid density at 60°F = 6.596 lb/gal
Kinematic viscosity at 100°F = 0.861 cSt
Ideal gas heat capacity at 60°F = 0.3260 Btu/(lb · °F)
Normal heat of vaporization = 131.5 Btu/lb

1.4 BASIC DEFINITIONS OF IMPORTANCE IN THERMODYNAMICS

During the study of thermodynamics many terms which are important to recognize appear and reappear repeatedly. This section briefly defines terms of importance, many of which will be further defined and amplified throughout the book.

1.4.1 Pressure, Volume, and Temperature

Pressure of a fluid acting on a surface is the normal force exerted by a fluid per unit area of the surface of the fluid. This force per unit area can be measured by various devices including a deadweight gauge, a Bourdon tube pressure gauge, or a manometer. The deadweight gauge is used to calibrate the other two devices. Manometers are used for high-accuracy work at relatively low pressures, while Bourdon tubes are used at higher pressures. Such devices are discussed in detail in many sources; Hala et al. (1967) give a reasonably comprehensive but brief treatment.

Volume is the amount of three-dimensional space occupied by a substance and is dependent on the mass of material. *Specific volume* is the volume per unit mass or volume per mole of material. *Density* is the reciprocal of the specific volume.

Temperature may be simply defined as a variable directly proportional to the degree of hotness of a substance. Alternatively, temperature might be defined as the property which has equal magnitude in systems where thermal flow does not exist.

Energy continuously passes between bodies in thermal contact. If two bodies A and B are brought into contact where the temperature of A is greater than the temperature of B, net energy will flow from A to B until the temperatures of the two bodies are equal. For three bodies in contact, consider two illustrative cases. If $T_A = T_B$ and $T_B = T_C$, then $T_A = T_C$. If $T_A > T_B$ and $T_B \geq T_C$, then $T_A > T_C$. Thus, the net energy transferred from a cold body to a hot body is always less than the energy which is simultaneously transferring from hot to cold. This principle has often been called the *zeroth law of thermodynamics*.

Thermodynamic temperature is defined as the property independent of any particular material which increases positively with increasing degree of hotness. A temperature scale can be defined by choosing a reference substance and noting the property which is a function of pressure and volume and remains constant as the substance changes equilibrium states. As will be shown later, fixing any two of the pressure, volume, and temperature variables for a pure substance will fix the other variable. Denbigh (1981) discusses these principles in more detail.

1.4.2 Intensive vs. Extensive Properties

An *intensive* property of a material is a property for which the value is independent of the size or extent of the system. Conversely, an *extensive* property depends on the extent or size of the system. For example, volume (m^3) is an extensive property, whereas specific volume ($m^3/kmol$) and density (kg/m^3) are intensive properties. Pressure, temperature, and most tabulated thermodynamic properties are intensive and not additive, while the important quantities of heat and work are extensive and additive. The distinction between the two will become important in the process of thermodynamic calculations.

1.4.3 Thermodynamic Systems

A thermodynamic *system* is that part of the universe we are considering. It is separated from the *surroundings*, the remainder of the universe, by a *boundary* which may be physically real or may be assumed for purposes of thermodynamic calculations. A system which is totally unaffected by changes in the surroundings and where neither energy nor matter can cross the boundary is called an *isolated* system. Such a system is hypothetical yet a useful tool in understanding thermodynamics. A system where the boundary prohibits the transfer of matter is known as a *closed* system. Conversely, a system where matter crosses through the boundary is known as an *open* system.

A batch vessel is a closed system as energy but not matter may cross the boundary. A tubular flow reactor is an open system as both matter and energy cross the boundary. Cyclical processes such as power and refrigeration cycles are flow processes and are closed when considered as a whole. However, each component of the cycle is open, e.g., the compressor, the pump, the expansion valve, and the heat exchanger.

Example 1.2 An adiabatic turbine extracts work from steam when attached to an electric generator. Draw a sketch of the process, indicating the system, the surroundings, and streams in and out. Is the system open or closed?

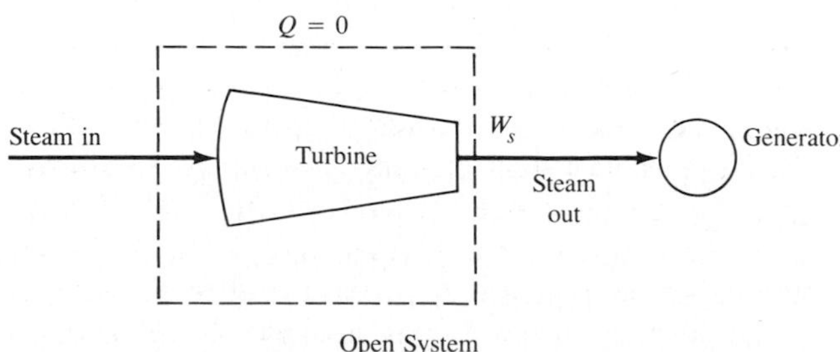

Open System

The dashed line shows the system which is open; the remainder of the universe is the surroundings.

Exercise 1.B Sketch the following processes, indicating the system, the surroundings, heat and work input and output, and whether the system is open or closed.

1. A motor and pump assembly pumping drinking water
2. An internal-combustion engine
3. A pressure cooker
4. A steam boiler fueled by gas
5. A storage battery
6. A reaction taking place in a high-pressure bomb
7. You

1.4.4 System States

A system is *isothermal* if the temperatures in all parts of the system are identical.
An *isobaric* system is at constant pressure throughout.
An *isometric* or *isochoric* system is at constant volume.
If no heat energy enters or leaves a system, the system is called *adiabatic*.

Exercise 1.C Specify as possible the system states for the processes in Exercise 1.B.

1.4.5 Process States

Any thermodynamic process producing or consuming energy in open or closed systems may be of two types: reversible or irreversible. A *reversible* process is one which can be changed in direction by an infinitesimal change in magnitude of the force exerted on the system. An *irreversible* process has a finite imbalance of forces or degradation of energy by friction and cannot be reversed by an infinitesimal change.

The concept of reversibility is difficult to define as it is an idealized process that cannot be attained in practice. The major purpose of the reversible process is to provide a standard of comparison such that real or irreversible processes can be evaluated. The maximum work can be accomplished in a reversible work-producing process, while the minimum work is required for a reversible work-consuming process. Thus, the efficiency of actual processes can be calculated. Since no process can be more efficient than the reversible process, engineers can determine when the law of diminishing returns will be applicable to improvement of a particular process.

Reversibility and irreversibility can best be defined by illustration. Consider the following processes:

1. *Expansion of a gas from high pressure to low pressure through a valve*. Once the gas expands it cannot be compressed to its original pressure without the addition of a device that puts work into the system, e.g., a compressor. Thus, the process is always irreversible. However, a cyclical process containing a valve, compressor, and heat exchange devices can be made reversible and will be discussed later.
2. *Transfer of heat between two bodies of fluid at different temperatures*. If transfer is allowed to go on indefinitely, both reservoirs will reach a common temperature. Since this process cannot be reversed without cooling one reservoir and heating the other with outside sources of energy, the process is irreversible. If the heat from one reservoir is alternately allowed to flow to a gas kept under pressure by a piston of suitable mass, the latter gas will expand and can be maintained at constant temperature by proper metering of the amount of heat allowed to transfer. This gas in the piston need initially only be at an infinitesimally lower temperature than the reservoir for the process to work. If the gas temperature increases infinitesimally, the flow of heat from the reservoir will stop; if the gas in the piston again increases infinitesimally in temperature, the flow of heat will reverse. Thus, the process will reverse. Such a process, although not totally attainable, will be a reversible process.
3. *Mixing of two gases*. Two gases can be mixed easily and cannot be separated without an outside energy source being imposed. The process is irreversible. How could this process be made reversible?
4. *A chemical reaction*. Chemical reactions such as combustion are normally irreversible. Chemical reactions in some cases can be made reversible by devices which maintain equilibrium, as discussed later, or by galvanic cells.

Reversibility arguments often raise as many questions as answers. However, as thermodynamics is studied further, it will be apparent that many real-life problems can be solved reasonably accurately by the application of the principle of reversibility. Both cyclic processes, i.e., processes in which the working fluid starts and ends in the same state, and noncyclic processes as discussed above will be considered in Chaps. 3 and 4.

Many thermodynamic calculations can be made by assuming that the process being analyzed is reversible as many thermodynamic properties are *state properties*. A state property is one that depends on the current condition of the system and is independent of the path by which the condition was reached or by the past history of the system. Usually temperature, pressure, volume, and mass of the components of each phase of a system are used as variables to define a state. As pressure, volume, and temperature are related, they usually may not all be specified independently. For example, for a pure gas, fixing any two of the three variables determines the other. The phase rule discussed in a later chapter shows how many variables may be specified. State properties such as internal energy, enthalpy, and entropy are discussed in Chaps. 3 and 4. It will be shown later that many thermodynamic calculations are made possible or at least easier by assuming a reversible process or series of processes in proceeding from an initial to a final state rather than by considering the process as it actually occurs.

1.4.6 Thermodynamic Equilibrium

Equilibrium can be most easily defined as a state that after any small, short mechanical disturbance of external conditions will return to its initial state. This return may be slow or rapid according to the substance. A low-viscosity liquid such as water will rapidly return to its original state after a perturbation, while a high-viscosity material may take a much longer time; this is a mechanical rather than thermodynamic equilibrium state. In the same manner a chemical reaction rather than a mechanical change may cause a disturbance in the composition of the chemical species present. The system, however, will proceed toward a steady-state composition at some finite although sometimes low rate. The chemical system attains thermodynamic equilibrium when it undergoes no apparent macroscopic changes and will not depart from this composition no matter what its mobility over any period of time. Obviously each molecule may be changing, but the net quantitative analysis of the mixture will not change. Physical or phase equilibrium is attained when components in a mixture of compounds are changing phases and a composition is reached where no change is noted with passage of time. For example, if a mixture of ethanol and methanol is heated in a constant-volume closed system to a given temperature which is maintained, the vapor and liquid phases will each be of constant composition and will not change with time. This is physical equilibrium. If the temperature, pressure, or volume are disturbed, the system will depart from the equilibrium state. A detailed discussion of equilibrium will be given in Chap. 8 for physical equilibria and in Chap. 9 for chemical equilibria.

Exercise 1.D For each process shown, define an equilibrium state, note whether the process is reversible or irreversible, and define the type of thermodynamic system.

1. Mixing gasoline and water in an open vessel at atmospheric pressure
2. Reaction of gasoline and oxygen to carbon dioxide and water in an internal-combustion engine
3. Crystallization of components from a mixture in a freezer
4. Reaction of gasoline and oxygen to carbon dioxide and water in a constant-temperature closed vessel

1.4.7 Phases

A phase may be defined as gaseous, liquid, or solid. While only one gaseous phase may exist in any system, several liquid or solid phases may be present. A homogeneous phase or system has properties which are identical in all parts or at least vary continuously from point to point. A heterogeneous system has two or more distinct homogeneous phases which appear to be separated from one another by surfaces of discontinuity.

1.5 BASIC TEXTUAL REFERENCES

Chemical engineering thermodynamics has been developed over the past half-century primarily emanating from chemical thermodynamics after the publication in 1923 of the classic work by G. N. Lewis and M. Randall, "Thermodynamics and the Free Energy of Chemical Substances," which is often cited as the basis on which chemical engineering thermodynamics was predicated. The book survives in a second expanded edition, published in 1961, adding K. S. Pitzer and L. Brewer as coauthors.

Some of the better texts published in the formative years of chemical engineering thermodynamics are:

Dodge, B. F.: "Chemical Engineering Thermodynamics," 1944 (an excellent intermediate book).

Hougen, O. A., K. M. Watson, and R. A. Ragatz: "Chemical Process Principles, Part II, Thermodynamics," 1st ed., 1947; 2d ed., 1959 (an excellent book for the intermediate student).

Weber, H. C.: "Thermodynamics for Chemical Engineers," 1939 (an excellent beginning book).

The preeminent elementary textbook for chemical engineering thermodynamics during the 1950s and 1960s was J. M. Smith, and later J. M. Smith and H. C. Van Ness, "Introduction to Chemical Engineering Thermodynamics," 1st ed., 1949; 2d ed., 1959. The third edition of this book was published in an expanded form in 1975, which changed the emphasis of the book to the intermediate level.

During the 1970s several other chemical engineering thermodynamics books were published—some primarily textbooks and others more for the reference of the practitioner. Some of the more prominent of these books are:

Balzhiser, R. E., M. R. Samuels, and J. D. Eliasen: "Chemical Engineering Thermodynamics," 1972 (an upper-intermediate textbook).

Chao, K. C., and R. A. Greenkorn: "Thermodynamics of Fluids, An Introduction to Equilibrium Theory," 1975 (an intermediate-level reference).

Edmister, W. C., and B. I. Lee: "Applied Hydrocarbon Thermodynamics, Volume 1," 2d ed., 1984 (primarily a professional reference book).

Modell, M., and R. C. Reid: "Thermodynamics and Its Applications," 1974 (an upper-intermediate or graduate general textbook).

Prausnitz, J. M.: "Molecular Thermodynamics of Fluid Phase Equilibria," 1969 (primarily an advanced-level book).

Van Ness, H. C., and M. M. Abbott: "Classical Thermodynamics of Non-Electrolyte Solutions," 1982 (primarily an advanced-level and professional reference book).

The selection of references given in this section is primarily illustrative and does not attempt to indicate the quality of the reference. In addition, omission of a reference does not indicate any value judgment.

References to the primary literature and other appropriate thermodynamic symposia books will be made throughout this text as the various topics are developed. As is the case in all textbooks, no attempt will be made to include a comprehensive set of references to the entire field of chemical engineering thermodynamics.

1.6 REFERENCES

American Institute of Chemical Engineers: *AIChE J.*, **17**:511 (1971).

American Petroleum Institute: "Technical Data Book—Petroleum Refining," 4th ed., T. E. Daubert and R. P. Danner (eds.), Washington (extant 1982).

Denbigh, K.: "The Principles of Chemical Equilibrium," 4th ed., Cambridge, New York, 1981.

Hala, E., J. Pick, V. Fried, and O. Vilim: "Vapour-Liquid Equilibrium," 2d ed., Pergamon, New York, 1967.

1.7 PROBLEMS

1 Convert all properties of sulfur dioxide listed in App. A to the English system of units.

2 Give an original example of the following systems:
- (*a*) Isothermal, isobaric, open
- (*b*) Isometric, isobaric, open
- (*c*) Adiabatic, open
- (*d*) Isolated
- (*e*) Reversible, isothermal

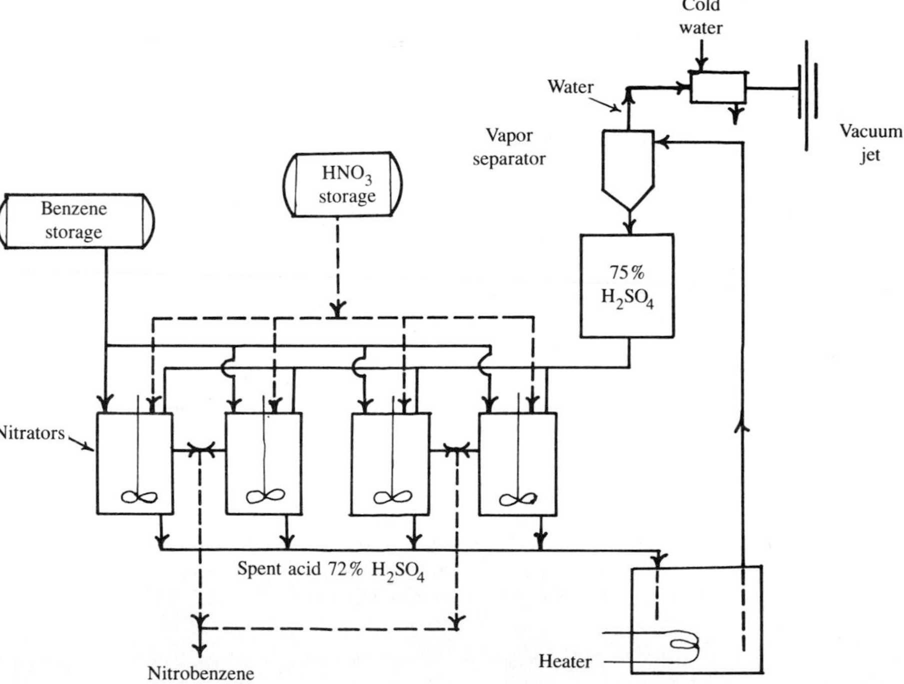

Figure 1.1 Nitration of benzene using nitric acid fortified spent acid.

3 A simplified flow diagram of a chemical plant is given in Fig. 1.1. Indicate for each piece of equipment in the process and for the plant as a whole, (*a*) whether the system is open or closed, (*b*) the system state, (*c*) the process state, (*d*) the equilibrium state, and (*e*) the phases present.

4 Repeat Prob. 3 for Fig. 1.2.

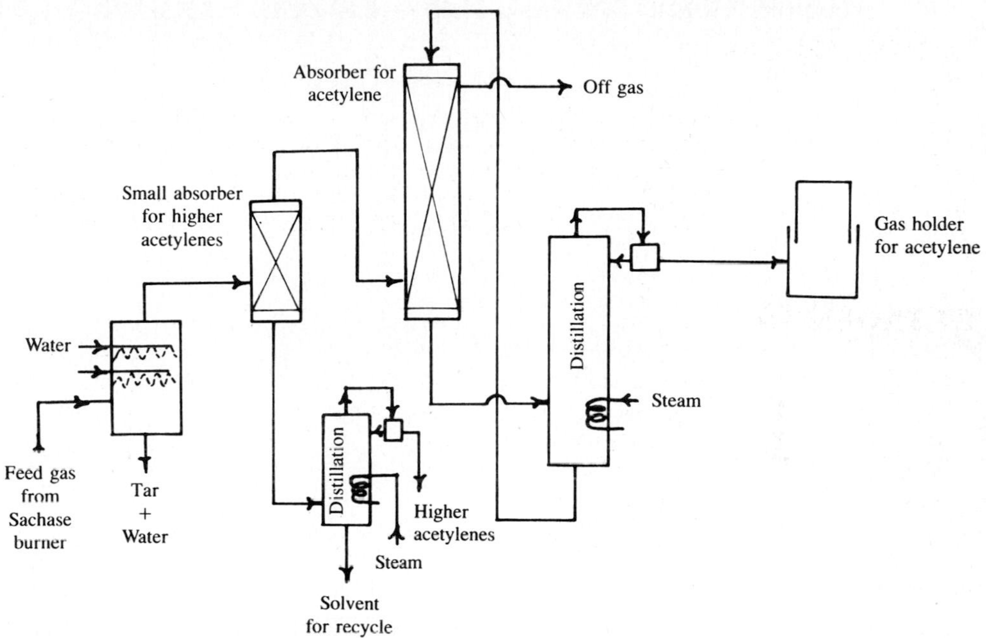

Figure 1.2 Simplified sketch of an absorption plant.

CHAPTER
TWO

PVT PROPERTIES OF FLUIDS—EQUATIONS OF STATE

A relationship describing how the *PVT* behavior of a chemical as the temperature, pressure, and/or volume varies is known as an *equation of state*. Such relationships are essential for the estimation of accurate thermodynamic properties of the compound. While the equation of state primarily describes gas-phase properties, many modern equations are also useful for prediction of liquid-phase properties of the substance. A thorough understanding of these methods is prerequisite to an understanding of classical thermodynamics.

Gases are divided into two types—ideal and nonideal based on the concept of the *PVT* behavior agreeing or not agreeing with the classical ideal-gas law. Nonideal gases or real gases have been treated by many different methods including a large number of general analytical equations of state or the generalized extended theorem of corresponding states. Each of these methods attempts to take into account deviations from ideality by introducing physical parameters other than the pressure, volume, and temperature of the gas.

For some common substances, such methods do not adequately describe the molecules to the desired degree of accuracy, and experimental data must be used directly. However, more complex formulations have made it possible to use semitheoretical methods to estimate properties of all hydrocarbon-type fluids, and methods are under development to improve the prediction of properties of polar fluids by introducing additional parameters. Molecules classified as polar are asymmetric with an uneven spatial distribution of charges about the positively charged nucleus. Thus, permanent dipole and higher moments exist. Symmetric molecules like argon and methane have zero dipole moments, hydrocarbons have low dipole moments, and most organic compounds containing oxygen, nitrogen, or sulfur are more polar.

2.1 THE IDEAL GAS

An ideal gas at the molecular level is a mixture of hard, elastic, and inert moving particles so far apart that the particles have no interactions with one another. No forces of attraction or repulsion among molecules exist. As such the ideal gas is one limiting condition of matter. Consequently, the ideal-gas law applies to no real substance. However, many simple gases under conditions of moderate and high temperature and at low and moderate pressures for all practical purposes follow the ideal-gas law shown by Eq. (2.1). At zero pressure all gases approach this law.

$$P\overline{V} = RT$$

or

$$PV = nRT \tag{2.1}$$

where n = number of moles of gas
R = universal gas constant

The ideal-gas law was developed by Boyle, Charles, and Gay-Lussac in the seventeenth and eighteenth centuries by noting for simple gases that (1) at constant temperature the volume of a gas decreases inversely with increasing pressure, $V \sim 1/P$ and (2) at constant pressure the volume of a gas increases directly with increasing temperature, $V \sim T$. Combining these two observations led to the proportionality $PV \sim T$. The constant of proportionality R for 1 mol of such a gas was determined to be a universal constant for simple, monatomic gases at low pressures.

Any interaction of molecules tends to cause a deviation from ideal-gas behavior. The more attractions and repulsions which cause interactions between molecules, the more nonideal the gas. Thus, the inert gases—helium, argon, krypton, neon—are the most ideal. Simple diatomic gases—nitrogen, oxygen, carbon monoxide—are ideal at high temperatures and low pressures. Triatomic gases—carbon dioxide, hydrogen sulfide, sulfur dioxide, water—do not follow the ideal-gas law well except at very high temperatures and very low pressures.

Ideal-gas thermodynamic properties only depend upon temperature and are independent of pressure. Use of the ideal-gas assumption can greatly simplify many calculations. Internal energy and enthalpy, two such properties, will be discussed in Chap. 3.

Example 2.1 Assuming that *n*-butane is an ideal gas, calculate its molar volume at the normal boiling point.

BASIS 1 kmol n-butane.

$$PV = nRT$$

$$P = 1.013 \times 10^5 \text{ N/m}^2$$

$$R = 8.3143 \times 10^3 \text{ J/(kmol} \cdot \text{K)}$$

$$T = -0.50°\text{C} = 272.6 \text{ K} \qquad \text{(from Appendix A)}$$

Therefore,

$$V = \frac{(1)(8.3143 \times 10^3)(272.6)}{1.013 \times 10^5} = 22.37 \text{ m}^3$$

Repeat for steam.

$$T = 100°\text{C} = 373.1 \text{ K} \qquad \text{(from App. A)}$$

$$V = \frac{(1)(8.3143 \times 10^3)(373.1)}{1.013 \times 10^5} = 30.62 \text{ m}^3$$

2.2 THE NONIDEAL OR REAL GAS

Many efforts have been made to improve the accuracy of the ideal-gas law for low temperatures, high pressures, or both. Two basic approaches have been used.

The first approach involves the addition of terms to the ideal-gas law solved for pressure, i.e., for 1 mol of gas,

$$P = \frac{RT}{\overline{V}} + C_1 f(T, \overline{V}) + C_2 f(T, \overline{V}) + \cdots \tag{2.2}$$

or

$$P = R\rho T + C_1' f(T, \rho) + C_2' f(T, \rho) + \cdots \tag{2.3}$$

These analytical equations of state vary appreciably in complexity and accuracy and fall into various categories such as the virial equations, the most simple cubic equation of van der Waals, and the numerous cubic equations of state to be discussed later. The constants are either specific or generalized depending on whether they must be developed for each compound by using experimental PVT data or whether they can be estimated for any compound by using a few readily available properties.

The second approach utilizes an empirical correction factor Z, the compressibility factor.

$$P = \frac{ZRT}{\overline{V}} = ZR\rho T \tag{2.4}$$

Compressibility factors may be tabulated separately for a compound as a function of temperature and pressure as derived from experimental data. Alternatively, the theorem of corresponding states postulates that different compounds at identical conditions of reduced temperature and reduced pressure, to be discussed later in this chapter, will exhibit identical reduced volumes. This allows a generalized plot of compressibility factor against these two parameters. However, this approach is only strictly applicable for spherical, nonpolar molecules. For more complex molecules additional parameter(s) are necessary. Three-parameter approaches have been determined to be successful in correlating properties for nonspherical, nonpolar molecules, and three alternate parameters have been proposed: critical compressibility factor (Lydersen, Hougen, and Greenkorn), acentric factor (Pitzer), and alpha (Riedel). Polar molecules add an additional complexity, and four-parameter models have been under development with varying degrees of success for the past 20 years.

Example 2.2 The molar volume of steam at its normal boiling point is 30.0 $m^3/kmol$. Determine the compressibility factor.

BASIS 1 mol steam.

$$P\bar{V} = ZRT$$

$$Z = \frac{P\bar{V}}{RT}$$

$$= \frac{(1.013 \times 10^5)(30.0)}{(8.3143 \times 10^3)(373.1)}$$

$$= 0.980$$

Exercise 2.A Determine compressibility factors for steam at the following sets of conditions:

Temperature, °C	Pressure, kPa
300	101.3
1000	101.3
1000	15,000
1000	25,000
700	25,000

Compare your answers with those derived from steam tables.

2.3 CRITICAL PROPERTIES

Vapor and liquid phases can be shown to coexist by a vapor pressure curve on a pressure-temperature diagram as shown in Fig. 2.1. For a pure substance the curve starts at the triple point (t), where vapor, liquid, and solid phases are in

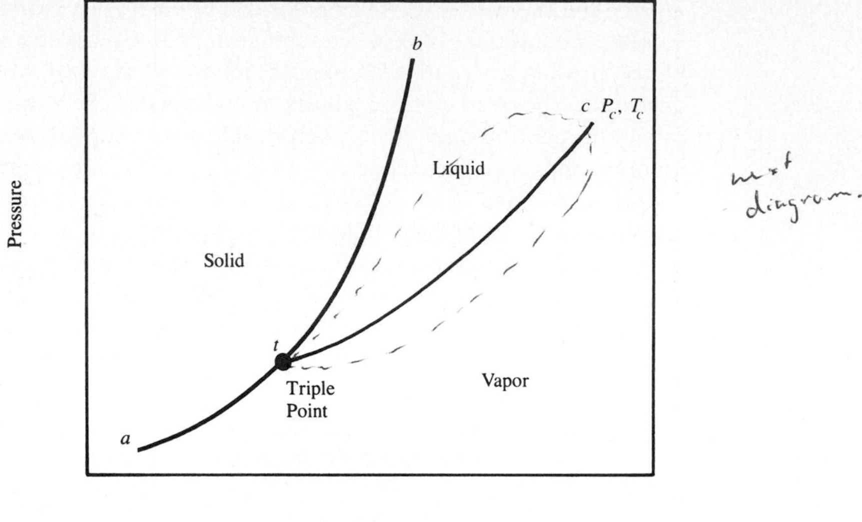

Temperature

Figure 2.1 Pressure-temperature diagram for a pure component.

equilibrium, and ends at the critical point (c). The critical point is the temperature-pressure point along the vaporization curve approached by the coexisting phases (vapor and liquid) where the phase properties become identical and a single homogeneous phase develops. At the critical point vapor and liquid have merged into a fluid region where vapor and liquid are not distinguishable. Above the critical temperature only a vapor phase can exist. The curve $\overline{at}$ is the sublimation curve, while $\overline{tb}$ is the melting curve.

Exercise 2.B Using the values of the critical parameters from App. A and the properties of steam from App. B, sketch a rough PV diagram for water showing the saturation curves for the liquid-vapor and solid-liquid regions from the triple point to the critical point.

For a mixture of liquids, vaporization at a constant pressure will occur over a range of temperatures. Thus, the vaporization requires two curves on the pressure-temperature diagram to define the vaporization characteristics. Figure 2.2 shows this behavior for a general binary mixture where an LLV (bubble point) and VVL line (dew point) are shown to meet at a common point, the critical point. This two-phase envelope represents a mixture of defined composition. Any change in the composition of the mixture is reflected in changes in the curves enclosing the two-phase region. The critical temperature and pressure will change as will the maximum temperature and pressure and slopes of the lines.

If a constant-pressure path indicated on Fig. 2.2 as case I is followed between points 1 and 2, the increase in temperature will cause the homogeneous liquid to begin vaporizing when the bubble point line is passed (A). As the temperature is

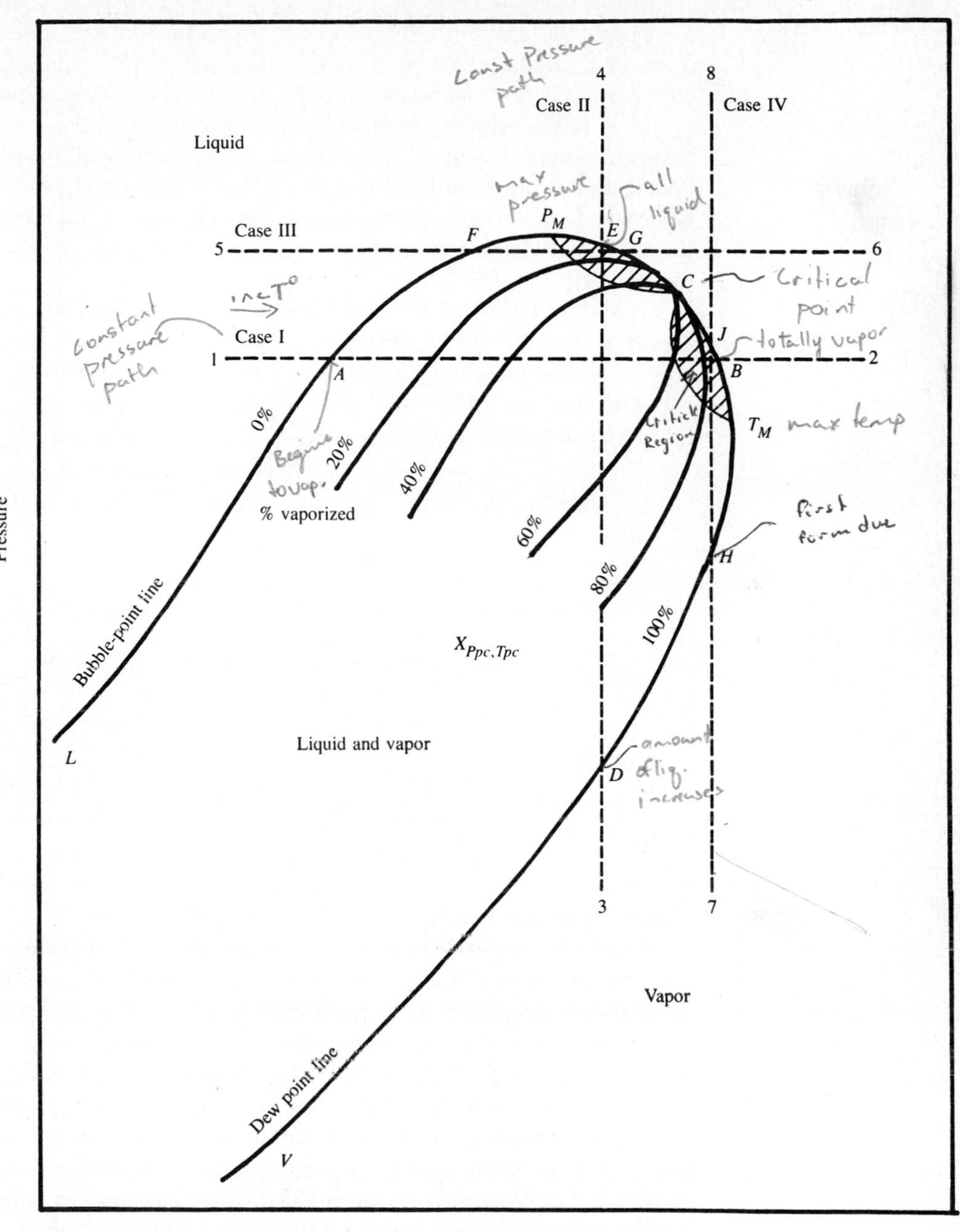

Figure 2.2 Pressure-temperature diagram for a mixture of constant composition near the critical point.

increased, the amount of liquid will decrease and the amount of vapor will increase until the mixture is totally vapor at the dew point line (B). As the temperature is further increased, only a homogeneous vapor phase exists.

If a constant-pressure path indicated on Fig. 2.2 as case II is followed between points 3 and 4, an increase in pressure will have no effect on the homogeneous vapor until the dew point line is passed (D). Between D and E the amount of vapor will decrease and the amount of liquid will increase until the mixture is totally liquid at the bubble point line (E). Above E a homogeneous liquid exists.

It should be noted that the two-phase region does not necessarily have its maximum pressure and temperature at the critical point. Figure 2.2 illustrates this point with the maximum pressure at P_{m} and the maximum temperature at T_{m}. These points are called the *critical condensation pressure* (*cricondenbar*) and *critical condensation temperature* (*cricondentherm*), respectively.

Exercise 2.C Postulate the process which occurs in cases III and IV shown on Fig. 2.2.

For multicomponent systems, envelopes will replace the vapor pressure lines, but the same principles will hold. More detailed descriptions of critical phenomena are summarized and referenced to the original literature in the "Technical Data Book—Petroleum Refining" (American Petroleum Institute) and in the "Data Prediction Manual" (American Institute of Chemical Engineers).

The supercritical state of a fluid is the fluid above its critical temperature and pressure. Use of fluids under supercritical conditions has recently gained prominence for contact with solid or liquid mixtures to selectively extract one or more components. Thus, knowledge of the critical state is required for calculation of equilibria in such systems, as well as in vapor-liquid equilibrium situations. Further discussions will be given in Chap. 8.

For most thermodynamic calculations, the critical properties of pure compounds are important as correlations for mixtures are based on formulations using critical properties as input parameters. True critical properties of mixtures are mainly required to determine operating temperatures and pressures for heat and mass transfer equipment and for chemical reactors. For purposes of this chapter further discussion will be limited to pure-compound properties.

We have defined the critical temperature of a pure material as the maximum temperature at which liquid and vapor phases can coexist at equilibrium. At this point the vapor pressure is called the *critical pressure*, and the volume per unit mass is called the *critical volume*. Considering a PV diagram as shown in Fig. 2.3, the critical-temperature isotherm must have a horizontal inflection point which mathematically reduces to

$$\left(\frac{\partial P}{\partial V}\right)_{T_c} = 0 \quad \text{and} \quad \left(\frac{\partial^2 P}{\partial V^2}\right)_{T_c} = 0 \tag{2.5}$$

As the average molecular kinetic energy equals the negative potential energy of attraction at the critical point, no liquid phase can possibly exist above this point. Investigators disagree as to whether there is a definite, precise critical point or whether we should discuss a critical region with a small range of P, V, and T. If the latter is more accurate, the disagreement is minor as the width of the region is within the range of experimental error in measuring critical properties.

Various methods have been proposed for estimating critical properties of pure compounds and mixtures. Recommended prediction methods are given in Chap. 6, which should be consulted for values not given in App. A or in the standard references listed.

As noted earlier the theorem of corresponding states requires the critical temperature and critical pressure of a pure compound for entry. For a mixture, a set of pseudocritical properties of the mixture is considered to have the same significance for the mixture as do the true critical conditions for a pure substance. The pseudocritical concept was introduced by Kay (1936) by graphically identifying temperatures and pressures for mixtures which had the same significance for *PVT* behavior as the true criticals have for pure compounds. These can be approximated as simple molar averages:

$$T_{pc} = \sum_{i=1}^{n} x_i T_{c_i} \qquad P_{pc} = \sum_{i=1}^{n} x_i P_{c_i} \tag{2.6}$$

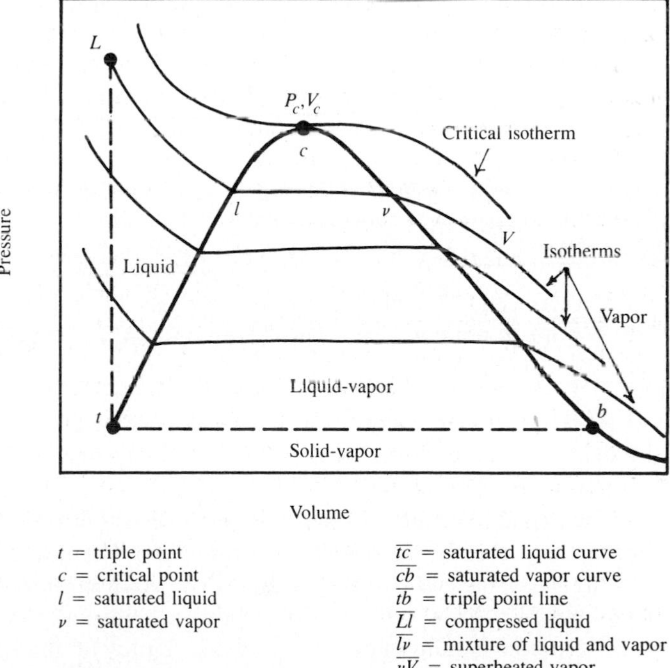

t = triple point
c = critical point
l = saturated liquid
v = saturated vapor
$\overline{tc}$ = saturated liquid curve
$\overline{cb}$ = saturated vapor curve
$\overline{tb}$ = triple point line
$\overline{Ll}$ = compressed liquid
$\overline{lv}$ = mixture of liquid and vapor
$\overline{vV}$ = superheated vapor

Figure 2.3 Pressure-volume diagram for a pure component.

where x = mole fraction
i = component
pc = pseudocritical
c = critical

This simple rule of postulating the behavior of a pseudocompound as a pure compound is actually incorrect, especially in the region between the true and the pseudocritical temperatures where liquids can exist but the theorem fails to recognize it. Nevertheless, the concept is extremely useful, simple, and accurate for prediction of volumetric, thermodynamic, and transport properties of mixtures in most regions of the phase diagram. For wide-boiling mixtures, the simple rule of molar average sometimes fails, and several sets of more complex equations have been developed for specific uses. However, for the great majority of cases, Kay's rule will suffice. For mixtures of unknown composition such as petroleum fractions, other correlations for pseudocritical properties are necessary, and predictive methods have been developed as functions of common inspection properties such as normal boiling point and relative density. Such correlations are given in Chap. 6.

Entrance to the methods of the theorem of corresponding states requires the reduced properties defined as

$$T_r = \frac{T}{T_c} \quad \text{and} \quad P_r = \frac{P}{P_c} \quad \text{for pure compounds} \tag{2.7}$$

and

$$T_r = \frac{T}{T_{pc}} \quad \text{and} \quad P_r = \frac{P}{P_{pc}} \quad \text{for mixtures} \tag{2.8}$$

Obviously, the reduced properties are identically unity at the critical point. Reduced temperatures above 1 indicate vapor phase.

Exercise 2.D Calculate the reduced pressure and reduced temperature for a 50-mass % aqueous solution of ammonia at any set of conditions.

As noted earlier, a third characterizing parameter is necessary for entering the theorem of corresponding states—the critical compressibility factor, the acentric factor, and Reidel's alpha.

The critical compressibility factor is merely defined as $Z_c = P_c\overline{V}_c/RT_c$ and can be calculated from the critical properties. The parameter varies from about 0.23 for water to 0.30 for light gases. Most polar compounds have factors in the range of 0.24 to 0.26, while hydrocarbons are normally in the 0.26 to 0.28 range. Hydrogen and helium cannot be treated. A plot of the saturation envelopes for various critical compressibilities is given in Fig. 2.4. Methods for prediction of Z_c will be discussed in Chap. 6.

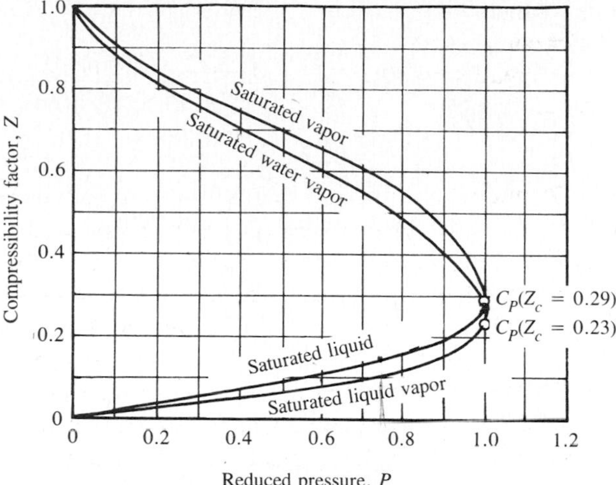

Figure 2.4 Saturation envelopes of compressibility factors at various critical compressibility factors.

Riedel's (1954) alpha is based on the slope of the reduced vapor pressure curve at the critical point, $\alpha = [d(\ln P)/d(\ln T)]_{T_c, P_c}$. This parameter has never been used to develop a complete corresponding-states approach.

Pitzer postulated that the slope of the reduced vapor pressure curve is the most sensitive property for a third-parameter base. Since vapor pressure can be measured with greater accuracy than critical properties, this approach should be superior to the critical compressibility factor. Thus, he defined the acentric factor ω as

$$\omega = -\log \frac{P_s}{P_c} - 1.000 \tag{2.9}$$

where P_s = the vapor pressure at $T_r = 0.700$. This particular form was chosen to ensure that $\omega = 0$ for simple spherical noble gases—argon, krypton, and xenon—while other fluids will have small positive values of ω. The acentric factor then measures the deviation of the intermolecular potential function from that of simple spherical molecules. The acentric factor serves as the primary third parameter in accurate three-parameter corresponding-states methods for normal fluids, primarily hydrocarbons. The critical compressibility factor is used more often for three-parameter correlation of polar fluids. Each of these methods will be discussed in more detail later in the chapter.

Example 2.3 Calculate the critical compressibility factor and the acentric factor for water.

$$P_c = 22.12 \times 10^6 \text{ Pa}$$

$$T_c = 374.2°\text{C} = 647.3 \text{ K}$$

$$\overline{V}_c = 0.05697 \text{ m}^3/\text{Kmol}$$

BASIS 1 kmol water.

$$Z_c = \frac{P_c V_c}{RT_c} = \frac{(22.12 \times 10^6)(0.05697)}{(8.3143 \times 10^3)(647.3)}$$

$$= 0.234$$

$$\omega = -\log\left(\frac{P_s}{P_c}\right) - 1.000$$

$$T_r = 0.70 = \frac{T}{647.3}$$

Therefore,

$$T = 453.1 \text{ K} = 180.0°\text{C}$$

$$P_s = 1.0 \times 10^6 \text{ Pa}$$

$$\omega = -\log\left(\frac{1.0 \times 10^6}{22.12 \times 10^6}\right) - 1.000$$

$$= 1.3448 - 1.000 = 0.3448$$

2.4 ANALYTICAL EQUATIONS OF STATE

Equations of state take many forms. This section will discuss various forms of the equations beginning with the analytical forms of the virial expansion, Benedict-Webb-Rubin formulation, van der Waals quadratic equation, and several important forms of the many other cubic equations such as those of Redlich-Kwong and Soave. Finally, the two generalized graphical forms formulated by Lydersen, Greenkorn, and Hougen and by Pitzer and Curl will be presented. An analytical form of the latter method useful for calculation by computer will also be noted. Each equation is derived on the basis of 1 mol.

2.4.1 Virial Equations

Considering the interaction of gas molecules by the methods of statistical mechanics, the virial equation of state can be derived to describe a real gas. This power series is

$$Z \equiv \frac{P\overline{V}}{RT} = 1 + \frac{B}{\overline{V}} + \frac{C}{\overline{V}^2} + \cdots \tag{2.10}$$

Equation (2.10) assumes that PV along an isotherm can be represented by a power series expansion.

This equation has also been written in an alternate form as a function of pressure to express specific volume explicitly as a function of pressure and to correlate more easily experimental data:

$$Z \equiv \frac{P\bar{V}}{RT} = 1 + B'P + C'P^2 + \cdots \tag{2.11}$$

where $B' = B/RT$ and $C' = (C - B^2)/R^2T^2$.

Virial coefficients are functions of temperature only and are calculated from experimental *PVT* data. Values of second virial coefficients are available for many simple gases. In general, they can be correlated in terms of intermolecular forces of attraction and repulsion between molecules expressed as a function of their intermolecular distances. However, this derivation is available only for simple materials such as rigid spherical molecules.

The virial coefficients can be estimated by corresponding-states techniques as shown later and have proved useful for describing gases accurately at low pressures. Virial equations truncated to two terms have been found useful up to 15 abs total pressure, while equations truncated after three terms have been used up to 50 abs total pressure.

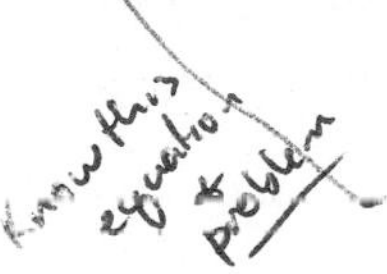

Exercise 2.E Given the first and second virial coefficients for sulfur dioxide at 431 K, $B = -0.159\ \mathrm{m^3/kmol}$ and $C = 9.0 \times 10^{-3}\ \mathrm{m^6/kmol^2}$, calculate the molar volume of sulfur dioxide at 7.5 MPa total pressure and 431 K.

2.4.2 Benedict-Webb-Rubin (1940) Equation

This equation shown below is empirical in nature and contains eight constants which have been derived from experimental data for light hydrocarbons:

$$P = \frac{RT}{V} + \frac{1}{\bar{V}^2}\left(B_0RT - A_0 - \frac{C_0}{T^2}\right) + \frac{1}{\bar{V}^3}(bRT - a) + \frac{a\alpha}{\bar{V}^6} + \frac{c}{\bar{V}^3T^2}\left(1 + \frac{\gamma}{\bar{V}^2}\right)e^{-\gamma/\bar{V}^2} \tag{2.12}$$

This equation has been determined to be accurate for the compounds for which constants are available and is thermodynamically consistent in form. Data have been correlated in this form in several compilations. A form of this equation is described later in this book for analytical representation of corresponding-states plots.

2.4.3 van der Waals (1873) Equation

Derivation of the ideal-gas law assumes that the volume of molecules is infinitesimal compared to the total volume. Actually, the volume in which a molecule can move is less than the total volume by some amount, say b, which takes into account the finite size of the molecules and the repulsion between molecules. The force of attraction between molecules, called the *van der Waals forces*, are also neglected in the simple-gas theory. This force reduces the pressure below the value calculated from ideal-gas principles by pulling molecules together. Kinetic theory shows this pressure to be inversely proportional to the molar volume squared.

Using these observations, the van der Waals equation has been derived as

$$\left(P + \frac{a}{\overline{V}^2}\right)(\overline{V} - b) = RT \tag{2.13}$$

where a and b are characteristics of each gas and are called the *van der Waals constants*. These constants, however, have been determined to be dependent on temperature and density of the fluid, limiting the use of the equation. However, this equation can be shown to support the theorem of corresponding states and has been useful in its development.

Values of a and b can be calculated from critical properties by applying the van der Waals equation to the critical point and using the fact that at the critical point the first and second derivatives of pressure with respect to volume equal zero. Rearranging Eq. (2.13):

$$P = \frac{RT}{\overline{V} - b} - \frac{a}{\overline{V}^2} \tag{2.13a}$$

These derivatives may be taken in three ways—T_c, V_c, or V_c, P_c, or P_c, T_c constant.

1. For example, if T_c is constant,

$$\left(\frac{\partial P}{\partial \overline{V}}\right)_{T_c} = \frac{-RT_c}{(\overline{V}_c - b)^2} + \frac{2a}{\overline{V}_c^3} = 0$$

$$\left(\frac{\partial^2 P}{\partial \overline{V}^2}\right)_{T_c} = \frac{2RT_c}{(\overline{V}_c - b)^3} - \frac{6a}{\overline{V}_c^4} = 0$$

which when solved lead to

$$a = \tfrac{9}{8}RT_c\overline{V}_c \qquad b = \frac{\overline{V}_c}{3} \tag{2.14}$$

2. If solved in terms of P_c and T_c,

$$a = \frac{27R^2T_c^2}{64P_c} \qquad b = \frac{RT_c}{8P_c} \tag{2.15}$$

3. If solved in terms of V_c and P_c,

$$a = 3P_cV_c^2 \qquad b = \frac{V_c}{3} \tag{2.16}$$

Isotherms drawn following these various constants best follow experimental data for the second formulation, that is, P_c and T_c constant, as shown in Fig. 2.5.

Usually values for the van der Waals constants are calculated from critical pressure and temperature by using these equations. Since experimental critical volumes are scarce and predicted critical volumes are often inaccurate, the use of a formulation in terms of P_c and T_c is also warranted on these grounds.

The van der Waals equation is a simple although somewhat inaccurate equation but provides a starting point for many more sophisticated formulations.

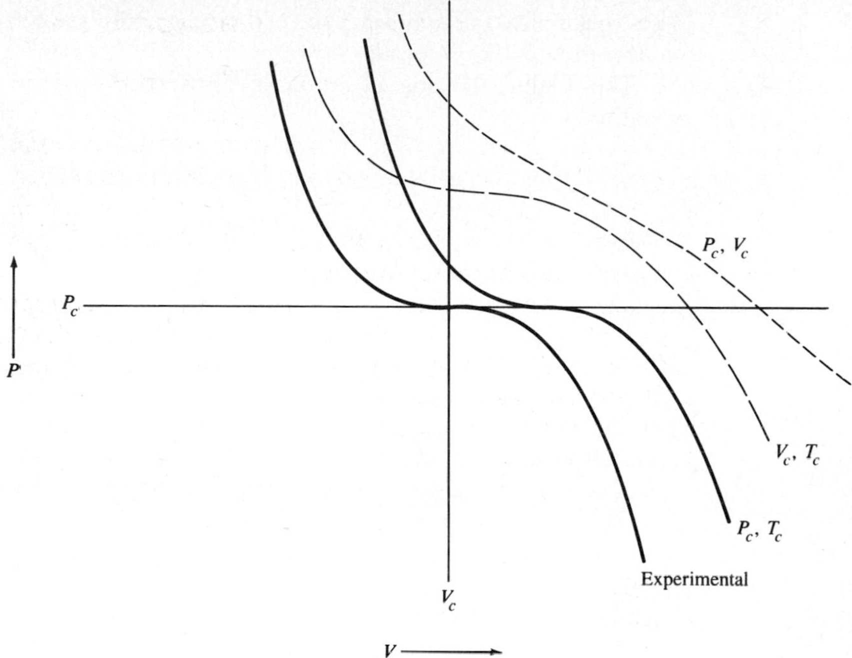

Figure 2.5 Isotherms from van der Waals formulation.

2.4.4 Other Cubic Equations

A large number of cubic equations of state have been developed and include those of Berthelot, Beattie and Bridgeman, Redlich and Kwong, and various modifications thereof, the most successful of which have been those of Soave and Peng and Robinson. Only the current commonly used Redlich-Kwong type of equations will be discussed in this section as the earlier equations have essentially been replaced by these formulations.

Redlich-Kwong (1949) equation The two-constant Redlich-Kwong equation of state has been determined to be one of the most useful equations which has been proposed in the century following van der Waals' work. The equation is

$$P = \frac{RT}{\overline{V} - b} - \frac{a}{T^{1/2}\overline{V}(\overline{V} + b)} \tag{2.17}$$

Constants a and b can be most accurately determined by fitting the constants, using experimental *PVT* data. However, as this information is not usually available, the criteria of the critical point [Eqs. (2.5)] are used to determine values for a and b as a function of the critical temperature and critical pressure of the material in question. The results of this derivation are

$$a = \frac{0.4278R^2T_c^{2.5}}{P_c} \qquad b = \frac{0.0867RT_c}{P_c} \tag{2.18}$$

These generalized parameters can then be used with the Redlich-Kwong equation to determine *PVT* properties.

The Redlich-Kwong equation is also often written as a compressibility equation:

$$Z = \frac{1}{1-h} - \frac{A}{B}\frac{h}{1+h} \tag{2.19}$$

where

$$h = \frac{b}{\overline{V}} = \frac{BP}{Z} \qquad B = \frac{b}{RT} \qquad \frac{A}{B} = \frac{a}{bRT^{1.5}}$$

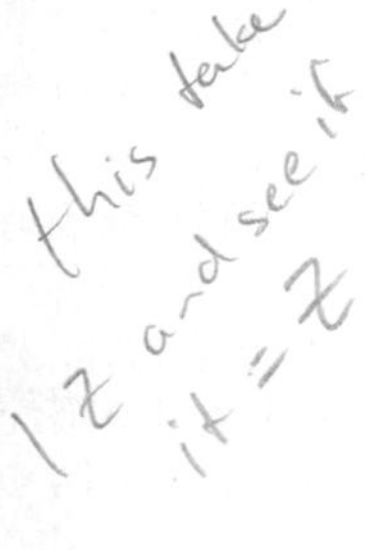

This equation will allow values of Z and h and thus Z to be determined by iterative solution of the equation. Such a procedure would set a value of Z (say 1), calculate h, then calculate a new value for Z. The process would be repeated until the assumed Z and calculated Z agree within certain limits.

Any cubic equation will give three roots for $\overline{V}$. The largest is the vapor volume, the smallest is the liquid volume, and the intermediate root has no physical significance.

Redlich-Kwong-Soave (1972) equation Soave (1972) proposed a modification to the Redlich-Kwong equation of state which introduced a third parameter, the acentric factor, and a temperature dependency into the cohesive energy term to account for the effect of nonsphericity on fluid *PVT* properties. This equation is

$$P = \frac{RT}{\overline{V}-b} - \frac{a\alpha}{\overline{V}(\overline{V}+b)} \tag{2.20}$$

where a and b are similar to the Redlich-Kwong a and b [Eqs. (2.18)] with the exception that $T_c^{2.5}$ becomes T_c^2 and $\alpha = f(T/T_c, \omega)$, an empirical function derived from pure-hydrocarbon vapor pressure data. Soave used the critical point and the vapor pressure at a reduced temperature of 0.7, that is, the acentric factor, to determine the function required to correlate vapor pressures. He determined that

$$\alpha = \left[1 + S\left(1 - \sqrt{T_r}\right)\right]^2 \tag{2.21}$$

where $S = a' + b'\omega + c'\omega^2$
$a', b', c' =$ constants

Graboski and Daubert (1978) refit Soave's original correlation for a', b', and c' and obtained

$$S = 0.48508 + 1.55171\omega - 0.15613\omega^2 \tag{2.22}$$

which for hydrocarbons allowed vapor pressure prediction over a wide range of conditions to be predicted for pure hydrocarbons within 3 percent. Further work showed that the Soave equation is accurate for pure hydrocarbons and nonhydrocarbon gases (except hydrogen) without modification (Graboski and Daubert, 1978*a*, 1978*b*, 1979).

For mixtures, the equation can be modified by using average a and b terms derived by Soave from the van der Waals one-fluid combining rules:

$$\overline{\alpha}\overline{a} = \sum_{i=1}^{n} \sum_{j=1}^{n} x_i x_j a_{ij} \alpha_{ij} \qquad \overline{b} = \sum_{j=i} x_j b_j \tag{2.23}$$

where $\alpha_{ij} a_{ij} = (1 - C_{ij})(\alpha_i a_i \alpha_j a_j)^{1/2}$.

C_{ij}, the Chueh and Prausnitz (1967) interaction coefficient, corrects for the effect of deviation from geometric-mean combining rules and must be obtained from binary-mixture data. It is generally independent of temperature, pressure, and composition. Soave suggested and Graboski and Daubert (1978*a*) verified that C_{ij} for hydrocarbon mixtures approached zero. Graboski and Daubert (1978*b*) suggested methods for predicting C_{ij} for hydrocarbon-nonhydrocarbon gas mixtures. Further modifications by Graboski and Daubert (1979) extended the equation to hydrogen.

This equation is currently considered to be the most general and simple analytical equation of state which can be used to accurately predict thermodynamic, especially vapor-liquid equilibrium, properties. Soave (1979, 1980) has recently extended this equation to polar compounds by redefinition of the α function.

Peng-Robinson (1976) equation This equation of state was developed by using the same bases as the Soave equation. The equation takes the form

$$P = \frac{RT}{\overline{V} - b} - \frac{\alpha a}{\overline{V}(\overline{V} + b) + b(\overline{V} - b)} \tag{2.24}$$

where

$$a = 0.45724 \frac{R^2 T_c^2}{P_c} \qquad \text{and} \qquad b = 0.07780 \frac{RT_c}{P_c}$$

$$\alpha = \left[1 + S\left(1 - \sqrt{T_r}\right)\right]^2 \tag{2.25}$$

$$S = 0.37464 + 1.54226\omega - 0.26992\omega^2$$

The equation for alpha was obtained by using vapor pressure data from the normal boiling point to the critical point and is of the same form as the Soave expression which used only the critical point and the calculated vapor pressure at $T_r = 0.7$ based on the value of the acentric factor. Only the constants in the S equation are different.

The Peng-Robinson equation is almost equivalent to the Soave equation for prediction of compressibility factors and vapor-liquid equilibria and has been used extensively for water systems, as will be discussed in Chap. 8 on equilibrium. However, the Soave equation is slightly easier to solve. Consider the Z form of the equations as an example of this.

Soave:

$$Z^3 - Z^2 + Z(A - B - B^2) - AB = 0$$

Peng-Robinson:

$$Z^3 - (1 - B)Z^2 + (A - 3B^2 - 2B)Z - (AB - B^2 - B^3) = 0$$

$$A = \frac{aP}{R^2T^2} \qquad B = \frac{bP}{RT}$$

Thus, the Soave equation has been more readily adopted because of its earlier introduction and comparative simplicity.

An excellent summary of methods for solution of cubic equations of state has been published by Edmister and Lee (1984).

Example 2.4 For ammonia at 100°C and 1 MPa pressure, calculate its molar volume by using (a) the van der Waals equation with $a = 4.19$ (atm · m^6)/kmol^2 and $b = 0.0373$ m^3/kmol, (b) the van der Waals equation with estimated values of a and b, and (c) the Redlich-Kwong-Soave equation.

BASIS 1 kmol ammonia.

(a)

$$\left(P + \frac{a}{\bar{V}^2}\right)(\bar{V} - b) = RT$$

$$\left(10^6 + \frac{4.19 \times 1.013 \times 10^5}{\bar{V}^2}\right)(\bar{V} - 0.0373) = (8314.3)(373.1)$$

If ideal, $\bar{V} = 3.10$ m^3/kmol. Therefore

$$\text{Trial and error } \bar{V} = 3.0 \qquad 3.102 \times 10^6 = 3.102 \times 10^6$$

$$= 3.0 \text{ m}^3\text{/kmol}$$

(b) From Eq. (2.15),

$$a = \frac{27R^2T_c^2}{64P_c} = \frac{27(8314.3)^2(405.5)^2}{64(11.28 \times 10^6)} = 425{,}116$$

$$b = \frac{RT_c}{8P_c} = \frac{(8314.3)(405.5)}{8(11.28 \times 10^6)} = 0.03736$$

Since these values are very close to those in (a), the answer would be essentially the same.

(c) For the Redlich-Kwong-Soave equation,

$$a = \frac{0.4278R^2T_c^2}{P_c} = 431{,}087$$

$$b = 0.0867\frac{RT_c}{P_c} = 0.02591$$

Using Eq. (2.20), where

$$P = \frac{RT}{\bar{V} - b} - \frac{a\alpha}{\bar{V}(\bar{V} + b)}$$

and using Eq. (2.21), where

$$\alpha = \left[1 + S\left(1 - \sqrt{T_r}\right)\right]^2$$

then

$$S = 0.48508 + 1.55171(0.2515) - 0.15613(0.2515)^2$$
$$= 0.86546$$
$$T_r = \frac{373.1}{405.7} = 0.91965$$
$$\alpha = \left[1 + 0.86546(1 - \sqrt{0.91965}\,)\right]^2 = 1.0355$$
$$10^6 = \frac{(8314.3)(373.1)}{\overline{V} - 0.02591} - \frac{(431{,}087)(1.0355)}{\overline{V}(\overline{V} + 0.02591)}$$

By trial and error,

If $\overline{V} = 3.0$: $\quad 10^6 = 1{,}043{,}030 - 49{,}174 = 993{,}856$

If $\overline{V} = 2.9$: $\quad 10^6 = 1{,}079{,}321 - 52{,}609 = 1{,}026{,}712$

Interpolating

$$\overline{V} = 2.98 \text{ m}^3/\text{kmol}$$

Exercise 2.F Derive Eq. (2.15).

Exercise 2.G Derive Eq. (2.19) starting with Eq. (2.17).

2.5 GENERALIZED EQUATIONS OF STATE

As discussed earlier the simple theorem of corresponding states as first postulated by van der Waals in 1873 suggested that all pure gases have the same compressibility factors at the same conditions of temperature and pressure. This method was extended to liquids by Young in 1899. This two-parameter theorem then states that $Z = f(P_r, T_r)$. If this theory is true, it logically follows that all pure fluids would have the same departure from ideality at equal reduced conditions. Generalized charts were developed for this theorem and were used for prediction of properties for many years although they were less than satisfactory. This led to the development of three-parameter methods to take into account the deviations in size and shape of molecules. As alluded to earlier, the most successful attempts at such methods were the methods of Lydersen, Greenkorn, and Hougen (1955) using the critical compressibility factor as the third parameter and the method of Pitzer and colleagues (1955, 1957, 1958) using the acentric factor as a third parameter. The former method is simple to use and fairly accurate for a wide variety of compounds, while the latter method was developed for and is more accurate for normal fluids, basically hydrocarbons. Normal fluids fall into a single family of curves which may be characterized by a single parameter when the reduced theoretical second virial coefficients (BP_c/RT_c) of these materials are plotted against T_r^{-2}. Both methods have been put into analytical form for computerized calculations; only the latter will be considered in this book.

Only the compressibility factor and reduced density will be developed in this chapter, while estimation of other thermodynamic properties will be considered as it is used in future chapters.

2.5.1 Lydersen et al. Method

The critical compressibility factor for substances varies from 0.20 to 0.30 and was used to derive this three-parameter approach:

$$Z = f(P_r, T_r, Z_c) \tag{2.26}$$

This parameter was chosen based on the fact that when compressibility factors at saturation are plotted against corresponding values of Z_c at a given reduced saturation pressure, the points for a given saturation pressure fall on a straight line almost exactly for liquids and with some scatter for gases. The correlation is best as the critical pressure is approached and decreases in accuracy at low reduced pressures. Based on 8000 values for 82 different liquids and gases, tables for compressibility factors and reduced densities were developed in four groups ranging from 0.23 to 0.30. Such plots are shown as Figs. 2.4 and 2.6, respectively. The data were generalized at a Z_c of 0.27 as most materials fell in the range of 0.26 to 0.28, and correction tables were developed to correct for a different value

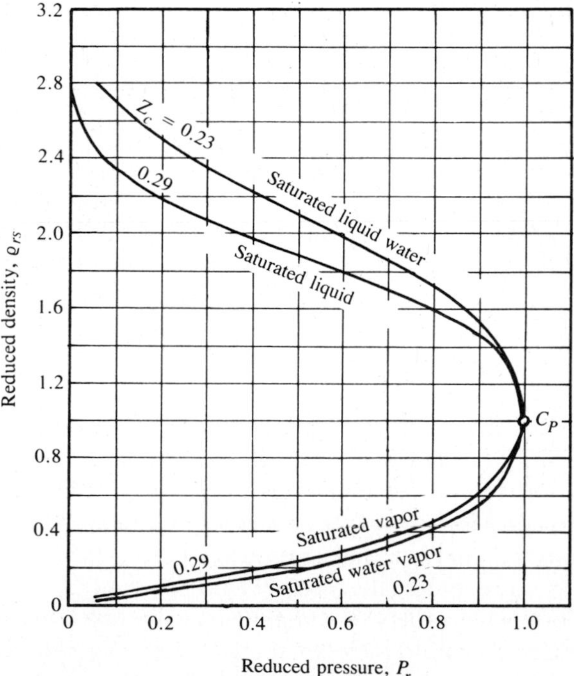

Figure 2.6 Saturation envelopes of reduced density at various critical compressibility factors.

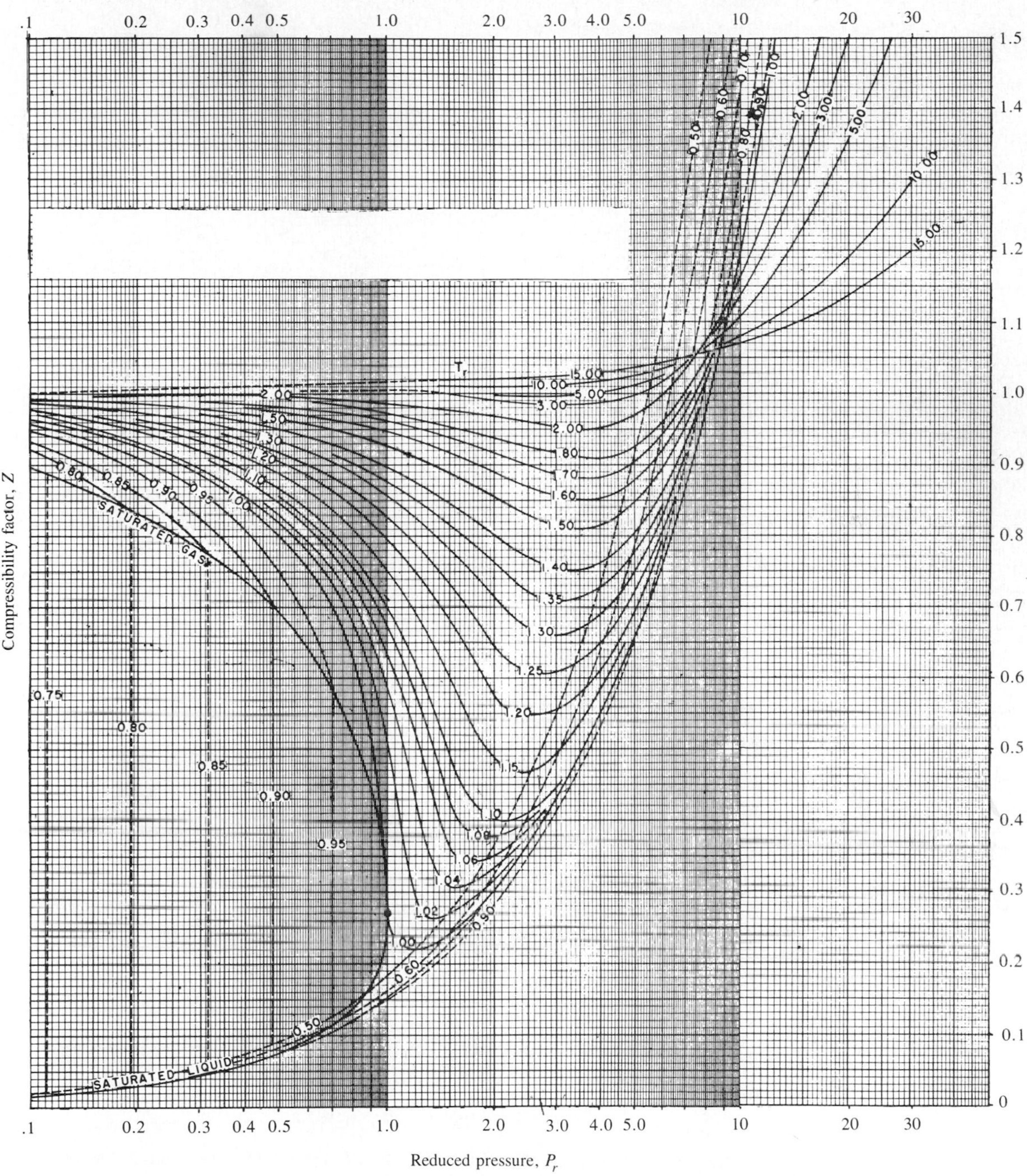

Figure 2.7 Generalized compressibility factors ($Z_c = 0.27$). *(Adapted from Hougen, Watson, and Ragatz, "Chemical Process Principles Charts," 3d ed., copyright 1964 by Wiley. Used with permission.)*

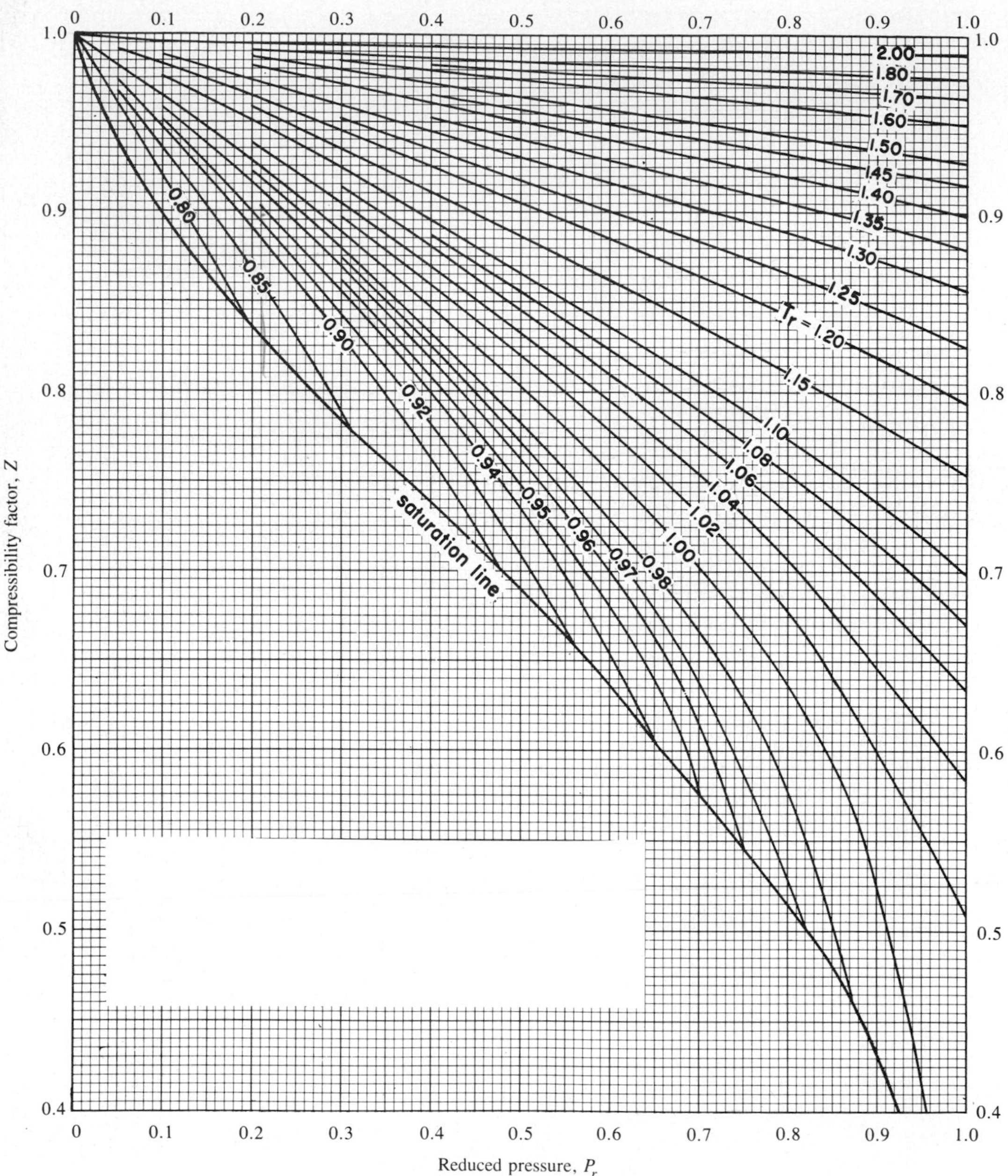

Figure 2.8 Generalized compressibility factors ($Z_c = 0.27$), low-pressure range. *(Adapted from Hougen, Watson, and Ragatz, "Chemical Process Principles Charts," 3d ed., copyright 1964 by Wiley. Used with permission.)*

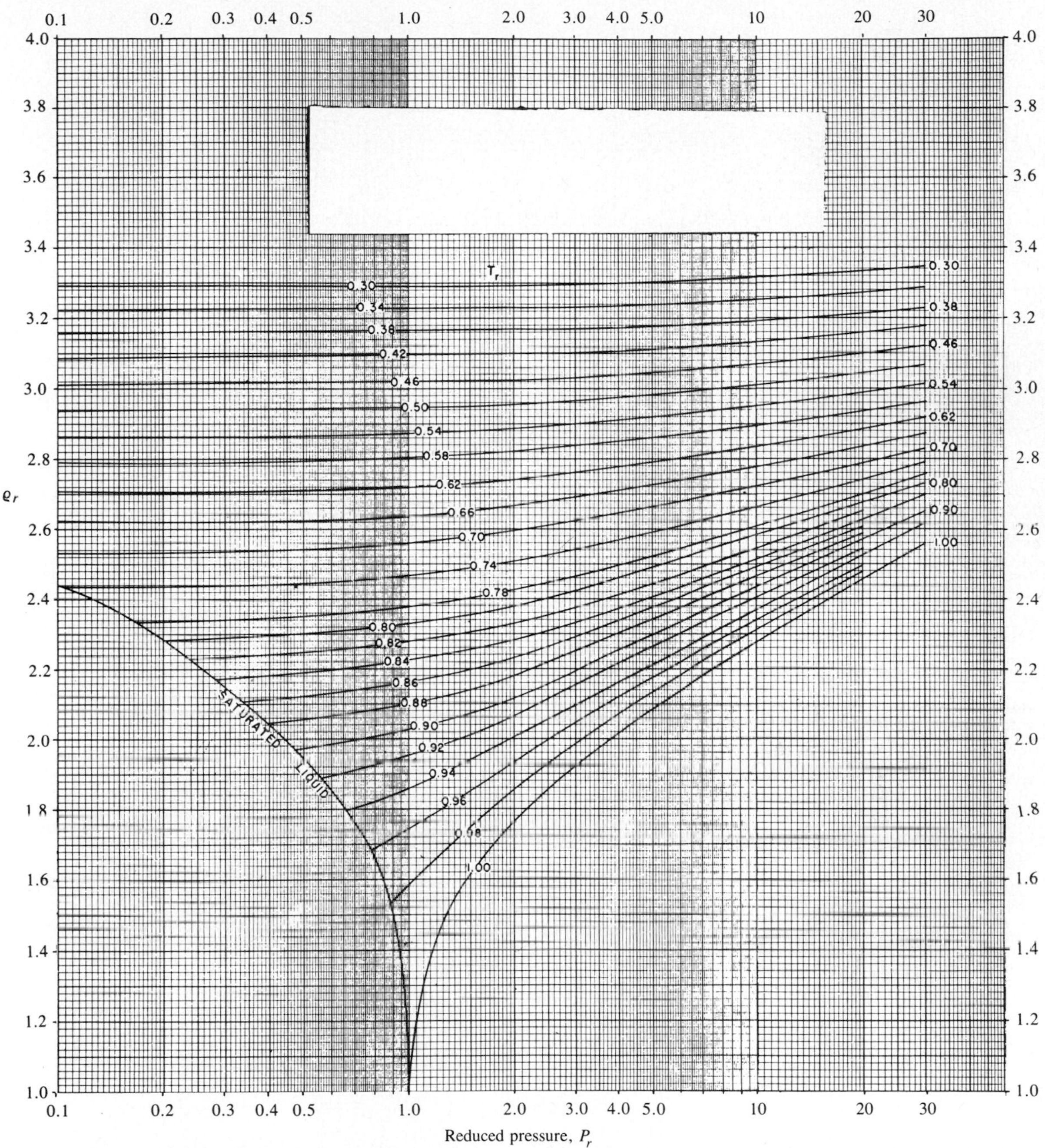

Figure 2.9 Generalized reduced densities of pure liquids ($Z_c = 0.27$). *(Adapted from Hougen, Watson, and Ragatz, "Chemical Process Principles Charts," 3d ed., copyright 1964 by Wiley. Used with permission.)*

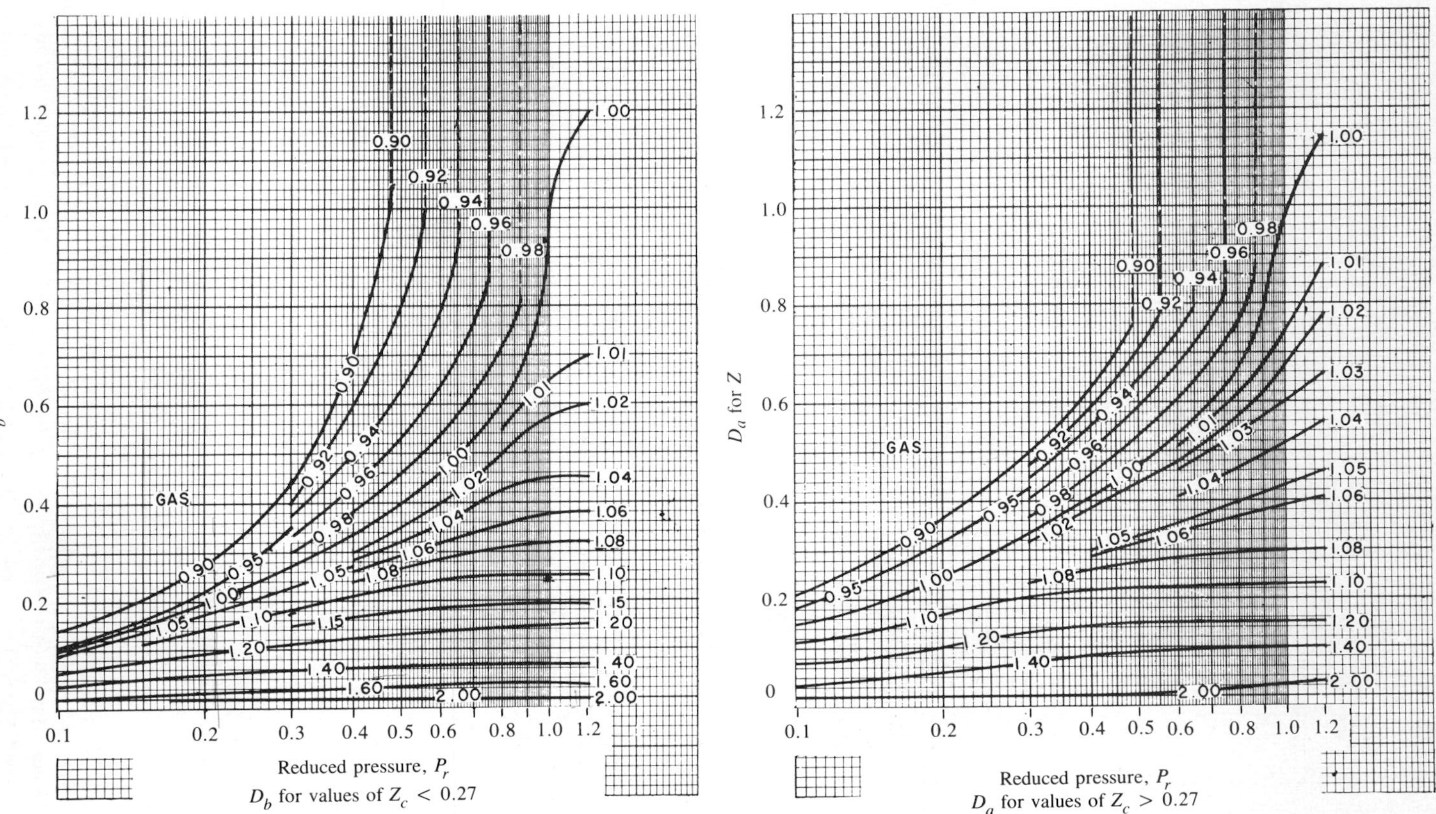

Figure 2.10 Generalized deviation of compressibility factor Z from values at $Z_c = 0.27$ (for vapor phase). *(Adapted from Hougen, Watson, and Ragatz, "Chemical Process Principles Charts," 3d ed., copyright 1964 by Wiley. Used with permission.)*

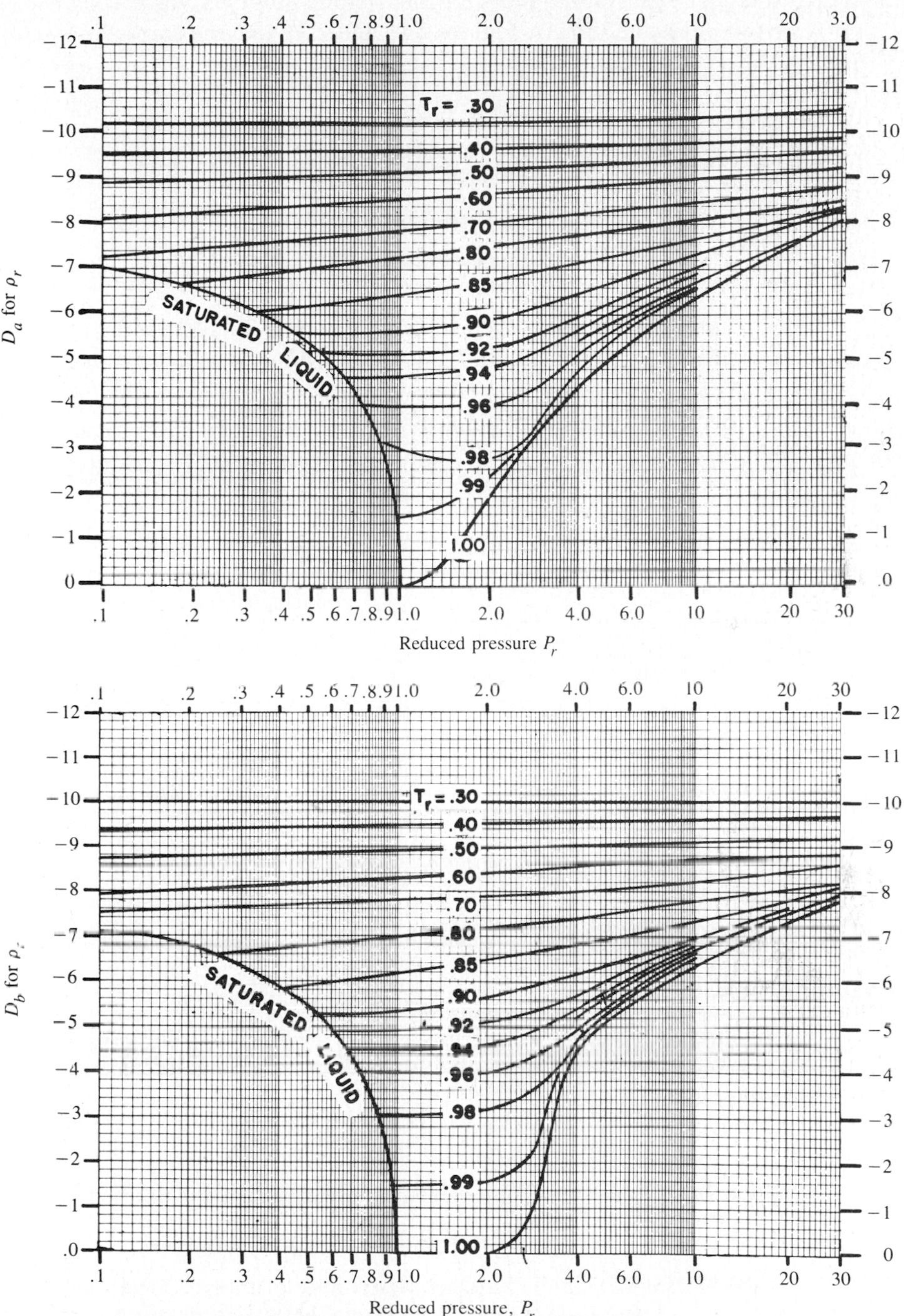

Figure 2.11 Generalized deviation of reduced densities of liquids from values at $Z_c = 0.27$. *(Adapted from Hougen, Watson, and Ragatz, "Chemical Process Principles Charts," 3d ed., copyright 1964 by Wiley. Used with permission.)*

of Z_c. The generalized plots for the compressibility factor and reduced density are given in Figs. 2.7 to 2.9. Figures 2.10 and 2.11 give the correction factors where

$$B' = B + D(Z_c - 0.27)$$

$$B = \text{general property at } Z_c = 0.27$$

$$B' = \text{general property at } Z_c \neq 0.27$$

$$D_a = \text{deviation term, } Z_c > 0.27$$

$$D_b = \text{deviation term, } Z_c < 0.27$$

Example 2.5 Using the results of Example 2.4, determine the compressibility factor of water at 700°C and 25 MPa pressure by using the Lydersen generalized technique. Compare your answer with the value of $Z = 0.97$ derived from experimental data.

$$P_c = 22.12 \text{ MPa}$$

$$T_c = 647.3 \text{ K}$$

$$Z_c = 0.234$$

$$T_r = \frac{973.1}{647.3} = 1.50 \qquad P_r = \frac{25}{22.12} = 1.13$$

From Fig. 2.7,

$$Z_{\text{at } Z_c=0.27} = 0.916$$

$$Z = Z_{\text{at } Z_c=0.27} + D_b(0.234 - 0.27)$$

From Fig. 2.10, $D_b = 0.05$. Therefore

$$Z = 0.916 + 0.05(-0.036)$$
$$= 0.914$$

2.5.2 Pitzer et al. Method

As discussed earlier, the method assumes that any property of a fluid in reduced form can be given as a function of three parameters as illustrated for the compressibility factor:

$$Z = Z(P_r, T_r, \omega)$$

It has been determined that a linear equation in ω is usually adequate:

$$Z = Z^{(0)}(P_r, T_r) + \omega Z^{(1)}(P_r, T_r) \tag{2.27}$$

where $Z^{(0)}$ is the simple fluid term and $Z^{(1)}$ is a correction term. Pitzer presented tabular and graphical values of the functions $Z^{(0)}$ and $Z^{(1)}$. Figures 2.12 and 2.13 give these values as extended and slightly revised by Lee and Kesler (1975). Extended range plots are given in the "Technical Data Book—Petroleum Refining."

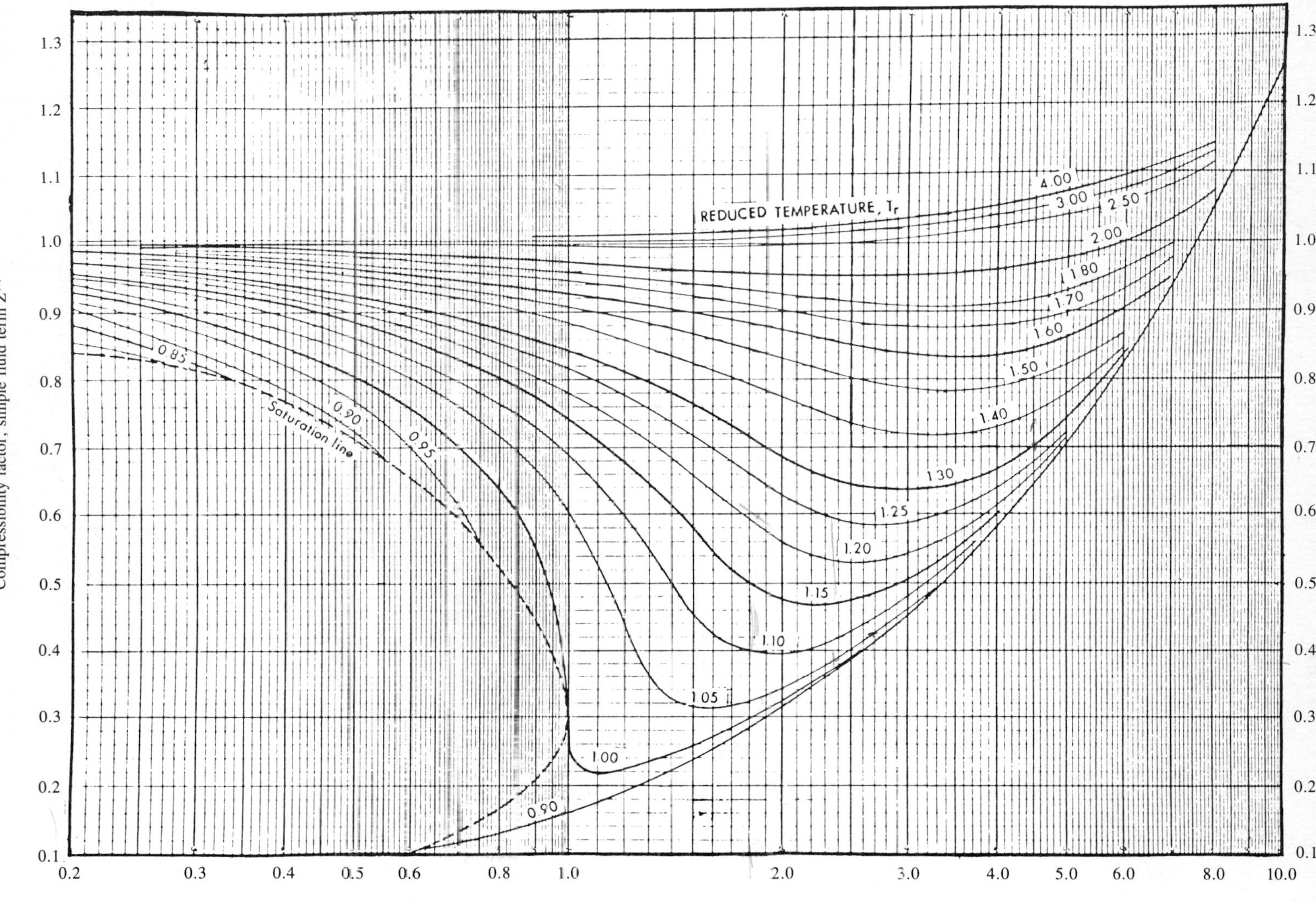

Figure 2.12 Generalized compressibility factors, simple fluid term. *(Adapted from "Technical Data Book—Petroleum Refining," 4th ed., copyright 1982 by American Petroleum Institute. Used with permission.)*

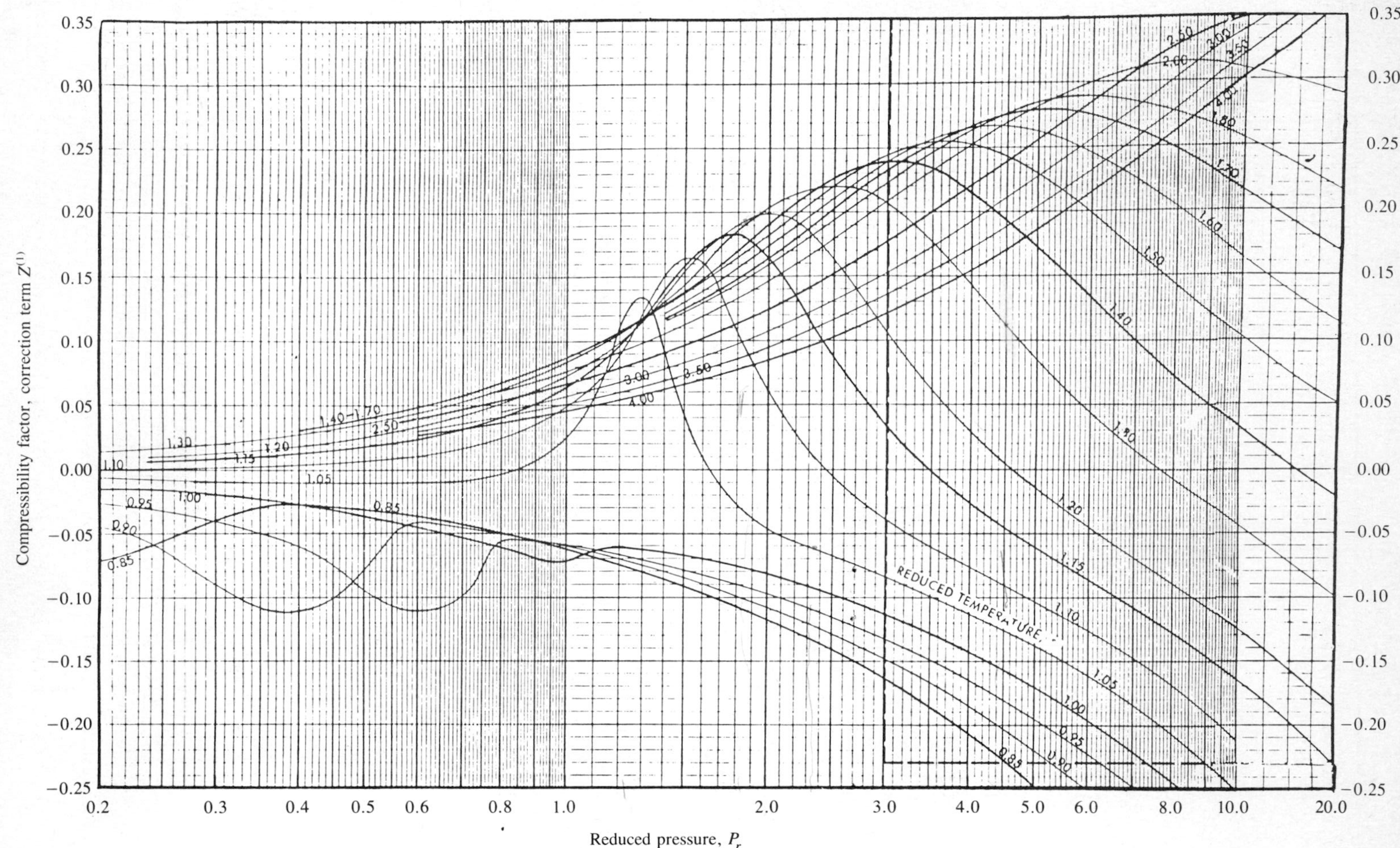

Figure 2.13 Generalized compressibility factors, correction term. *(Adapted from "Technical Data Book—Petroleum Refining," 4th ed., copyright 1982 by American Petroleum Institute. Used with permission.)*

Example 2.6 Repeat Example 2.5 by using the Pitzer generalized technique. Compare your answer with the value of $Z = 0.97$ derived from experimental data.

$$T_r = 1.50 \qquad P_r = 1.13 \qquad \omega = 0.3448$$

By Eq. (2.27),

$$Z = Z^{(0)} + \omega Z^{(1)}$$

From Fig. 2.12:

$$Z^{(0)} = 0.898$$

From Fig. 2.13:

$$Z^{(1)} \approx 0.08$$

Thus

$$Z = 0.898 + (0.08)(0.3448) = 0.926$$

Exercise 2.H Determine the density of water at 1000°C and 25 MPa pressure by using the Lydersen generalized technique.

Exercise 2.I Determine the compressibility factor for 2, 2, 4-trimethylpentane at 30 MPa total pressure and 800 K by using both the Lydersen and Pitzer generalized techniques. Compare the results.

Use of this method to estimate compressibility factors should give values for normal fluids within 1 percent of the true value except in the critical region as shown on the plots where errors up to 25 percent can occur. The plots may also be extrapolated below reduced pressures of 0.2 by noting that $Z^{(0)}$ approaches unity and $Z^{(1)}$ approaches zero as the pressure approaches zero.

Lee and Kesler (1975) have also fit the Pitzer tables by considering a reference fluid, *n*-octane, and fitting the equations by a Benedict-Webb-Rubin type of equation. This method is completely suited for digital-computer use and consists of the equations and constants shown below.

The following method is recommended for estimating the density of pure hydrocarbon and nonpolar gases by a digital computer:

$$Z = Z^{(0)} + \frac{\omega}{\omega^{(h)}}\left(Z^{(h)} - Z^{(0)}\right) \tag{2.28}$$

where Z= compressibility factor, dimensionless

$Z^{(0)}$= compressibility factor for the simple fluid, which is obtained from Eq. (2.29)

$Z^{(h)}$= compressibility factor for the heavy reference fluid (*n*-octane), which is obtained from Eq. (2.29)

ω= acentric factor of the compound for which the density is sought

$\omega^{(h)}$= acentric factor for the heavy reference fluid (*n*-octane) = 0.3978

The compressibility factors for the simple fluid $Z^{(0)}$ and the heavy reference fluid

$Z^{(h)}$ are obtained from the following equation:

$$Z^{(i)} = \frac{P_r V_r}{T_r}$$

$$= 1 + \frac{B}{V_r} + \frac{C}{V_r^2} + \frac{D}{V_r^5} + \frac{c_4}{T_r^3 V_r^2}\left(\beta + \frac{\gamma}{V_r^2}\right) e^{-\gamma/V_r^2} \qquad (2.29)$$

$$B = b_1 - \frac{b_2}{T_r} - \frac{b_3}{T_r^2} - \frac{b_4}{T_r^3}$$

$$C = c_1 - \frac{c_2}{T_r} + \frac{c_3}{T_r^3}$$

$$D = d_1 + \frac{d_2}{T_r}$$

where $Z^{(i)} = Z^{(0)}$ when the constants in the equation correspond to the simple fluid
$Z^{(i)} = Z^{(h)}$ when the constants in the equation correspond to the heavy reference fluid
P_r = reduced pressure, P/P_c
P = pressure, kPa
P_c = critical pressure of the compound whose density is sought, kPa
$V_r = P_c V/RT_c$
V = molar volume of the simple fluid or of the heavy reference fluid, as the case may be, m³/kmol
R = gas constant = 8.3140 (m³·kPa)/(kmol·K)
T_c = critical temperature of the compound whose density is sought, K
T_r = reduced temperature, T/T_c
T = temperature, K

Two sets of the constants b_1, b_2, b_3, b_4, c_1, c_2, c_3, c_4, d_1, d_2, β, and γ are given below, one set for the simple fluid and the other set for the heavy reference fluid:

Constant	Simple fluid	Heavy reference fluid
b_1	0.1181193	0.2026579
b_2	0.265728	0.331511
b_3	0.154790	0.027655
b_4	0.030323	0.203488
c_1	0.0236744	0.0313385
c_2	0.0186984	0.0503618
c_3	0.0	0.016901
c_4	0.042724	0.041577
$d_1 \times 10^4$	0.155488	0.48736
$d_2 \times 10^4$	0.623689	0.0740336
β	0.65392	1.226
γ	0.060167	0.03754

The method is used by solving the equation for $Z^{(i)}$ for V_r using the constants for the simple fluid by an iterative procedure, making certain that the volume converges on the correct root. This is quite simple since incorrect values will deviate substantially from expected values. This same procedure is used for the heavy reference fluid to obtain $Z^{(h)}$. Then the compressibility factor is calculated by using the first equation. The procedure is accurate within the same limits as the graphical or tabular approach. For saturated vapors both temperature and pressure must be used as input parameters, and the calculation is made as if the point were in the homogeneous region. Subroutines for implementing this procedure for compressibility factors are given in App. E.

For nonpolar gas mixtures, the same methods are applicable by using pseudocritical properties, the molar averages of the component true critical properties, for T_c, P_c, and ω. Errors in such use do not normally exceed 2 percent except in the critical region where 15 percent errors are common and 50 percent errors can occur. More complicated and slightly more accurate methods especially in the immediate critical region are applied by using more complex mixing rules. The "Technical Data Book—Petroleum Refining" (procedures 6B2.1 and 6B2.2) discusses such improvements. A subroutine for utilizing the method for mixtures is given in App. E.

2.6 POSTSCRIPT

Equations of state are quite useful for the estimation of thermodynamic properties. However, they have basically only been perfected for normal fluids, hydrocarbons, and some nonhydrocarbon gases. Care should be used in attempting to use such methods for polar compounds with only known critical properties. However, the methods can be used with somewhat more confidence for interpolating and extrapolating a limited amount of experimental data. For certain common nonideal materials such as water, plots and tabulations of experimental data are recommended.

2.7 REFERENCES

Benedict, M., G. Webb, and L. C. Rubin: *J. Chem. Phys.*, **8**:334 (1940).

Chueh, P. L., and J. M. Prausnitz: *AIChE J.*, **13**:1099 (1967).

Edmister, W. C., and B. I. Lee, "Applied Hydrocarbon Thermodynamics, Volume 1," 2d ed., Gulf, Houston, 1984.

Graboski, M. S., and T. E. Daubert: *Ind. Eng. Chem. Process Des. Dev.*, **17**:443 (1978*a*); **17**:448 (1978*b*); **18**:300 (1979).

Kay, W. B.: *Ind. Eng. Chem.*, **28**:1014 (1936).

Lee, B. I., and M. G. Kesler: *AIChE J.*, **21**:510 (1975).

Lewis, G. N., M. Randall, K. S. Pitzer, and L. Brewer: "Thermodynamics," appendix 1, McGraw-Hill, New York, 1961. Also *J. Am. Chem. Soc.*, **77**:3433 (1955) and **79**:2369 (1957) and *Ind. Eng. Chem.*, **50**:265 (1958).

Lydersen, A. L., R. A. Greenkorn, and O. A. Hougen: *Univ. Wis. Eng. Exp. Stn Rep.*, **4** (Oct. 1955). Also in Hougen, O. A., K. M. Watson, and R. A. Ragatz: "Chemical Process Principles, Part Two: Thermodynamics," Wiley, New York, 1959.

Peng, D. Y., and D. B. Robinson: *Ind. Eng. Chem. Fundam.*, **15**:59 (1976).
Redlich, O., and J. N. S. Kwong: *Chem. Rev.*, **44**:233 (1949).
Riedel, L.: *Chem.-Ing.-Tech.*, **26**:83, 259 (1954).
Soave, G.: *Chem. Eng. Sci.*, **27**:1197 (1972).
______: *Int. Chem. Eng. Symp. Ser.*, no. 56, p. 1.2 (1979).
______: *Chem. Eng. Sci.*, **35**:1725 (1980)
van der Waals, J. D.: Doctoral Dissertation, Leiden, Holland (1873).
Webb, G.: *Chem. Eng. Progr.*, **47**:419 (1951).

The literature on equations of state continues to expand at a rapid rate, with symposia being held to review developments almost annually. The most recent symposia which includes both new developments and reviews of previous work that have been published are listed below for reference.

Chao, K. C., and D. B. Robinson (eds.): "Equations of State in Engineering and Research," *Adv. Chem. Ser.*, no. 182 (1979).
Knapp, H., and S. I. Sandler: "Proceedings of the Second International Conference on Phase Equilibria and Fluid Properties in the Chemical Industry," Dechema Frankfurt/Main, 1980 (second international conference on fluid properties and phase equilibria).
Newman, S. A. (ed.): "Chemical Engineering Thermodynamics," Ann Arbor Science Publishers, Ann Arbor, 1983 (thermodynamic sessions of second World Congress of Chemical Engineering, Montreal, October 1981).
O'Connell, J. P., and A. A. Fredenslund (eds.): "Proceedings of the Third International Conference on Fluid Properties and Phase Equilibria for Chemical Process Design," *Fluid Phase Equilibria*, **13** and **14**, 800 pp. (1983).
Storvich, T. S., and S. I. Sandler (eds.): "Phase Equilibria and Fluid Properties in the Chemical Industry," Symposium Series no. 60, American Chemical Society, Washington, 1977 (first international conference on fluid properties and phase equilibria sponsored by Engineering Foundation, Asilomar, Calif.).

2.8 PROBLEMS

1 Using both the ideal-gas law and the van der Waals equation of state, calculate:

(*a*) The pressure in atmospheres of 10 lb mol of nitrogen stored in a 15-ft^3 container at 200°C.

(*b*) The temperature in degrees Fahrenheit of 150 standard ft^3 (measured at 0°C and 1 atm) of nitrogen stored in a 2-ft^3 container at 50 atm total pressure.

2 Estimate the compressibility factor and the density of naphthalene at 3 MPa total pressure by using the corresponding-states method illustrated in Figs. 2.4 and 2.6.

3 Write the van der Waals equation in terms of Z, P_r, and T_r only. Comment on the fact that Z depends only on P_r and T_r.

4 Compare the van der Waals, Redlich-Kwong, Redlich-Kwong-Soave, and Peng-Robinson equations of state in similar forms. Qualitatively analyze how the progression of authors may have arrived at the form of their equations.

5 A cylinder contains air at 25°C and 4000 lb/in^2 atm and has a volume of 8 ft^3. Estimate the amount of air by using the ideal-gas law and the van der Waals equation of state.

6 Calculate the molar volume and the compressibility factor for isopropyl alcohol at 200°C and 1.0 MPa total pressure by using the van der Waals, Redlich-Kwong, and Lydersen equations of state. Compare the results.

7 Calculate the compressibility factor for isobutane at 154.5°C and 8.62 MPa total pressure by using the Redlich-Kwong-Soave and the Pitzer equations of state.

8 Given the extensive data in steam tables for water, state qualitatively how the virial coefficients might be obtained. Comment on the range of applicability of the equation.

9 The second virial coefficient for isopropanol at 200°C is $-388\ \text{cm}^3/\text{g}$. Determine its density at 1 atm and at 10 atm. Repeat if you also know the third virial coefficient to be $-26{,}000\ \text{cm}^6/(\text{g} \cdot \text{mol}^2)$.

10 The successful three-parameter corresponding-states methods use the critical compressibility factor or the acentric factor as a third parameter. For polar fluids a four-parameter method may be necessary. Suggest additional possible third parameters for size-shape characterization and fourth parameters which may take into account polar effects. Comment on each.

11 Chloroform exists at 370°C and 10,000 kPa total pressure. What is its density?

12 Outline in stepwise form the most accurate procedure on how you would proceed to complete the following problem, including estimation of all properties not given in the problem statement.

Estimate the specific volume of 2-methyl-2-butene at 150°C and 3 MPa total pressure by using the Redlich-Kwong-Soave analytical equation of state. The normal boiling point is 38.5°C and the relative density at 25°C is 0.65.

13 Consider the Berthelot equation of state given below. Show how to calculate the constants a and b for any substance by applying only the conditions at the critical point.

$$P = \frac{RT}{\overline{V} - b} - \frac{a}{T\overline{V}^2}$$

14 Consider a thick-walled tube that is to be filled with benzene—liquid and vapor—at 0.1 MPa, after which it is sealed and heated such that it passes through the critical state. What fraction of the volume of the tube must be filled with liquid?

15 Compressibility factors for methane at 182 K have been measured as given below. Calculate the second and third virial coefficients from the data.

P, atm	6.62	9.62	13.67	18.82	24.74	30.58
Z	0.943	0.915	0.875	0.819	0.745	0.652

16 Write the van der Waals equation in the virial equation form.

17 Calculate the saturated-vapor and saturated-liquid densities of Freon 12 by using the Redlich-Kwong-Soave equation and the Pitzer corresponding-states method at 250 lb/in^2 abs. Compare the results with those given in Table 5.2.

18 The Beattie-Bridgeman equation of state is

$$P\overline{V}^2 = RT\left[\overline{V} + B\left(1 - \frac{b}{\overline{V}}\right)\right]\left(1 - \frac{c}{\overline{V}T^3}\right) - A\left(1 - \frac{a}{\overline{V}}\right)$$

Convert this equation to the virial form.

19 Make a table of all equations of state studied and list for each (a) areas of applicability with respect to range and compound type, (b) relative advantages, and (c) relative disadvantages. Finally, develop a short guide for their use.

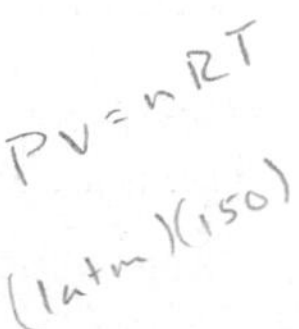

CHAPTER

THREE

CONSERVATION OF ENERGY—FIRST LAW OF THERMODYNAMICS

Clausius summed up the first law of thermodynamics a century ago in a single statement: "Die Energie der Welt ist konstant." This chapter will attempt to develop this law in the various ways necessary to carry out practical chemical engineering calculations. Starting with defining the various types of energy which exist, the energy balances for various types of processes will be generally derived. Following this development, thermodynamic properties relating to the energy content of substances and to the energy involved in various processes will be discussed. Finally, applications to ideal and real gases and to various types of processes will be given.

3.1 TYPES OF ENERGY

Energy can be divided into several types: energy of external position or potential energy, energy of external motion or kinetic energy, energy intrinsically a part of any substance or internal energy, and energy due to processes taking place within a system—heat and work.

3.1.1 Potential Energy

Potential energy is defined as the energy an object contains due to its position above some arbitrary reference plane. For example, consider a mass m suspended

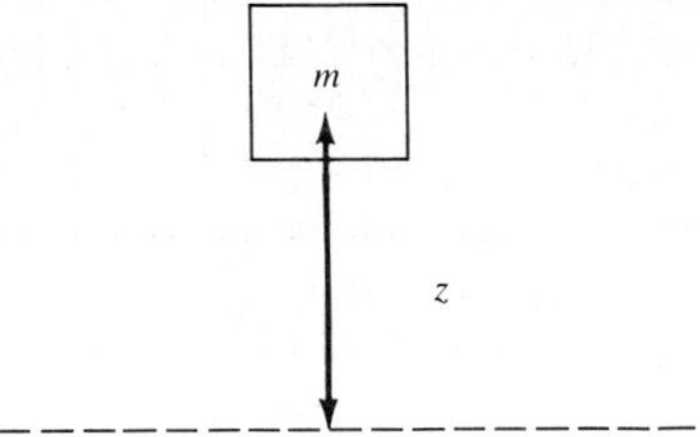

z m above the ground, which will be chosen as a reference plane. The potential energy of the mass is then defined as mgz, where g is the acceleration of gravity, 9.8 m/s^2, in the earth's gravitational field. Thus, the potential energy has the units (kg)(m/s^2)(m), or N · m. An 80-kg diver standing on a 10-m diving platform would have a potential energy equal to (80)(9.8)(10), or 7840 N · m. This energy is independent of temperature and pressure.

3.1.2 Kinetic Energy

External kinetic energy can be defined as the amount of external energy a substance possesses as a result of motion. Thus, if a mass m is moving at a velocity u, its kinetic energy is $m(u^2/2)$. The units would be identical to those of potential energy, kg(m^2/s^2), or N · m. Thus, the diver as he stands on the platform has zero kinetic energy. As he falls through the air, his potential energy is converted to kinetic energy. By the time he hits the water, his potential energy will all be converted to kinetic energy. Thus, his velocity at that point can be calculated by equating the two energies; that is, $7840 = 80(u^2/2)$, or $u = 14$ m/s. External kinetic energy is not considered to be a function of temperature or pressure.

Exercise 3.A A 2-kg weight is dropped from a 16-m platform in a plant. What is its potential energy at the time of release and its kinetic energy as it strikes the ground? At what velocity does it strike the ground?

3.1.3 Internal Energy

Internal energy can be defined as the total energy a substance possesses as a result of the presence of molecules and atoms and their translational, rotational, and vibrational motion as well as the attractions and repulsions between each atomic part of the substance. Although the total absolute internal energy of any substance cannot be calculated, the amount of energy relative to a given temperature and pressure or base state can be calculated. The scientific base for zero internal energy is normally the perfectly ordered crystalline solid at 0° absolute, while the engineering base varies. Often 0°C (or 0°F) and 1 atm pressure is used as a base. Internal energy is given the symbol E and can be calculated from the first law of thermodynamics or estimated as shown later. It is a state property and only depends on the conditions of the system: temperature and pressure. The path of a process occurring between the same two states is immaterial. The internal-energy change between any T_1, P_1 and T_2, P_2 is the same no matter how the change occurs.

The dependence of internal energy of a gas on pressure arises from the forces between molecules. No such forces exist for the ideal gas. Thus, if pressure or volume changes take place at constant temperature and no molecular interactions occur, the internal energy of the ideal gas would not change. For the ideal gas where no interactions occur and for the real gas approaching ideality at zero pressure, the internal energy is independent of pressure and volume and is only a function of temperature.

3.1.4 Work

Work is physically defined as a force acting through a distance. Work cannot be stored as such but is the result of one form of energy being converted to another:

$$dW = F\,dS$$

where W = work
F = force
S = distance

Thermodynamic work as discussed in this chapter does not include surface, electric, or magnetic work since the only form of work of any magnitude in normal applications is pressure-volume work.

Consider a piston of cross-sectional area A compressing a gas of pressure P and volume V. The force of the piston on the gas is PA, while the distance is V/A.

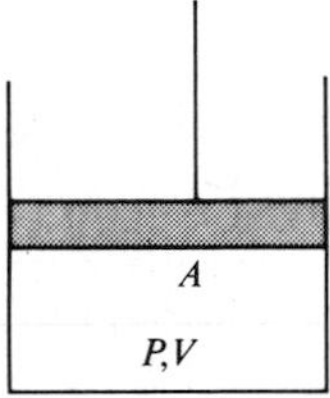

Thus

$$dW = PA\,d\frac{V}{A} = P\,dV$$

Integrating,

$$W = \int_{V_1}^{V_2} P\,dV$$

This is the general expression for a finite compression or expansion process and can be calculated by graphical integration of PV data unless an analytical expression can be substituted for P as a function of V and directly integrated. Work is a path function and will vary with the manner in which a process is carried out. For example, it obviously will require more work to move a piano from the first floor to the second by carrying it outside, hoisting it, and then carrying it inside than to merely raise it through a hole in the ceiling. Standard chemical engineering convention is to consider work done by a system as positive and work done on a system as negative. Physical chemists today use the opposite convention.

Example 3.1 A piston is held in place by a latch which is suddenly released. One kilogram of carbon monoxide is contained in the cylinder initially at 800 kPa and 30°C. The piston has a mass of 4000 kg/m² of surface area. The piston generates friction with the walls. The gas expansion is stopped at triple its initial volume.

Determine the work done on the surroundings, the work done by the gas, and the reversibility of the system.

SOLUTION Taking the gas as the system and the piston, cylinder, and all outside as the surroundings, the work done on the surroundings is

$$W = \int P_S \, dV$$

The pressure on the surroundings is

$$\begin{aligned} P_S &= P_{\text{atm}} + P_{\text{piston}} \\ &= 1.013 \times 10^5 + (4000)(9.807) \\ &= 140{,}528 \text{ N/m}^2 \end{aligned}$$

If carbon monoxide is an ideal gas,

$$\begin{aligned} V_1 &= \frac{nRT}{P_1} = \frac{\frac{1}{28}(8.3143 \times 10^3)(303)}{800 \times 10^3} \\ &= 0.1125 \text{ m}^3 \end{aligned}$$

The work done on the surroundings is

$$\begin{aligned} W &= (140{,}528)(3V_1 - V_1) \\ &= (140{,}528)(2)(0.1125) \\ &= 31{,}609 \text{ J} \end{aligned}$$

The work done by the system is

$$W = \int P_s \, dV$$

However, the pressure of the system changes and the movement of the piston causes frictional heating; thus, the variation of pressure cannot be calculated. The average pressure cannot be used since the pressure of importance is that at the piston face. Some work is lost as heat. As a result the work done by the system is not calculable although it obviously must be greater than the work done on the surroundings.

The process is definitely not reversible as the pressure difference between the system and the surroundings is finite. A differential increase in pressure of the surroundings will not return the system to its original state.

3.1.5 Heat

Heat represents energy transfer caused by a temperature difference. It may cause a change in internal energy of a substance, produce work, or cause changes in potential or kinetic energy. Heat cannot be stored as such and is a path property. It is readily apparent that it would require more heat to keep your electric blanket (the system) at 30°C with a room (the surroundings) at 16°C rather than at 20°C.

Heat added to a system is considered positive by standard chemical engineering convention, while heat taken from a system is considered negative. Again, physical chemists use the opposite convention.

3.1.6 Enthalpy

Enthalpy is a derived thermodynamic quantity which consists of the sum of the internal energy and the pressure-volume product of the system:

$$H = E + PV$$

where H = enthalpy. The PV product has units of energy but only truly represents energy in flow processes.

Enthalpy is a state function and can be tabulated as a function of only temperature and pressure of a substance. It becomes a very useful parameter for many thermodynamic applications and can be estimated from other properties, as will be shown later.

3.2 LAW OF CONSERVATION OF ENERGY—ENERGY BALANCES

The postulate that the energy of the world is constant manifests itself in an expression known as an energy balance. A general energy balance based on unit time can be set up and reduced for flow systems at steady or nonsteady state and for batch or nonflow systems. The general energy balance is equivalent to the general mass balance:

$$\text{Energy input} = \text{energy output} + \text{energy accumulation}$$

Energies to be considered may include any of the following list according to the type of system being analyzed.

Energy carried by the fluid:

1. Internal energy, E.
2. Pressure energy carried by the fluid at the expense of the surroundings, i.e., the energy added in forcing a stream of materials into a system under the restraint of pressure or in forcing a stream from the system, PV.

$$\text{Enthalpy} = (1) + (2) = E + PV$$

3. External potential energy, mgZ.
4. External kinetic energy, $mu^2/2$.
5. Surface energies of materials, $E\sigma$. (These are usually considered negligible except for sprays and emulsions where extremely large surface areas exist.)
6. Net energy change within the system as a result of changes in other energies is often referred to as accumulation of energy. In a steam-state flow process this term would normally be zero, while in a nonflow process the term is quite important.

Energy transferred between the system and surroundings:
1. Net work done by the system including both mechanical and electrical sources, W
2. Net heat added to the system from all sources, Q

In equation form the general energy balance could then be written

$$\text{Energy input} - \text{energy output} = \text{energy accumulation}$$

or

$$\begin{matrix}\text{Difference in energy flow} \\ \text{into and out of the system}\end{matrix} = \begin{matrix}\text{accumulation of energy} \\ \text{in the system}\end{matrix}$$

Based on a unit mass of material,

$$\text{Accumulation in system} = d\left(E + \frac{u^2}{2} + gz\right)$$

$$\text{Energy input} - \text{energy output} = \Delta\left(H + \frac{u^2}{2} + gz\right) + Q' - W'$$

Thus, in general,

$$Q' - W' = \underbrace{d\left(E + \frac{u^2}{2} + gz\right)}_{\text{System}} - \underbrace{\Delta\left(H + \frac{u^2}{2} + gz\right)}_{\text{Flow}} \tag{3.1}$$

In this general equation the work term includes any work of contraction or expansion against the surroundings plus the shaft work W_s' done by the system. For example, for a turbine the work includes W_s' plus $\Delta(PV)$ for the flowing fluid. Equation (3.1) can be simplified for any particular system being studied.

Exercise 3.B Write and simplify the energy balance for the following situations:

1. Nitrogen escaping from a cylinder at high pressure to the atmosphere
2. Gas being charged to an evacuated cylinder from a source at 15 MPa
3. Discharge of oil from a barge to a pipeline

Two important special cases of the overall energy balance follow.

3.2.1 General Energy Balance for a Constant-Volume Nonflow Process

$$\text{Energy input} - \text{energy output} = \text{energy accumulation}$$

Assuming that there is no pressure-volume work caused by streams entering or leaving the system and that changes in potential and kinetic energy are negligible,

$$\text{Energy input} = Q$$
$$\text{Energy output} = W$$
$$\text{Energy accumulation} = \Delta E = E_{\text{out}} - E_{\text{in}}$$

Thus, the first law for a constant-volume nonflow process is

$$\boxed{Q - W = \Delta E} \tag{3.2}$$

The units of all terms on a common mass or mole basis are joules in the SI system and Btu in the English system.

3.2.2 General Energy Balance for a Constant-Pressure Flow Process at Steady State

Consider a constant-pressure, steady-state flow process with one stream (1) entering and one stream (2) leaving the system. Using a basis of a unit of time and including all possible energy terms,

$$\text{Energy inputs} = E_1\dot{m} + \dot{m}\frac{u_1^2}{2} + \dot{m}gz_1 + \dot{m}P_1V_1 + Q$$

$$\text{Energy outputs} = E_2\dot{m} + \dot{m}\frac{u_2^2}{2} + \dot{m}gz_2 + \dot{m}P_2V_2 + W_s$$

$$\text{Energy accumulation} = 0$$

Therefore

$$E_1\dot{m} + \dot{m}\frac{u_1^2}{2} + \dot{m}gz_1 + \dot{m}P_1V_1 + Q = E_2\dot{m} + \dot{m}\frac{u_2^2}{2} + \dot{m}gz_2 + \dot{m}P_2V_2 + W_s \tag{3.3}$$

where $\dot{m}$ = mass flow rate, kg/time
E = internal energy, $\text{N}\cdot\text{m/kg} = \text{J/kg}$
u = velocity, m/time
g = acceleration of gravity, m/time^2
z = distance above reference phase, m
P = pressure, $\text{N/m}^2 = \text{Pa}$
V = volume, $\text{m}^3\text{/kg}$
Q = heat flow, J/time
W_s = shaft work, J/time

This equation can then be written as

$$E_1 + P_1V_1 + \frac{u_1^2}{2} + gz_1 + Q' = E_2 + P_2V_2 + \frac{u_2^2}{2} + gz_2 + W_s' \tag{3.4}$$

where $Q' = Q/\dot{m}$ and $W_s' = W_s/\dot{m}$, and is often called the *total energy balance*. If by definition H is substituted for $E + PV$, the equation simplifies to

$$H_1 + \frac{u_1^2}{2} + gz_1 + Q' = H_2 + \frac{u_2^2}{2} + gz_2 + W_s' \tag{3.5}$$

Equations (3.4) and (3.5) can then be used to evaluate any steady-state flow process. According to the process, certain terms will be important and other terms may be small or negligible. For example, for a water nozzle, the kinetic-energy change will be very large, while the internal-energy, heat, and potential-energy terms will be negligible, reducing the equation to

$$W_s' = \frac{u_1^2 - u_2^2}{2} + P_1V_1 - P_2V_2$$

For a chemical reaction the kinetic-energy, work, and potential-energy terms will normally be negligible, reducing the equation to

$$H_2 - H_1 = \Delta H = Q'$$

For hydroelectric power generation the potential-energy term will be very large.

It should be noted that often practitioners will make the mistake of utilizing a reduced form of the equation in the wrong situation. Thus, any calculation should begin with the entire energy balance, deleting those terms that are negligible for the specific process being analyzed. For example, in nonflow processes at constant pressure where work other than that of expansion is negligible, the heat added to the process is equal to the increase in enthalpy. That is,

$$E_2 - E_1 = Q - W$$
$$= Q - \int_{V_1}^{V_2} P\,dV = Q - P(V_2 - V_1)$$

Therefore

$$\Delta H = Q$$

However, in a nonflow process at constant volume, this formulation is obviously not accurate as $W = 0$ and therefore $\Delta E = Q$. However, the pressure change is no longer zero.

Example 3.2 Consider the following properties of steam:

P, MPa	20	20	10	10	20
T, K	1000	700	600	700	639 (sat)
V, m^3/kg	0.02188	0.01156	0.02008	0.02825	0.00584
H, kJ/kg	3879.3	2961.0	2818.3	3175.6	2409.5
E, kJ/kg	3441.8	2729.9	2617.5	2893.1	2292.8

Go over this

(a) If the steam is cooled at constant volume from 20 MPa and 1000°K to 10 MPa, determine the heat transfer requirement per kilogram of steam.

$$\Delta E = Q - W$$

If $W = 0$,

$$\Delta E = Q$$

If V is constant at 0.02188,

$$T_f = 600 + \frac{0.02188 - 0.02008}{0.02825 - 0.02008}(100) = 622 \text{ K}$$

$$E_f = 2617.5 + 0.22(2893.1 - 2617.5)$$

$$= 2678.1$$

Therefore

$$Q = \Delta E = 2678.1 - 3441.8 = -763.7 \text{ kJ/kg}$$

(b) If the steam in (a) is cooled at constant pressure to 622 K, determine the heat transfer requirement per kilogram of steam.

$$\Delta H = Q \qquad H_f = H_{\text{sat}} - mC_P \Delta T$$

$$H_f = 2409.5 - (1)(1)(4.186)(639 - 622)$$

$$= 2409.5 - 71.2 = 2338.3$$

Therefore

$$Q = \Delta H = 2338.3 - 3879.3 = -1541 \text{ kJ/kg}$$

Exercise 3.C If the chemical reaction shown below occurs in a constant-volume reactor, what will be the direction of the change in internal energy and enthalpy? The feed is stoichiometric. Molar volumes are about equal.

$$P + Q = R$$

Repeat for a constant-pressure reactor.

The total energy balance will be applied to various types of systems in Sec. 3.5.

The *total energy balance* on a steady-state flow system is an external balance only considering inputs and outputs to the system. The system is a black box. For fluid mechanics calculations, it has become useful to design an internal balance of the system, normally called the *mechanical energy balance* or expanded Bernoulli equation. This equation can be easily derived from the differential form of the steady-state total energy balance based on one mass of material by taking into account the following facts. For a reversible process in which only PV work is done, $dE = \delta Q_{\text{rev}} - P\,dV$. Since the reversible work is equal to the sum of the heat absorbed from the surroundings Q' and the heat added to the fluid due to friction h_f,

$$dE = \left(Q' + h_f\right) - P\,dV$$

From Eq. (3.1) and the fact that $W' = W_s' + d(PV)$,

$$g\,dz + dE + d(PV) + u\,du = Q' - W_s'$$

$$g\,dz + h_f - P\,dV + d(PV) + u\,du = -W_s'$$

Integrating between points 1 and 2, the entrance and the exit of the system,

$$gz_1 + P_1V_1 + \frac{u_1^2}{2} + \int_{V_1}^{V_2} P\,dV = gz_2 + P_2V_2 + \frac{u_2^2}{2} + h_f + W_s' \qquad (3.6)$$

This balance is quite useful for fluid mechanics calculations and is discussed in detail in Hougen, Watson, and Ragatz (1959) from the thermodynamic viewpoint and in McCabe and Smith (1976) from the fluid mechanics viewpoint.

One other modification of the mechanical-energy balance would substitute the following identity:

$$\int_1^2 PV = \int_1^2 P\,dV + \int_1^2 V\,dP$$

Therefore

$$\int_1^2 V\,dP = P_2V_2 - P_1V_1 - \int_1^2 P\,dV$$

Thus

$$gz_1 + \frac{u_1^2}{2} = gz_2 + \frac{u_2^2}{2} + h_f + \int_1^2 V\,dP \qquad (3.7)$$

where $\int P\,dV$ = reversible, nonflow work
$-\int V\,dP$ = reversible, shaft work

3.3 HEAT CAPACITY OF VAPORS, LIQUIDS, AND SOLIDS

The heat capacity of a substance is defined as the amount of heat required to increase its temperature one degree. Specific heat is the ratio of the heat capacity of a substance to the heat capacity of an equal mass of water. Specific heat has no units but depends on both the temperature of the substance and the temperature of water. Normally water at 15°C is taken as the reference.

$$C = \frac{dq}{dT} \qquad (3.8)$$

where C = heat capacity, J/deg
dq = heat added to produce a change in temperature dT

If a substance is heated at constant volume in a nonflow process, the energy balance reduces to $dq = dE$. Thus, the heat capacity at constant volume can be defined as the change in internal energy with temperature:

$$C_V = \left(\frac{\partial E}{\partial T}\right)_V \qquad (3.9)$$

If a substance is heated at constant pressure, then $dq = dE + P\,dV$. Since $dH = dE + P\,dV$ for a constant-pressure process, the heat capacity at constant pressure can be defined as the change in enthalpy with temperature:

$$C_P = \left(\frac{\partial H}{\partial T}\right)_P \tag{3.10}$$

For an ideal gas, $PV = nRT$. Since the internal energy is independent of volume or pressure, $(\partial E/\partial T)_P = (\partial E/\partial T)_V$. Differentiating Eq. (3.2) at constant pressure,

$$\left(\frac{\partial H}{\partial T}\right)_P = \left(\frac{\partial E}{\partial T}\right)_P + P\left(\frac{\partial V}{\partial T}\right)_P$$

Differentiating the ideal-gas law at constant pressure for 1 mole of gas,

$$\left(\frac{\partial V}{\partial T}\right)^* = \frac{R}{P}$$

Combining the three equations,

$$\left(\frac{\partial H}{\partial T}\right)_P = \left(\frac{\partial E}{\partial T}\right)_V + R$$

or

$$\boxed{C_P = C_V + R} \tag{3.11}$$

For ideal monatomic gases, only translational energy is important. The kinetic theory of gases states that $E = \frac{1}{2}mnu^2 = \frac{3}{2}RT$. Thus $(\partial E/\partial T)_V = C_V = \frac{3}{2}R$. For ideal diatomic gases, both translational and rotational contribution are important, and $E = \frac{5}{2}RT$ or $(\partial E/\partial T)_V = C_V = \frac{5}{2}R$. For all other gases, additional rotational and vibrational contributions will increase the heat capacities.

Heat capacities of ideal gases have been tabulated by API Research Project 44 (1983) and are available from that source. Common ideal-gas heat capacities are given in Table 3.1 as a function of temperature.

The most generally used method for correlating ideal-gas heat capacities is by a power series in temperature. A consistent set of heat capacities for common gases and all hydrocarbons for which heat capacity data are reported in API Research Project 44 was developed by Passut and Danner (1972), with additional compounds added by Huang and Daubert (1974). Their correlation equation is

$$C_P^* = B + 2CT + 3DT^2 + 4ET^3 + 5FT^4 \tag{3.12}$$

The constants B to F for this equation are given in Table 3.2 for a subset of the compounds correlated. The equations reproduce the tables within an average error of 0.5 percent. In addition, these same constants can be used for a completely consistent set of ideal-gas enthalpy data with an equation of the form

$$H^* = A + BT + CT^2 + DT^3 + ET^4 + FT^5 \tag{3.13}$$

using only the additional constant A given in Table 3.2.

Table 3.1 Common ideal-gas heat capacities C_P, kJ / (kg · K)

	T, °C					
	0	25	100	200	400	800
Oxygen	0.912	0.916	0.933	0.962	1.024	1.098
Hydrogen	14.322	14.358	14.437	14.499	14.591	15.109
Water	1.857	1.864	1.893	1.945	2.076	2.334
Hydrogen sulfide	0.999	1.006	1.034	1.080	1.186	1.379
Nitrogen	1.037	1.038	1.043	1.055	1.092	1.183
Ammonia	2.067	2.101	2.221	2.406	2.798	3.411
Carbon monoxide	1.037	1.038	1.045	1.060	1.105	1.201
Carbon dioxide	0.822	0.846	0.915	0.993	1.114	1.254
Sulfur dioxide	0.611	0.625	0.664	0.713	0.789	0.857
Methane	2.187	2.240	2.444	2.785	3.540	4.706
Ethane	1.660	1.753	2.050	2.464	3.237	4.192
Propane	1.571	1.675	1.997	2.424	3.180	4.060
n-Butane	1.603	1.712	2.023	2.442	3.144	4.036
n-Hexane	1.554	1.660	1.988	2.415	3.107	4.242
n-Decane	1.537	1.645	1.974	2.398	3.071	4.047
Cyclopentane	1.035	1.172	1.560	2.023	2.770	3.691
Cyclohexane	1.107	1.248	1.650	2.137	2.935	3.997
Methylcyclopentane	1.156	1.293	1.675	2.123	2.836	3.726
Methylcyclohexane	1.210	1.355	1.759	2.236	2.992	3.904
Ethylene	1.461	1.546	1.800	2.128	2.706	3.451
Propylene	1.422	1.524	1.817	2.172	2.764	3.547
1-Butene	1.413	1.525	1.840	2.213	2.817	3.603
Isobutylene	1.487	1.591	1.885	2.238	2.824	3.611
1, 3-Butadiene	1.351	1.467	1.779	2.122	2.623	3.188
Benzene	0.945	1.050	1.340	1.672	2.181	2.765
Toluene	1.020	1.125	1.418	1.756	2.283	2.911
Ethylbenzene	1.102	1.210	1.511	1.858	2.396	3.033
o-Xylene	1.159	1.255	1.527	1.851	2.376	3.017
m-Xylene	1.096	1.199	1.487	1.824	2.360	3.018
p-Xylene	1.099	1.197	1.474	1.808	2.349	3.007

The effect of pressure on heat capacity of gases is negligible below 1 atm pressure and below the critical temperature. At higher pressures, the constant-pressure heat capacity increases with increasing pressure and reaches infinity at the critical point. At pressures above the critical pressure, pressure effects decrease with increase in temperature. Such effects can be estimated by corresponding-states principles, as will be discussed in a later chapter.

To simplify an energy balance, it is often desirable to use a mean heat capacity over the temperature range desired rather than to integrate a polynomial expression. Since

$$\left(\frac{\partial H}{\partial T}\right)_P = C_P \tag{3.14}$$

Table 3.2 Coefficients for the ideal-gas enthalpy, entropy, and heat capacity equations*

CPD no.	Compound name	A	B	$C \times 10^3$	$D \times 10^6$	$E \times 10^9$	$F \times 10^{13}$	G	Range C
					Nonhydrocarbons				
1	Oxygen	−2.283574	0.952440	−0.281140	0.655223	−0.452316	1.087744	2.080310	−175 to 1200
2	Hydrogen	28.671997	13.396156	2.960131	−3.980744	2.661667	−6.099863	−11.801371	−175 to 1200
3	Water	−5.729915	1.915007	−0.395741	0.876232	−0.495086	1.038613	0.702815	−175 to 1200
4	Hydrogen sulfide	−1.437049	0.998865	−0.184315	0.557087	−0.317734	0.636644	1.394812	−175 to 1200
5	Nitrogen	−2.172507	1.068490	−0.134096	0.215569	−0.078632	0.069850	1.805409	−175 to 1200
6	Ammonia	−2.202606	2.010317	−0.650061	2.373264	−1.597595	3.761739	0.990447	−175 to 1200
7	Carbon[†]	9.572700	−0.199901	1.535456	0.267516	−0.811532	2.726889	1.687611	−175 to 1200
8	Carbon monoxide	−2.269176	1.074015	−0.172664	0.302237	−0.137533	0.200365	2.018445	−175 to 1200
9	Carbon dioxide	11.113744	0.479107	0.762159	−0.359392	0.084744	−0.057752	2.719180	−175 to 1200
10	Sulfur dioxide	3.243188	0.461650	0.248915	0.120900	−0.188780	0.568232	2.086924	-175 to 1200
					Paraffins				
11	Methane	−16.228549	2.393594	−2.218007	5.740220	−3.727905	8.549685	−0.339779	−175 to 1200
12	Ethane	−0.049334	1.108992	−0.188512	3.965580	−3.140209	8.008187	1.995889	−175 to 1200
13	Propane	−1.717565	0.722648	0.708716	2.923895	−2.615071	7.000545	2.289659	−175 to 1200
14	*n*-Butane	17.283134	0.412696	2.028601	0.702953	−1.025871	2.883394	2.714861	−75 to 1200
15	2-Methylpropane	26.744208	0.195448	2.523143	0.195651	−0.772615	2.386087	3.466595	−75 to 1200
16	*n*-Pentane	63.201677	−0.011701	3.316498	−1.170510	0.199648	−0.086652	4.075275	−20 to 1200
17	*n*-Hexane	−17.191071	0.959226	−0.614725	6.142101	−6.160952	20.868190	−0.207040	−75 to 700
18	*n*-Heptane	−0.153725	0.754499	0.261728	4.366358	−4.484510	14.842099	0.380048	−75 to 700
19	*n*-Octane	2.604725	0.724670	0.367845	4.142833	−4.240199	13.734055	0.327588	−75 to 700
20	*n*-Nonane	4.000278	0.707805	0.438048	3.969342	−4.043158	12.876028	0.257265	−75 to 700
21	*n*-Decane	−6.962020	0.851375	−0.263041	5.521816	−5.631733	18.885443	−0.412446	−75 to 700
22	*n*-Eicosane	59.163624	−0.095147	3.456592	−1.360776	0.267410	−0.147933	2.963523	−20 to 1200
					Naphthenes				
23	Cyclopentane	134.396280	−0.730818	3.676931	−1.071943	−0.063241	0.823378	7.295290	−20 to 1200
24	Methylcyclopentane	127.244412	−0.684541	4.005691	−1.681762	0.357838	−0.218737	6.766396	−20 to 1200

25	Cyclohexane	108.312586	−0.627383	3.446132	−0.525507	0.437375	1.667307	6.150081	−20 to 1200
26	Methylcyclohexane	107.598271	−0.705014	4.103364	−1.528647	0.183407	0.266355	6.267710	−20 to 1200
27	Ethylcyclohexane	60.932524	−0.355703	3.224107	−0.384025	−0.541312	2.002343	4.583946	−20 to 1200
				Olefins					
28	Ethene	60.093536	0.606930	1.288788	1.033636	−1.099537	2.929326	4.489853	−20 to 1200
29	Propene	56.369991	0.128994	2.646910	−0.671019	−0.055255	0.494690	5.117553	−20 to 1200
30	1-Butene	76.155333	−0.077537	3.213039	−1.275922	0.261813	−0.153755	5.138521	−20 to 1200
31	*Cis*-2-Butene	131.751917	−0.179174	3.040359	−0.928242	0.032840	0.386271	5.696056	−20 to 1200
32	*Trans*-2-Butene	40.122265	0.155046	2.676286	−0.760249	0.038694	0.195440	3.908880	−20 to 1200
33	2-Methylpropene	34.814312	0.138200	2.850686	−0.994755	0.170328	−0.076840	3.876160	−20 to 1.200
34	1-Hexene	63.931831	−0.017846	3.162701	−1.196596	0.225939	−0.118897	3.941543	−20 to 1200
35	1-Decene	60.316832	−0.055226	3.292792	−1.286107	0.251849	−0.138976	3.374782	−20 to 1200
36	1-Hexadecene	59.389339	−0.084092	3.384826	−1.354812	0.273895	−0.157003	3.089307	−20 to 1200
37	Cyclopentene	83.920312	−0.250908	2.228924	0.451583	−0.948697	2.850605	5.325994	−20 to 1200
				Diolefins and acetylenes					
38	1, 3-Butadiene	94.816692	−0.421207	4.259386	−2.880527	1.179493	−2.082612	6.362204	−20 to 1200
39	Ethyne (Acetylene)	81.982824	0.094773	4.114197	−4.037767	2.133692	−4.415085	6.190109	−20 to 1200
				Aromatics					
40	Benzene	84.467062	−0.513560	3.248740	−1.543913	0.365037	−0.248222	5.631041	−20 to 1200
41	Methylbenzene	74.164254	−0.423499	3.184606	−1.439865	0.326620	−0.212757	5.167448	−20 to 1200
42	Ethylbenzene	70.553907	−0.392022	3.308891	−1.527851	0.356058	−0.238744	4.849365	−20 to 1200
43	1,2-Dimethylbenzene	32.516503	−0.062591	2.518936	−0.656669	−0.112362	0.749621	3.271024	−20 to 1200
44	1,3-Dimethylbenzene	59.286530	−0.288481	3.010730	−1.254453	0.258772	−0.151750	4.379096	−20 to 1200
45	1,4-Dimethylbenzene	43.042514	−0.125980	2.487353	−0.534952	−0.200507	0.955170	3.634082	−20 to 1200
46	*n*-Propylbenzene	75.604397	−0.418290	3.517918	−1.721802	0.431672	−0.305210	4.835570	−20 to 1200
47	Isopropylbenzene	58.761110	−0.354917	3.335217	−1.509732	0.341841	−0.223950	4.458348	−20 to 1200

* $H = 0$ kJ/kg for the ideal gas at 0 K. $S = 1$ kJ/(kg · K) for the ideal gas at 0 K and 1 kPa pressure. Units are kJ/kg for enthalpy, kJ/(kg · K) for heat capacity and entropy.

† Solid graphite (not ideal gas).

Therefore

$$dH = C_P \, dT$$

$$\Delta H = \int C_P(T)\, dT$$

The mean heat capacity is then merely defined as

mean heat capacity

$$C_{Pm} = \frac{\int_{T_1}^{T_2} C_P(T)\, dT}{T_2 - T_1} \tag{3.15}$$

Values for common gases for simplified calculations are given in Table 3.3.

This method is useful over narrow temperature ranges where the heat capacity does not vary appreciably. Use of the table avoids the necessity of integrating heat capacity data.

Exercise 3.D Using the mathematical definitions of C_P and C_V justify why these quantities are appreciably different for gases and essentially the same for liquids and solids.

Exercise 3.E Show that Eqs. (3.12) and (3.13) are completely consistent thermodynamically.

Exercise 3.F Compare the value of the mean molal heat capacity between 25 and 1000°C for nitrogen in Table 3.3 with that derived from the data of Table 3.2.

Heat capacities of solids vary considerably according to the molecular structure, homogeneity or heterogeneity of the mixture, and phase transitions

Table 3.3 Mean constant-pressure heat capacities between 25°C and T(°C), kJ / (kg · K)

T	O_2	H_2	H_2O	H_2S	N_2	NH_3	CO	CO_2	SO_2	CH_4	C_2H_6	C_3H_8	C_2H_4	C_6H_6
100.0	0.924	14.400	1.877	1.020	1.040	2.160	1.053	0.881	0.644	2.339	1.900	1.836	1.673	1.199
200.0	0.937	14.440	1.901	1.041	1.045	2.247	1.053	0.924	0.670	2.493	2.103	2.051	1.841	1.377
300.0	0.952	14.469	1.928	1.064	1.051	2.341	1.059	1.961	0.693	2.667	2.309	2.261	2.001	1.536
400.0	0.967	14.495	1.958	1.089	1.059	2.437	1.068	0.995	0.714	2.850	2.508	2.459	2.154	1.679
500.0	0.982	14.522	1.990	1.115	1.069	2.532	1.079	1.025	0.733	3.033	2.696	2.644	2.296	1.805
600.0	0.996	14.557	2.023	1.141	1.079	2.625	1.090	1.052	0.749	3.211	2.869	2.811	2.426	1.918
700.0	1.009	14.600	2.056	1.167	1.089	2.712	1.101	1.076	0.763	3.379	3.025	2.960	2.546	2.018
800.0	1.019	14.653	2.088	1.192	1.100	2.794	1.113	1.097	0.775	3.536	3.165	3.092	2.654	2.108
900.0	1.029	14.717	2.119	1.215	1.111	2.872	1.124	1.116	0.784	3.681	3.291	3.211	2.752	2.188
1000.0	1.037	14.790	2.150	1.238	1.121	2.945	1.134	1.134	0.793	3.818	3.407	3.320	2,842	2.056
1100.0	1.046	14.868	2.180	1.259	1.131	3.017	1.144	1.150	0.801	3.950	3.520	3.425	2.926	2.328
1200.0	1.055	14.948	2.210	1.280	1.139	3.090	1.152	1.164	0.809	4.083	3.637	3.536	3.008	2.389

Table 3.4 Heat capacities C_P of representative solids, J / (mol · K), at 25°C, 0.1 MPa

Compound	Formula	MW	C_P
Iodine	I_2	253.8	54.438
Sulfur (rhombic)	S	32.064	22.64
Ammonium chloride	NH_4Cl	53.492	84.1
Phosphorus, white	P	30.974	23.84
Phosphoric acid	H_3PO_4	98.00	106.06
Carbon (graphite)	C	12.011	8.527
Carbon (diamond)	C	12.011	6.113
Urea	$CO(NH_2)_2$	60.056	93.14
Silicon	Si	28.086	20.00
Silicon dioxide (quartz)	SiO_2	60.085	44.43
Lead	Pb	207.19	26.44
Aluminum	Al	26.982	24.35
Aluminum oxide, α	Al	101.961	79.04
Aluminum chloride	$AlCl_3$	133.341	91.84
Aluminum sulfate	$Al_2(SO_4)_3$	342.148	259.41
Mercuric oxide, red	HgO	216.589	44.06
Copper	Cu	63.54	24.435
Cupric oxide	CuO	79.539	42.30
Cuprous oxide	Cu_2O	143.079	63.64
Cupric chloride	$CuCl_2$	134.446	71.88
Cupric sulfate	$CuSO_4$	159.602	100.0
Silver	Ag	107.87	25.351
Silver oxide	Ag_2O	231.739	65.86
Silver chloride	AgCl	143.323	50.79
Silver nitrate	$AgNO_3$	169.875	93.05
Gold	Au	196.967	25.418
Nickel	Ni	58.710	26.07
Nickel oxide	NiO	74.709	44.31
Iron	Fe	55.847	25.10
Ferrous chloride	$FeCl_2$	126.753	76.65
Ferric oxide	Fe_2O_3	159.692	103.85
Ferrous sulfate	$FeSO_4$	151.909	100.58
Palladium	Pd	106.40	25.98
Platinum	Pt	195.09	25.86
Manganese, α	Mn	54.938	26.32
Chromium	Cr	51.996	23.35
Tungsten	W	183.85	24.27
Vanadium pentoxide	V_2O_5	181.881	127.65
Titanium dioxide, rutile	TiO_2	79.899	55.02
Magnesium	Mg	24.312	24.89
Magnesium chloride	$MgCl_2$	95.218	71.38
Magnesium sulfate	$MgSO_4$	120.374	96.48
Calcium, α	Ca	40.08	25.31
Calcium oxide	CaO	56.079	42.80
Calcium hydroxide	$Ca(OH)_2$	74.095	87.49
Calcium chloride	$CaCl_2$	110.986	72.59
Sodium hydroxide	NaOH	39.997	59.54
Sodium chloride	NaCl	58.443	50.50
Sodium sulfate	Na_2SO_4	142.041	128.20
Sodium nitrate	$NaNO_3$	84.995	92.88
Sodium carbonate	Na_2CO_3	105.989	112.30
Sodium bicarbonate	$NaHCO_3$	84.007	87.61
Potassium hydroxide	KOH	56.109	64.9
Potassium chloride	KCl	74.555	51.30
Potassium sulfate	K_2SO_4	174.266	131.46

Abstracted from Wagman et al., "The NBS Tables of Chemical Thermodynamic Properties," *J. Phys. Chem. Ref. Data, Suppl. 2*, **11** (1982).

which have occurred. Generally experimental data are used since no good predictive methods are available. Heat capacities of solids increase moderately with temperature, and pressure has no appreciable effect. A representative selection of solid heat capacities is given as Table 3.4. Additional data are given by Perry (1973).

Heat capacities of liquids are usually greater than that of the corresponding solid or vapor. They normally increase with increasing temperature. Pressure has moderate effects on the properties. Some representative values are given in Table 3.5. The method of Hadden (1964) is recommended for prediction of liquid heat capacities of pure hydrocarbons below the normal boiling point, while corresponding-states methods to be discussed later are recommended for predictions above the normal boiling point. Weight averages of pure-component liquid heat capacities are used for determining liquid heat capacities of defined mixtures.

For aqueous solutions, the heat capacities normally decrease with increasing solute concentration. Experimental data are normally utilized.

Table 3.5 Selected liquid heat capacities* C_P, [kJ / (kmol · K)] 10^3; T = K

Liquid	C_P	T	C_P	T
Dichlorodifluoromethane	97.68	160	148.6	370
Carbon tetrachloride	124.2	233	260.3	450
Chloroform	107.8	210	134.2	450
Methyl chloride	74.55	177	93.34	360
Methanol	70.45	175	127.9	425
Acetic acid	123.0	293		
Acetone	116.7	193	129.8	320
Isopropyl alcohol	109.9	185	277.9	400
1,2-Propylene glycol	153.0	213	298.5	540
Glycerol	215.4	291	341.0	540
n-Butane	112.9	134	184.9	380
Diethyl ether	145.5	157	180.4	330
n-Pentane	140.8	149	169.2	303
Cyclohexane	148.1	279	203.2	400
Benzene	132.1	279	176.4	440
n-Hexane	169.4	178	226.2	366
n-Octane	232.5	216	297.8	390
Biphenyl	229.0	303	410.1	623
n-Hexadecane	498.4	291	527.1	335
Nitric acid	111.9	238	109.7	320
Water	76.14	273	78.41	400
Sulfuric acid	139.1	298	177.5	1500
Ammonia	75.96	195	116.9	373
Phosphoric acid	168.2	375	371.5	1050
Sulfur dioxide	84.29	253	152.6	413
Sulfur trioxide	258.1	303		

* Liquid heat capacities are given at 1 atm pressure below the boiling point and at saturation pressure above the normal boiling point.

3.4 HEATS OF FUSION, VAPORIZATION, FORMATION, REACTION, COMBUSTION, AND SOLUTION

3.4.1 Latent Heats of Fusion and Vaporization

As a pure substance changes phases an amount of heat is either consumed or released at constant temperature. A schematic plot of enthalpy of a pure compound vs. temperature illustrates this phenomenon (Fig. 3.1).

When a crystalline solid fuses at its melting point to form a liquid at constant temperature, heat is absorbed, causing an increase in enthalpy of the compound. This heat is called the *heat of fusion* of the substance. Methods have been proposed to estimate the heat of fusion and are available in compilations such as Reid, Prausnitz, and Sherwood (1978). However, no method is very accurate. Some examples of heats of fusion as reported by various investigators are given in Table 3.6.

It should be noted that many crystalline materials will exhibit transition points where changes in crystalline structure occur; e.g., iron exists in four crystalline states α, β, γ, and δ. Each of these points has an accompanying absorption or evolution of heat. The transition from a crystalline state at low temperature to a state at a higher temperature requires absorption of heat. For example, iron absorbs 363, 313, and 106 kcal/kmol of heat in changing from $\alpha \rightarrow \beta \rightarrow \gamma \rightarrow \delta$ at temperatures of 770, 910, and 1400°C, respectively.

As the temperature of a liquid is raised, a point is reached where the liquid vaporizes accompanied by an absorption of heat necessary to overcome the intermolecular forces of attraction in the liquid as well as to supply the energy required to allow the vapor to expand against an external pressure. This heat absorption is called the *latent heat of vaporization* of the liquid and is a function of temperature and pressure.

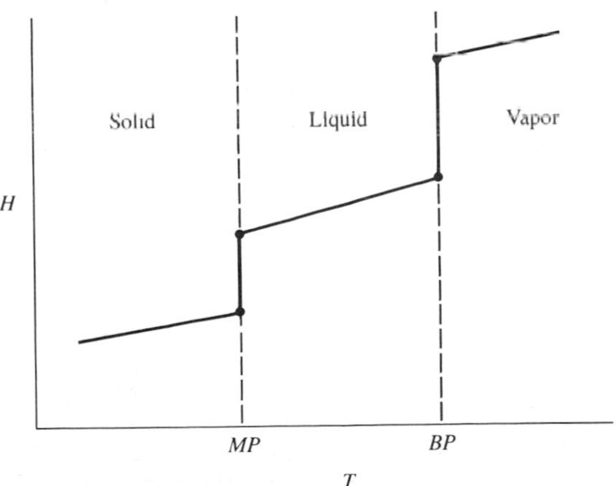

Figure 3.1 Schematic plot of enthalpy vs. temperature for a pure compound.

Clapeyron (1834) developed an exact relationship between vapor pressure and temperature:

$$\frac{dP^\circ}{dT} = \frac{\lambda}{T(\overline{V}_G - \overline{V}_L)} \tag{3.16}$$

where P° = vapor pressure
T = absolute temperature
λ = latent heat of vaporization at T
$\overline{V}_G$ = gas volume
$\overline{V}_L$ = liquid volume

Use of experimental vapor pressure and molar volume data allow exact calculation of the latent heat at any temperature. However, the paucity of data on accurate liquid molar volumes as a function of temperature makes use of this equation limited.

Table 3.6 Common heats of fusion*
All heats are given at the melting point in kJ/kmol.

	Hydrocarbons		
Ethylene	3,351	Toluene	6,636
Propylene	3,003	Ethylbenzene	9,184
1,3-Butadiene	7,985	Styrene	10,950
Cyclohexane	2,677	Cumene	7,786
Benzene	9,866		
	Organic compounds		
Methanol	3,205	Ethylene glycol	11,623
Ethanol	4,819	Vinyl chloride	4,744
Isopropanol	5,300	1,2-Dichloroethane	8,828
Urea	14,853	Phenol	~ 11,580
Ethylene oxide	5,171	Acetic acid	11,728
Propylene oxide	6,531	Acetic anhydride	390
	Gases		
Carbon dioxide	8,326	Hydrogen chloride	1,998
Carbon monoxide	841	Chlorine	6,406
Nitrogen	720	Ammonia	5,657
Oxygen	444		
	Inorganic compounds		
Nitric acid	10,473	Sulfuric acid	9,874
Sodium hydroxide	6,694	Phosphoric acid	12,979
Water	6,008		

* Values for heats of fusion are given for many hydrocarbons in the API Research Project 44 (1982) compilation and for common nonhydrocarbons in the TRC compilation.

Clausius modified this equation to derive the well-known Clausius-Clapeyron equation by assuming that the volume of the liquid was negligible in comparison to the vapor and that the vapor follows the ideal-gas law; that is, $\overline{V}_G = RT/P°$. Thus,

$$\frac{dP°}{dT} = \frac{\lambda}{T(RT/P°)}$$

or

$$\frac{dP°}{P°} = \frac{\lambda}{R}\frac{dT}{T^2} \tag{3.17}$$

These assumptions are normally made at moderate temperatures and pressures. If, in addition, λ is assumed to be independent of temperature over a small temperature range, the equation can be integrated as

$$\int_{P_1}^{P_2}\frac{dP°}{P°} = \frac{\lambda}{R}\int_{T_1}^{T_2}\frac{dT}{T^2}$$

and

$$\ln\frac{P_2°}{P_1°} = -\frac{\lambda}{R}\left(\frac{1}{T_2} - \frac{1}{T_1}\right) \tag{3.18}$$

This assumption should be used with caution.

Example 3.3 Consider water at 500 K. Calculate its latent heat by using the Clapeyron equation and the Clausius-Clapeyron equation and compare the results. Use vapor pressure data at 490 and 510 K.

SOLUTION The Clapeyron equation (3.16) is

$$\frac{dP°}{dT} = \frac{\lambda}{T(\overline{V}_G - \overline{V}_L)}$$

T, K	490	500	510
Vapor pressure, MPa	2.181	2.637	3.163
$\overline{V}_G$, m^3/kg	0.09150	0.07585	0.06323
$\overline{V}_L$, m^3/kg	0.00118	0.00120	0.00122

$$P_2° - P_1° = \frac{\lambda}{(\overline{V}_G - \overline{V}_L)}\ln\frac{T_2}{T_1}$$

$$3163 - 2181 = \frac{\lambda}{(0.07585 - 0.00120)(18)}\ln\frac{510}{490}$$

$$= \frac{\lambda}{0.07465(18)}\ln\frac{510}{490}$$

$$\lambda = 32{,}983 \text{ kJ/kmol}$$

From the Clausius-Clapeyron equation (3.18),

$$\ln\frac{3.163}{2.181} = \frac{\lambda}{8.3143}\left(\frac{1}{490} - \frac{1}{510}\right)$$

$$\lambda = 38{,}618 \text{ kJ/kmol}$$

Clausius-Clapeyron is in error by 17 percent as water is especially nonideal.

Exercise 3.G Repeat Example 3.3 for benzene at 500 K.

Many empirical or semiempirical methods for estimating latent heats of vaporization have been developed. Available methods have been noted in Chap. 6. A listing of some common heats of vaporization for a variety of compounds is given in App. A.

A more general method of estimating latent heats of vaporization is by use of the theorem of corresponding states.

An accurate method is to rearrange the Clapeyron equation in reduced form, use an accurate vapor pressure equation for dP_r/dT_r, and use the compressibility charts to estimate molar volumes:

$$\lambda = P_c\left(\frac{dP_r}{dT_r}\right)(\overline{V}_G - \overline{V}_L)T_r \tag{3.19}$$

where

$$\overline{V}_G - \overline{V}_L = (Z_G - Z_L)\frac{RT}{P}$$

$$= (Z_G - Z_L)\frac{T_r T_c}{P_r P_c}$$

Lydersen, Greenkorn, and Hougen reported successful use of this method.

A simpler way to use the corresponding-states techniques for estimating latent heat of vaporization and simultaneously take into account the effect of pressure is to use the definition of latent heat in terms of enthalpy:

$$\lambda = (H_V - H_L)_{T,P} \tag{3.20}$$

The enthalpy of the vapor and the liquid can be determined by the methods given in Chap. 4, after which the latent heat can be easily calculated. This method has been determined to be very accurate for hydrocarbon compounds using the Pitzer approach and is a reasonable estimate for polar materials using both the Pitzer and the Lydersen approaches.

As the temperature and pressure of a compound are increased, the heat of vaporization becomes smaller. As the critical point is approached, the kinetic energy of translation of a molecule overcomes the attractive potential-energy forces and the molecule vaporizes without any energy requirement. At the critical point, the liquid and vapor phases have equal thermodynamic properties; thus, the latent heat of vaporization is zero.

Watson (1931, 1943) determined that the following equation is accurate in predicting latent heat at one temperature if the latent heat at another temperature is known:

$$\frac{\lambda_2}{\lambda_1} = \left(\frac{1 - T_{r2}}{1 - T_{r1}}\right)^{0.38} \tag{3.21}$$

where subscripts 1 and 2 refer to different states. Often T_1 is taken as the normal boiling point where most data are available. Obviously the effect of pressure on latent heat is assumed to be negligible.

3.4.2 Heat of Formation

Heat of formation of any compound is defined as the heat required to form the compound from the elements at some standard state conditions normally defined at 25°C and 1 atm pressure. The reactants are only stable elements at these conditions, and the compound is the sole product formed. Elements by definition have zero heats of formation.

As discussed earlier for flow situations with negligible kinetic- and potential-energy changes and no external mechanical work *and* for nonflow situations operating at constant pressure, $Q = \Delta H$. For nonflow situations at constant volume, $Q = \Delta E$. These are the laws of thermochemistry.

Thus, since heats of formation are fixed at constant pressure, $Q = \Delta H$, and the heat evolved or absorbed is equal to the change in enthalpy of the reaction. If the enthalpy increases when a compound is formed from the elements, its heat of formation is positive and it is called an *endothermic compound*. The opposite situation forms *exothermic compounds*. Values of heats of formation must be specified as to what state the elements and the compound will exist, i.e., crystalline, liquid, gas. Values for typical compounds are given in App. A. Heats of formation are normally derived from experimental combustion data. Group contribution methods such as those of Benson (1976) can be used to estimate heats of formation.

3.4.3 Heat of Reaction

Based on the discussion for heat of formation, it follows that the heat of a chemical reaction is the heat absorbed or rejected by the reaction. The standard heat of reaction for any chemical reaction is defined as the change in enthalpy which results when any compounds are reacted at a constant pressure of 1 atm and constant temperature of 25°C. Many real processes cannot actually be carried out at 25°C; however, if a higher temperature is required, data are reduced to a form where all initial reactants and final products are at 25°C as shown in Fig. 3.2. An example of a reaction would be

$$A + B \rightarrow C$$

$$\underset{25^\circ\text{C}}{A + B} \rightarrow \underset{T}{A + B} \xrightarrow{\Delta H_R} \underset{T}{C} \rightarrow \underset{25^\circ\text{C}}{C}$$

Basis: 1 mol product

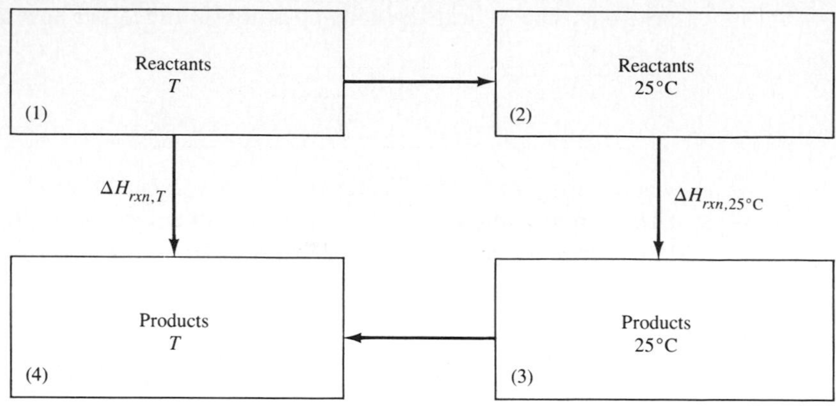

Using eq. (3.14) $H_2 - H_1 = \int_T^{298} C_p \text{ (reactants) } dT$

$$H_3 - H_2 = \Delta H^\circ_{rxn}$$

$$H_4 - H_3 = \int_{298}^{T} C_p \text{ (products) } dT$$

$$\Delta H_{rxn,T} = H_4 - H_1 = (H_4 - H_3) + (H_3 - H_2) + (H_2 - H_1)$$

$$\Delta H_{rxn,T} = \int_{298}^{T} [C_p \text{ (products)} - C_p \text{ (reactants)}]\, dT + \Delta H^\circ_{rxn}$$

Figure 3.2 Calculation of heat of reaction.

Thus, the standard heat of reaction at 25°C is specified as ΔH_{25}. If heat is evolved in the reaction, the reaction is called *exothermic*. If heat is absorbed, the reaction is called *endothermic*.

The standard heat of reaction can be calculated if the heats of formation of all compounds taking part in the reaction are known. If the reference states of both heats of formation as well as heats of reaction are both taken at 25°C, the standard heat of reaction is equal to the algebraic sum of the standard heats of formation of the products minus the algebraic sum of the standard heats of formation of the reactants:

$$\Delta H^\circ_{rxn} = (\Sigma n\, \Delta H_f)_{\text{Pr}} - (\Sigma n\, \Delta H_f)_{\text{Re}} \tag{3.22}$$

where n = stoichiometric number of moles. Thermodynamic convention dictates that endothermic heats of reaction are positive, while exothermic heats of reaction are negative.

Example 3.4 Determine the heat of reaction at 25°C for the reaction

Ethylbenzene → styrene + hydrogen

SOLUTION From App. A,

	Heat of formation at 25°C, kJ/kmol
Ethylbenzene	29,790
Styrene	147,360
Hydrogen	0

From Eq. (3.22),

$$\begin{aligned}\Delta H_{rxn} &= \Delta H_f(\text{styrene}) - \Delta H_f(\text{ethylbenzene}) - \Delta H_f(\text{hydrogen}) \\ &= 147{,}360 - 29{,}790 - 0 \\ &= 117{,}570 \text{ kJ/kmol} \qquad (28.1 \text{ kcal/mol})\end{aligned}$$

Exercise 3.H For Example 3.4, show how to determine the heat of reaction if the reaction is to be carried out at 800 K.

3.4.4 Heat of Combustion

The heat of combustion of any compound is the heat of reaction resulting from oxidizing the compound with oxygen. Since combustion processes are without exception exothermic, all heats of combustion are negative. The standard heat of combustion follows the same definition as the standard heat of reaction with all reactants and products at 25°C and 1 atm total pressure. Standard heats of combustion of hydrocarbons correspond to complete oxidation of all carbon and hydrogen to carbon dioxide and liquid water. Sometimes heats of combustion specify water vapor, and this must be noted. For compounds containing oxygen as well as carbon and hydrogen, the same convention applies. If a halogen, nitrogen, or sulfur is present, the data must specify which products are formed and whether they form solutions with water. Some examples of heats of combustion are given in App. A.

Heats of combustion can be calculated from heats of formation in the way any heat of reaction can be calculated. Conversely, heats of formation can be calculated from heats of combustion if heats of formation of all but one of the substances taking part in the reaction(s) are known. For example, assume the heat of combustion of formaldehyde is known but its heat of formation is unknown.

$$CH_2O + O_2 \rightarrow CO_2 + H_2O$$

where ΔH_C known
$\Delta H_f(CO_2, g)$ known
$\Delta H_f(H_2O, g)$ known

Therefore

$$\Delta H_C = \Delta H_f(CO_2, g) + \Delta H_f(H_2O, g) - \Delta H_f(CH_2O, g)$$

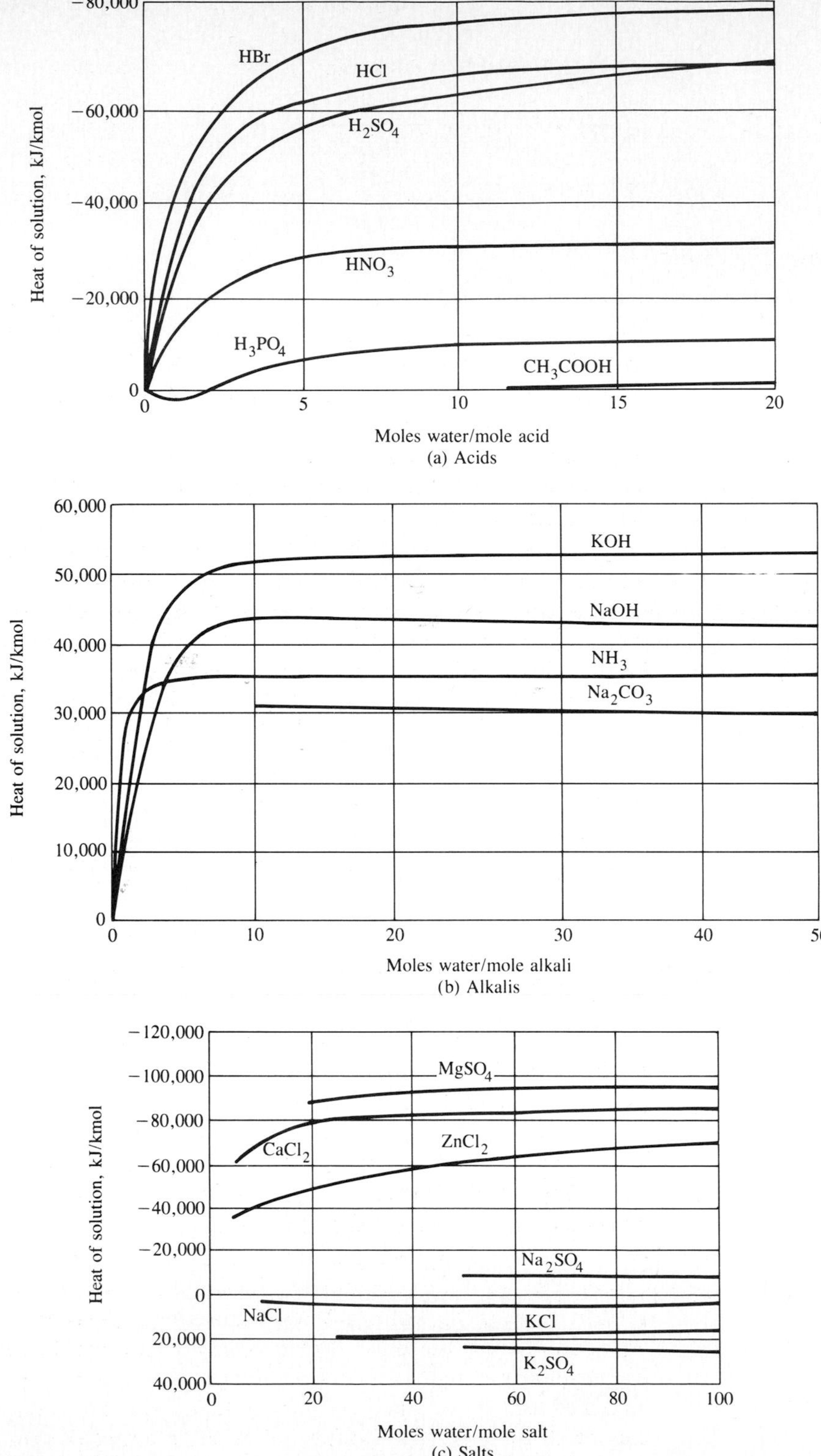

Figure 3.3 Integral heats of solution at 25 °C for common substances.

The equation can be solved for the heat of formation of formaldehyde. The same methods can be used for calculating heats of formation where multiple reactions are involved by algebraic manipulation of the equations.

In addition, heats of reactions can be calculated in the same manner as heats of formation by using heats of combustion and algebraic manipulations.

3.4.5 Heat of Solution

The heat of solution is defined as the enthalpy change which accompanies the dissolution of a compound. When chemical combination occurs, the heat of hydration or heat of solvation will be included. If ionization occurs, the energy of ionization is included. Neutral salts normally have positive heats of solution, i.e., absorb heat from the surroundings. Heats of solution during hydration are normally highly negative and evolve large quantities of heat. The temperature, solution concentrations, type of solvent, and amount of solvent must be specified for each heat of solution.

The standard integral heat of solution is the change in enthalpy of a system when 1 mol of solute is dissolved in n mol of solvent at a constant temperature of 25°C and a pressure of 1 atm. This heat of solution approaches a maximum at infinite dilution. Values of heats of solution are available (Hougen, Watson, and Ragatz). Figure 3.3 shows some examples for some common substances. If m mol of solute is dissolved in n mol of solvent, the enthalpy of the solution will be

$$H_{\text{sol}} = mH_m + nH_n + n\,\Delta H_{\text{sol}}$$

where all enthalpies and the heat of solution are taken at the standard temperatures and the base of each enthalpy is at the same reference temperature. If both solute and solvent are liquids, the heat of solution is often called the *heat of mixing*.

3.5 APPLICATIONS OF ENERGY BALANCES TO PROCESSES

This section will attempt to illustrate the application of the first law of thermodynamics to both flow and nonflow situations for the cases of ideal gases, nonideal gases, and liquid systems. The equations of energy will be reduced for various situations useful in chemical engineering calculations, namely, constant temperature (isothermal), constant volume (isometric), constant pressure (isobaric), and cases in which no heat enters or leaves the system (adiabatic). Processes may be reversible, i.e., frictionless and doing maximum work or requiring minimum work, or irreversible, i.e., having a finite imbalance or with friction. The concept of reversibility allows calculations to be simplified since the system can always be identified at some uniform temperature and pressure for application of an equation of state. Since the system is only infinitesimally removed from mechanical equilibrium with the surroundings, the internal and external forces are balanced and the internal pressure is that which is producing work. Thus, work can be defined as $\int P\,dV$, where P is the internal pressure. For an irreversible process, this assumption cannot be made.

3.5.1 Ideal Gas—Nonflow Reversible Processes

Consider 1 mol of a gas contained in a frictionless piston, either expanding or being compressed. This system is considered to be a nonflow process where the first law is

$$\Delta E = Q - W \tag{3.2}$$

If the process can be said to be reversible, then $W = \int P\,dV$. If, in addition, the ideal-gas law holds, then

$$P = \frac{RT}{\overline{V}} = \frac{nRT}{V}$$

Isothermal processes If the system can be assumed to be isothermal, i.e., the temperature remains constant throughout the entire process while expanding from state 1 to state 2, the first-law equation can be reduced. As was stated earlier, the internal energy of an ideal gas is only a function of temperature. Thus, the change in internal energy for an isothermal process will be zero.

If $\Delta E = 0$, then $Q = W = \int P\,dV$, and

$$Q = W = \int_{V_1}^{V_2} \frac{nRT}{V}\,dV$$

Since the temperature is constant, the integral can be evaluated. Therefore

$$Q = W = nRT \ln \frac{V_2}{V_1} \tag{3.23}$$

Since $P_1V_1 = P_2V_2$ at constant temperature, an equivalent expression is

$$Q = W = nRT \ln \frac{P_1}{P_2} \tag{3.24}$$

Since $V_2 > V_1$ and $P_1 > P_2$, the isothermal work of expansion is positive.

Isobaric processes If the system pressure remains constant throughout the entire process as an ideal gas expands from state 1 to state 2 as caused by the addition of heat, then the work term will reduce to

$$W = \int_{V_1}^{V_2} P\,dV = P(V_2 - V_1)$$

As the gas is ideal, $PV = nRT$, and therefore

$$W = nR(T_2 - T_1) \tag{3.25}$$

The internal energy will change as the temperature is not constant and thus

$$\Delta E = Q - nR(T_2 - T_1)$$

During expansion since

$$T_2 - T_1 = T_1\left(\frac{V_2}{V_1} - 1\right)$$

is positive, the work is positive.

As was derived earlier,

$$\Delta H = \Delta E + \Delta(PV)$$

Therefore for an isobaric process $\Delta E = \Delta H - P(\Delta V)$. Since $\Delta E = Q - P(\Delta V)$, then

$$\Delta H = Q$$

Since $C_P = (\partial H/\partial T)_P$ and $C_{\bar{V}} = (\partial E/\partial T)_V$ and for ideal gases $C_P = C_V + R$,

$$\Delta H = Q = C_P\, dT = nC_P\, \Delta T \qquad \text{and} \qquad \Delta E = nC_V\, \Delta T \tag{3.26}$$

Isometric processes If the gas in the cylinder remains at constant volume, no expansion or compression occurs, and $W = \int P\, dV$ is zero. Thus,

$$\Delta E = Q \qquad \text{and} \qquad \Delta H = Q + \Delta(PV)$$

As discussed under isobaric processes, then

$$\Delta E = nC_V\, \Delta T \qquad \text{and} \qquad \Delta H = nC_P\, \Delta T \tag{3.27}$$

Adiabatic processes Assume the cylinder is perfectly insulated such that no heat can enter or leave the system; that is, $Q = 0$ as it expands from state 1 to state 2. Thus, the first law reduces to $\Delta E = -W$ or $W = -\Delta E$. Since the pressure, temperature, and volume of the gas will change, the $\int P\, dV$ cannot be evaluated directly. However, since the change in internal energy by definition can be related to the heat capacity at constant volume, $\Delta E = nC_V(T_2 - T_1)$ and $W = -nC_V(T_2 - T_1)$. Thus, the work of expansion is produced by decreasing the internal energy of the gas. Since the temperature decreases as the volume increases, the pressure of the ideal gas must also decrease.

Relationships can easily be derived between the initial and final pressures, temperatures, and volumes for adiabatic ideal-gas processes. Since $C_P = C_{\bar{V}} + R$ and defining $k = C_P/C_{\bar{V}}$,

$$W = P\, dV = -nC_V\, dT \tag{3.28}$$

Also since $PV = nRT$, differentiating yields $P\, dV + V\, dP = nR\, dT$. Substituting to reduce the equation to one pair of variables, PV, PT, or VT, and integrating leads to the equations

$$P_1V_1^k = P_2V_2^k \tag{3.29}$$

$$\frac{T_2}{T_1} = \left(\frac{P_2}{P_1}\right)^{(k-1)/k} \tag{3.30}$$

$$\frac{T_2}{T_1} = \left(\frac{V_1}{V_2}\right)^{k-1} \tag{3.31}$$

These expressions can then be substituted into the definition of work to give

$$W = \frac{-nRT_1}{k-1}\left[\left(\frac{P_2}{P_1}\right)^{(k-1)/k} - 1\right] \tag{3.32}$$

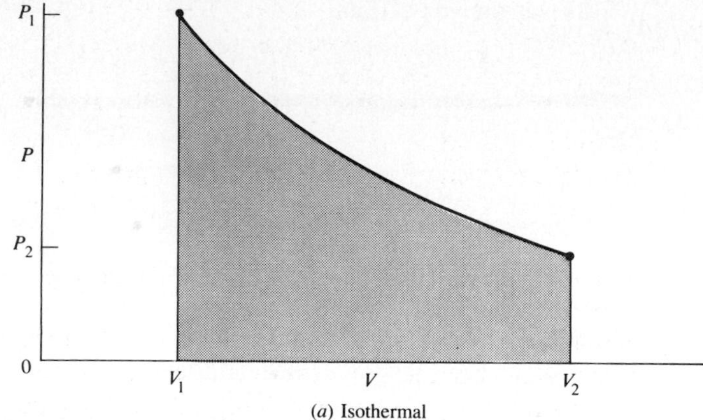

(a) Isothermal

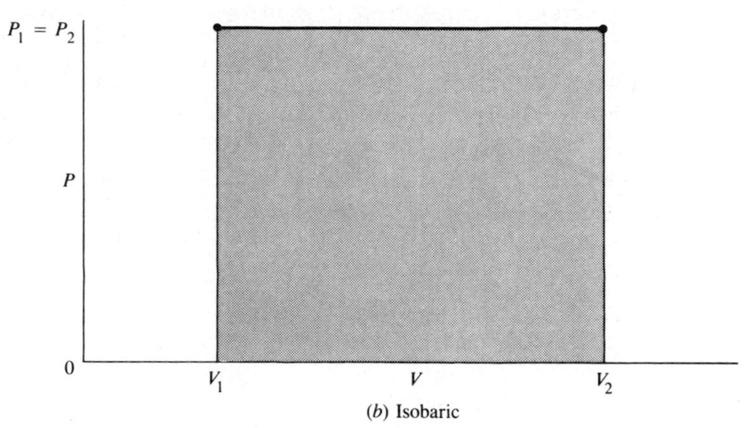

(b) Isobaric

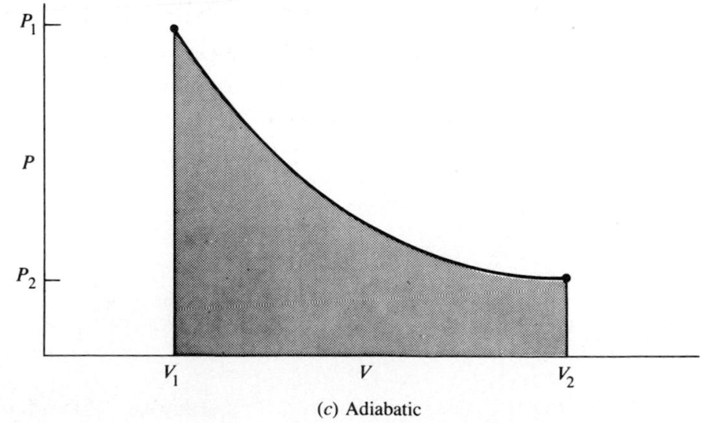

(c) Adiabatic

Figure 3.4 Pressure-volume diagrams for ideal-gas processes.

Equation (3.32) is the standard equation used to determine the work and hence ideal power requirements for the industrial design of compressors. The standard assumption of an adiabatic process is reasonable for the compression of many gases. Nonidealities are taken into account by use of adiabatic efficiencies derived from previous experimental data. Efficiency averages about 75 to 80 percent and usually ranges from 65 to 85 percent. Thus, the actual work required for compression is normally greater than that of the adiabatic process.

Polytropic processes Polytropic processes follow the same derivation as adiabatic processes except that $k \neq C_P/C_{\bar{V}}$ due to nonidealities in the gas. Thus, k becomes an empirical parameter which must be determined from experiment.

For example, k for an ideal, diatomic gas is 1.4, while k for a polytropic process may vary considerably but is normally greater than the ideal k, say 1.5 to 1.6 for a nonideal diatomic gas. This will increase the work of compression.

***PV* diagrams** It is often desirable to graphically represent the path of a process undergoing work to aid in understanding and to compare various modes of operation. In addition, it should be noted that the work can be calculated directly from the area under the curve.

Consider an ideal gas being expanded from V_1 to V_2 under isothermal, isobaric, and adiabatic conditions as shown in Fig. 3.4.

It is readily apparent that the isobaric process produces more work than the isothermal process than does the adiabatic process when expanded between the same P and V.

Example 3.5 One kilomole of carbon monoxide considered to be an ideal gas with a constant isobaric heat capacity of 29.3 kJ/(kmol · K) is subjected to the following process.

Carbon monoxide at 2.758 MPa and 700 K (state 1) is expanded isothermally to 0.552 MPa (state 2), is cooled at constant volume to 437.5 K (state 3), is cooled at constant pressure to 350 K (state 4), is compressed adiabatically to 2.758 MPa (state 5), and is heated at constant pressure to 700 K.

Compute Q, W, ΔH, and ΔE for each step of the process if reversibility can be assumed. Sketch the path on a PV diagram.

SOLUTION

(a) State 1 → state 2 (isothermal):

$$\Delta H = 0 \qquad \Delta E = 0$$

$$Q = W = -RT \ln \frac{P_2}{P_1}$$

$$= -(8314.3)(700) \ln \frac{0.552}{2.758}$$

$$= 9.363 \times 10^6 \text{ J/kmol}$$

$$= 9363 \text{ kJ/kmol}$$

(b) State 2 → state 3 (isometric):

$$W = 0 \qquad Q = \Delta E = C_V \Delta T \qquad \Delta H = C_P \Delta T$$

$$C_V = C_P - R = 29.3 - 8.3 = 21.0 \text{ kJ/(kmol} \cdot \text{K)}$$

$$Q = \Delta E = (21.0)(437.5 - 700) = -5513 \text{ kJ/kmol}$$

$$\Delta H = (29.3)(437.5 - 700) = -7691 \text{ kJ/kmol}$$

$$P_3 = \frac{P_2 T_3}{T_2} = 0.345 \text{ MPa}$$

(c) State 3 → state 4 (isobaric):

$$Q = \Delta H = C_P \Delta T = (29.3)(350 - 437.5) = -2564 \text{ kJ/kmol}$$

$$\Delta E = C_V \Delta T = (21.0)(350 - 437.5) = -1838 \text{ kJ/kmol}$$

$$W = Q - \Delta E = -2564 - (-1838) = -726 \text{ kJ/kmol}$$

(d) State 4 → state 5 (adiabatic):

$$\frac{T_5}{T_4} = \left(\frac{P_5}{P_4}\right)^{(k-1)/k} = \left(\frac{2.758}{0.345}\right)^{0.2857} = 1.811$$

$$T_5 = (1.811)(350) = 634\ K$$

$$Q = 0 \qquad \Delta H = C_P \Delta T = (29.3)(634 - 350) = 8321 \text{ kJ/kmol}$$

$$W = -\Delta E = C_V \Delta T = (-21.0)(634 - 350) = -5964 \text{ kJ/kmol}$$

(e) State 5 → state 1 (isobaric):

$$Q = \Delta H = C_P \Delta T = 29.3(700 - 634) = 1934 \text{ kJ/kmol}$$

$$\Delta E = C_V \Delta T = 21.0(700 - 634) = 1386 \text{ kJ/kmol}$$

$$W = Q - \Delta E = 1934 - 1386 = 548 \text{ kJ/kmol}$$

Overall Process As ΔE and ΔH are point functions, the sum of each around the path must equal zero. Thus, it follows that for the overall process the sum of the Q's equals the sum of the W's as

$$\Delta E = Q - W = 0$$

$$\Sigma \Delta E = 0 - 5513 - 1838 + 5964 + 1386 = -1 \approx 0$$

$$\Sigma \Delta H = 0 - 7691 - 2564 + 8321 + 1934 = 0$$

$$\Sigma Q = 9363 - 5513 - 2564 + 0 + 1934 = 3220 \text{ kJ/kmol}$$

$$\Sigma W = 9363 + 0 - 726 - 5964 + 548 = 3221 \text{ kJ/kmol}$$

A diagrammatic sketch of the process is given in Fig. 3.5.

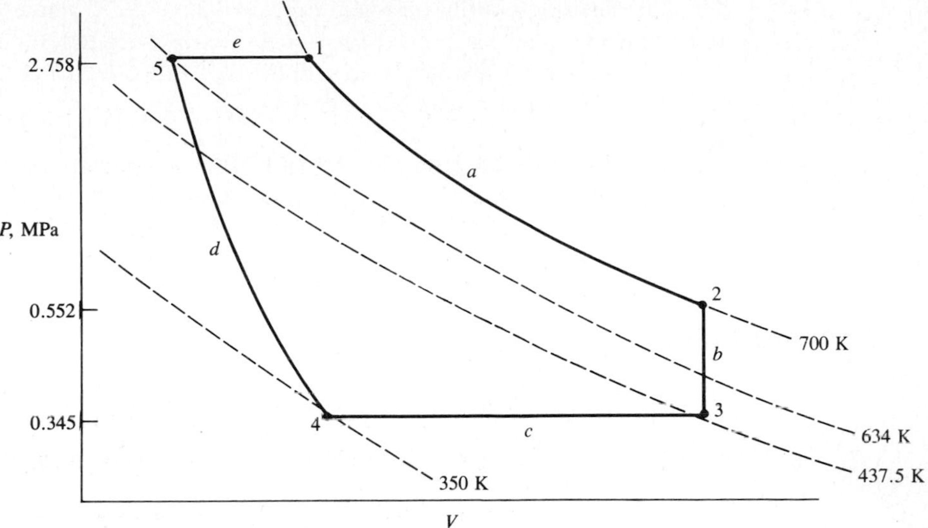

Figure 3.5 Process of Example 3.5.

Exercise 3.I Determine Q, W, ΔE, and ΔH for each step of a process in which argon is compressed isothermally at 500 K from 0.2 to 3.0 MPa, is cooled isobarically to 300 K, is expanded adiabatically to 1.0 MPa, is expanded isobarically to its initial specific volume, and is then compressed isometrically to its initial state.

3.5.2 Ideal Gas—Nonflow Irreversible Processes

As discussed for reversible processes, the first law $\Delta E = Q - W$ holds as does the expression $\Delta E = nC_V \Delta T$. However, although work is still defined as $\int P\,dV$, the internal pressure cannot be used directly.

The simplest form of such a process is a *free expansion*. Consider an ideal gas contained in one-half of a perfectly insulated container with a diaphragm in the center. The other half of the container is a perfect vacuum. The system is defined as the gas. If the diaphragm is broken, the gas will expand and fill the entire container. Since the process is adiabatic, $Q = 0$. Since no work is done on the surroundings, $W = 0$. Therefore, $\Delta E = 0$. Since for an ideal gas the internal-energy change is only a function of temperature, $\Delta E = nC_V \Delta T$, then $\Delta T = 0$ and the process is also isothermal. Obviously, $\Delta E = 0$ for any free expansion process whether or not the gas is ideal; however, for real gases $\Delta T \neq 0$.

For irreversible processes in general the work cannot be evaluated directly from $\int P\,dV$ as explained earlier and can only be calculated by using the first law. Obviously, if process conditions both initially and finally are known, the change in internal energy, a state function, can be calculated by devising a reversible process operating between these conditions. For example, an irreversible process operating between P_1T_1 and P_2T_2 can be modeled as a series of two reversible processes operating between $P_1T_1 \xrightarrow{1} P_1T_2 \xrightarrow{2} P_2T_2$ where one process is isobaric and the other isothermal for calculation of ΔE. ΔE for the irreversible process will be merely the algebraic sum of ΔE_1 and ΔE_2. Once ΔE is calculated and if Q is known or calculable, then $W = Q - \Delta E$.

Exercise 3.J Consider 1 kmol of an ideal diatomic gas which is expanded from 2.0 MPa and 300°C to 0.2 MPa and 100°C. Calculate ΔE, W, and Q for the process. Calculate the same variables for a free expansion between the same initial and final state.

3.5.3 Nonideal Gases—Nonflow Processes

The same derivation methods as applied to the ideal gas apply to nonideal gases except that a different equation of state or tabular data must be used to evaluate the $\int P\,dV$ for reversible processes. The work of a real gas is not equal to the work of an ideal gas since the final volumes are not equal.

For example, consider a van der Waals gas where

$$P = \frac{RT}{\overline{V} - b} - \frac{a}{\overline{V}^2}$$

For an isothermal process,

$$W = \int P\,d\overline{V} = RT\int_{V_1}^{V_2} \frac{d\overline{V}}{\overline{V} - b} - a\int_{V_1}^{V_2} \frac{d\overline{V}}{\overline{V}^2}$$

or

$$W = -RT\ln\left(\frac{\overline{V}_2 - b}{\overline{V}_1 - b}\right) + a\left(\frac{1}{\overline{V}_2} - \frac{1}{\overline{V}_1}\right) \tag{3.33}$$

For an isobaric process, obviously there is no change in the results as the pressure is a constant and $W = P\Delta V$.

Other than making the mathematics slightly more complex, any equation of state can be used to develop all the relations between ΔE, Q, and W as was previously discussed for the ideal gas.

Sometimes, an equation of state is not available for a nonideal gas, and tabular data must be used. Again the same definitions are maintained, but work terms must be evaluated from the data available. For isobaric processes, the actual specific volumes must be used. For isothermal processes, a graphical integration of the pressure-volume plot must be made. Internal-energy changes can usually be calculated by utilization of the tabulated PVT and enthalpy data.

Steam is the most important example of a nonideal gas where such methods are normally required. Only at extremely high temperatures and very low pressures does steam behave as an ideal gas. As it is a very important material, manipulation of steam tables is very important. Such tables in condensed form are given in App. B.

Example 3.6 One kilogram of steam is subjected to the following process.

Steam at 2.75 MPa and 440°C (state 1) is expanded isothermally to 0.50 MPa (state 2), is cooled at constant volume to 170°C (state 3), is cooled at constant pressure to 140°C (state 4), is compressed adiabatically to 2.75 MPa and 410°C (state 5), and is heated at constant pressure to 440°C.

Compute Q, W, ΔH, and ΔE for each step of the process if reversibility can be assumed. Sketch the path on a PV diagram.

BASIS 1 kg steam. Since steam is nonideal, values for $\overline{V}$ and H must be obtained from the steam tables (App. B).

State	1	2	3	4	5
Pressure, MPa	2.75	0.5	0.31	0.31	2.75
T, °C	440	440	170	140	410
$10^3\overline{V}$, m³/kg	116.17	654.8	654.8	597	110.65
H, kJ/kg	3325.0	3356.0	2802.6	2738.5	3257.7
E, kJ/kg	3005.6	3028.6	2602.6	2553.6	2953.4

(*a*) State 1 → state 2

$$\Delta H = 3356.0 - 3325.0 = 31.0$$
$$\Delta E = 3028.6 - 3005.6 = 23.0$$

W can be calculated by graphical integration of $\int P\,dV$. Since $Q - W = 23.0$, thus $Q = 23.0 + W$.

(*b*) State 2 → state 3

$$W = 0 \qquad Q = \Delta E = 2602.6 - 3028.6 = -426$$
$$\Delta H = 2802.6 - 3356.0 = -553.4$$

(*c*) State 3 → state 4

$$\Delta H = 2738.5 - 2802.6 = -64.1$$
$$\Delta E = 2553.6 - 2602.6 = -49.0$$
$$W = P\Delta\overline{V} = (0.31 \times 10^{-6})(597 - 654.8)10^{3}10^{-3} \text{ kJ/(kg} \cdot \text{m}^2 \cdot \text{s}^2)$$
$$= -17.9 \text{ kJ/kg}$$
$$Q = \Delta E + W = -49.0 - 17.9 = -66.9$$

(*d*) State 4 → state 5

$$Q = 0$$
$$\Delta H = 3257.7 - 2738.5 = 519.2$$
$$\Delta E = -W = 2953.4 - 2553.6 = 399.8$$

(*e*) State 5 → state 1

$$\Delta H = 3325.0 - 3257.7 = 67.3$$
$$\Delta E = 3005.6 - 2953.4 = 52.2$$
$$W = P\Delta\overline{V} = (2.75)(116.17 - 110.65) = 15.2$$
$$Q = \Delta E + W = 52.2 + 15.2 = 67.4$$

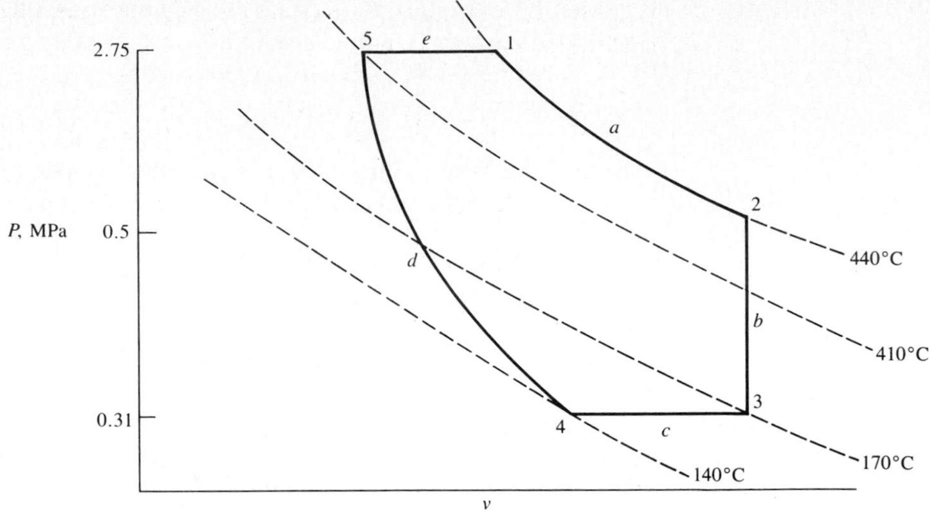

Figure 3.6 Process of Example 3.6.

Overall Process By definition for the closed cycle $\Sigma \Delta H = 0$,

$$\Sigma \Delta E = 0 \quad \text{and} \quad \Sigma Q = \Sigma W$$

$$\Sigma \Delta H = 0$$

$$\Sigma Q = Q_a - 425.5 \qquad \Sigma W = W_a - 402.5$$

$$Q_a - W_a = +23.0 \qquad \text{Why?}$$

A diagrammatic sketch of the process is given in Fig. 3.6.

Exercise 3.K Determine Q, W, ΔE, and ΔH for each step of a process in which steam is compressed isothermally at 130°C from 0.2 to 3.0 MPa, is cooled isobarically to 30°C, is expanded adiabatically to 1.0 MPa, is expanded isobarically to its initial specific volume, and is then compressed isometrically to its initial state.

3.5.4 Flow Processes

In Sec. 3.2 energy balances for constant-pressure flow processes were derived for a single stream flowing between points 1 and 2:

$$E_1 + P_1V_1 + \frac{u_1^2}{2} + gz_1 + \frac{Q}{\dot{m}} = E_2 + P_2V_2 + \frac{u_2^2}{2} + gz_2 + \frac{W_s}{\dot{m}} \tag{3.4}$$

or

$$H_1 + \frac{u_1^2}{2} + gz_1 + \frac{Q}{\dot{m}} = H_2 + \frac{u_2^2}{2} + gz_2 + \frac{W_s}{\dot{m}} \tag{3.5}$$

The equation was also shown to be reducible to a mechanical-energy balance. Based on one mass of material,

$$gz_1 + P_1V_1 + \frac{u_1^2}{2} + \int P\,dV = gz_2 + P_2V_2 + \frac{u_2^2}{2} + h_f + W_s' \qquad (3.6)$$

or

$$gz_1 + \frac{u_1^2}{2} = gz_2 + \frac{u_2^2}{2} + h_f + \int_1^2 -V\,dP + W_s'$$

where $\int P\,dV$ = reversible, nonflow work
$-\int V\,dP$ = reversible, shaft work

The mass balance for a steady flow process (often called the *continuity equation*) is

$$\dot{m}_1 = \dot{m}_2$$

or

$$u_1A_1\rho_1 = u_2A_2\rho_2 \qquad (3.34)$$

or

$$\frac{u_1A_1}{\overline{V}_1} = \frac{u_2A_2}{\overline{V}_2}$$

where $\dot{m}$ = mass flow rate
u = average velocity
ρ = density of flowing fluid
$\overline{V}$ = specific volume of flowing fluid
A = cross-sectional area of conduit

This equation is of importance in solving an energy balance in a flow system.

This section will show some examples of reduction of these equations for typical flow systems.

Ideal gas For an ideal gas, it has been shown earlier that $\Delta E = nC_V\Delta T$ and $\Delta H = nC_P\Delta T$. Therefore, the total energy balance will reduce to

$$Q' - W_s' = g(z_2 - z_1) + \left(\frac{u_2^2}{2} - \frac{u_1^2}{2}\right) + (P_2V_2 - P_1V_1) + nC_V(T_2 - T_1) \qquad (3.35)$$

or

$$Q' - W_s' = g(z_2 - z_1) + \left(\frac{u_2^2}{2} - \frac{u_1^2}{2}\right) + nC_P(T_2 - T_1)$$

These equations are quite useful for any flow process involving ideal gases. For example a turbine operating on an ideal gas will produce work in the amount W_s, while a compressor will require $-W_s$ amount of work to compress an ideal gas.

Referring to the mechanical-energy balance, expressions for the reversible shaft work term can be derived:

$$W_s' = -\int V\,dP \qquad (3.36)$$

At constant volume: $W_s' = -V\int_{P_1}^{P_2} dP = -V(P_2 - P_1)$

At constant pressure: $W_s' = 0$

At constant temperature: $W_s' = -nRT \ln \dfrac{P_2}{P_1}$

Adiabatic process assuming no potential- or kinetic-energy effects:

$$W_s' = -\Delta H = -nC_P(T_2 - T_1)$$

$$W_s' = \frac{-knRT_1}{k-1}\left[\left(\frac{P_2}{P_1}\right)^{(k-1)/k} - 1\right]$$

$$= -\frac{k}{k-1}(P_2V_2 - P_1V_1)$$

Nonideal gases and liquids The total energy balance can be applied to nonideal gases by utilization of more advanced equations of state or by utilization of tabulated values of P, V, E, and H as is usually required for steam and for many industrially important gases, especially refrigerants such as SO_2, NH_3, Freons, etc.

For most liquids, the material can be assumed to be incompressible, implying a constant density. Often other major simplifying assumptions can be made such as in a horizontal system where the potential-energy change is negligible; an incompressible system where velocities are constant and, thus, the kinetic-energy change is negligible; a completely insulated system where $Q = 0$; or an isothermal low-pressure system where ΔE and ΔH are essentially negligible. For example, if all the assumptions are made,

$$W_s' = (P_1 - P_2)\bar{V}$$

or

$$-W_s' = \frac{\Delta P}{\rho} \tag{3.37}$$

which could be used as a first estimate of pumping power required in frictionless pipelines. For an actual pipeline with heat degradation due to friction but still horizontal and isothermal with negligible kinetic-energy changes, the mechanical-energy balance yields

$$-W_s' = h_f + \frac{P_2 - P_1}{\rho} \tag{3.38}$$

Examples of situations utilizing the various forms of these equations follow.

Example 3.7 A compressor is fed with 0.25 m^3/s of dry air at 0.1 MPa total pressure and 30°C which is flowing in an entrance pipe 0.154 m in diameter. Downstream from the compressor a cooler removes 2.764×10^8 J/h from the gas. In the exit line of 0.028 m in diameter, the air is at 43°C and 0.55 MPa total pressure. Assuming the gas is ideal with a molal heat capacity of 29.3 kJ/(kg · K), determine the power input to the compressor.

BASIS 1 kg air.

Use Eqs. (3.35) and (3.34):

$$Q' - W_s' = g(z_2 - z_1) + \frac{u_2^2 - u_1^2}{2} + nC_P(T_2 - T_1)$$

$$u_1 A_1 \rho_1 = u_2 A_2 \rho_2$$

From the ideal-gas law,

$$\rho_1 = \frac{PM}{RT} = \frac{(0.1 \times 10^6)(28.84)}{(8314.4)(303.1)} = 1.144 \text{ kg/m}^3$$

$$u_1 = 0.25 \frac{1}{(\pi/4)(0.154)^2} = 13.42 \text{ m/s}$$

$$\rho_2 = \frac{(0.55 \times 10^6)(28.84)}{(8314.4)(316.1)} = 6.035 \text{ kg/m}^3$$

$$u_2 = \frac{u_1 A_1 \rho_1}{A_2 \rho_2} = \frac{(13.42)(0.154)^2(1.144)}{(0.028)^2(6.035)} = 76.95 \text{ m/s}$$

$$\frac{-2.764 \times 10^8}{(0.25)(1.144)(3600)} - W_s' = 0 + \frac{76.95^2 - 13.42^2}{2} + \frac{29.3 \times 10^3}{28.84}(316.1 - 303.1)$$

$$-268{,}454 - W_s' = 2871 + 13{,}207$$

$$-W_s' = 284{,}532 \text{ J/kg}$$

From Eq. (3.34),

$$\begin{aligned} \dot{m} &= u_1 A_1 \rho_1 \\ &= (13.42 \text{ m/s})\left[\frac{\pi}{4}(0.154)^2 \text{m}^2\right](1.144 \text{ kg/m}^3) \\ &= 0.2860 \text{ kg/s} \end{aligned}$$

$$\begin{aligned} -W_s &= \dot{m}(-W_s') = 81{,}376 \text{ J/s} = 81{,}376 \text{ W} \\ &= (81{,}376 \text{ W})\frac{\text{hp}}{745.7 \text{ W}} = 109 \text{ hp} \end{aligned}$$

Example 3.8 A pump transferring a solution of 1.21 relative density from a mixing vessel to a storage tank travels through a pipe of 0.078 m in diameter at 1.1 m/s. The level of liquid in the tank is 18 m above that in the mixing vessel; both are open to the atmosphere. Pipe friction and entrance and exit effects amount to 4 m of solution pressure drop. Determine the power input to the pump and the increase in pressure over the pump. A sketch of the process is given in Fig. 3.7.

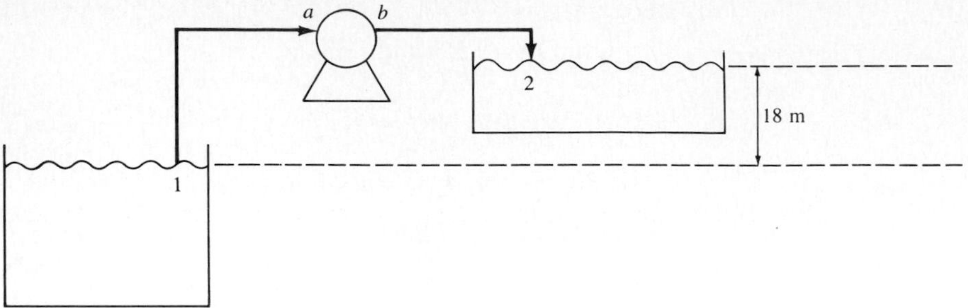

Figure 3.7 Physical process of Example 3.8.

BASIS 1 kg solution.

$$gz_1 + P_1V_1 + \frac{u_1^2}{2} + \int P\,dV = gz_2 + P_2V_2 + \frac{u_2^2}{2} + h_f + W_s'$$

$$P_1 = P_2 = \text{atmospheric} \qquad V_1 = V_2 \qquad \text{(incompressible)}$$

$$u_1 \to 0 \qquad \text{and} \qquad u_2 \to 0 \qquad \text{(very large cross-sectional area)}$$

$$\int P\,dV = 0 \qquad \text{(incompressible)}$$

Therefore

$$-W_s' = h_f + g(z_2 - z_1)$$

Using Eq. (3.34),

$$\begin{aligned} \dot{m} &= u_1\rho_1A_1 \\ &= (1.1\text{ m/s})(1.21 \times 10^3\text{ kg/m}^3)\left[\frac{\pi}{4}(0.078)^2\text{ m}^2\right] \\ &= 6.36\text{ kg/s} \end{aligned}$$

$$-W_s' = 4 + (18 - 0) = 22\text{ m solution}$$

$$\begin{aligned} -W_s &= (22)\dot{m}g = (22)(6.36)(9.806) \\ &= 1372\text{ W} \end{aligned}$$

For the pump, points a and b, the balance will reduce to $-W_s = (P_b - P_a)V$. Therefore

$$P_b - P_a = \frac{-W_s}{V} = W_s\frac{\rho}{\dot{m}}$$

$$\begin{aligned} \Delta P &= \frac{1372(1.21 \times 10^3)}{6.36} \\ &= 0.261\text{ MPa} \end{aligned}$$

Exercise 3.L Air is flowing at a rate of 0.025 kg/s through a horizontal tube with an inside diameter of 0.023 m. At one point the air has a bulk temperature of 21°C and a pressure of 0.04 MPa. The tube is heated electrically with a net input of 130 W between the first point and a second point where the pressure is 0.02 MPa. Estimate the temperature at the second point.

Exercise 3.M Consider the flow of an incompressible fluid with the properties of water in a system such as the system of Example 3.8 except that the tanks are closed with $P_1 = 0.2$ MPa and $P_2 = 0.4$ MPa. How far apart vertically are the tanks if the pump inputs 5000 W and the liquid flows at 1 m/s?

3.6 APPLICATIONS OF THERMOCHEMISTRY

Section 3.4 defined the various important quantities of thermochemistry. The principles to be discussed in this section are necessary for the analysis of almost any chemical engineering process and consist of applying the material already discussed in this chapter to such an analysis.

If an energy balance is carried out on a flow process where potential- and kinetic-energy effects are negligible and no work is done on or by the system, it reduces to $Q = \Delta H$. (Similarly, for a nonflow process, $Q = \Delta E$.)

A thermochemical analysis of a process then consists of analyzing the heat inputs and outputs of each stream entering or leaving a process, hence the pseudonym—heat balance. All sensible (no phase change) heat effects, latent (phase change) heat effects, heats of reaction, heats of combustion, and heats of mixing must be taken into account. Each stream must be identified as to its composition, temperature, pressure, and phase. Also the reactions, mixing processes, and external sources of heat must be specified.

To carry out such a balance, enthalpies and heats of reaction must be known at the correct temperature and pressure all referred to the same state of matter and the same base temperature. Since this base temperature is constant and only ΔH is important, selection of the base is arbitrary but must be uniform for any analysis. Thus, this section begins with an outline of the methods necessary to convert heats of reaction to varying pressures and temperatures. A general method for analyzing processes follows. The important cases of adiabatic processes, $\Delta H_T = Q = 0$, and flame temperatures conclude the discussion. The basic material in this section will be especially useful when studying the material of Chap. 9.

3.6.1 Heat of Reaction—Variable Effects

Pressure The standard heat of reaction was defined at 25°C (298.16 K) and 1 atm total pressure. To analyze a process at a different pressure, the heat of reaction at that pressure must be calculated. Consider the *PH* diagram in Fig. 3.8*a*.

$$\Delta H_{R1} = \text{heat of reaction at 1 atm}$$

$$\Delta H_{RP} = \text{heat of reaction at pressure } P$$

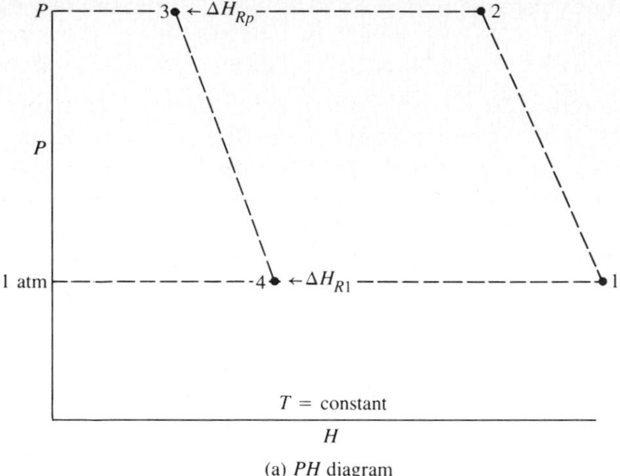

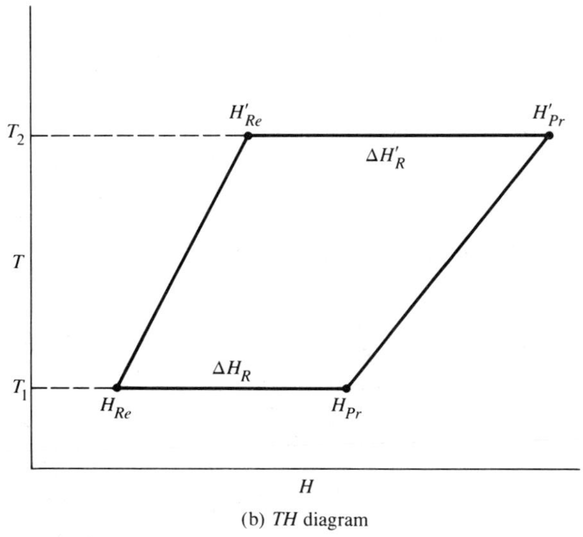

Figure 3.8 Schematic diagrams for heat of reaction calculations. (*a*) *PH* diagram; (*b*) *TH* diagram.

Since enthalpy is a state function, the change in enthalpy in going from point 1 to point 3 by paths 1-2-3 and 1-4-3 must be identical. Therefore

$$(H - H_1)_{\text{Re}} + \Delta H_{RP} = (H_3 - H_4)_{\text{Pr}} + \Delta H_{R1}$$

where Re = reactants and Pr = products. Thus

$$\Delta H_{RP} = \Delta H_{R1} + (H - H_1)_{\text{Pr}} - (H - H_1)_{\text{Re}} \tag{3.39}$$

This is the general equation for the effect of pressure which is applicable to any system. However, for ideal gases it can be simplified by noting that enthalpy is only a function of temperature for the ideal gas. Therefore, $(H - H_1)_{\text{Pr}} = 0$ and $(H - H_1)_{\text{Re}} = 0$, and hence $\Delta H_P = \Delta H_1$. For solid or liquid reactants, enthalpy does not change with pressure over a moderate pressure range. Thus, the same simplification can be made.

At higher pressures for nonideal gases such as water vapor, the various enthalpies must be calculated to determine ΔH_P. For liquids and solids, if enthalpies cannot be easily calculated, the assumption that the fluid is incompressible and that the internal energy will be essentially constant with change in pressure allows the enthalpy change to be estimated as $\Delta H = \bar{V}\Delta P$. This correction is often quite small.

Temperature Actual processes usually occur at varying conditions with reactants entering at a certain temperature, reactions occurring at another temperature, and products leaving at yet a different temperature. The *TH* diagram in Fig. 3.8*b* illustrates such a process.

The enthalpy change in going from reactants at T_1 to products at T_2 is independent of path. Thus

$$\Delta H_R + (H'_{\text{Pr}} - H_{\text{Pr}}) = \Delta H'_R + (H'_{\text{Re}} - H_{\text{Re}})$$

Therefore,

$$\Delta H'_R - \Delta H_R = (H'_{\text{Pr}} - H_{\text{Pr}}) - (H'_{\text{Re}} - H_{\text{Re}})$$

$$H'_{\text{Pr}} - H_{\text{Pr}} = C_{P,\text{Pr}}(T_2 - T_1) \qquad \text{and} \qquad H'_{\text{Re}} - H_{\text{Re}} = C_{P,\text{Re}}(T_2 - T_1)$$

when there are no phase changes and C_P is constant. If these expressions are substituted in the balance,

$$\Delta H'_R - \Delta H_R = C_{P,\text{Pr}}(T_2 - T_1) - C_{P,\text{Re}}(T_2 - T_1)$$

If

$$\Delta C_P = \sum_{\text{Pr}} n_{\text{Pr}} C_{P,\text{Pr}} - \sum_{\text{Re}} n_{\text{Re}} C_{P,\text{Re}}$$

then

$$\Delta H'_R - \Delta H_R = \Delta C_P(T_2 - T_1) \tag{3.40}$$

Often this equation is written in differential form in order to use heat capacities given as a polynomial function of temperature as discussed earlier; for example, $C_P = a + bT + cT^2 + \cdots$. In this form,

$$\Delta H'_R = \Delta H_R + \int_{T_1}^{T_2} \Delta C_P \, dT \tag{3.41}$$

where $\Delta C_P = \Sigma_{\text{Pr}} n_{\text{Pr}} C_{P,\text{Pr}} - \Sigma_{\text{Re}} n_{\text{Re}} C_{P,\text{Re}}$.

If phase changes are also involved, terms corresponding to the latent heats of vaporization and/or fusion must be added to the right side of the equation.

Volume Normally heats of reaction are calculated at constant pressure. For nonflow, constant-volume systems these must be converted to constant volume, whereas for the constant-pressure case, $Q_P = \Delta H, Q_{\bar{V}} = \Delta E_{\bar{V}}$.

Assume a reaction is carried out at the final temperature of the constant-volume case and a constant pressure equal to the starting pressure of the constant-volume case. The first law states that $Q_P = \Delta E_P + P\Delta V$. Thus, if the difference is taken,

$$Q_P - Q_{\bar{V}} = \Delta E_P - \Delta E_{\bar{V}} + P\Delta V \tag{3.42}$$

This equation can then be used to calculate the difference between the heat requirement under the two sets of conditions. For the ideal gas, $\Delta E_P = \Delta E_{\bar{V}}$. Thus $Q_P - Q_{\bar{V}} = P\Delta V = \Delta nRT$. For liquids and solids, the assumption that $\Delta E_P = \Delta E_{\bar{V}}$ is possible. For solid-gas reactions, for example, $S(s) + O_2(g) \rightarrow SO_2(g)$, the volume of the solid is usually neglected. Thus, $\Delta n = \Sigma n_{\mathrm{Pr}} - \Sigma n_{\mathrm{Re}}$ only considering the gases. Thus, for the example given, $\Delta n = 0$ and $Q_P = Q_{\bar{V}}$. For liquid-liquid reactions, the change in volume is usually so small that it can be assumed to be negligible.

Exercise 3.N Determine the approximate error in a heat of reaction calculation which assumes that the heat of reaction for a combustion reaction at 10 MPa is the same as that at 0.1 MPa (atmospheric pressure) if the enthalpy of the steam is about 50 percent of the total enthalpy and the combustion is carried out at 1000°C.

Exercise 3.O Calculate the heat of reaction of ethane when pyrolyzed to ethylene and hydrogen at 600°C.

3.6.2 General Method for Simplified Energy Balances

A form of the energy balance where changes in kinetic, potential, and work energies are negligible is often used in process analysis.

Application of the general energy balance yields the general equation

$$q = \Delta H$$

where

$$\Delta H = \Sigma H_{\mathrm{Pr}} + \Sigma \Delta H_{R,T_0} - \Sigma H_{\mathrm{Re}}$$

and T_0 is an arbitrary temperature base, usually 25°C.

The enthalpies of the products and reactants must also be related to this same temperature base. Enthalpy cannot be assumed to be conserved as heat is gained from friction and other sources within a process. It cannot be stressed too often that this balance does not include work effects, and a total-energy or mechanical-energy balance must be used if work is present.

For any process, each entering and exiting stream plus all work and heat terms must be tabulated. Heat into the system is positive, while work done on the system is positive. Negative heats of reaction are inputs, while positive heats of reaction are outputs. Such an analysis is necessary for the design of any new process or analysis of any other process.

Examples illustrate these methods for processes where work is not involved and for processes with work terms.

Example 3.9 Carbon monoxide and water vapor react in stoichiometric amounts to form carbon dioxide and hydrogen. The feed enters at 25°C and the products exit at 540°C with a carbon monoxide conversion of 75 percent. Determine the total amount of heat which must be added or removed in the reactor per 1000 kg of hydrogen produced.

For purposes of the problem assume:

	Heat of formation at 25°C, kJ/kmol	Heat capacity, kJ/(kmol · °C)
CO	−110,600	30.35
H_2O	−241,980	36.00
CO_2	−393,770	45.64
H_2	0	29.30

BASIS 1 kmol CO feed.

$$q = \Delta H = \Sigma H_{\mathrm{Pr}} - \Sigma \Delta H_{R,T_0} - \Sigma H_{\mathrm{Re}}$$

$$\Delta H_{R25} = \Delta H_{f,CO_2} + \Delta H_{f,H_2} - \Delta H_{f,CO} - \Delta H_{f,H_2O}$$

$$= -41{,}190 \text{ kJ/kmol}$$

$$\Sigma H_{\mathrm{Re}} = 0$$

$$\Sigma H_{\mathrm{Pr}} = \left[0.75\left(C_{P,CO_2} - C_{P,H_2}\right) + 0.25\left(C_{P,CO} + C_{P,H_2O}\right)\right](540 - 25)$$

$$= [0.75(45.64 + 29.30) + 0.25(30.35 + 36.00)](515)$$

$$= 37{,}488 \text{ kJ/kmol}$$

$$\Delta H_{R,T_0} = (0.75)(-41{,}190) = -30{,}893$$

$$q = -30{,}893 + 37{,}488 = 6595 \text{ kJ/kmol}$$

$$\text{Heat} = 6595 \frac{1 \text{ kmol CO fed}}{0.75 \text{ kmol } H_2 \text{ produced}} 500 \text{ kmol } H_2 \text{ produced}$$

$$= 4.397 \times 10^6 \text{ kJ}$$

Example 3.10 If a compressor transfers 10^6 kJ of work to the process in Example 3.9, recalculate the answer.

SOLUTION Using Eq. (3.5),

$$q = \Delta H + W_s$$

$$= 4.397 \times 10^6 - 10^6$$

$$= 3.397 \times 10^6 \text{ kJ}$$

Exercise 3.P Saturated steam at 100°C is blown into a 3000-kg-mass bed of coke at 1300°C until the bed falls to a temperature of 1000°C. The gases leaving are at 1000°C and analyze by volume as 3.1% CO_2, 45.35% CO, and 51.55% H_2. All water vapor is reacted and no heat losses occur. Determine the steam requirement.

Exercise 3.Q Repeat Exercise 3.P if 20 percent of the steam passes through undecomposed.

3.6.3 Specific Cases of General Importance

Adiabatic reactions The general energy balance is applicable to any reaction process. Heat in or out of the system by means of heat exchangers or mass transfer equipment are taken into account. However, if only a reactor is analyzed and it can be assumed to neither gain heat nor lose heat with respect to the surroundings, then the reactor will reach a steady-state temperature called the *adiabatic reaction temperature*. The balance would be

$$\text{Enthalpy input} = \text{enthalpy output}$$

or

$$\text{Enthalpy of reactant stream} = \text{enthalpy of product stream} + \text{standard heat of reaction}$$

This equation assumes constant-pressure operation if a nonflow process and no effect of pressure on enthalpy for a flow process. In addition the basis of the calculation is that reaction occurs at the base temperature (normally 25°C) after which the products and any unreacted materials are heated to the adiabatic-reaction temperature. Obviously the reactants must be heated or cooled to the base temperature (T_0). Thus,

$$\Sigma H_{\text{Re}} = \Sigma H_{\text{Pr}} + \Delta H_{R,T_0} \tag{3.43}$$

where

$$\Sigma H_{\text{Re}} = \Sigma n_{\text{Re}} \int_{T_0}^{T_i} C_{P,\text{Re}}\, dT + \Sigma n_{\text{Re}} \lambda_{\text{Re}} \tag{3.44a}$$

T_i = initial temperature

T_0 = base temperature

$$\Sigma H_{\text{Pr}} = \Sigma n_{\text{Pr}} \int_{T_0}^{T} C_{P,\text{Pr}}\, dT + \Sigma n_{\text{Pr}} \lambda_{\text{Pr}} \tag{3.44b}$$

$\Delta H_{R,T_0}$ = standard heat of reaction adjusted for actual extent of reaction, i.e., number of moles of base product actually reacting

This equation can then be solved for the adiabatic-reaction temperature, although trial and error is often required as heat capacities are normally polynomials in temperature.

Many reactions which reach thermodynamic equilibrium are treated in this manner. A complete composition analysis of all reactants and products including inerts must be known or able to be estimated. The procedure is trial and error when the extent of reaction is a strong function of temperature. The former simpler case where compositions are known is illustrated here, while treatment of the latter case where chemical equilibrium must be considered will be deferred to Chap. 9.

Example 3.11 In the manufacture of sulfuric acid, FeS_2 is burned with 100% excess air over that required to form 100% Fe_2O_3 + SO_2. No SO_3 is formed, and reaction is complete. In a catalytic converter, 75% of the SO_2 is oxidized to SO_3 with the oxygen remaining in the gases. If the gases enter the adiabatic converter at 400°C, calculate the temperature of the exit gases.

BURNER

$$4FeS_2 + 11O_2 \rightarrow 2Fe_2O_3 + 8SO_2$$

BASIS 4 kmol FeS_2.

	Entering burner	Leaving burner
Oxygen	$11 \times 2 = 22$	$22 - 11 = 11$
Nitrogen	$22 \times \frac{79}{21} = 82.76$	82.76
SO_2	0	8

CONVERTER

$$SO_2 + \tfrac{1}{2}O_2 \rightarrow SO_3$$

$$SO_3 \text{ formed} = (0.75)(8) = 6.0 \text{ kmol}$$

$$O_2 \text{ used} = (6.0)(\tfrac{1}{2}) = 3.0 \text{ kmol}$$

GASES LEAVING

$$SO_2 = 2 \qquad O_2 = 8$$

$$SO_3 = 6 \qquad N_2 = 82.76$$

Using Eq. (3.43) on converter,

$$\Sigma H_{Re} = \Sigma H_{Pr} + \Delta H_{R,T_0}$$

Take $T_0 = 25°C$:

$$\Delta H_f \; SO_2 = -296{,}840 \text{ kJ/kmol}$$

$$\Delta H_f \; SO_3 = -395{,}720 \text{ kJ/kmol}$$

Therefore

$$\Delta H_{R,T_0} = 6.0[-395{,}720 - (-296{,}840)]$$

$$= -593{,}260 \text{ kJ}$$

Using mean molal heat capacities, kJ/kmol from Table 3.3 for gases entering converter,

$$nC_P(N_2) = (1.059)(28)(82.76) = 2454 \text{ kJ/K}$$
$$nC_P(O_2) = (0.967)(32)(11) = 340 \text{ kJ/K}$$
$$nC_P(SO_2) = (0.714)(64)(8) = 366 \text{ kJ/K}$$
$$3160 \text{ kJ/K}$$

$$\Sigma H_{Re} = (3160)(400 - 25) = 1{,}185{,}000 \text{ kJ}$$

Therefore

$$\Sigma H_{Pr} = 1{,}185{,}000 - (-593{,}260)$$
$$= 1{,}778{,}260 \text{ kJ}$$

The temperature of the gases exiting the converter can then be calculated by guessing the exit temperature, determining the mean molal heat capacity, and solving the equation:

$$1{,}776{,}345 = [(2)C_{P,SO_2} + (6)C_{P,SO_3} + (8)C_{P,O_2} + (82.76)C_{P,N_2}](4.186)(T - 25)$$

At 600°C, C_P's for SO_2, SO_3, O_2, and N_2 are 0.749(64), 0.855(80), 0.996(32), and 1.079(28), respectively.

If $T = 600$°C,

$$1{,}778{,}260 = 3261.6(T - 25)$$
$$T = 570\text{°C}$$

The heat capacities can be modified with a new estimated temperature and the calculation redone to obtain a more accurate anwer. A good estimate would be 575°C; why?

It should be noted that a more accurate procedure is to calculate heat capacities from Table 3.2 and utilize Eq. (3.44*a*) and (3.44*b*) for calculating ΣH_{Re} and ΣH_{Pr}.

Flame temperatures If a combustible material is burned in air or oxygen under adiabatic conditions, a temperature is reached which is known as the *theoretical flame temperature*. Again it is assumed that no potential- or kinetic-energy effects and that no mechanical work are present in the system. The exact same method of analysis as given in the previous section applies to the calculation of flame temperature. The maximum adiabatic flame temperature is that which results when the material is burned with the required stoichiometric amount of pure oxygen for complete combustion, consisting of carbon dioxide and water for materials containing only carbon, hydrogen, and oxygen and requiring specification of the oxide if elements such as nitrogen or sulfur are present. As excess air or oxygen is normally used, actual adiabatic flame temperatures will be lower than the maximum because of the dilution effect. If the stoichiometric amount of air is used, a maximum adiabatic flame temperature in air can be obtained which

Table 3.7 Heating values and flame temperatures of gases

HV = total heating value, kJ/m^3 measured at 15.6°C and 1 atm saturated with water vapor (assuming ideal-gas behavior)

			Maximum flame temperatures with dry air at 25°C		
Gas	Formula	HV	Theoretical (assuming complete combustion), °C	Calculated (allowing for equilibrium conditions), °C	Experi-mental, °C
Carbon monoxide	CO	11,774	2342		
Hydrogen	H_2	11,886	2217		
Paraffins:					
Methane	CH_4	37,036	2012	1918	1885
Ethane	C_2H_6	64,907	2065	1949	1900
Propane	C_3H_8	92,368	2356	1967	1930
n-Butane	C_4H_{10}	119,791	2084	1973	1905
n-Pentane	C_5H_{12}	147,140	2087		
Olefins:					
Ethylene	C_2H_4	58,722	2250	2072	1980
Propylene	C_3H_6	85,661	2180	2050	1940
Butylene	C_4H_8	113,121	2158	2033	1935
Amylene	C_5H_{10}	140,470	2204		
Acetylene	C_2H_2	54,064	2586		
Aromatics:					
Benzene	C_6H_6	137,378	2211		
Toluene	$C_6H_5CH_3$	164,279	2187		

will obviously be lower than the corresponding oxygen value. Such temperatures for common hydrocarbons are given in Table 3.7.

Actual flame temperatures are always lower than theoretical flame temperatures as heat loss occurs, 100 percent complete combustion usually does not occur, and an equilibrium between CO, CO_2, and oxygen may occur especially at very high temperatures. In addition, the other terms in the energy balance may have some effect on the results. Table 3.7 also illustrates these effects. This table also gives the total heating value (HV) of the substance which is merely the standard heat of combustion with the opposite sign, assuming that any water present in the combustible material and water present in the products are in the liquid state. This total HV is also called gross HV. Net HV assumes that the water present in the products after combustion is vapor at 25°C. Thus

$$\text{Gross HV} = \text{net HV} + \text{latent heat contribution of water formed in the reaction and subsequently vaporized} \quad (3.45)$$

Example 3.12 Calculate the theoretical flame temperature of a gas initially at 25°C containing 20% C_2H_6 and 80% oxygen when completely burned after dilution with double the original volume of CO_2.

SOLUTION

$$C_2H_6 + \tfrac{7}{2}O_2 \rightarrow 2CO_2 + 3H_2O$$

Since $\Sigma H_{\text{Re}} = 0$ (base temperature = 25°C),

$$\Sigma H_{\text{Pr}} = -\Delta H_{R,25°\text{C}}$$

BASIS 1 kmol C_2H_6 + 4 kmol O_2 in feed, 10 kmol CO_2 added (assuming ideal gases).

$$-\Delta H_{R,25°\text{C}}\ (\text{App. A}) = 1.478 \times 10^6\ \text{kJ/kmol}\ C_2H_6$$

PRODUCT 12 kmol CO_2, 3 kmol H_2O, 0.5 kmol O_2. Therefore

$$1.478 \times 10^6 = 4.186\left(12C_{P,CO_2} + 3C_{P,H_2O} + 0.5C_{P,O_2}\right)(T - 25)$$

Assume T = 1200°C and using mean heat capacities from Table 3.3,

$$C_{P,CO_2} = 1.164(44) \qquad C_{P,H_2O} = 2.210(18) \qquad C_{P,O_2} = 1.055(32)$$

$$1.278 \times 10^6 = (750{,}812)(T - 25)$$

$$T = 1927°\text{C}$$

Assume T = 1800°C:

$$C_{P,CO_2} = 1.24(44) \qquad C_{P,H_2O} = 2.39(18) \qquad C_{P,O_2} = 1.11(32)$$

$$1.4278 \times 10^6 = (801.54)(T - 25)$$

$$T = 1806°\text{C}$$

Since heat capacities will change very little in 6°C,

Theoretical temperature ≈ 1806°C

Note again that the use of heat capacities from Table 3.2 with trial and error for T would be a more accurate procedure.

Exercise 3.R Repeat Example 3.11 by using exact heat capacity relations from Table 3.2.

Exercise 3.S Accurately calculate the theoretical flame temperature of a gas containing 25% CO and 75% CO_2 when burned with 50% excess air. The gas is initially at 25°C. Repeat if the gas is initially at 800°C. Assume in both cases that burning is complete.

3.7 REFERENCES

The first law of thermodynamics has been treated in many ways in various material and energy balance (stoichiometry) and thermodynamics texts. Some of the elementary treatments are given in the texts noted which are recommended for reference.

Felder, R. M., and R. W. Rousseau: "Elementary Principles of Chemical Processes," Wiley, New York, 1978.

Hougen, O. A., K. M. Watson, and R. A. Ragatz: "Chemical Process Principles, Part I, Material and Energy Balances," 2d ed., Wiley, New York, 1954 (classic treatment).

Smith, J. M., and H. C. Van Ness: "Introduction to Chemical Engineering Thermodynamics," 3d ed., McGraw-Hill, New York, 1975.

Sussman, M. V.: "Elementary General Thermodynamics," Addison-Wesley, Reading, Mass., 1972.

References cited are given below:

American Petroleum Institute: "Technical Data Book—Petroleum Refining," 4th ed., T. E. Daubert and R. P. Danner (eds.), Washington (extant 1982).

______: Research Project 44, "Selected Values of Physical and Thermodynamic Properties of Hydrocarbons and Related Compounds," Thermodynamics Research Center, Texas A & M, College Station, Texas (extant 1982).

Benson, S. W.: "Thermochemical Kinetics," 2d ed., Wiley, New York, 1976.

Clapeyron, J.: L'Ecole Polytechnique, **14**(23):153 (1834).

Hadden, S. T.: Private communication (1964) as recorded in "Technical Data Book—Petroleum Refining," American Petroleum Institute, Washington.

Hougen, O. A., K. R. Watson, and R. A. Ragatz: "Chemical Process Principles, Part I," 2d ed., Wiley, New York, 1954.

Huang, P. K., and T. E. Daubert: *Ind. Eng. Chem. Process Des. Dev.*, **13**:193 (1974).

McCabe, W. L., and J. C. Smith: "Unit Operations of Chemical Engineering," 3d ed., McGraw-Hill, New York, 1976.

Passut, C. A., and R. P. Danner: *Ind. Eng. Chem. Process Des. Dev.*, **11**:543 (1972).

Perry, J. H., et al.: "Chemical Engineers Handbook," 5th ed., McGraw-Hill, New York, 1973.

Reid, R. C., J. M. Prausnitz, and T. K. Sherwood: "Properties of Liquids and Gases," 3d ed., McGraw-Hill, New York, 1978.

Watson, K. M.: *Ind. Eng. Chem.*, **23**: 360 (1931); **35**:398 (1943).

3.8 PROBLEMS

1 One cubic foot of a diatomic ideal gas is contained in a vertical cylinder by a piston of 1 ft^2 cross section held in place by a latch. A perfect vacuum exists above the piston. The latch is released and the 1000-lb piston rises. Determine the piston's velocity and the heat input required to maintain the system isothermal at 100°F when the piston has risen 9 ft. The initial gas pressure is 100 lb/in^2 abs and the system is vertical and can be assumed frictionless.

2 Two tanks of equal volume are connected by a valve. One contains steam at 68 atm and 370°C, while the other is fully evacuated. The valve is fully opened and isothermal expansion occurs. Determine the work done by the steam, the enthalpy change in the system, and the heat transferred to or from the steam in the process.

3 Repeat Prob. 2 for a monatomic ideal gas. Comment on the reversibility of the process.

4 Ideally how much work is required to bring a 2000-lb automobile traveling on a level road at 55 mi/h to a stop?

5 One pound mole of a diatomic, ideal gas is to undergo a reversible expansion from 12 atm and a volume of 50 ft^3 to 1 atm. Determine the change in internal energy, the change in enthalpy, the work done by the process, and the heat required if the process is carried out (*a*) isothermally, (*b*) isometrically, (*c*) adiabatically, and (*d*) polytropically if $k = 1.2$. Compare the processes on a *PV* plot of the process.

6 The heat of reaction of the gaseous methanol synthesis ($CO + 2H_2 \rightarrow CH_3OH$) at 1000 K is -21.61 kcal. What is its heat of reaction at 750°?

7 Derive an expression analogous to Eq. (3.32) for the adiabatic compression of a van der Waals gas.

8 The adiabatic work of compression of an ideal gas is a function of the constant k. Compare the work required for a monatomic and a diatomic gas for a compression ratio of 10, as well as for a value of k of unity. What type of compression does the latter k correspond to?

9 How far will a 10-lb frictionless piston of 1 in^2 area fall if suspended 1 ft from the bottom of a cylinder containing air at 70°F and 1 atm? Assume no heat losses.

10 Derive a general expression for the mean heat capacity of an ideal gas between temperature T_1 and T_2 in terms of temperatures and the constants A, B, C, and D if $C_P = A + BT + CT^2 + D/T$.

11 Calculate Q, W, and ΔE for the following two processes for raising 2 lb mol of an ideal, diatomic gas from 1 atm and 0°C to 50 atm and 1000°C.

Path 1: The gas is compressed isothermally and reversibly and then heated isobarically.
Path 2: The gas is heated isobarically and then compressed isothermally.

12 Repeat the processes of Prob. 11 if the gas is nitrogen and can be assumed to follow the Soave equation of state.

13 Repeat the processes of Prob. 11 if the fluid is liquid water changing to superheated steam.

14 An oil flows at 2000 lb/min from an open reservoir on a hill 1000 ft high into another open reservoir at the bottom of the hill. The oil with mean heat capacity of 0.8 Btu/(lb · °F) is heated to assure rapid flow at a rate of 100,000 Btu/h and a 1-hp pump assists in moving the oil. What is the change in the temperature of the oil?

15 Calculate the standard heat of reaction of the reaction

$$C_4H_{10}(g) + 4S(s) \rightarrow C_4H_4S(g) + 3H_2S(g)$$

if $\Delta H_c(C_4H_4S)_l = -670.5$ kcal/g mol for products of liquid water and gaseous SO_2 and CO_2 and $\Delta H_{\bar{V}}(C_4H_4S)_{25°C} = 99.1$ cal/g. Data of App. A may also be used.

16 If 1 kmol of HCl is added to 5 kmol of water at 25°C, estimate the heat that must be transferred to maintain the system isothermal.

17 A waste gas at 426°F contains 5 mol % HCN and 95 mol % N_2. It is to be disposed of by reaction with 150 percent of the air needed for complete combustion to CO_2, H_2O, and N_2. Before mixing with the waste gas the air is preheated in a countercurrent heat exchanger by the combustion gases leaving the reactor. The entire system is well-insulated. The air enters the heat exchanger at 25°C. The combustion gases leave the reactor at 2012°F.

(*a*) At what temperature does the air enter the reactor?

(*b*) At what temperature do the combustion gases leave the heat exchanger?

Note: C_P values can be considered constant as follows: for H_2, N_2, O_2, $H_2O(g)$, CO_2, and $HCN(g)$, 7, 7, 7, 10.5, 10.5, 10.5 cal/(g mol · °C), respectively. The standard heat of formation of HCN is 31.2 kcal/(g mol).

18 Ethylbenzene can be converted to styrene in two ways—dehydrogenation with evolution of hydrogen and oxidative dehydrogenation with formation of water. Compare the heating requirements for reaction of 1 kmol of ethylbenzene.

19 A gas containing 4.8% CO_2, 27.0% CO, 0.4% O_2, 1.9% CH_4, 14.1% H_2, and 51.8% N_2 enters a combustion chamber dry at 25°C. Combustion is complete to CO_2 and H_2O using 50 percent excess air. Determine the theoretical flame temperature, assuming the system is adiabatic.

20 A gas stream composed of 9.1 mol % NO, 4.7% O_2, 69.5% N_2, and 16.7% H_2O leaves an ammonia oxidation unit at 1200 K and is to be cooled in a heat exchanger with air entering at 25°C. The air is supplied in the ratio of 1 mol dry air/mol gas.

How much heat could be recovered from the gases if it were possible to cool them to 25°C?

How much heat could be absorbed by the air if it could be heated to 1200 K?

Determine the minimum temperature to which the gases can actually be cooled.

21 It is proposed to use waste gas [C_P (mean) = 8.2 Btu/(lb · mol · °F)] being discarded at 750°F to produce saturated steam at 100°F from feedwater at 70°F in a waste heat boiler. Under adiabatic conditions, with unlimited heat transfer surface, what is the maximum possible production of steam, lb steam/lb mol gas?

22 Estimate the latent heat of vaporization of benzene at 200°C, using only data of App. A where necessary.

23 Estimate the vapor pressure of benzene at 200°C. Note any assumptions.

24 Normal heats of reaction are tabulated at constant pressure. What correction must be applied to calculate a heat of reaction at constant volume?

CHAPTER

FOUR

THE SECOND LAW OF THERMODYNAMICS AND ITS APPLICATIONS

The essence of thermodynamics is the second law and its applications. The basis for analyzing important cyclical processes for propulsion, compression, and refrigeration for use in motor vehicles, steam plants, electrical generating facilities, and low-temperature heat exchange lies in an understanding of the second-law principles and is important to almost all engineers. Of particular consequence to the chemical engineer and almost totally carried out by chemical engineers are three areas which comprise the intrinsic nature of chemical engineering.

1. Complete thermodynamic analyses of entire chemical process plants or subsections of plants necessary in the design stage and subsequent operating states of almost every facility. Such analyses have increased in importance with the necessity for energy conservation and will increasingly be a necessary part of any process analysis.
2. Calculation of the physical equilibrium relationships between materials necessary for any design or analysis of mass transfer equipment dependent on equilibrium between phases including vapor-liquid, liquid-liquid, vapor-solid, liquid-solid, and multiphase systems. Basic thermodynamic equilibrium data must be available for design and improvement of equipment for distillation, extraction, adsorption, absorption, etc.
3. Calculation of the chemical equilibrium state which may be approached in any reacting system whether actually at equilibrium or approaching equilibrium.

Such calculations are necessary for almost all kinetic calculations leading to reactor design and operation and are usually prerequisite to an understanding of reaction kinetics as well as necessary for design of reactors operating at equilibrium.

This chapter begins with the concept of entropy and its applications to cyclical processes. An outline for thermodynamic analysis of processes is given. In addition, methods for the calculation or estimation of free energy, the parameter necessary to calculate both physical and chemical equilibrium constants, is presented. The application of free energy and its derived properties for physical and chemical equilibrium calculations are deferred to Chaps. 7 to 9, where extensive treatments of such situations are considered.

4.1 STATEMENTS OF THE SECOND LAW AND DEFINITIONS OF TERMS

The second law of thermodynamics has been quoted in many ways, all of which are equivalent. Some of these statements refer to heat flow, some to cyclical processes, and others to entropy.

Some equivalent statements are:

Heat flow:

"Heat cannot of itself pass from a colder to a hotter body." (Steiner quoting Clausius)

"Any process, the sole net result of which is equivalent to the transfer of heat from a low-temperature level to a high one is impossible." (Weber)

"Heat cannot flow from a body at a low temperature to another at a higher temperature unless energy is supplied by some external system to bring about the change." (Young and Young) Airconditioner

Process:

"It is impossible for a self-acting machine, unaided by an external agency, to convey heat continuously from one body to another at a higher temperature." (Kiefer and Stuart quoting Clausius)

"It is impossible to construct an engine which, working in a complete cycle, will produce no effect other than the raising of a weight and the cooling of a heat reservoir." (Zemansky quoting Planck)

"It is impossible by means of inanimate material agency to derive mechanical effect from any portion of matter by cooling it below the temperature of the coldest of its surroundings." (Zemansky quoting Clausius)

"No actual or ideal heat engine which operates on a closed cycle can convert all the heat supplied to it into work." (Faires)

Entropy:

"Every system which is left to itself will, on the average, change toward a system of maximum probability." (Lewis and Randall)

"All spontaneous processes are to some extent irreversible and are accompanied by a degradation of energy." (Hougen and Watson)

$$\Delta S = \sum \frac{dQ_{\text{rev}}}{T} \geq 0$$

where ΔS = change in entropy
T = absolute temperature
dQ_{rev} = reversible heat transferred

"Die Entropie der Welt strebt einem maximum zu." (Clausius)

Although each of these statements is equivalent, a close analysis of each is necessary to prove that any device which violates one of these statements will violate all others. This chapter shall attempt to develop the principles which will allow such analyses. The reader should continually refer back to these statements and eventually satisfy the criterion that all statements are equal.

Certain auxiliary definitions are necessary throughout any consideration of the second law. Most of these are contained in the preceding statements.

A *heat reservoir* is a very large system that can be assumed to be at constant temperature which serves as a heat source or heat sink for another system.

A *heat engine* is a mechanical device which produces work by transferring heat from a high-temperature source to a low-temperature source. In such an engine the *maximum* transformation of heat to work will occur in a reversible process, with all heat transfer taking place over infinitesimal temperature differences so that an infinitesimal change will change the direction of the process. The development of the Carnot cycle in Sec. 4.3.2 shows the validity of this statement.

Thermal efficiency is defined as the work obtained from a certain process divided by the heat input to that process. The maximum thermal efficiency occurs for a reversible engine operating between a high-temperature heat source reservoir and a low-temperature heat sink reservoir, each of which is so large that its temperature is constant. (Such an engine is called a $2T$ engine.)

4.2 ENTROPY

Entropy is one of the most difficult concepts in thermodynamics for many to grasp as it is intangible and cannot be measured calorimetrically or any other way. This section will attempt by means of several equivalent definitions of entropy to dispel the "fear" of entropy, one of the most useful and indispensable properties of thermodynamics. After defining the term, methods for calculation of the entropy changes occurring for reversible and irreversible processes, ideal and nonideal gases, and other fluids will be discussed. Finally, the application of entropy in the calculations necessary for complete thermodynamic process analysis and lost-work calculations will be addressed and illustrated. Full proofs and derivations are beyond the scope of this text; Denbigh (1981) gives the most lucid analysis.

4.2.1 Definitions and Their Interpretation

Clausius defined entropy as follows: "When a system passes in a reversible way from one state to another state, at a constant absolute temperature T, its change in entropy ΔS is equal to the heat Q it absorbs, divided by T."

As pointed out in an excellent discourse on entropy by Darrow in a Bell Laboratories publication pamphlet (1942), not only entropy and absolute temperature are contained in Clausius' statement. The second law enters the definition by the phrase "in a reversible way." Thus, to understand entropy, some understanding of absolute temperature and reversibility is required.

Another definition of entropy as quoted from Hougen, Watson, and Ragatz (1959) is related to the unavailability or degradation of energy in a system: "Entropy is an intrinsic property of matter so defined that an increase in the unavailability of the total energy of a system is quantitatively expressed by a corresponding increase in its entropy."

If entropy is defined as a measure of unavailability of internal energy, any degradation of this energy to *heat* or any other form of energy will obviously lead to an increase in entropy. As this property is not affected by position or motion, the cause of such degradation is immaterial and any heat generation will lead to an entropy gain.

Lewis and Randall (1923), after discussing the Clausius definition of entropy, relate entropy to the probability that if any system is left to itself, it will, on the average, change toward a condition of maximum probability; i.e., if change is spontaneous, the entropy will increase. A complete discussion of the statistical thermodynamic treatment of entropy is beyond the scope of this book. Lewis and Randall (1923, 1961) give a very understandable description of such statistical concepts. Hill (1960) may be consulted for a more detailed treatment.

Since entropy is defined as

$$\Delta S \equiv \frac{dQ_{\text{rev}}}{T} \tag{4.1}$$

it is obvious that the lower the *temperature* of the system the greater the increase in entropy. [This thermodynamic temperature can be shown to be equal to the ideal-gas temperature (Denbigh, 1981; Lewis and Randall, 1961; Hougen, Watson, and Ragatz, 1959) satisfying the equation $PV = RT$.] Thus, the entropy increase resulting from the addition of a certain amount of heat to a system is larger, the lower the temperature at which heat is added.

Entropy, like internal energy and enthalpy, is defined as a state property and does not depend on its position or motion with respect to any other body at the same temperature and pressure. If entropy can be proven to be a function of state, then the process actually occurring is immaterial, and a reversible process can be assumed to calculate entropy.

Consider a reversible cyclical process consisting of two isothermal processes at T_1 and T_2 and two adiabatic processes joining the isothermal processes. This process is called the Carnot cycle, derived in detail in Sec. 4.3.2. By definition the changes in entropy of the adiabatic processes are each zero, while the changes in entropy for the isothermal processes are Q_1/T_1 and Q_2/T_2. Q_1 and Q_2 must also be of opposite sign or the process could not be cyclical. If entropy is a state function, the sum of the isothermal entropy changes must be zero.

$$dS = \frac{Q_1}{T_1} + \frac{Q_2}{T_2} = 0$$

This can only be true if temperatures are chosen such that $-Q_1/Q_2 = T_1/T_2$. This is the basis on which the thermodynamic absolute temperature scale is founded which as alluded to earlier can be equated to the ideal-gas-law temperature. Since the cycles are reversible, as long as the heat sink and heat source temperatures are the same, it makes no difference in what direction the cycles are carried out, nor does it matter what particular pressures and volumes occur in the cycle as long as the cycle is between the same pair of isothermals. Thus, the ratio of Q's is only a function of the temperatures T_1 and T_2.

These arguments can be extended to any cyclical process by assuming the process to be a series of adiabatic-isothermal cycles as derived in detail by Denbigh (1981), proving that entropy is truly a function of state.

Irreversibilities may occur whether or not any heat is transferred or work is done. Thus energy is degraded, and by definition an increase in entropy takes place. However, in such a process the defining equation cannot be used as it assumed reversibility. However, calculation of an irreversible process is quite simple if one remembers that entropy is a state function rather than a path function like heat and work. Thus, the change in entropy for an irreversible process proceeding between any two states can be calculated by devising a reversible process for accomplishing the same change and evaluating the $\Sigma(dQ_{\text{rev}}/T)$ for the entire process. This will then give the entropy change for the irreversible process.

It is important to compare reversible and irreversible processes occurring between a given system and a given surroundings. If the identical change takes place in the system by reversible or irreversible processes, the change in entropy of the system in both cases will be equal. If the process occurring in the surroundings, e.g., gain or loss of heat, is transferred reversibly, its entropy change will be equal in magnitude and opposite in sign to that of the system, yielding a net change in the entropy of the universe as zero. However, if the process occurs irreversibly, the net increase in the entropy of the universe will be positive.

$$\Delta S_{\text{univ or total}} = \Delta S_s + \Delta S_S \tag{4.2}$$

where the subscripts s and S refer to system and surroundings, respectively. The restriction of reversibility does not apply to the process used to degrade higher forms of energy or to the way in which heat is added to a process. Heating an object by a constant-temperature heat source is irreversible; however, $\Delta S = dQ/T$. Only if the properties of matter change must the process be reversible. Thus, if only heat transfer is occurring, no matter whether the process is reversible or irreversible, the change in entropy equals the heat gain or loss divided by the temperature. However, if a pressure difference or other transport process occurs, a reversible path must be developed.

Entropy is an extensive property like internal energy or enthalpy and depends on the mass of the system. Entropy, like internal energy, must be conserved.

Example 4.1 A reservoir at 500 K receives 5×10^6 kJ of heat from a constant-temperature heat source at 600 K. Determine the change in entropy of the system, the surroundings, and the universe.

SOLUTION Taking the reservoir as the system,

$$\Delta S_s = \frac{Q_{\text{rev}}}{T} = \frac{5 \times 10^6}{500} = 10{,}000 \text{ kJ/K}$$

(Note that by the definition of a reservoir it is so large that it will not change in temperature.)

$$\Delta S_S = \frac{Q_{\text{rev}}}{T} = \frac{-5 \times 10^6}{600} = -8333 \text{ kJ/K}$$

$$\Delta S_{\text{univ}} = \Delta S_s + \Delta S_S$$

$$= 10{,}000 + (-8333) = +1667 \text{ kJ/K}$$

A general entropy balance analogous to the energy balance discussed in Chap. 3 can be set up.

Consider a system with flow of material in and out and flow of entropy across the system boundary not dependent on mass flow. A general entropy balance is

$$S_{\text{in}} - S_{\text{out}} + S_{\text{generated}} = S_{\text{accumulated}}$$

or

$$(\bar{S}\delta m)_{\text{in}} - (\bar{S}\delta m)_{\text{out}} + \delta S_{\text{flow}} + \delta S_{\text{generated}} = d(m\bar{S})_s$$

The generation term includes all generation of entropy in the system volume including lost work (defined in Sec. 4.2.5) and is zero (for a reversible process) or positive (for an irreversible process). The flow term includes all entropy crossing the system boundary excluding that carried by the flow of material.

For an isothermal system the balance can be written as

$$(\bar{S}\delta m)_{\text{in}} - (\bar{S}\delta m)_{\text{out}} + \frac{\delta Q}{T} + \frac{\delta W_{\text{lost}}}{T} = d(m\bar{S})_s$$

For a closed, isothermal system,

$$\frac{\delta Q}{T} + \frac{\delta W_{\text{lost}}}{T} = m\,\Delta\bar{S}_s$$

For an open, isothermal, steady-state system,

$$(\bar{S}_{\text{in}} - \bar{S}_{\text{out}})\,\delta m + \frac{\delta Q}{T} + \frac{\delta W_{\text{lost}}}{T} = 0$$

These formulations will result in the same expressions derived in Sec. 4.2.5 for lost work and process analyses. The use of the entropy balance is more general and can be applied to any situation, steady or unsteady state.

Exercise 4.A Write entropy balances for the situations of Exercise 3.B.

The following sections treat entropy change calculations for various types of common situations.

4.2.2 Entropy Changes for Ideal-Gas Systems

As discussed in Chap. 3, Sec. 3.5.1, for the calculation of heat, work, changes in internal energy, and changes in enthalpy of ideal-gas systems, a similar analysis can be done for the calculation of changes in entropy in such systems, both for reversible and irreversible systems.

For *reversible* systems, starting with the basic definition $\Delta S = dQ_{\text{rev}}/T$, an expression can be derived for the various cases.

Isothermal As discussed earlier, the change in internal energy for an ideal gas is only a function of temperature. Thus, $\Delta E = 0$ and $dQ_{\text{rev}} = W$. As derived earlier for the ideal gas, $Q = W = nRT\ln(V_2/V_1) = nRT\ln(P_1/P_2)$. Thus

$$\Delta S_T = \frac{Q}{T} = nR\ln\frac{V_2}{V_1} = nR\ln\frac{P_1}{P_2} \tag{4.3}$$

Isobaric By definition,

$$\Delta S_P = \left(\frac{dQ_{\text{rev}}}{T}\right)_P = \frac{dH_P}{T} = \int_{T_1}^{T_2}\frac{C_P\,dT}{T} \tag{4.4}$$

If C_P is a function of temperature, the expression must be integrated according to the function. If C_P can be assumed to be an average constant value of heat capacity over the temperature range under consideration,

$$\delta S_P = \bar{C}_P\ln\frac{T_2}{T_1} \tag{4.5}$$

Isometric By definition,

$$\Delta S_V = \left(\frac{dQ_{\text{rev}}}{T}\right)_V = \frac{dE_V}{T} = \int_{T_1}^{T_2}\frac{C_V\,dT}{T} \tag{4.6}$$

Again, if C_V can be assumed to be an average constant value,

$$\Delta S_V = \bar{C}_V\ln\frac{T_2}{T_1} \tag{4.7}$$

Adiabatic By definition, $dQ_{\text{rev}} = 0$ for an adiabatic system. Thus $\Delta S = 0$.

Nonisothermal-nonadiabatic If an ideal-gas system changes from $P_1V_1T_1$ to $P_2V_2T_2$, none of the previous cases can be applied directly. The fact that change in entropy is a state property and does not depend on path can be used to calculate the change in entropy as discussed earlier for internal-energy calculations. Consider two reversible processes, an isobaric process followed by an isothermal process, that is,

$$P_1V_1T_1 \xrightarrow{1} P_1VT_2 \xrightarrow{2} P_2V_2T_2$$

Thus, the change in entropy for the actual process is merely the sum of the changes for the two assumed processes.

$$\Delta S = \int_{T_1}^{T_2} \frac{C_P\,dT}{T} - R\ln\frac{P_2}{P_1} \tag{4.8}$$

Obviously the same result would occur from reversing the order of the assumed processes. An equivalent expression would result from a series of processes which assume an isometric process followed by an isothermal process.

$$\Delta S = \int_{T_1}^{T_2} \frac{C_V\,dT}{T} + R\ln\frac{V_2}{V_1} \tag{4.9}$$

Isothermal mixing Such a process can be treated by devising a reversible process to carry out the mixing. Consider two separate ideal gases A and B at P and T and mixed with mole fractions y_A and y_B, respectively. Assume that each gas is expanded isothermally and reversibly from P to a pressure equal to its partial pressure in the mixture. Therefore

$$\Delta S_A = -y_A R\ln\frac{y_A P}{P} \qquad \Delta S_B = -y_B R\ln\frac{y_B P}{P} \tag{4.10}$$

if Dalton's law applies. Therefore

$$\Delta S = -y_A R\ln y_A - y_B R\ln y_B \tag{4.11}$$

Then, reversibly force each pure component into a volume of gas at P and T, where the ratio of A/B is y_A/y_B. For this operation, $\Delta S = 0$. Therefore, the total charge in entropy of mixing of the ideal gases will be equal to the entropy charge resulting from the first step of the assumed process.

For *irreversible* ideal-gas systems, the entropy changes can be calculated in the same way as will be discussed in Sec. 4.2.3. However, the case of a free expansion, the adiabatic expansion of a gas into a vacuum in an isolated system, is a special case which should be discussed. Consider any gas, totally isolated from the surroundings at $P_1V_1T_1$, expanding into a vacuum in the same isolated system reaching a state of $P_2V_2T_2$. Since the system is adiabatic, $Q = 0$. Since no work is done by or on the gas, $W = 0$. Therefore, $\Delta E = 0$. This fact is true for any gas. If, in addition, the gas is ideal,

$$\Delta E = C_V(T_2 - T_1) = 0 \tag{3.26}$$

Since C_V cannot be zero, $T_2 - T_1$ must equal zero. Thus, $\Delta T = 0$ and the gas expands from $P_1V_1T_1$ to $P_2V_2T_1$, which is an isothermal process. As the change in entropy does not depend on path, ΔS_s can be calculated as a reversible, isothermal expansion. Thus, $\Delta S_s = R \ln (P_1/P_2)$, which must be a positive value. Since the system is isolated, $\Delta S_S = 0$, and the change in entropy of the universe is the same as the change in entropy of the system and is positive in value.

Example 4.2 Reconsider Example 3.5. Compute the change in entropy for each step in the process and for the entire process.

SOLUTION

(*a*) *Isothermal*:

$$P_1 = 2.758 \text{ MPa} \rightarrow P_2 = 0.552 \text{ MPa}$$

$$T_1 = 700 \text{ K} \qquad T_2 = 700 \text{ K}$$

By Eq. (4.3),

$$\Delta S = nR \ln \frac{P_1}{P_2} = 8.3143 \ln \frac{2.758}{0.552}$$

$$\Delta S_a = +13.375 \text{ kJ/(kmol} \cdot \text{K)}$$

(*b*) *Isometric*:

$$T_2 = 700 \text{ K} \rightarrow T_3 = 437.5 \text{ K}$$

By Eq. (4.7),

$$\Delta S_V = \overline{C}_V \ln \frac{T_3}{T_2} = 21.0 \ln \frac{437.5}{700}$$

$$\Delta S_b = -9.870 \text{ kJ/(kmol} \cdot \text{K)}$$

(*c*) *Isobaric*:

$$T_3 = 437.5 \text{ K} \rightarrow T_4 = 350 \text{ K}$$

By Eq. (4.5),

$$\Delta S_P = C_P \ln \frac{T_4}{T_3} = 29.3 \ln \frac{350}{437.5}$$

$$\Delta S_c = -6.538 \text{ kJ/(kmol} \cdot \text{K)}$$

(*d*) *Adiabatic*:

$$T_4 = 350 \text{ K} \rightarrow T_5 = 634 \text{ K}$$

By definition, $Q_{\text{rev}} = 0$, therefore $\Delta S_d = 0$.

(*e*) *Isobaric*:

$$T_5 = 634 \text{ K} \rightarrow T_6 = 700 \text{ K}$$

By Eq. (4.5),

$$\Delta S_P = \overline{C}_P \ln \frac{T_6}{T_5} = 29.3 \ln \frac{700}{634}$$

$$\Delta S_e = +2.902 \text{ kJ/(kmol} \cdot \text{K)}$$

OVERALL PROCESS

$$\Delta S_T = 13.375 - 9.870 - 6.538 + 0 + 2.902$$

$$\Delta S_T = (-0.131)\ \text{kJ/(kmol} \cdot \text{K)} \to 0$$

Since S is a state function, S_T must be equal to zero. The actual number is not zero because of roundoff errors in temperatures and heat capacities.

Example 4.3 Helium and nitrogen at 300 K in a proportion of 1 : 2 are mixed. What is the entropy of mixing?

$$\begin{aligned}\Delta S &= -y_A R \ln y_A - y_B R \ln y_B \\ &= -0.333(8.314) \ln 0.333 - 0.667(8.314) \ln 0.667 \\ &= 3.044 + 2.246 = 5.290\ \text{kJ/(kmol} \cdot \text{K)}\end{aligned}$$

If this mixture of gases is caused to go through a free expansion from 0.5 to 0.1 kPa, what is the change in entropy of the universe?

$$\Delta S_{\text{univ}} = \Delta S_s + \Delta S_S$$

Since in a free expansion for ideal gases $\Delta T = 0$, and since a free expansion implies an isolated system, $\Delta S_S = 0$. Thus

$$\begin{aligned}\Delta S_{\text{univ}} = \Delta S_s &= R \ln \frac{P_1}{P_2} \\ &= 8.314 \ln \frac{0.5}{0.1} \\ &= 13.38\ \text{kJ/(kmol} \cdot \text{K)}\end{aligned}$$

Exercise 4.B Calculate ΔS for each step of the process in Exercise 3.I as well as for the overall process.

Exercise 4.C Calculate the entropy change in a heat exchanger where 3 mol of hydrogen and 1 mol of nitrogen at 1 MPa pressure are fed separately at 100°C, are mixed, and then are heated in the exchanger to 500°C with a corresponding increase in pressure.

4.2.3 Entropy Changes for Nonideal Gas and Heat Transfer Systems

Obviously for nonideal gases where an equation of state is available to describe the gas, the principles of Sec. 4.2.2 can be used to derive equations for calculating entropy changes. If an equation of state if not applicable, tabular or graphical data are required at the end points to calculate the change in entropy. Alternatively, the theorem of corresponding states as discussed later can be used to estimate such changes.

If isothermal phase changes under constant-pressure conditions are carried out slowly so that they can be assumed reversible and at equilibrium, $\Delta S = \Delta H_{\text{phase change}}/T$. For fusion and vaporization equilibrium processes as discussed in detail later,

$$\Delta S_{\text{fus}} = \left(\frac{\lambda_{\text{fus}}}{T_{\text{fus}}}\right)_{P,T_{\text{fus}}}$$

and
$$\Delta S_{\text{vap}} = \left(\frac{\lambda_{\text{vap}}}{T_{\text{vap}}}\right)_{P,T_{\text{vap}}} \tag{4.12}$$

If a system is considered where only heat transfer is occurring at constant pressure and no work is done, it can be easily shown that the entropy change for the system, if heating is done irreversibly with high temperature gradients, will be the same as if the process were carried out reversibly with infinitesimal temperature differences. Thus, the change in entropy of the system will be

$$\Delta S = n \int_{T_1}^{T_2} \frac{C_P\,dT}{T}$$

Obviously in a heat exchanger, this difference can be calculated for each of the process streams. If the process is carried out reversibly, the algebraic sum of the entropy changes for each process stream will be zero. For the practical situation, the process will be somewhat irreversible, and the entropy change in the universe will be positive. It can be shown relatively easily that the total entropy change in a countercurrent exchanger is smaller than that in a cocurrent exchanger.

An adiabatic mixing process between hot and cold fluids at constant pressure would require identical calculations to those required for heat exchangers.

Example 4.4 Consider the process given in Example 3.6. Calculate ΔS for each step in the process and for the overall process.

SOLUTION Interpolating from the steam tables in App. B,

$$S_1 = 7.096 \text{ kJ/(kg}\cdot\text{K)} \qquad S_4 = 7.014 \text{ kJ/(kg}\cdot\text{K)}$$

$$S_2 = 7.915 \text{ kJ/(kg}\cdot\text{K)} \qquad S_5 = 6.999 \text{ kJ/(kg}\cdot\text{K)}$$

$$S_3 = 7.16 \text{ kJ/(kg}\cdot\text{K)}$$

$$\Delta S_a = S_2 - S_1 = 0.819$$

$$\Delta S_b = S_3 - S_2 = -0.755$$

$$\Delta S_c = S_4 - S_3 = -0.16$$

$$\Delta S_d = S_5 - S_4$$

$$= -0.015 = 0 \qquad \text{by definition of adiabatic process}$$

$$\Delta S_e = S_1 - S_5 = +0.097$$

$$\Delta S_T = +0.001 \rightarrow 0 \qquad \text{(state function)}$$

Example 4.5 Assume that 5000 kg/h of oil with a heat capacity of 3.2 kJ/(kg · K) is to be cooled from 220°C as far as possible if a 10°C temperature approach is required with water available at 20°C in a countercurrent heat exchanger. Determine the change in entropy of the universe per hour.

SOLUTION

$$\Delta S = \left(mC_P \ln\frac{T_2}{T_1}\right)_{\text{oil}} + \left(mC_P \ln\frac{T_2}{T_1}\right)_{H_2O}$$

$$10^\circ \text{ approach} < \begin{matrix} \text{oil } 220^\circ \rightarrow 30^\circ\text{C} \\ 210^\circ \leftarrow \text{water } 20^\circ\text{C} \end{matrix} > 10^\circ \text{ approach}$$

$$(mC_P\Delta T)_{\text{oil}} = (mC_P\Delta T)_{H_2O}$$

Assuming the water is pressurized so vaporization does not occur,

$$(5000)(3.2)(220-30) = m_{H_2O}(4.186)(210-20)$$

$$m_{H_2O} = 3822 \text{ kg/h}$$

$$\begin{aligned}\Delta S &= (5000)(3.2)\ln\frac{303.1}{493.1} + (3822)(4.186)\ln\frac{483.1}{293.1} \\ &= -7786 + 7995 \\ &= +209 \text{ kJ/(h}\cdot\text{K)}\end{aligned}$$

Exercise 4.D Calculate ΔS for each step of the process and for the overall process given in Exercise 3.K.

Exercise 4.E The oil of Example 4.5 is to be cooled in a cocurrent exchanger with the same water flow and cooling as far as possible with the water 10°C below the oil at the outlet. Calculate the overall change in entropy. Repeat for the same temperatures for a countercurrent exchanger.

4.2.4 Entropy Changes for Chemical Reactions—The Third Law of Thermodynamics

Entropy changes occurring in chemically reacting systems can be calculated in the same manner as enthalpy changes as discussed in Chap. 3.

$$\Delta S_{\text{rxn}} = \sum n_{\text{Pr}} S_{\text{Pr}} - \sum n_{\text{Re}} S_{\text{Re}} \tag{4.13}$$

Such a calculation requires the determination of the absolute entropies of the substances involved which require the third law of thermodynamics as a basis. Nernst first advanced the concept of the third law, which states that the entropy of any perfect crystal, that is, a crystalline substance with no random arrangements, is zero at absolute zero temperature. This has been proven experimentally and allows absolute values of entropy to be calculated from extrapolation of experimental latent heats and specific heats taken under cryogenic conditions down to 0 K. In comparison with internal energy and enthalpy which are calculated relative to an arbitrary reference state, entropy is a reference property and is absolute as are pressure, volume, and temperature. The absolute entropy is then calculated usually with respect to a reference of 25°C as

$$S^\circ_{\text{abs}} = \int_0^{298} \frac{C_P\, dT}{T} + \Delta S_{\text{phase changes}} \tag{4.14}$$

Table 4.1 lists values of S° for common substances. It is apparent that the absolute entropy will increase as the disorder of the substance increases. Thus, entropy increases as a substance goes from crystalline solid to amorphous solid to liquid to gas.

Once an absolute entropy is available for any compound, the entropy at any other temperature can be calculated by adding the entropy contributions due to

Table 4.1 Standard molal entropies S°, kJ / (kmol · K), at 298.16 K

Standard states:
Gases: The ideal-gaseous state at 1 atm pressure
Liquids: The pure liquid at 1 atm pressure
Solids: The pure solid at 1 atm pressure
Aqueous solutions: The hypothetical ideal 1.0 molal solution in which $a/m = 1.0$
The ionic entropies are relative to zero for the hydrogen ion H^+.

Abbreviations:
c = crystalline state g = gaseous state
l = liquid state aq = dilute aqueous solution

	Carbon		
C(graphite)	5.697	$CBr_4(g)$	358.3
C(diamond)	2.440	$CCl_4(g)$	309.55
$C(g)$	158.068	$CCl_4(l)$	214.53
$C_2(g)$	200.47	$CF_4(g)$	262.5
$CO(g)$	198.00	$COCl_2(g)$	289.38
$CO_2(g)$	213.74	$CH_4(g)$	186.28
$CS(g)$	211.0	$CO_3^{2-}(aq)$	−53.2
$CS_2(l)$	151.11	$HCO_3^-(aq)$	55.0
$CS_2(g)$	237.93	$H_2CO_3(aq)$	191.3
$COS(g)$	231.65		
	Chlorine		
$Cl_2(g)$	223.055	$Cl_2O(g)$	266.65
$Cl(g)$	165.167	$ClO_2(g)$	249.5
$Cl^-(aq)$	55.13		
	Oxygen		
$O_2(g)$	205.127	$OH(g)$	183.72
$O(g)$	161.031	$OH^-(aq)$	−10.54
$O_3(g)$	237.8		

Table 4.1 Continued

Alcohols			
Methyl alcohol(*g*)	237.8	Ethyl alcohol(*l*)	160.7
Methyl alcohol(*l*)	126.8	Isopropyl alcohol(*l*)	180.0
Ethyl alcohol(*g*)	282.1	Isopropyl alcohol(*g*)	307.3
Hydrogen			
$H_2(g)$	130.649	$H_2S(g)$	205.74
$D_2(g)$	144.844	$HBr(g)$	198.571
$H(g)$	114.667	$HCl(g)$	186.767
$H^+(aq)$	0 by definition	$HF(g)$	173.59
$H_2O(l)$	69.973	$HI(g)$	206.428
$H_2O(g)$	188.814	$HCN(l)$	112.70
$D_2O(l)$	76.026	$HCN(g)$	201.69
$D_2O(g)$	198.328		
Iron			
$Fe(c)$	27.17	$Fe_3O_4(c)$	146.5
$Fe(g)$	180.46	$Fe_2O_3(c)$	90.0
$FeO(c)$	56.1	FeS_2 (pyrites)(*c*)	53.2
Nitrogen			
$N_2(g)$	191.581	$NH_4Cl(c)$	94.6
$N(g)$	153.268	$NH_4HCO_3(c)$	118.5
$N_2O(g)$	220.10	$NH_4HS(c)$	113.4
$NO(g)$	210.72	$NH_4OH(aq)$	179.2
$NO_2(g)$	240.57	$NH_4^+(aq)$	112.7
$N_2O_4(g)$	304.45	$NO_2^-(aq)$	125.2
$N_2O_5(c)$	153.2	$NO_3^-(aq)$	146.5
$N_2O_5(g)$	343.	$HNO_3(l)$	155.7
$NH_3(g)$	192.6		
Sodium			
$Na(c)$	51.1	$NaOH(c)$	59.4
$Na(g)$	153.69	$Na_2CO_3(c)$	136.0
$Na_2(g)$	230.31	$NaHCO_3(c)$	102.1
$Na+(aq)$	60.3	$NaNO_3(c)$	116.4
$NaCl(c)$	72.42	$Na_2SO_4(c)$	149.57
$NaCl(g)$	232.3	$Na_2SO_4 \cdot 10H_2O(c)$	593.2
Aldehydes, ketones, ethers			
Formaldehyde(*g*)	218.8	Dimethyl ether(*g*)	266.7
Acetone(*g*)	304.3	Dimethyl ether(*l*)	188.3
Acetone(*l*)	200.5		
Acids			
Formic acid(*g*)		Formic acid(*l*)	129.0
(monomer)	129.0	Acetic acid(*l*)	159.9

Table 4.1 Continued

Hydrocarbons			
Methane(*g*)	186.28	Benzene(*g*)	269.33
Ethyne(*g*)	200.92	Benzene(*l*)	172.88
Ethene(*g*)	219.56	Cyclohexane(*g*)	298.38
Ethane(*g*)	229.60	Cyclohexane(*l*)	204.49
Propyne	248.23	Methylcyclopentane(*l*)	248.06
Propadiene(*g*)	244.04	Methylcyclohexane(*l*)	248.06
Propene(*g*)	267.07	Toluene(*g*)	319.89
Propane(*g*)	270.04	Toluene(*l*)	219.68
2-Methylpropene	293.73	*o*-xylene(*g*)	352.92
cis-2-Butene(*g*)	300.97	*o*-xylene(*l*)	246.60
trans-2-Butene(*g*)	296.62	*p*-xylene(*g*)	352.59
1-Butene(*g*)	305.75	*p*-xylene(*l*)	247.48
n-Butane(*g*)	310.27	Styrene(*g*)	345.26
n-Butane(*l*)	231.1	Styrene(*l*)	237.68
2-Methylpropane	294.78	Ethylbenzene(*l*)	255.30
2-Methylpropane	218.05	*n*-Decane(*g*)	544.89
n-Pentane(*g*)	349.11	Naphthalene(*c*)	167.0
2-Methylbutane	343.75	Diphenyl(*c*)	206.0
Cyclopentane(*g*)	293.02	Anthracene(*c*)	207.6
Cyclopentane(*l*)	204.36	Phenanthrene(*c*)	211.8
n-Hexane(*g*)	388.59		
Halogen compounds			
Carbon tetrachloride(*g*)	309.55	Methyl chloride(*l*)	153.79
Carbon tetrachloride(*l*)	214.53	Ethylene dichloride(*l*)	208.63
Chloroform(*g*)	296.62	Ethylene dichloride(*g*)	304.49
Methylene chloride(*g*)	270.75	Phosgene(*g*)	289.38
Methyl chloride(*g*)	234.29		
Sulfur compounds			
Methyl mercaptan(*g*)	254.93	Dimethyl sulfide(*l*)	196.49
Dimethyl sulfide(*g*)	285.82		
Nitrogen compounds			
Methylamine(*g*)	241.66	Methyl cyanide(*g*)	243.54
Methylamine(*l*)	150.28	Methyl isocyanide(*g*)	246.05
Dimethylamine(*g*)	273.35	Urea(*c*)	104.65
Dimethylamine(*l*)	182.43	Cyanogen(*g*)	242.20
Sulfur			
S(rhombic)	31.90	$H_2S(g)$	205.74
S(monoclinic)	32.57	$H_2S(aq)$	122.2
S(*g*)	167.80	$SO_3^{2-}(aq)$	43.5
$SO_2(g)$	248.65	$HSO_4^-(aq)$	126.92
$SO_3(g)$	256.35	$SO_4^{2-}(aq)$	17.2
$SO_3(l)$	132.7		

temperature changes and any additional phase changes.

$$S_T = S^\circ + \int_{298}^{T} \frac{C_P\,dT}{T} + \sum \Delta S_{\substack{\text{phase changes}\\ \text{above 298 K}}} \tag{4.15}$$

These entropies can then be used to calculate ΔS_{rxn} by Eq. (4.13).

Example 4.6 Determine the entropy change for the reaction of methanol and oxygen to formaldehyde and water in the vapor phase at 25°C.

SOLUTION

$$CH_3OH + \tfrac{1}{2}O_2 \rightarrow CH_2O + H_2O$$

$$\Delta S = S^\circ_{CH_2O} + S^\circ_{H_2O} - S^\circ_{CH_3OH} - \tfrac{1}{2}S^\circ_{O_2}$$

$$= 218.8 + 188.85 - 237.8 - \tfrac{1}{2}(205.2)$$

$$= 67.25\ \text{kJ/(kmol} \cdot \text{K)}$$

Exercise 4.F Determine at 200°C the absolute entropy of water vapor above 0°C base at atmospheric pressure, using the values of the heat capacity from Table 3.2 and the values of heat of vaporization and fusion from the steam tables. Compare your result with the values recorded in the steam tables in App. B.

4.2.5 Classical Lost Work and Processes Analysis

The *ideal work* done by a process is the maximum amount of work which can be done by the process by operating reversibly within the system and by transferring heat between the system and the surroundings reversibly. Obviously this ideal process does not exist but merely serves as a basis for calculation of the thermodynamic efficiency of the process. The *lost work* can then be calculated as the difference between the ideal, reversible work done by the process and the actual work done by the process:

$$W_{\text{lost}} = W_{\text{ideal}} - W_{\text{actual}}$$

Consider a *nonflow process* where $W = Q - \Delta E$. If the process is reversible, then by Eq. (4.2), $\Delta S_s = -\Delta S_S$. Assuming the surroundings are a constant-temperature heat reservoir at T_S, $\Delta S_S = Q_S/T_S$. Since $Q_S = -Q_s$, then $Q_s = T_S \Delta S_s$. Thus, $W_{\text{rev}} = Q_{\text{rev}} - \Delta E = T_S \Delta S_s - \Delta E$. The ideal work is different from the reversible work by the amount of work done by the system on the surroundings at pressure P_S by expansion of the system fluid or done by the surroundings on the system by contraction of the system fluid. Thus

$$W_{\text{ideal}} = W_{\text{rev}} - P_S \Delta V_s$$

or

$$W_{\text{ideal}} = T_S \Delta S_s - \Delta E - P_S \Delta V_s \tag{4.16}$$

The lost work is then the difference between the ideal work and the actual work:

$$W_{\text{lost}} = W_{\text{ideal}} - W_{\text{actual}} = T_S \Delta S_s - Q - P_S \Delta V_s \tag{4.17}$$

If the industrially more important *steady-state flow process* is considered, Eq. (3.5) shows $\Delta H + \Delta(\text{KE}) + \Delta(\text{PE}) = Q - W_s'$. If reversible, $Q = T_S \Delta S_s$. Thus

$$W_{\text{ideal}} = T_S \Delta S_s - \Delta H - \Delta(\text{KE}) - \Delta(\text{PE}) \tag{4.18}$$

The lost work, $W_{\text{lost}} = W_{\text{ideal}} - W_s'$, is calculated by combining Eqs. (3.5) and (4.18):

$$W_{\text{lost}} = T_S \Delta S_s - Q \tag{4.19}$$

where Q is the heat transferred to the system.

If the same formulation is derived with respect to the heat transferred to the surroundings Q_S, it is apparent since $Q_S = -Q$ that

$$W_{\text{lost}} = T_S \Delta S_s + Q_S \tag{4.20}$$

The entropy change in the surroundings, ΔS_S, a large constant-temperature heat reservoir, is $\Delta S_S = Q_S / T_S$. Thus

$$W_{\text{lost}} = T_S \Delta S_s + T_S \Delta S_S$$

Since $\Delta S_{\text{univ}} = \Delta S_s + \Delta S_S$,

$$W_{\text{lost}} = T_S \Delta S_{\text{univ}} \tag{4.21}$$

Since the second law of thermodynamics requires that the entropy change in the universe is zero for a reversible process and is positive for an irreversible process, it naturally follows that the lost work for an irreversible process is positive. By definition the lost work in a completely reversible process is zero.

The concept of lost work is a measure of the inefficiency of a process. The thermodynamic work efficiency of a process doing work is then merely the actual work produced divided by the ideal work which could be produced:

$$\eta = \left(\frac{W}{W_{\text{ideal}}} \right)_{\text{nonflow}} \qquad \text{or} \qquad \eta = \left(\frac{W_s'}{W_{\text{ideal}}} \right)_{\text{flow}} \tag{4.22}$$

Example 4.7 Steam enters a turbine at 1.5 MPa and 500°C and exhausts at 0.1 MPa. The turbine delivers 85 percent of the shaft work of a reversible, adiabatic turbine although it is neither reversible nor adiabatic. Heat losses to the surroundings at 20°C are 6 kJ/kg steam. Determine the lost work in the process.

BASIS: 1 kg.

$$W_{\text{ideal}} = T_S \Delta S_s - \Delta H$$

$$W_{\text{lost}} = T_S \Delta S_s - Q \tag{4.19}$$

Ideal case	In	$\xrightarrow[\text{adiab}]{\text{rev}}$	Out	
T	500°C		140.8°C	
P	1.5 MPa		0.1 MPa	
H	3473.1 kJ		2758.1 kJ	Interpolating from App. B
S	7.5698 kJ/K		7.5698 kJ/K	

$$\Delta S_s = 0 \qquad W_{\text{ideal}} = -\Delta H$$

$$W_{\text{ideal}} = 3473.1 - 2758.1 = 715 \text{ kJ}$$

Since the efficiency is 0.85, $W_s = (0.85)(715) = 607.7$ kJ. Since we are neglecting kinetic and potential energy,

$$\Delta H = Q - W_s$$

$$= -6 - 607.7 = -613.7 \text{ kJ}$$

$$\Delta H = H_2 - H_1 = -613.7 = H_2 - 3473.1$$

Therefore

$$H_2 = 2859.4$$

At 0.1 MPa, $T = 192$°C and $S = 7.8005$ (from the steam tables). Therefore

$$\Delta S_s = 7.8005 - 7.5698 = 0.2307$$

Thus

$$W_{\text{lost}} = (293.1)(0.2307) - (-6)$$

$$= 73.6 \text{ kJ}$$

For heat transfer systems, the ideal work is simply the amount of heat transferred from the system which could be converted to work $(-Q)$, while the actual work done by the system is the reversible work done on the surroundings $(-T_S \Delta S_s)$.

Example 4.8 Assume that 5000 kg/h of oil with a heat capacity of 3.2 kJ/(kg · K) is to be cooled from 220 to 40°C, using a large quantity of water which can be assumed to be at a constant temperature of 30°C. Determine the lost work in the process and the thermodynamic efficiency of the process.

SOLUTION Use Eq. (4.19):

$$W_{\text{lost}} = T_S \Delta S_s - Q$$

Assume that $P_S \Delta V = 0$. Then

$$Q = -5000(3.2)(180) = -2.88 \times 10^6 \text{ kJ}$$

$$\Delta S_s = mC_P \ln \frac{313.1}{493.1}$$

$$= (5000)(3.2) \ln \frac{313.1}{493.1}$$

$$= -7267 \text{ kJ}$$

$$T_S = 303.1$$

Therefore

$$W_{\text{lost}} = (303.1)(-7267) - (-2.88 \times 10^6)$$

$$= -2.20 \times 10^6 + 2.88 \times 10^6$$

$$= 680{,}000 \text{ kJ}$$

$$\text{Efficiency} = \frac{2.20}{2.88} = 0.764$$

For processes containing several units, lost-work calculations can be made for each unit and summed to determine the overall value. The overall process efficiency would then be the ratio of $\Sigma W_s / \Sigma W_{\text{ideal}}$, taking into account each process step.

For systems in which work must be done on the system (e.g., compression), the efficiency of the process would be defined as the ideal work required to operate the process (e.g., adiabatic, reversible work of compression) divided by the heat transferred during the process (e.g., the charge in enthalpy of the gas being compressed).

Example 4.9 Assume that 100 kg of methane gas/h is adiabatically compressed from 0.5 MPa and 300 K to 3.0 MPa and 500 K after which it is cooled isobarically to 300 K by a large amount of water available at 290 K.

Determine the efficiency of the compressor. If the surroundings are assumed to be at 290 K, do a thermodynamic analysis of the process.

Assume ideal gas and use Eq. (3.32).

BASIS: 1 h.

COMPRESSION

$$W_{\text{ideal}} = \frac{nRT_1}{k-1}\left[\left(\frac{P_2}{P_1}\right)^{(k-1)/k} - 1\right]$$

$$= \frac{-\frac{100}{16}(8.314)(300)}{1.305 - 1}\left[\left(\frac{3.0}{0.5}\right)^{0.2337} - 1\right]$$

$$Q_{\text{rev}} = 0 \qquad \Delta S = 0$$

$$C_P \text{ at } 25°\text{C (App. A)} = 35.58$$

$$C_V = 35.58 - 8.31 = 27.27$$

Therefore

$$k = \frac{C_P}{C_V} = 1.305$$

$$W_{\text{ideal}} = -26{,}582 \text{ kJ} = -\Delta H_{\text{ideal}}$$

(This is the work required for compression if adiabatic and reversible.) However, the actual amount of heat transferred is

$$\Delta H_{\text{act}} = \frac{100}{16}(35.58)(500 - 300)$$
$$= +44{,}475 \text{ kJ}$$

Therefore, the adiabatic efficiency is

$$\eta = \frac{26{,}582}{44{,}475} = 0.598$$

Considering the change in entropy,

$$\Delta S_s = C_P \ln\frac{T_2}{T_1} - R\ln\frac{P_2}{P_1}$$
$$= 35.58 \ln\frac{500}{300} - 8.314 \ln\frac{3.0}{0.5}$$
$$= 18.175 - 14.897 = 3.278 \text{ kJ/K}$$
$$\Delta S_S = 0 \qquad \text{as} \quad Q = 0$$

Therefore

$$\Delta S_{\text{univ}} = \Delta S_s$$
$$W_{\text{lost},1} = T_S \Delta S_s = (290)(3.278) = 950.6 \text{ kJ}$$

Isobaric Cooling Process

$$\Delta S_s = nC_P \ln\frac{T_2}{T_1} = \frac{100}{16} 35.58 \ln\frac{300}{500}$$
$$= -113.595 \text{ kJ/K}$$
$$\Delta S_S = \frac{Q}{T_S} = \frac{\Delta H}{T_S} = \frac{44{,}475}{290}$$
$$= 153.362 \text{ kJ/K}$$
$$\Delta S_{\text{total}} = -113.595 + 153.362 = +39.767 \text{ kJ/K}$$
$$W_{\text{lost},2} = T_S \Delta S_s - Q = (290)(-113.595) - (-44{,}475)$$
$$= 11{,}532.5 \text{ kJ}$$

or

$$W_{\text{lost},2} = T_S \Delta S_{\text{univ}} = 11{,}532.5 \text{ kJ}$$
$$\Sigma W_{\text{lost}} = W_{\text{lost},1} + W_{\text{lost},2}$$
$$= 950.6 + 11{,}532.5 = 12{,}483 \text{ kJ}$$
$$\Sigma \Delta S = 3.278 + 39.767 = 43.045 \text{ kJ/K}$$
$$\Sigma W_{\text{lost}} = T_S \Sigma \Delta S = (290)(43.045) = 12{,}483 \text{ kJ}$$

The concept of lost work can also be approached by using the general entropy balance as illustrated in the introduction to this section. Energy or availability analysis, briefly described in Sec. 4.3.4, is an alternate approach for process analysis.

Exercise 4.G If 1 kg of liquid water is heated at atmospheric pressure from 20°C to vapor at 100°C in a flow process, determine the ideal work done on the system, the thermodynamic efficiency of the process, and the entropy change in the system, the surroundings at 20°C, and the universe. If the energy came from a furnace at 260°C, determine the change in entropy of the universe and the lost work in the process.

Exercise 4.H Do a thermodynamic process analysis of the process of Example 4.2, including calculation of all lost-work terms if the surroundings are at 300 K.

4.3 APPLICATIONS OF THE SECOND LAW OF THERMODYNAMICS TO IDEAL CYCLICAL AND NONCYCLICAL PROCESSES

A treatment of ideal mechanical devices will be given to show how the second law applies in the limit of frictionless devices. Although no such devices can practically exist, calculations for these ideal processes are necessary to determine the thermodynamic limit of efficiency that is possible. A short discussion of an ideal noncyclical process will be followed by a more-thorough treatment of a closed, ideal-work-producing cycle, the familiar Carnot cycle and its cognate, the Carnot engine. This discussion will be followed by a discussion of the reverse Carnot cycle, commonly called a refrigeration cycle. Discussion of most practical cycles will be deferred to Sec. 4.6.

4.3.1 Noncyclical Process

Consider a frictionless, perfectly insulated cylinder-piston arrangement, shown in Fig. 4.1, containing an ideal gas connected to a large heat reservoir by a very high capacity conducting plate adjacent to the cylinder head. The frictionless piston is attached to a frictionless receiver of work. If this arrangement can be considered to be isothermal, a small amount of heat can be taken from the reservoir and

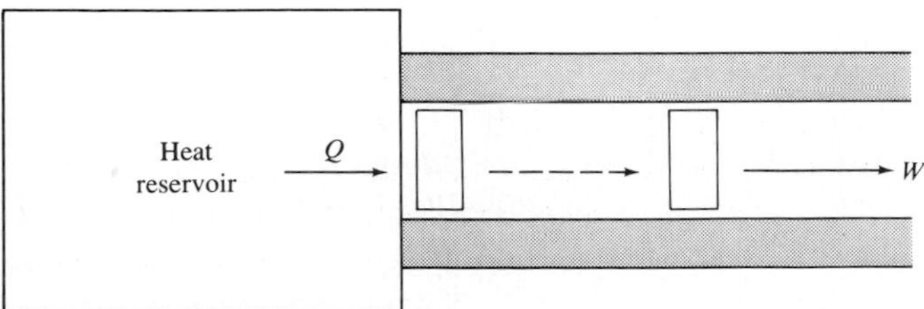

Figure 4.1 Cylinder-piston arrangement.

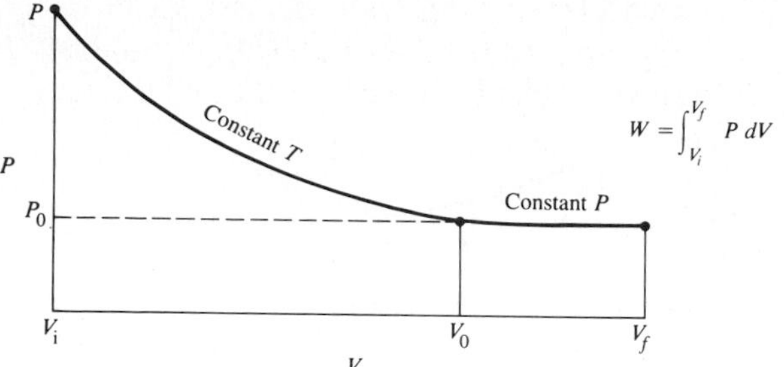

Figure 4.2 Noncyclical ideal-gas expansion.

converted to work. Since $\Delta E = 0$, for the isothermal expansion of an ideal gas, $Q = W$. If the gas is initially at a pressure P above atmospheric pressure P_0, it can expand isothermally and reversibly to a pressure P_0 after which it can continue to expand at constant pressure only if the gas temperature increases. Such a process is shown in Fig. 4.2. Since the temperature of the heat reservoir is only infinitesimally greater than the temperature of the gas in the cylinder, to be a reversible process the temperature of the heat source would have to continue to rise. Such a process cannot continue indefinitely unless heat is rejected to a sink at a lower temperature than the heat reservoir in some sort of a cyclical process.

Obviously the efficiency of heat conversion to work is unity during the isothermal process. However, during the isobaric process, ΔE will also increase and the work will be less than the heat inputted.

This analysis provides a starting point for the consideration of cyclical processes.

Exercise 4.1 Using the noncyclical process diagrammed, show why it is impossible to convert the heat indefinitely fed into the heat engine into work. In addition, determine and contrast the ratio of W/Q for the isothermal and the isobaric parts of the process.

4.3.2 Cyclical Processes—The Carnot Cycle and the Carnot Engine

The Carnot ideal engine is an imaginary device which outputs the maximum possible amount of shaft work when operating reversibly with no friction in a cyclic process. The efficiency of such an engine is the standard, therefore, with which all practical engines operating between the same temperatures can be compared with respect to their ability in changing internal energy and/or heat to mechanical work. The cycle is shown on a PV diagram in Fig. 4.3

The *first* stage of a Carnot engine is identical to the isothermal portion of the noncyclical process considered in the previous section. In general, if an amount of heat Q_1 is supplied to the expanding gas at a constant temperature T_1 when expanding from state A to state B, the work of expansion is

$$W_{AB} = \int_A^B P\,dV$$

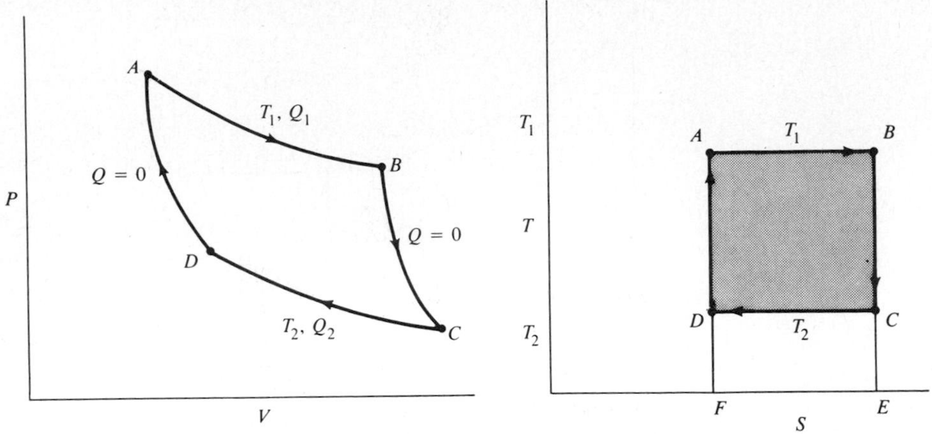

Figure 4.3 PV and TS diagrams for Carnot cycle.

Consider a *second* stage in which a nonconducting head is attached to the cylinder where the gas is expanded reversibly and adiabatically, hence isentropically, by decreasing the internal energy of the gas from state B to state C. The work of expansion is

$$W_{BC} = \int_B^C P\,dV$$

and the temperature drops to T_2.

A *third* stage would consist of isothermally compressing the fluid by placing a conducting head on the cylinder which inputs heat at T_2 to the head of the cylinder. Heat is then removed from the cylinder in an amount of $-Q_2$ as heat flows to the receiver at T_2. The work done by the gas,

$$W = \int_C^D P\,dV$$

in being compressed from state C to state D is obviously negative.

A *fourth* stage consists of again replacing the cylinder head by a nonconductor and compressing adiabatically and reversibly, isentropically, from state D to the initial state A. The negative work done by the gas in this process is

$$W = \int_D^A P\,dV$$

For the complete cycle the overall change in the internal energy must be zero. Thus, the net work done by the process must be the difference between the heat input from the reservoir at T_1 during the first stage of the process and the heat output to the reservoir at T_2 during the third stage of the process. Thus, $W_{\text{overall cycle}} = Q_1 + Q_2$, where

$$W_{\text{overall cycle}} = \int_A^B P\,dV + \int_B^C P\,dV + \int_C^D P\,dV + \int_D^A P\,dV$$

The change in entropy over the entire process and over stages 2 and 4 are zero; thus $\Delta S = 0 = \Delta S_1 + \Delta S_3$.

Since $S_1 = Q_1/T_1$ and $S_3 = Q_2/T_2$, then

$$\frac{Q_1}{T_1} + \frac{Q_2}{T_2} = 0 \tag{4.23}$$

Since the work of the entire cycle is equal to the algebraic sum of the heat terms, the thermodynamic efficiency of the cycle can be determined.

$$\text{Efficiency} = \eta = \frac{W_{\text{cycle}}}{Q_{\text{input}}} = \frac{Q_1 + Q_2}{Q_1} \tag{4.24}$$

Rearranging Eq. (4.23),

$$\frac{Q_2}{Q_1} = -\frac{T_2}{T_1}$$

Substituting in Eq. (4.24),

$$\frac{Q_1 + Q_2}{Q_1} = \frac{T_1 - T_2}{T_1} \quad \text{and} \quad \eta = \frac{T_1 - T_2}{T_1} \tag{4.25}$$

This same conclusion can be reached by remembering that the cycle operates on an ideal gas and that Q_1 and Q_2 enter the cycle isothermally. The efficiency can only reach unity if T_2 approaches absolute zero or T_1 is infinite, both of which are impossible.

It is often desirable to express the Carnot cycle on a *TS* diagram as also shown in Fig. 4.3. The *TS* diagram graphically shows the efficiency of a cycle. The area *ABCD* is the work of the cycle as calculated from

$$Q_1 + Q_2 = \int T_1\,dS + \int T_2\,dS$$

As the heat input is the area *ABEF*, the efficiency is the ratio of the areas *ABCD*/*ABEF*.

The Carnot cycle and engine then show conclusively the validity of several of the statements of the second law of thermodynamics as presented earlier and validate the important principle that regardless of the fluid or the system, the Carnot efficiencies are equal as long as the source and sink temperatures T_1 and T_2 are equal. Rather than expanding and compressing an ideal gas, a Carnot engine could first vaporize a saturated liquid to start the cycle. The analysis would be identical.

Example 4.10 A cyclic engine is alleged to produce 0.5 kJ of work/kJ of heat extracted from a hot reservoir. The cycle operates between reservoirs at 400 and 20°C. Is this quantitatively possible?

SOLUTION The maximum efficiency of any cycle is that of the Carnot cycle. Using Eq. (4.25),

$$\eta = \frac{W_{\text{cycle}}}{Q_{\text{input}}} = \frac{T_1 - T_2}{T_1} = \frac{673 - 293}{673} = 0.565$$

Since the maximum efficiency is 0.565, an efficiency of 0.5 is possible.

The terms energy availability or unavailability have been used. The definitions given below are related to the Carnot cycle and are classical definitions. A discussion of the more recent use of these terms is given in Sec. 4.3.4. The first term is merely the fraction of heat energy available which can be converted to mechanical work or the efficiency of the Carnot cycle:

$$\text{Classical availability} = \frac{Q_1 + Q_2}{Q_1} = \frac{T_1 - T_2}{T_1} \tag{4.26}$$

The second term is the fraction of heat energy which cannot be converted to work:

$$\text{Classical unavailability} = \frac{-Q_2}{Q_1} = \frac{T_2}{T_1} \tag{4.27}$$

The sink temperature T_2 is often the cooling water or air temperature for actual processes, while the heat source temperature T_1 is often the available steam or other heat transfer fluid temperature. The efficiency declines as these temperatures approach one another.

The Carnot efficiency of more-complicated processes where the heat source temperature is not constant can be calculated by using the same general equation in differential form and noting that

$$dQ_1 = -mC_P\,dT \tag{4.28}$$

where m = mass of heat source material
C_P = heat capacity of heat source material
dT = change in temperature of heat source

Coupling of these equations will yield Carnot efficiencies for the so-called differential Carnot engines, where to preserve reversibility only a differential amount of work is done per cycle, $dW/dQ_1 = (T - T_2)/T$.

Exercise 4.J Consider a Carnot engine in which 1 kg of water at atmospheric pressure and 80°C is to be used to produce work exhausting to an atmosphere heat sink at 20°C. Determine the maximum amount of work that can be produced. How much work would be required to heat the water from 30 to 80°C with the same surroundings?

4.3.3 The Reverse Carnot or Refrigeration Cycle

A cycle operating in the same mode as the Carnot cycle in the reverse direction is of great practical importance for refrigeration. The net effect of such a cycle is the absorption of work energy and the transfer of heat energy from a low-temperature heat reservoir to a high-temperature heat reservoir. Such a cycle, depicted on a PV diagram in Fig. 4.4, consists of an adiabatic expansion ($A \rightarrow D$), followed by an isothermal expansion ($D \rightarrow C$), an adiabatic compression ($C \rightarrow B$), and finally an isothermal compression ($B \rightarrow A$).

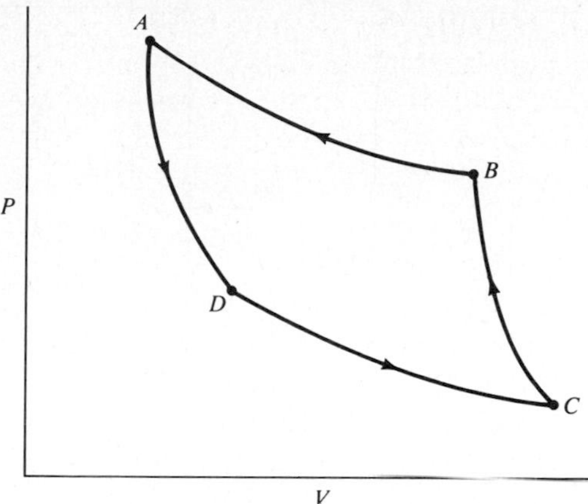

Figure 4.4 Reverse Carnot or refrigeration cycle.

As for the Carnot engine the work of the cycle is equal to the algebraic sum of the heat absorption in the isothermal expansion (Q_1) and the heat rejection in the isothermal compression (Q_2).

$$W_{\text{cycle}} = Q_1 + Q_2 \tag{4.29}$$

The difference between the Carnot engine and the reverse Carnot engine is that in the first case work is done by the system and W_{cycle} is positive, while in the second case work is done on the system and W_{cycle} is negative.

The efficiency of a Carnot cycle is equal to the ratio of the work of the cycle divided by the heat input. Since the object of the reverse cycle is to remove heat, the efficiency of any refrigeration cycle is the reciprocal of this ratio, the heat absorption or input divided by the net work of the cycle, and is called the *coefficient of performance* (COP). While the efficiency of a Carnot cycle is always less than 1, the coefficient of performance is always greater than 1, with the largest COP being the most desirable.

A refrigerator consists of a vapor in a closed loop which absorbs heat from the low-temperature cold box and rejects heat to the higher-temperature surroundings, thus requiring work. A heat pump, currently very popular as a supplementary heating source for homes, also uses mechanical work to transfer heat from the cold surroundings and reject heat to the interior of a building. The heat rejected will be both the heat absorbed from the surroundings and the heat generated by mechanical work. A schematic comparison of the Carnot engine and heat pump is given in Fig. 4.5.

Practical refrigeration systems, primarily vapor recompression cycles used in refrigerators and air conditioners as well as other refrigeration cycles of more limited use, will be discussed in Sec. 4.6.

Example 4.11 The coefficient of performance is defined as the heat absorbed (Q_1) divided by the work obtained. For an ideal-gas refrigeration system operating between a low temperature of 7°C and surroundings at 27°C, determine the coefficient of performance of the machine.

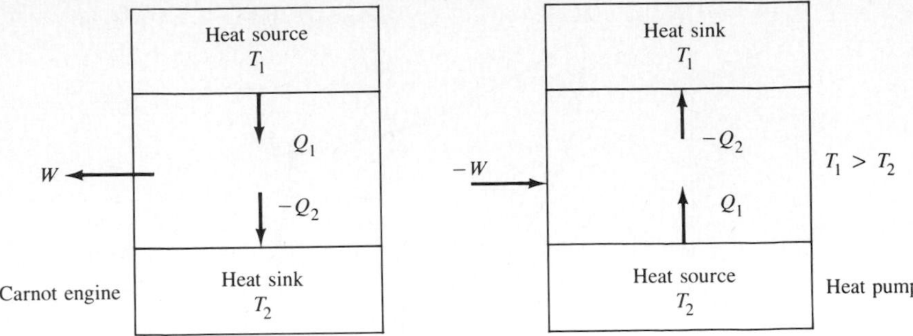

Figure 4.5 Comparison of Carnot engine and heat pump.

SOLUTION Analogous to the Carnot cycle:

$$\text{COP} = \frac{Q_1}{W_{\text{net}}} = \frac{T_1}{T_2 - T_1} = \frac{280}{300 - 280} = 14.0$$

Exercise 4.K Compare an ideal Carnot engine for air conditioning in summer and an ideal heat pump for heating in winter, assuming the following:

Average summer outdoor temperature = 29°C
Average winter outdoor temperature = −7°C
Average summer indoor temperature = 24°C
Average winter indoor temperature = 20°C

Use the same *basis* for each.

4.3.4 Availability or Exergy Analysis

With the recent interest in energy conservation the use of availability analysis, closely related to lost-work analysis, has become prevalent for steady-state flow systems; Sussman (1980) gives a complete discussion of availability analyses.

If a base state of temperature T_0, usually taken as 25°C and 1 atm, is selected, the maximum amount of shaft work that can be extracted from a unit mass of matter as it flows into equilibrium with the atmosphere at the base state is given as

$$-W'_{s,\max} = \Delta H_{1\to 0} - T_0\,\Delta S_{1\to 0} = \text{exergy} \tag{4.30}$$

This work is defined as the exergy or availability of energy and is related to the base state. The change in availability or exergy from any initial state to any final state is then

$$\Delta\ \text{exergy} = \Delta H - T_0\,\Delta S \tag{4.31}$$

This difference is a state property and is analogous to the ideal work (excluding kinetic- and potential-energy effects) shown in Eq. (4.18). It depends on the base state and is the potential recovery of work from the process.

Example 4.12 Calculate the exergy of saturated steam at 2 MPa total pressure relative to a base state of liquid water at 25°C and low pressure.

State	P, MPa	T, °C	H, kJ/kg	S, kJ/(kg · K)
1	2.0	212.4	2799.5	6.3409
0	...	25	104.89	0.3674

$$\text{Exergy} = \Delta H_{1\to 0} - T_0 \Delta S_{1\to 0}$$

$$\Delta H_{1\to 0} = 2799.5 - 104.89 = 2694.6$$

$$\Delta S_{1\to 0} = 6.3409 - 0.3674 = 5.9735$$

$$\text{Exergy} = 2694.6 - (298.15)(5.9735)$$

$$= 913.6 \text{ kJ/kg}$$

The second-law efficiency for a process could then be defined as the ratio of the sum of the changes in exergy for the output streams to the sum of the changes in exergy for the input streams:

$$\text{Efficiency} = \frac{(\Delta \text{ exergy})_{\text{out}}}{(\Delta \text{ exergy})_{\text{in}}} \tag{4.32}$$

or

$$\text{Efficiency} = \frac{(\Delta \text{ exergy})_{\text{in}} + (\Delta \text{ exergy})_{\text{process}}}{(\Delta \text{ exergy})_{\text{in}}}$$

This type of efficiency is an alternate to efficiencies for processes defined as $\Sigma W_s/\Sigma W_{\text{ideal}}$, calculated by taking into account each process step. Analyses of complete processes are given by Sussman (1980).

4.4 FREE ENERGY

Any real, i.e., irreversible, process causes an increase in the total entropy of the universe. A state of equilibrium is a state of rest and thus every infinitesimal process taking place is reversible. Therefore, the entropy is a constant, and the change in entropy in the process is zero. Thus, to define a state of equilibrium the most general way would be to calculate changes in entropy for all systems involved in a particular process and identify the condition of zero change in entropy to define the equilibrium state. This is a difficult task; thus two derived thermodynamic properties have been proposed and accepted to alleviate this problem by providing quantities which are state functions and hence are much more convenient to use. The quantities are the Helmholtz free energy or work function A and the Gibbs free energy G, each bearing the name of their proposer. The significance of these functions and their relationship to other thermodynamic properties are the major topics of this section. A brief introduction to the criteria for equilibrium in terms of free energy and entropy is also given together with a discussion of the prediction of whether reactions will or will not occur spontaneously. Detailed discussions of the use of free energy to predict physical equilibrium composition of mass transfer systems or chemical equilibrium resulting from reactions are deferred to Chaps. 8 and 9, respectively.

4.4.1 Helmholtz Free Energy or Work Function

The Helmholtz free energy is defined as

$$A \equiv E - TS \tag{4.33}$$

This function is useful in that for any process for which the initial and final temperatures are equal, the change in work function represents the total work done *on* the system.

Consider a system operating reversibly at constant temperature:

$$\Delta A = \Delta E - T\Delta S \tag{4.34}$$

As $\Delta E = Q_{rev} - W_{rev} = T\Delta S - W_{rev}$, therefore

$$\Delta A = -W_{rev} \tag{4.35}$$

Thus, the change in ΔA represents the maximum total work for a reversible process at constant temperature. If only expansion work is done by the system, then $\Delta A = -W_{rev} = \int -P\,dV$.

This free-energy term has limited utility in actual practice and is normally only utilized for certain calculations in constant-volume systems. However, it can be manipulated and utilized by the same methods as can the Gibbs free energy, which has gained far more utility and will be discussed next.

4.4.2 Gibbs Free Energy

The Gibbs free energy is defined as

$$G \equiv H - TS \tag{4.36}$$

Since $H = E + PV$,

$$G = E + PV - TS \tag{4.37}$$

Since $A = E - TS$,

$$G = A + PV \tag{4.38}$$

All forms are equivalent; however, the first is the most-used form of the definition. The Helmholtz free energy represented the maximum total work available for an isothermal, reversible process. Some of this work, $\int P\,dV$, is the work done against a constant-pressure atmosphere. The remainder of the work, or net work, is available for other purposes. Thus

$$W_{net} = W_{max\,rev} - \int P_{ext}dV = -dA - \int P_{ext}dV$$

Since $G = A + PV$, for a constant-pressure reversible process, $dG = dA + P\,dV$.

Therefore

$$dG = -W_{net} - \int P\,dV + \int P\,dV = -W_{net}$$

or

$$W_{net} = -\Delta G \tag{4.39}$$

This then indicates that at constant temperature and pressure, $-\Delta G$ represents the maximum amount of useful work which can be extracted from a specific process for any purpose, hence the name *free energy*. Since more processes are carried out at constant pressure rather than at constant volume, it is readily apparent why G (as well as H and C_P) is more used than A (as well as E and C_V). Clearly, in cases where the work against external pressure is small, both free energies will be little different in magnitude.

If a reaction (or transfer of material between phases) is considered where the only external forces on the reaction are the constant pressure of the atmosphere and no work is accomplished by the process, then the net work done by the process is zero. If this is true, $\Delta G = 0$. Also, for a reaction to occur spontaneously at constant temperature, the change in free energy must be negative (or zero) or work would have to be done on the system to make the reaction proceed. Alternatively, if the change in free energy is positive then the reaction cannot proceed without external aid.

Whether a reaction will or will not proceed spontaneously is an important application of free-energy calculations. It should be remembered that thermodynamics only indicates the possibility of reaction, and although possible a reaction may only proceed at an infinitesimal rate. Thus the negative free-energy change is the driving force for reaction and provides a test of equilibrium. If the net work equals zero, then for any isothermal and isobaric process, equilibrium can be defined as $\Delta G = 0$. This concept will be discussed in detail in Chaps. 7 and 8.

Consider the definition of Gibbs free energy: $dG = dH - d(TS) = dE + d(PV) - d(TS)$. Therefore

$$dG = dE + P\,dV + V\,dP - T\,dS - S\,dT$$

Since $dE = TdS - PdV$ for a reversible nonflow process only doing PV work, then

$$dG = V\,dP - S\,dT \tag{4.40}$$

This general formulation is limited to nonflow processes with no work other than electrical work and to pure substances with no compositional mixing effects. However, since none of the properties are path-dependent, reversibility need not be assumed. The equation can be reduced for various situations:

At constant pressure,

$$dG_P = -S\,dT \tag{4.40a}$$

At constant temperature,

$$dG_T = V\,dP \tag{4.40b}$$

At constant temperature and pressure,

$$dG_T = 0 \tag{4.40c}$$

Ideal gases For *ideal gases*, Eq. (4.40a) and (4.40b) can be easily reduced.
At constant temperature,

$$\int dG_T = \int V\,dP = \int_{P_1}^{P_2} \frac{nRT}{P}\,dP$$

$$\Delta G_T = nRT \ln \frac{P_2}{P_1} \tag{4.41}$$

At constant pressure,

$$\int dG_P = -\int S\,dT$$

$$\Delta G_P = -\int_{T_1}^{T_2} S\,dT \tag{4.42}$$

As discussed earlier, for an ideal gas

$$S = \int_0^T C_P \frac{dT}{T} + \sum \left(\frac{\Delta H}{T}\right)_{\text{phase changes}}$$

and S at any T can be calculated at 1 atm pressure. Since S is a state function, the change in entropy to go to any other state can be calculated easily for an ideal gas, e.g., from P_1T_1 to P_2T_2:

1. Expand isothermally from P_1 to P_2:

$$\Delta S = R \ln \frac{P_1}{P_2}$$

2. Heat at P_2 from T_1 to T_2:

$$\Delta S = \int_{T_1}^{T_2} C_P \frac{dT}{T}$$

Thus, the absolute entropy can be determined as a function of temperature for substitution into the previous equation.
At constant volume,

$$dG_V = \int V\,dP - \int S\,dT$$

$$\Delta G_V = V\Delta P - \int S\,dT \tag{4.43}$$

S can be determined as above.

Example 4.13 Determine the change in Gibbs free energy for each of the steps of the process given in Example 3.5.
(a) *Isothermal*: Using Eq. (4.41),

$$\Delta G_T = nRT \ln \frac{P_2}{P_1} = -W = -9363 \text{ kJ}$$

(*b*) *Isometric*: Using Eq. (4.43),

$$\Delta G_V = V\Delta P - \int_{T_2}^{T_1} S\,dT$$

$$P_2 = 0.552 \text{ MPa} \qquad P_3 = 0.345 \text{ MPa}$$

$$T_2 = 700 \text{ K} \qquad T_3 = 437.5 \text{ K}$$

$$S_{\text{at 1 atm}, T_1} = S_0 + \int_{T_0}^{T_1} C_P \frac{dT}{T}$$

$$S_{P_1, T_1} = S_0 + \int_{T_0}^{T_1} C_P \frac{dT}{T} - R \ln \frac{P_1}{P_0}$$

Assuming C_P constant at 29.3,

$$S = S_0 + C_P \ln \frac{T_1}{T_0} - R \ln \frac{P_1}{P_0}$$

From Table 4.1,

$$S_{CO(g)} \text{ at } 25°\text{C, 1 atm} = 198.0 \text{ kJ/(kmol} \cdot \text{K)}$$

$$V_{700} = \frac{RT}{P} = \frac{(8314.3)(700)}{0.552 \times 10^6} = 10.543 \text{ m}^3$$

$$\Delta G_V = 10.543(345 - 552) - \int_{700}^{437.5} S\,dT$$

Evaluate integral by using Simpson's rule:

$$\int_{700}^{437.5} S\,dT = \frac{437.5 - 700}{6}(S_{437.5} + 4S_{568.8} + S_{700})$$

$$S_{700, 0.552 \text{ MPa}} = 198.0 + 29.3 \ln \frac{700}{298} - R \ln \frac{0.552}{0.103} = 209.0$$

Similarly

$$S_{437.5, 0.345 \text{ MPa}} = 199.2 \qquad S_{568.8, 0.449 \text{ MPa}} = 204.7$$

Therefore

$$\int_{700}^{437.5} S\,dT = \frac{-262.5}{6}[209.0 + 4(204.7) + 199.2] = -53{,}681$$

and $$\Delta G_V = -2182 + 53{,}681 = +51{,}499 \text{ kJ}$$

(*c*) *Isobaric*: Using Eq. (4.42),

$$\Delta G_P = -\int_{T_1}^{T_2} S\,dT$$

$$T_3 = 437.5 \qquad T_4 = 350$$

$$P_3 = 0.345 \text{ MPa} \qquad P_4 = 0.345 \text{ MPa}$$

From the isothermal equations of (*b*),

$$S_{437.5, 0.345 \text{ MPa}} = 199.2 \qquad S_{350, 0.345 \text{ MPa}} = 192.6$$

Using the average S,

$$\Delta G_P = 195.9(350 - 437.5)$$
$$= +17{,}141 \text{ kJ}$$

(*d*) *Adiabatic*: Using Eq. (4.40),

$$\Delta G = V\,dP - S\,dT = \frac{RT}{P}\,dP - S\,dT$$
$$= \int_{P_1}^{P_2} V\,dP - \int_{T_4}^{T_5} S\,dT$$

$$P_4 = 0.345 \text{ MPa} \qquad P_5 = 2.758 \text{ MPa}$$
$$T_4 = 350 \qquad T_5 = 634$$

From the isothermal equations of (*b*),

$$S_{350} = 192.6 \qquad S_{634} = 192.8 \qquad S_{av} = 192.7$$

From ideal-gas law,

$$V_4 = 8.435 \text{ m}^3 \qquad V_5 = 1.911 \text{ m}^3$$

Again, to use Simpson's rule, if $P_{mid} = 1.5515$

$$T_{mid} = \left(\frac{1.5515}{0.345}\right)^{0.2857}(350) = 538 \text{ K}$$

Therefore

$$V_{mid} = 2.88$$

$$\int_{P_1}^{P_2} V\,dP = \frac{2758 - 345}{6}[8.435 + 4(2.88) + 1.911] = 8794 \text{ kJ}$$

$$\int S\,dT = 192.7(634 - 350) = +54{,}727 \text{ kJ}$$

$$\Delta G = 8794 - 54{,}727 = -45{,}933 \text{ kJ}$$

(*e*) *Isobaric*

$$\Delta G_P = -\int_{T_5}^{T_1} S\,dT$$

$$T_5 == 634 \qquad T_1 = 700$$
$$P_5 = 2.758 \text{ MPa} = P_1$$
$$S_{700} = 195.7 \qquad S_{634} = 192.8 \qquad S_{av} = 194.25$$
$$\Delta G_P = -194.25(700 - 634) = -12{,}821 \text{ kJ}$$
$$\Sigma\Delta G = -9363 + 51{,}499 + 17{,}141 - 45{,}933 - 12{,}821$$
$$= +523 \text{ kJ}$$

This value is not identically equal to zero as approximations are made in the evaluation of several steps.

Exercise 4.L For the process of Exercise 3.I, calculate the Gibbs free-energy change for each step of the process and for the entire process.

Exercise 4.M For the process of Example 3.6, calculate the Gibbs free-energy change for each step of the process and for the entire process.

The defining equations can be reduced for any equation of state including the corresponding-states principle which will be discussed in Sec. 4.5.

Flow processes For *flow* processes the defining equation must be modified to take into account kinetic- and potential-energy effects.

Consider the total energy balance for a flow process discussed in Chap. 3 for an isothermal, reversible, steady-state process carrying out shaft work:

$$Q - W_s' = \Delta E + \Delta(P\bar{V}) + \frac{\Delta u^2}{2} + g\Delta Z$$

Since $\Delta G = \Delta E + \Delta(P\bar{V}) - \Delta(TS)$,

$$Q - W_s' = \Delta G + T\Delta S - S\Delta T + \frac{\Delta u^2}{2} + g\Delta Z$$

Since the process is isothermal, $S\Delta T = 0$. Since the process is reversible, $Q = T\Delta S$. Thus

$$\Delta G = -W_s' - \frac{\Delta u^2}{2} - g\Delta Z \tag{4.44}$$

where $-W_s'$ is the reversible shaft work done on the system. ΔS, then, is a measure of the maximum, reversible shaft work possible.

Effect of pressure The effect of pressure on vapor pressure and hence on the Gibbs free energy is called the *Poynting effect*. Consider an isothermal system containing noncondensable gases and vapors at a total pressure greater than the vapor pressure of the vapor at that temperature. If the mixture is compressed to a high pressure, some of the vapor condenses, and the gas phase will contain some condensable vapor as determined by the vapor pressure of the condensable.

$$\Delta G_T = V\Delta P = \bar{V}(P_F - P_I) \tag{4.45}$$

where P_F = final total high pressure
P_I = initial pressure equal to the vapor pressure at the temperature
$\bar{V}$ = specific volume of liquid

Since the vapor and the liquid are in equilibrium, $\Delta G = 0$, or $G_{\text{vap}} = G_{\text{liq}}$. Thus ΔG increases equally in each phase so that the vapor pressure must have increased slightly. If the gas phase is ideal, then

$$nRT \ln \frac{P}{P_I} = \bar{V}(P_F - P_I) \tag{4.46}$$

where P is the vapor pressure of the condensable at the higher-pressure level. This pressure can then be used to calculate free energies or other desirable properties. This effect is only appreciable at high pressures of at least 1000 kPa. As written, the Poynting correction applies to pure components. Mixtures will be considered in Chap. 7.

Example 4.14 Assuming an ideal vapor phase, estimate the vapor pressure of ethylene at 0°C and 10 MPa total pressure. At 0°C and 0.1 MPa total pressure, the vapor pressure is 4.08 MPa. Repeat if the total pressure is 1 MPa.

SOLUTION Use Eq. (4.46):

$$nRT \ln \frac{P}{P_I} = \bar{V}(P_F - P_I)$$

Assume the liquid density is approximately 430 kg/m³ and take as a basis 1 kg of ethylene.

At $P_F = 10$ MPa:

$$\frac{1}{28}(8314.3)(273.1) \ln \frac{P}{4.08} = \frac{1}{430}(10 - 0.1)10^6$$

$$\ln \frac{P}{4.08} = 0.2839$$

$$\frac{P}{4.08} = 1.3283$$

$$P = 5.42 \text{ MPa}$$

At $P_F = 1$ MPa:

$$\frac{1}{28}(8314.3)(273.1) \ln \frac{P}{4.08} = \frac{1}{430}(1 - 0.1)10^6$$

$$\ln \frac{P}{4.08} = 0.02581$$

$$\frac{P}{4.08} = 1.026$$

$$P = 4.19 \text{ MPa}$$

4.4.3 Properties at Equilibrium—G, A, S

Although the major discussion of equilibrium properties will be deferred to Chaps. 7 and 8 for physical equilibria and to Chap. 9 for chemical equilibria, it is instructive to combine our previous discussions of the criteria for equilibrium in terms of the various properties.

Two terms not previously defined come into play: the concept of a spontaneous process and the concept of an isolated system. An *isolated system* is simply a system which is its own universe, thus having no surroundings. A *spontaneous process* is simply a process which will occur in some measurable amount of time either in an isolated system or between a system and its surroundings as long as the surroundings have no effect on the process. Clearly an isolated system can only support spontaneous processes. Thus the entropy of an isolated system must either remain constant or must increase.

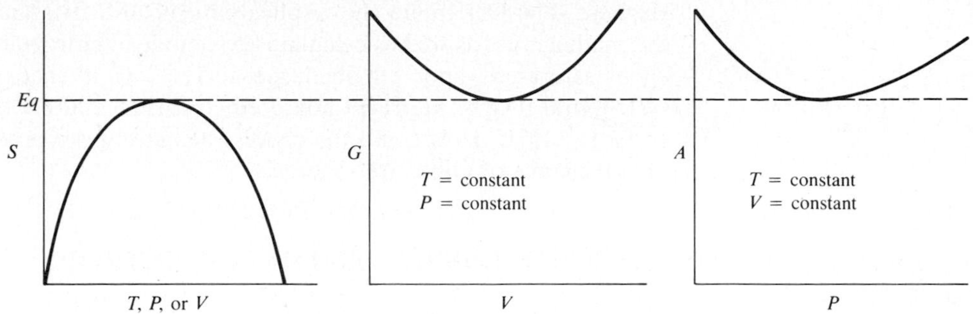

Figure 4.6 Entropy and free-energy behavior of pure compounds at equilibrium.

As discussed in the previous section, the free-energy change for a spontaneous process is negative at constant temperature and pressure. Similarly the change in entropy of the universe for a spontaneous process is positive. Movement toward equilibrium increases the entropy of the universe, and a system at equilibrium results in the fact that the change in the entropy of the universe must be zero or negative at equilibrium. A corollary of this fact is that the entropy of the universe must be at a maximum at equilibrium. If this is true, the Gibbs free energy of a system at constant temperature and pressure must then be a minimum at equilibrium. A similar analysis of the Helmholtz free energy shows that it is also a minimum at equilibrium at constant temperature and volume. This is shown graphically by Fig. 4.6.

Such an analysis is in congruence with the second law of thermodynamics in that the production of entropy in a spontaneous process is analogous to the production of entropy when heat flows from a warm body to a cold body.

Example 4.15 One kilogram of ethylbenzene is reversibly vaporized at its normal boiling point and atmospheric pressure when exposed to a heat source at 136.2°C. Determine the change in entropy and Gibbs free energy of the system and the universe.

SOLUTION From App. A,

$$\Delta H_{\text{vap}} = 35.910 \text{ kJ/kmol} = 338.14 \text{ kJ/kg}$$

$$\Delta S_{\text{univ}} = 0 = \Delta S_s + \Delta S_S$$

$$\Delta S_s = \frac{338.14}{409.3} = 0.826 \text{ kJ/(kg} \cdot \text{K)}$$

$$\Delta S_S = -0.826 \text{ kJ/(kg} \cdot \text{K)}$$

$$\Delta G = 0 = \Delta H_{\text{vap}} - T\Delta S_{\text{vap}}$$

$$= \Delta H_{\text{vap}} - T\frac{\Delta H_{\text{vap}}}{T} = 0$$

Could this process be carried out in an isolated *system*?

Exercise 4.N Determine the applicability of the first and second laws of thermodynamics as well as calculate the change in entropy of the universe if 1 kg of saturated vapor ethylbenzene at 136.2°C is changed in an isolated system to 0.1 kg of saturated liquid ethylbenzene and 0.9 kg of ethylbenzene vapor at 140°C. How could the process be carried out practically? Comment on the change in Gibbs free energy.

4.5 ESTIMATION OF INTERNAL ENERGY, ENTHALPY, ENTROPY, AND GIBBS FREE ENERGY

In Chap. 2, Secs. 2.4 and 2.5, analytical and generalized equations of state were described and used to estimate compressibility factors and densities. In this section, the same methods will be used to estimate internal energy, enthalpy, entropy, and Gibbs free energy. The general topic of departure functions, i.e., the departure of the thermodynamic functions from ideal-gas behavior due to pressure, will be discussed. In addition, the presentation of tabular data by means of free-energy functions will be illustrated.

4.5.1 Analytical Equations of State

The use of an equation of state to estimate the changes in thermodynamic properties has previously been illustrated only for the ideal gas. This section will extend the method to nonideal gases where the variation of pressure may also be important.

As will be discussed in Chap. 5, relationships between thermodynamic properties and PVT properties can be derived from basic definitions. For purposes of this discussion, we will consider the following expressions as starting points and leave their derivation to Chap. 5.

Each of these relationships will be of the form $\Delta P = \int f(1)\,dP + \int f(2)\,dT$, where ΔP is the change in the property represented as the sum of a pressure and a temperature effect.

$$\Delta\bar{S} = \int_{P'}^{P} -\left(\frac{\partial\bar{V}}{\partial T}\right)_P dP + \int_{T'}^{T} C_P\frac{dT}{T} \tag{4.47}$$

where P', T' are an initial reference state and P, T are any final state. P' and T' are chosen at a very low pressure where ideal-gas behavior can be assumed.

$$\Delta\bar{H} = \int_{P'}^{P}\left[\bar{V} - T\left(\frac{\partial\bar{V}}{\partial T}\right)_P\right] dP + \int_{T'}^{T} C_P\,dT \tag{4.48}$$

$$\Delta\bar{E} = \int_{P'}^{P}\left[-T\left(\frac{\partial\bar{V}}{\partial T}\right)_P - P\left(\frac{\partial\bar{V}}{\partial P}\right)_T\right] dP + \int_{T'}^{T} C_V\,dT \tag{4.49}$$

To illustrate the fact that each of these expressions will reduce to the ideal-gas form previously discussed, consider entropy. For an ideal gas,

$$\bar{V} = \frac{RT}{P}$$

Therefore

$$\left(\frac{\partial \bar{V}}{\partial T}\right)_P = \frac{R}{P}$$

Substituting into Eq. (4.47),

$$\Delta \bar{S} = \int_{P'}^{P} -\left(\frac{R}{P}\right) dP + \int_{T'}^{T} C_P \frac{dT}{T}$$

$$\Delta \bar{S} = -R \ln \frac{P}{P'} + \int_{T'}^{T} C_P \frac{dT}{T} = \Delta \bar{S}^* \tag{4.50}$$

which is identical to the expression obtained earlier for the ideal gas. To eliminate confusion any ideal-gas property will be asterisked for the remainder of the book. The enthalpy and internal-energy functions can be reduced to the ideal-gas state in the same manner.

van der Waals equation This equation of state is one of the simplest and will be used to illustrate the method.

For entropy,

$$\Delta \bar{S} = \int_{P'}^{P} -\left(\frac{\partial \bar{V}}{\partial T}\right)_P dP + \int_{T'}^{T} C_P \frac{dT}{T} \tag{4.47}$$

The van der Waals equation states that

$$P = \frac{RT}{\bar{V} - b} - \frac{a}{\bar{V}^2} \tag{2.13a}$$

This equation, as well as the other equations, is implicit in V and thus is difficult to differentiate. As will be discussed in Chap. 5, the following identity exists:

$$\left(\frac{\partial \bar{V}}{\partial T}\right)_P dP = -\left(\frac{\partial P}{\partial T}\right)_V d\bar{V}$$

This important relation allows Eq. (4.47) to be rewritten as

$$\Delta \bar{S} = \int_{\bar{V}'}^{\bar{V}} \left(\frac{\partial P}{\partial T}\right)_V d\bar{V} + \int_{T'}^{T} C_P \frac{dT}{T} \tag{4.51}$$

Differentiating the van der Waals equation with respect to temperature at constant volume,

$$\left(\frac{\partial P}{\partial T}\right)_V = \frac{R}{\bar{V} - b}$$

Thus

$$\Delta \bar{S} = \int_{\bar{V}'}^{\bar{V}} \frac{R}{\bar{V} - b} d\bar{V} + \int_{T'}^{T} C_P \frac{dT}{T}$$

or

$$\Delta \bar{S} = R \ln \frac{\bar{V} - b}{\bar{V}' - b} + \int_{T'}^{T} \frac{C_P}{T} dT \tag{4.52}$$

This is then the general equation for $\Delta \bar{S}$ of a gas which follows the van der Waals equation.

For the ideal gas,

$$\Delta \bar{S}^* = -R \ln \frac{P}{P'} + \int_{T'}^{T} C_P \frac{dT}{T} \tag{4.8}$$

Thus,

$$\Delta \bar{S}^* - \Delta \bar{S} = -R \ln \frac{P}{P'} - R \ln \frac{\bar{V} - b}{\bar{V}' - b} \tag{4.53}$$

This is called the departure function for entropy based on the van der Waals equation.

For enthalpy the same analysis can be undertaken. Making the same substitution in the defining equation for enthalpy change,

$$\Delta \bar{H} = \int_{P'}^{P} \bar{V}\, dP + \int_{V'}^{V} T \left(\frac{\partial P}{\partial T} \right)_V d\bar{V} + \int_{T'}^{T} C_P\, dT \tag{4.54}$$

Since $d(P\bar{V}) = \bar{V}\,dP + P\,d\bar{V}$,

$$\bar{V}\,dP = d(P\bar{V}) - P\,d\bar{V}$$

Thus

$$\Delta \bar{H} = \int_{P'}^{P} d(PV) + \int_{V'}^{V} \left[T \left(\frac{\partial P}{\partial T} \right)_V - P \right] d\bar{V} + \int_{T'}^{T} C_P\, dT$$

Since

$$\left(\frac{\partial P}{\partial T} \right)_V = \frac{R}{\bar{V} - b} \qquad \text{and} \qquad P = \frac{RT}{\bar{V} - b} - \frac{a}{\bar{V}^2}$$

$$\Delta \bar{H} = (P\bar{V} - P'\bar{V}') + \left(\frac{a}{\bar{V}'} - \frac{a}{\bar{V}} \right) + \int_{T'}^{T} C_P\, dT$$

or

$$\Delta \bar{H}^* - \Delta \bar{H} = (P'\bar{V} - P\bar{V}) + \left(\frac{a}{\bar{V}} - \frac{a}{\bar{V}'} \right) \tag{4.55}$$

This is the general equation for enthalpy change. Since the reference pressure is very low, $\bar{V}'$ will be very large and the term $a/\bar{V}'$ may often be neglected. In addition $P'V'$ can usually be assumed to be equal to RT, the ideal-gas volume. Thus

$$\Delta \bar{H}^* - \Delta \bar{H} = RT - P\bar{V} + \frac{a}{\bar{V}} \tag{4.56}$$

This is the departure function for enthalpy based on the van der Waals equation. A similar analysis for internal energy is possible.

Example 4.16 For ammonia at 100°C, calculate the change in entropy and the change in enthalpy above that of the ideal gas at 100°C and 1 atm if the pressure is raised to 5 MPa at constant temperature. Use the van der Waals formulation.

BASIS 1 kmol ammonia.

Use Eq. (4.53):

$$\Delta\bar{S}^* - \Delta\bar{S} = -R\ln\frac{P}{P'} - R\ln\frac{\bar{V}-b}{\bar{V}'-b}$$

$$P = 10 \text{ MPa} \qquad P' = 0.1013 \text{ MPa}$$

$$\bar{V}' = \frac{RT}{P'} = \frac{(8314.3)(373.1)}{0.1013\times10^6} = 30.6226 \text{ m}^3/\text{kmol}$$

$\bar{V}$ by similar calculation to Example 2.4 is

$$\left(5\times10^6 + \frac{424{,}447}{\bar{V}^2}\right)(\bar{V}-0.0373) = \frac{(8314.3)(373.1)}{3.102\times10^6}$$

Trial-and-error $\bar{V} = 0.50$

$$\Delta\bar{S}^* - \Delta\bar{S} = -8.3143\left(\ln\frac{5}{0.1013} + \ln\frac{0.5-0.0373}{30.6226-0.0373}\right)$$

$$= -8.3143(3.899 - 4.191)$$

$$= +2.4294 \text{ kJ/(kmol}\cdot\text{K)}$$

Using Eq. (4.56),

$$\Delta\bar{H}^* - \Delta\bar{H} = RT - P\bar{V} + \frac{a}{\bar{V}}$$

$$= (8.3143)(373.1) - (5000)(0.50) + \frac{424.447}{(0.50)}$$

$$= 3102 - 2500 + 1698 = 2300 \text{ kJ/kmol}$$

Other analytical equations of state The methods discussed for the van der Waals equation may be applied to any other equation of state, although the more complex the equation, the more difficult the algebraic manipulations. For example, consider the Redlich-Kwong-Soave equation described in Chap. 2:

$$P = \frac{RT}{\bar{V}-b} - \frac{\alpha a}{\bar{V}(\bar{V}+b)} \tag{2.20}$$

where $\alpha a = f(T)$.

Considering enthalpy as a representative property and using the same definitions given earlier,

$$\Delta\bar{H} - \Delta\bar{H}^* = \frac{-f + (\partial f/\partial T)_V T}{b}\ln\left(1+\frac{b}{V}\right) + ZRT - RT \tag{4.57}$$

where Z = compressibility factor

$$f = f_c\alpha$$

$$f_c = 0.42747\frac{R^2T_c^2}{P_c}$$

$$\left(\frac{\partial f}{\partial T}\right)_V = \frac{-f^{0.5}f_c^{0.5}S}{T_cT_r^{0.5}}$$

and α and S are given by Eqs. (2.21) and (2.22). Similar expressions can be derived for the other properties.

Exercise 4.0 Derive Eq. (4.57) and an equation for entropy departure following the Redlich-Kwong-Soave equation.

4.5.2 Departure Functions

The calculation of changes in thermodynamic properties usually is referred to the ideal gas, thus the term *departure function.*

Consider a process whereby a material is taken from its initial condition P_1T_1 to a final condition P_2V_2. This process can be visualized as a series of steps. From the initial conditions P_1T_1:

1. Move to ideal-state pressure $P'T_1$.
2. Increase or decrease temperature isobarically to $P'T_2$.
3. Increase pressure isothermally to P_2T_2.

Thus, the change in property for any process will be the sum of the three processes. In equation form, for any property which in its ideal-gas state has no pressure effect,

$$\Delta\overline{P} = \overline{P}_2 - \overline{P}_1 = \left(\overline{P}_1^* - \overline{P}_1\right) + \left(\overline{P}_2^* - \overline{P}_1^*\right) - \left(\overline{P}_2^* - \overline{P}_2\right) \tag{4.58}$$

where
$$\overline{P}_1^* - \overline{P}_1 = \text{departure at initial condition}$$
$$\overline{P}_2^* - \overline{P}_1^* = \text{change for ideal gas}$$
$$P_2^* - \overline{P}_2 = \text{departure at final condition}$$

For enthalpy,

$$\Delta\overline{H} = \left(\overline{H}_1^* - \overline{H}_1\right) + \left(\overline{H}_2^* - \overline{H}_1^*\right) - \left(\overline{H}_2^* - \overline{H}_2\right) \tag{4.59}$$

where
$$\overline{H}_2^* - \overline{H}_1^* = \Delta\overline{H}^* = \int_{T_1}^{T_2} C_P^*\,dT$$

For entropy the ideal-gas state has a pressure effect, so terms are necessary to take into account the ideal-gas pressure effect in going from P_1 to P' and in going from P' to P_2. Thus

$$\Delta\overline{S} = \left(\overline{S}_1^* - \overline{S}_1\right) + \left(\overline{S}_1'^* - \overline{S}_1^*\right) + \left(\overline{S}_2'^* - \overline{S}_1'^*\right) + \left(\overline{S}_2^* - \overline{S}_2'^*\right) - \left(\overline{S}_2^* - \overline{S}_2\right) \tag{4.60}$$

where
$$\overline{S}_1'^* - \overline{S}_1^* = R\ln\frac{P_1}{P'}$$
$$\overline{S}_2^* - \overline{S}_2'^* = R\ln\frac{P'}{P_2}$$
$$\overline{S}_2'^* - \overline{S}_1'^* = \int_{T_1}^{T_2} C_P^*\frac{dT}{T}$$

Substituting,

$$\Delta \bar{S} = \left(\bar{S}_1^* - \bar{S}_1\right) + R \ln \frac{P_1}{P_2} + \int_{T_1}^{T_2} C_P^* \frac{dT}{T} - \left(\bar{S}_2^* - \bar{S}_2\right) \tag{4.61}$$

The ideal-gas heat capacities may be calculated by using the methods of Chap. 3 with the constants of Table 3.2. A similar equation for ideal-gas entropy is

$$S^* = B \ln T + 2CT + \tfrac{3}{2}DT^2 + \tfrac{4}{3}ET^3 + \tfrac{5}{4}FT^4 + G \tag{4.62}$$

Constants for this equation are also given in Table 3.2.

The expression for internal energy could be derived in a similar way to that of enthalpy.

Free energy has not been discussed and is normally not treated in this manner but is calculated from enthalpy and entropy by its definition: $G \equiv H - TS$. Since free energies are normally useful only at constant-temperature conditions, $\Delta \bar{G} = \Delta \bar{H} - T\Delta \bar{S}$.

The departure functions just derived can be calculated by using an analytical equation of state as discussed in Sec. 4.5.1 or by a generalized equation of state to be discussed in Sec. 4.5.3.

Before leaving departure functions, it is important to note that such functions are general and can be used for any purpose whether or not phase boundaries are crossed. The ideal-gas temperature portion of the expression for any property change will equal zero for such changes as isothermal expansion or compression of a gas or for an isothermal vaporization or condensation. In these cases, the overall change in property will be merely the difference in the two departure functions, except for entropy where the ideal-gas pressure effect must be taken into account.

Example 4.17 Derive the entropy departure function for a gas that follows the Redlich-Kwong equation.

SOLUTION Using Eq. (4.47),

$$\Delta \bar{S} = \int_{P'}^{P} \left[-\left(\frac{\partial \bar{V}}{\partial T} \right)_P \right] dP + \int_{T'}^{T} C_P \frac{dT}{T}$$

Since

$$\left(\frac{\partial \bar{V}}{\partial T} \right)_P = -\left(\frac{\partial P}{\partial T} \right)_V d\bar{V}$$

$$\Delta \bar{S} = \int_{\bar{V}'}^{\bar{V}} \left(\frac{\partial P}{\partial T} \right)_V d\bar{V} + \int_{T'}^{T} C_P \frac{dT}{T}$$

Using Eq. (4.50),

$$\Delta \bar{S}^* = -R \ln \frac{P}{P'} + \int_{T'}^{T} C_P \frac{dT}{T}$$

$$\text{Departure} = \Delta \bar{S}^* - \Delta \bar{S}$$

The Redlich-Kwong equation is given by Eq. (2.17):

$$P = \frac{RT}{\bar{V} - b} - \frac{a}{T^{1/2}\bar{V}(\bar{V} + b)}$$

Taking

$$\left(\frac{\partial P}{\partial T}\right)_V = \frac{R}{\bar{V} - b} + \frac{a}{2\bar{V}(\bar{V} + b)T^{3/2}}$$

$$\int_{\bar{V}'}^{\bar{V}} \left(\frac{\partial P}{\partial T}\right)_V dV = \int_{\bar{V}'}^{\bar{V}} \frac{R\,d\bar{V}}{\bar{V} - b} + \frac{a}{2T^{3/2}} \int_{\bar{V}'}^{\bar{V}} \frac{d\bar{V}}{\bar{V}(V + b)}$$

$$= R \ln \frac{\bar{V} - b}{\bar{V}' - b} - \frac{a}{2bT^{3/2}} \ln \frac{\bar{V}'(\bar{V} + b)}{\bar{V}(\bar{V}' + b)}$$

Thus

$$\Delta\bar{S} = R \ln \frac{\bar{V} - b}{\bar{V}' - b} - \frac{a}{2bT^{3/2}} \ln \frac{\bar{V}'(\bar{V} + b)}{\bar{V}(\bar{V}' + b)} + \int_{T'}^{T} C_P \frac{dT}{T}$$

Subtracting this expression from Eq. (4.50),

$$\Delta\bar{S}^* - \Delta\bar{S} = R \ln \frac{P'}{P} - R \ln \frac{\bar{V} - b}{\bar{V}' - b} + \frac{a}{2bT^{3/2}} \ln \frac{\bar{V}'(\bar{V} + b)}{\bar{V}(\bar{V}' + b)}$$

Exercise 4.P Write the expression for internal energy that would be analogous to Eq. (4.59) for enthalpy. Write the equation for the process of Example 4.16.

4.5.3 Generalized Equations of State

Generalized equations of state have been introduced in Chap. 2. The discussion here will be limited to the Lydersen and Pitzer type of methods and the analogous Lee and Kesler analytical representation of the Pitzer method.

Lydersen method The method of Lydersen has been extended to enthalpy, entropy, internal-energy, heat capacity, and fugacity (to be discussed in Chap. 7) departures. The equations utilized are quite simple to derive based on the three-parameter theorem $Z = f(P_r, T_r, Z_c)$. Combining Eqs. (4.48) (real gas) and (3.14) (ideal gas),

$$\Delta\bar{H}^* - \Delta\bar{H} = \int_0^P \left[T\left(\frac{\partial\bar{V}}{\partial T}\right)_P - \bar{V}\right]_T dP$$

and

$$P\bar{V} = ZRT$$

Differentiating

$$\left(\frac{\partial\bar{V}}{\partial T}\right)_P = \frac{RZ}{P} + \frac{RT}{P}\left(\frac{\partial Z}{\partial T}\right)_P$$

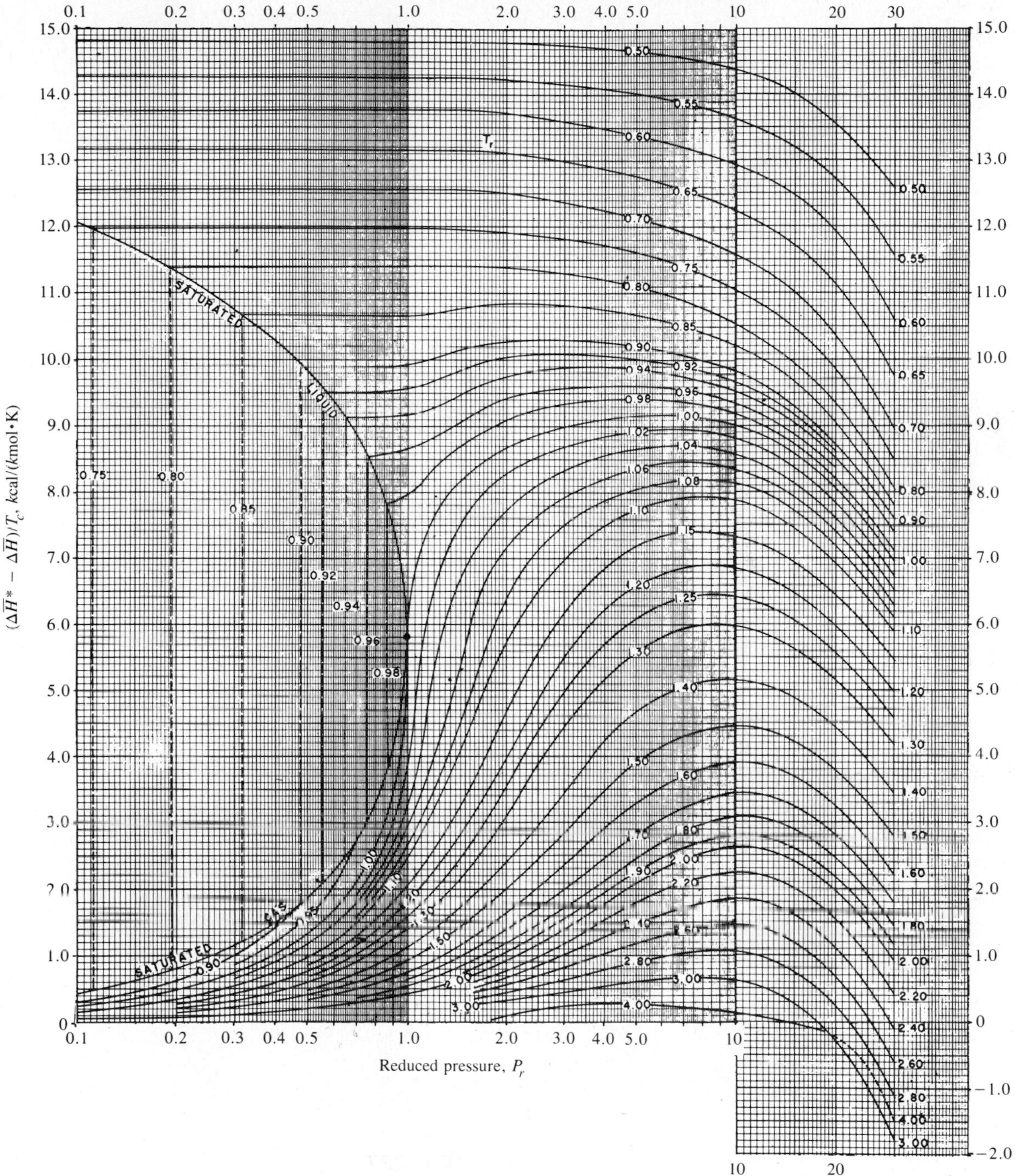

Figure 4.7 Generalized enthalpy departures from ideal-gas behavior at $Z_c = 0.27$ (molar basis). *(Adapted from Hougen, Watson, and Ragatz, "Chemical Process Principles Charts," 3d ed., copyright 1964 by Wiley. Used with permission.)*

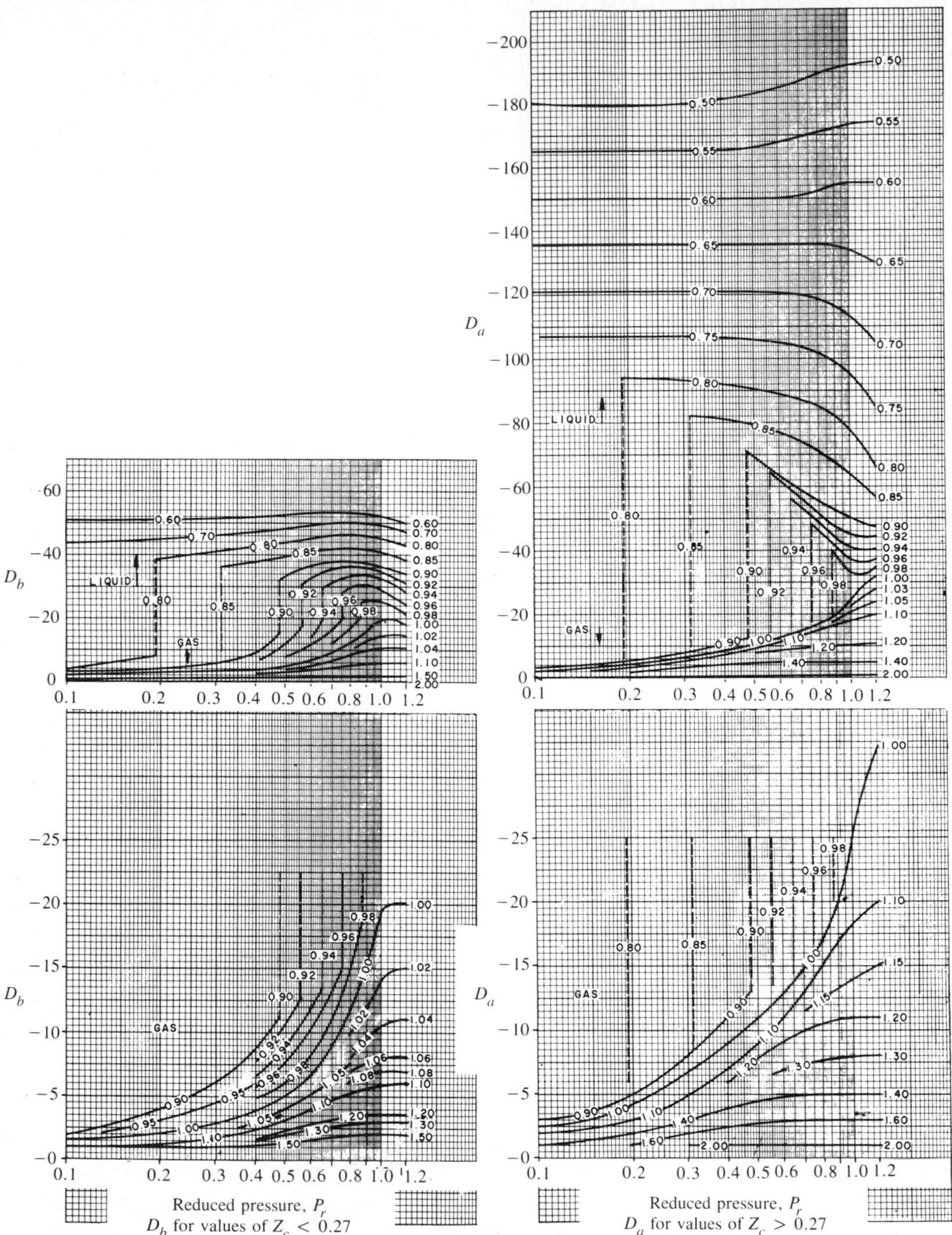

Figure 4.8 Generalized deviation of enthalpy departure from values at $Z_c = 0.27$. *(Adapted from Hougen, Watson, and Ragatz, "Chemical Process Principles Charts," 3d ed., copyright 1964 by Wiley. Used with permission.)*

Substituting

$$\Delta\overline{H}^* - \Delta\overline{H} = \int_0^P \frac{RT^2}{P}\left(\frac{\partial Z}{\partial T}\right)_P dP \tag{4.63}$$

in terms of reduced quantities,

$$\frac{\Delta\overline{H}^* - \Delta\overline{H}}{T_c} = \int_0^{P_r} + \frac{RT_r^2}{P_r}\left(\frac{\partial Z}{\partial T_r}\right)_{P_r} dP_r$$

$$\left(\frac{\Delta\overline{H}^* - \Delta\overline{H}}{T_c}\right)_T = \left[RT_r^2 \int_0^{P_r}\left(\frac{\partial Z}{\partial T_r}\right)_{P_r} \frac{dP_r}{P_r}\right]_T \tag{4.64}$$

The left side of the equation was calculated over a wide range of reduced temperature and reduced pressure for gases, liquids, and saturated materials at values of Z_c from 0.23 to 0.30. These were plotted for values of Z_c of 0.27 as given in Fig. 4.7. Correction factors given in Fig. 4.8 are used together with the equation $B' = B + D(Z_c - 0.27)$ described in Chap. 2 to calculate departures at other critical compressibility factors. It should be noted on Fig. 4.7 that the discontinuities in the vapor-liquid two-phase region are merely the latent heat of vaporization divided by the critical temperature of the material. Thus, a given value of enthalpy of an ideal gas may be converted to any gaseous or liquid state by using this method. Above the critical pressure, no heat of vaporization or condensation is involved and no discontinuities exist in the plots. A change in slope is caused by the increase in values of heat capacity as vapor goes to liquid.

Example 4.18 Calculate the enthalpy departure for water at 700°C and 25 MPa pressure by using the Lydersen generalized technique. Compare your value with that read from the steam tables. The system is isothermal with a base pressure of 0.1 MPa.

$$P_c = 22.12 \text{ MPa} \qquad T_c = 647.3 \text{ K} \qquad Z_c = 0.234$$

$$T_r = \frac{973.1}{647.3} = 1.50 \qquad P_r = \frac{25}{22.12} = 1.13$$

From Fig. 4.7 at $Z_c = 0.27$,

$$\frac{\Delta\overline{H}^* - \Delta\overline{H}}{T_c} = 1.1$$

From Fig. 4.8, $D_b^{\circ} = -2$. Therefore

$$\frac{\Delta\overline{H}^* - \Delta\overline{H}}{T_c} = 1.1 - 2(0.234 - 0.27) \doteq 1.17$$

$$\Delta\overline{H}^* - \Delta\overline{H} = (1.17)(647.3) = 757.3 \text{ kcal/kmol}$$

Since $\Delta\overline{H}^* = 0$ (isothermal),

$$\Delta\overline{H} = -3170 \text{ kJ/kmol} = -176 \text{ kJ/kg}$$

From the steam tables,

$$H_{700^\circ\text{C},25\text{ MPa}} = 3777.5 \text{ kJ/kg}$$

$$H_{700^\circ\text{C},0.1\text{ MPa}} = 3928.2 \text{ kJ/kg}$$

$$\Delta H = -150.7 \text{ kJ/kg}$$

A 16 percent error results from the comparison. As water is very nonideal, this is a rigorous test of the method.

Considering entropy, the definition of the departure function can be written by combining Eqs. (4.47) and (4.50):

$$(\Delta\bar{S}^* - \Delta\bar{S})_T = \int_P^0 \left[\frac{R}{P} - \left(\frac{\partial \bar{V}}{\partial T}\right)_P\right]_T dP \tag{4.65}$$

Using the previously derived equation,

$$\left(\frac{\partial \bar{V}}{\partial T}\right)_P = \frac{RZ}{P} + \frac{RT}{P}\left(\frac{\partial Z}{\partial T}\right)_P$$

$$(\Delta\bar{S}^* - \Delta\bar{S})_T = \int_P^0 \left[\frac{1-Z}{P} - \frac{T}{P}\left(\frac{\partial Z}{\partial T}\right)_P\right]_T dP \tag{4.66}$$

Writing this equation in reduced form,

$$(\Delta\bar{S}^* - \Delta\bar{S})_T = -R\int_0^{P_r} \frac{(1-Z)\,dP_r}{P_r} + RT_r\int_0^{P_r}\left(\frac{\partial Z}{\partial T_r}\right)_{P_r}\frac{dP_r}{P_r} \tag{4.67}$$

Plots of the departure function and the correction factors are given in Figs. 4.9 and 4.10.

Internal-energy departures have been calculated from enthalpy departures and compressibility data by using the definition of enthalpy: $H \equiv E + PV$. Therefore

$$E = H - PV = H - ZRT$$

Thus

$$\frac{\Delta\bar{E}^* - \Delta\bar{E}}{T_c} = \frac{\Delta\bar{H}^* - \Delta\bar{H}}{T_c} - (1 - Z)RT_r \tag{4.68}$$

These are plotted in Fig. 4.11, with correction factors in Fig. 4.12.

Heat capacity departures are functions of enthalpy or internal energy. The constant-pressure heat capacity C_P equals $(\partial H/\partial T)_P$. Thus

$$C_P - C_P^* = \frac{\partial\left(\dfrac{\Delta\bar{H} - \Delta\bar{H}^*}{T_c}\right)}{\partial T_r} \tag{4.69}$$

Figure 4.13 gives values over a wide range of reduced conditions. C_P^* can be calculated or estimated as discussed earlier.

In summary, Lydersen's method can be used to predict enthalpy, internal-energy, entropy, and isobaric heat capacity departures by using input parameters

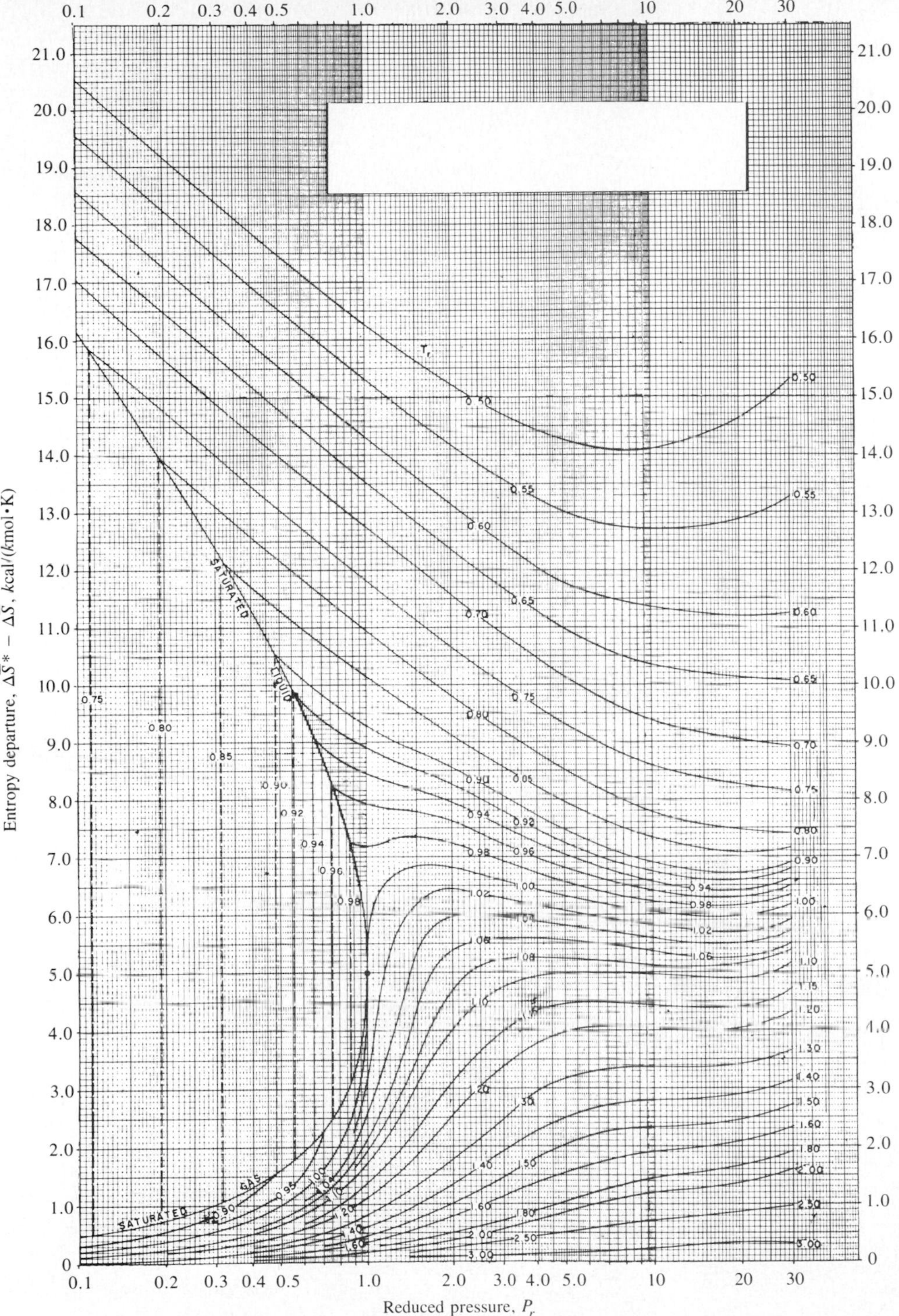

Figure 4.9 Generalized entropy departures from ideal-gas behavior at $Z_c = 0.27$ (molar basis). *(Adapted from Hougen, Watson, and Ragatz, "Chemical Process Principles Charts," 3d ed., copyright 1964 by Wiley. Used with permission.)*

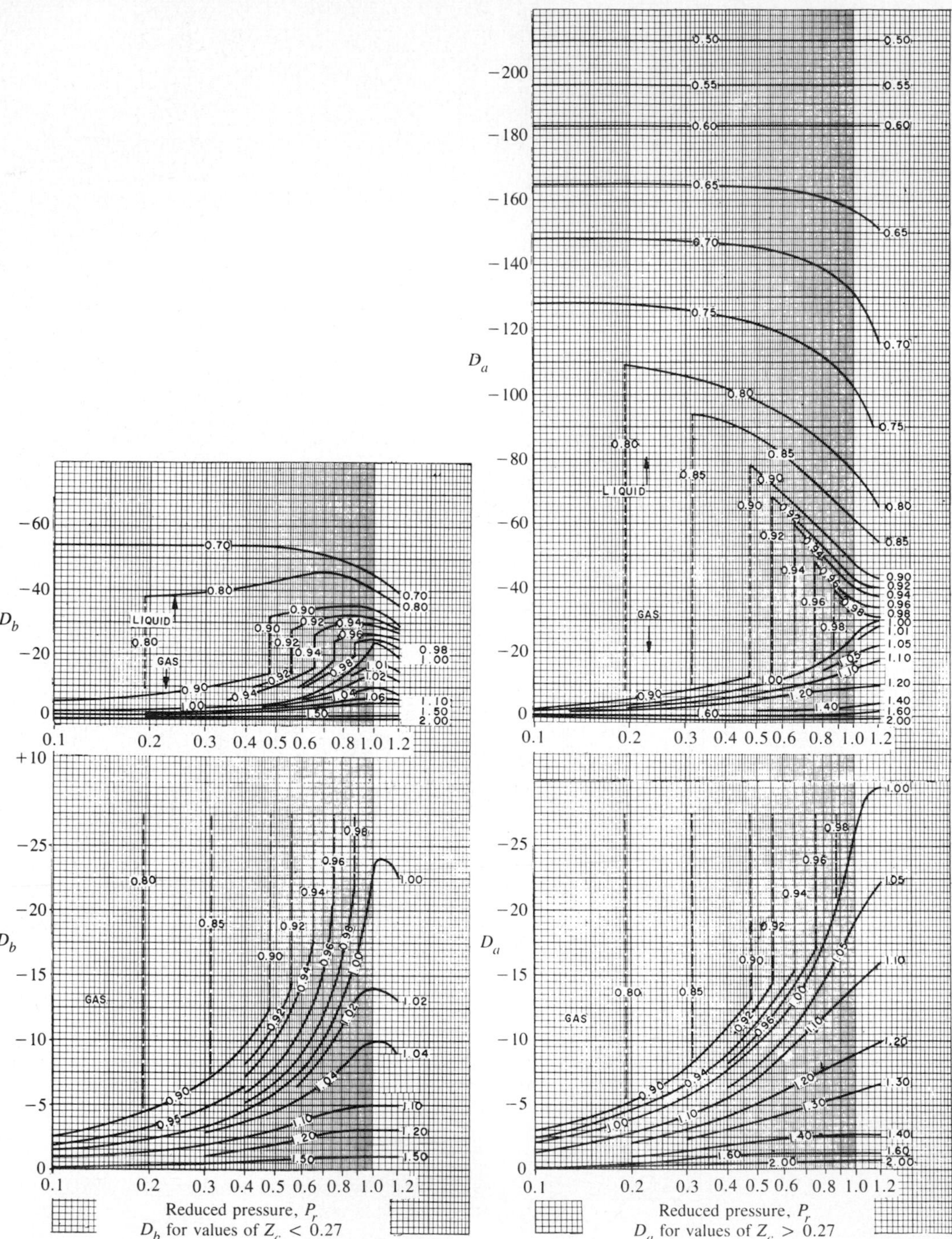

Figure 4.10 Generalized deviation of entropy departure from values at $Z_c = 0.27$. *(Adapted from Hougen, Watson, and Ragatz, "Chemical Process Principles Charts," 3d ed, copyright 1964 by Wiley. Used with permission.)*

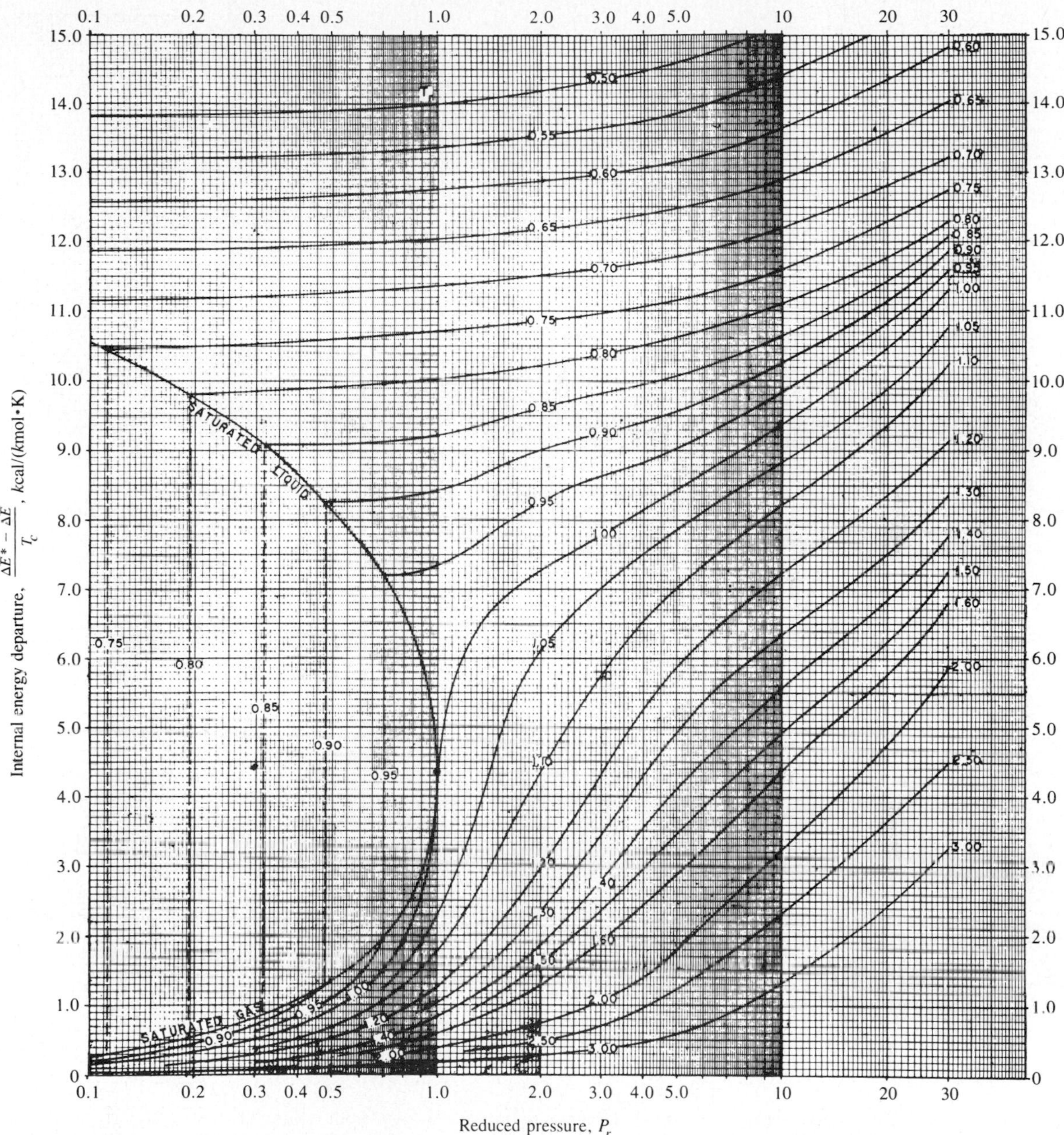

Figure 4.11 Generalized internal-energy departures from ideal-gas behavior at $Z_c = 0.27$ (molar basis). *(Adapted from Hougen, Watson, and Ragatz, "Chemical Process Principles Charts," 3d ed., copyright 1964 by Wiley. Used with permission.)*

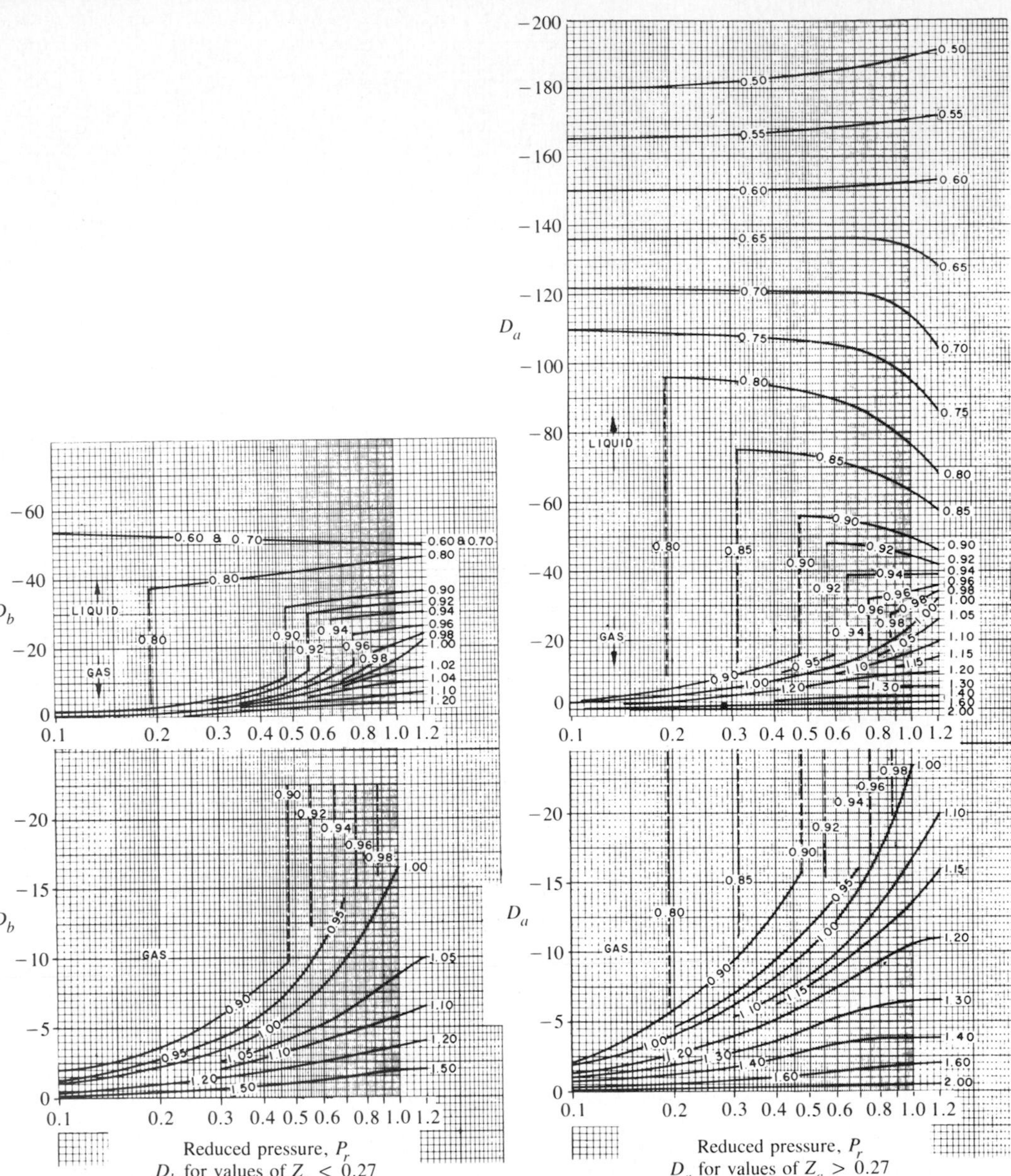

Figure 4.12 Generalized deviation of internal energy from values at $Z_c = 0.27$. *(Adapted from Hougen, Watson, and Ragatz, "Chemical Process Principles Charts," 3d ed., copyright 1964 by Wiley. Used with permission.)*

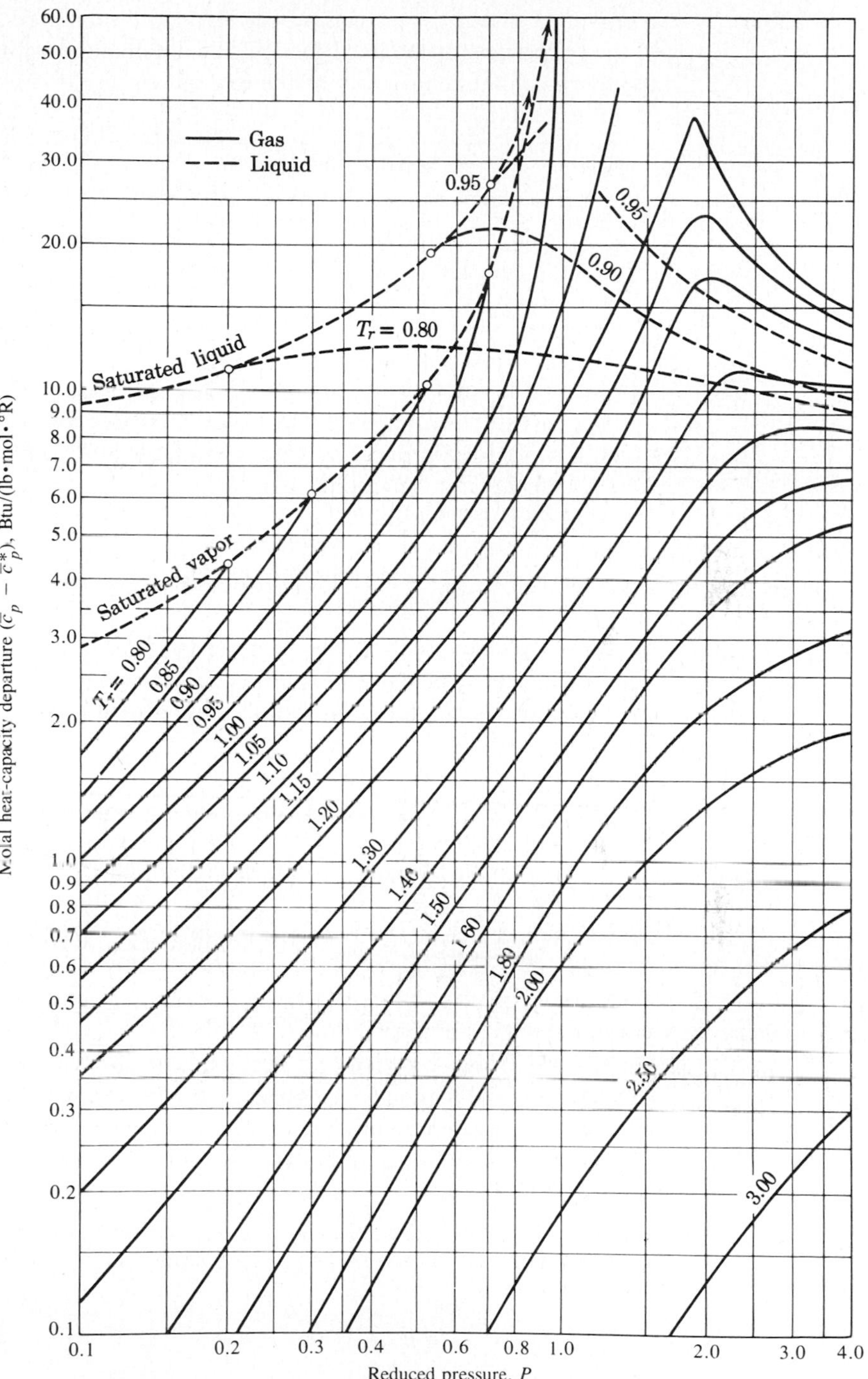

Figure 4.13 Departure of molar heat capacities of gases and liquids from ideal-gas behavior at $Z_c = 0.27$. *(Adapted from Hougen, Watson, and Ragatz, "Chemical Process Principles Charts," 3d ed., copyright 1964 by Wiley. Used with permission.)*

consisting only of the critical properties of the substance in question. If changes in the properties for the ideal gas are calculated and combined with the departure, the thermodynamic property of the real gas or liquid is obtained.

Property	$\Delta \bar{P}^* - \Delta \bar{P}$	$\Delta \bar{P}^*$
Enthalpy	Fig. 4.7	Eq. (3.14)
Entropy	Fig. 4.9	Eq. (4.8)
Internal energy	Fig. 4.11	Sec. 3.5.1
Heat capacity	Fig. 4.13	Eq. (3.12)

Application to mixtures of compounds will be discussed later.

Given experimental data can be extrapolated by using the departure functions; e.g., if an experimental value is available at T_1, P_1 and a value is desired at T_2, P_2, departure functions can be determined at T_{r1}, P_{r1}, $Z_c(C_1)$ and T_{r2}, P_{r2}, $Z_c(C_2)$. Then the value of the property can be estimated as

$$\text{Prop}_{T_2, P_2} = \text{Prop}_{T_1, P_1} \frac{C_2}{C_1} \tag{4.70}$$

This Lydersen method has been used extensively for estimation of properties and enhancement of known experimental data for a wide variety of compounds. Construction of the thermodynamic diagrams using these methods will be discussed in Chap. 5.

Exercise 4.Q Calculate enthalpy, entropy, and internal-energy departures for methane gas at 600°C and 6 MPa pressure from a base of 600°C and 0.1 MPa pressure by using Lydersen's techniques.

Pitzer method The generalized method of Pitzer which has been determined to be more accurate than the Lydersen method for normal fluids, hydrocarbons, can be extended from the discussion given in Chap. 2. In this case, $Z = Z(P_r, T_r, \omega)$. Pitzer used the same defining equations in reduced form as did Lydersen for enthalpy, entropy, fugacity, and heat capacity. Internal-energy functions could be derived in a like manner. Using the form that the departure from an ideal gas is composed of a simple fluid term and a correction term, Pitzer tabulated the two terms given below in Eq. (4.71) as functions of reduced temperature and reduced pressure.

Enthalpy

$$\left|\frac{\Delta \bar{H}^* - \Delta \bar{H}}{RT_c}\right| = \left|\frac{\Delta \bar{H}^* - \Delta \bar{H}}{RT_c}\right|^{(0)} + \omega \left|\frac{\Delta \bar{H}^* - \Delta \bar{H}}{RT_c}\right|^{(1)} \tag{4.71}$$

where superscripts (0) and (1) refer to the simple fluid term (spherical molecules) and correction term (correction for size and shape), respectively. Pitzer's original tables extended from $0.2 \le P_r \le 9$ and $0.80 \le T_r \le 4.0$. These limits were later

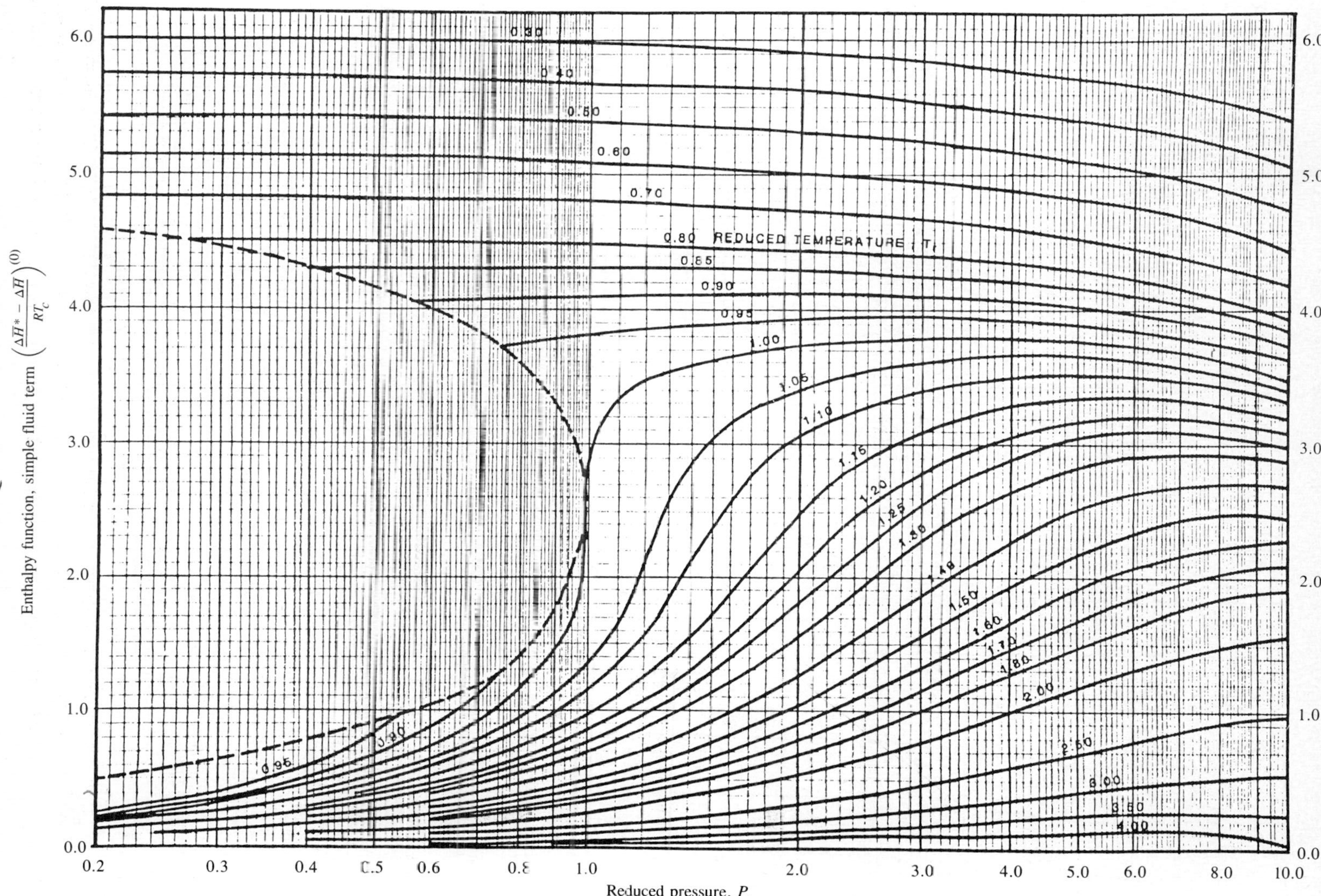

Figure 4.14 Effect of pressure on enthalpy, simple fluid term. *(Adapted from "Technical Data Book—Petroleum Refining," 4th ed., copyright 1982 by American Petroleum Institute. Used with permission.)*

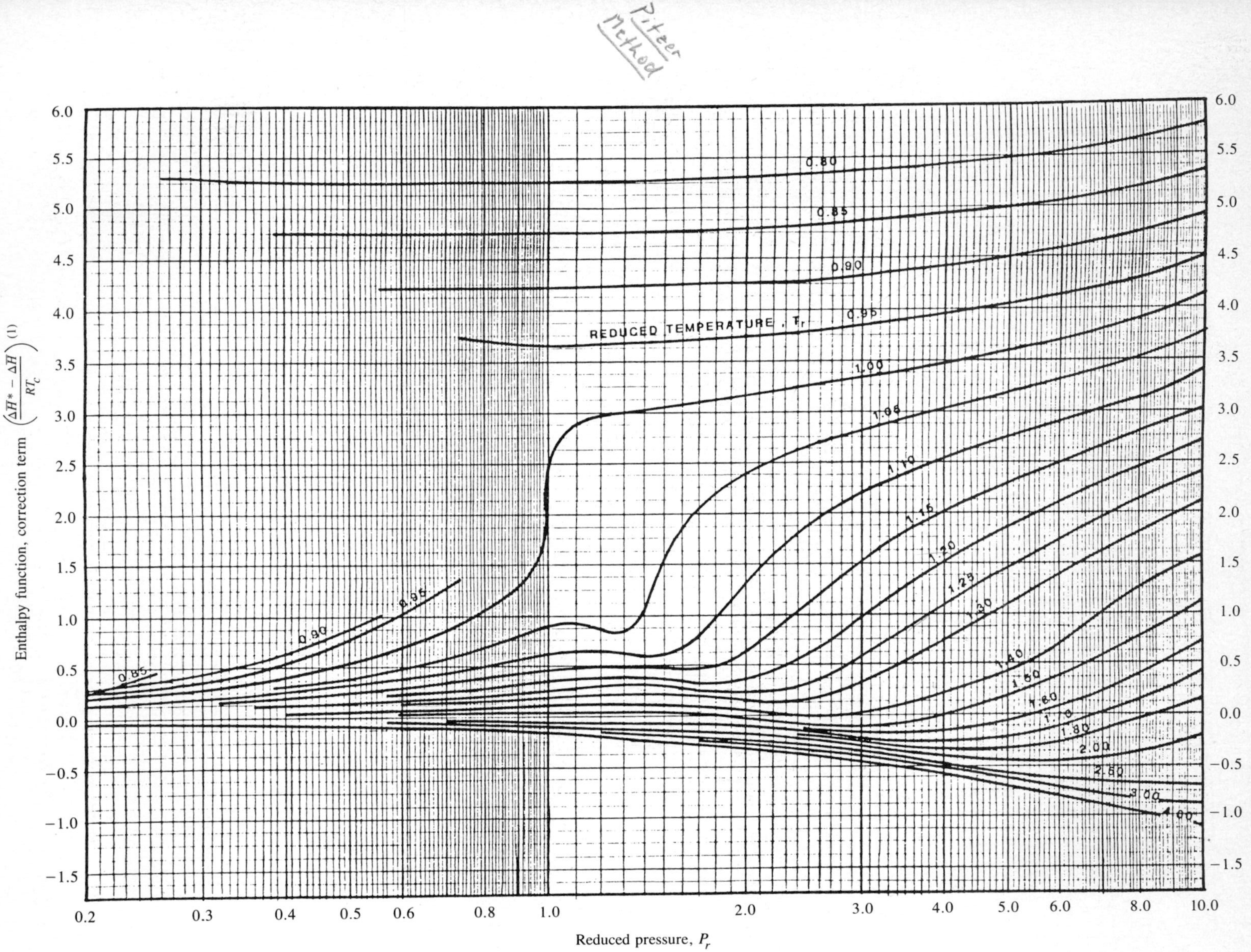

Enthalpy function, correction term $\left(\frac{\Delta H^* - \Delta H}{RT_c}\right)^{(1)}$

REDUCED TEMPERATURE, T_r

0.80, 0.85, 0.90, 0.95, 1.00, 1.05, 1.10, 1.15, 1.20, 1.25, 1.30, 1.40, 1.50, 1.60, 1.70, 1.80, 2.00, 2.50, 3.00, 4.00

Reduced pressure, P_r

Figure 4.15 Effect of pressure on enthalpy, correction term. *(Adapted from "Technical Data Book—Petroleum Refining," 4th ed., copyright 1982 by American Petroleum Institute. Used with permission.)*

increased in reduced-pressure and reduced-temperature ranges to $0.2 \leq P_r \leq 10$ and $0.30 \leq T_r \leq 4.0$ by various investigators. Below a reduced pressure of 0.2, the terms may be extrapolated, remembering that both terms must go to zero at zero pressure.

Figures 4.14 and 4.15 give the values for the two terms. Tables are available in the "Technical Data Book—Petroleum Refining." For normal fluids the error in pressure effect on enthalpy when predicted by this method averages 7 kJ/kg, although errors as high as 35 kJ/kg can occur in the critical region.

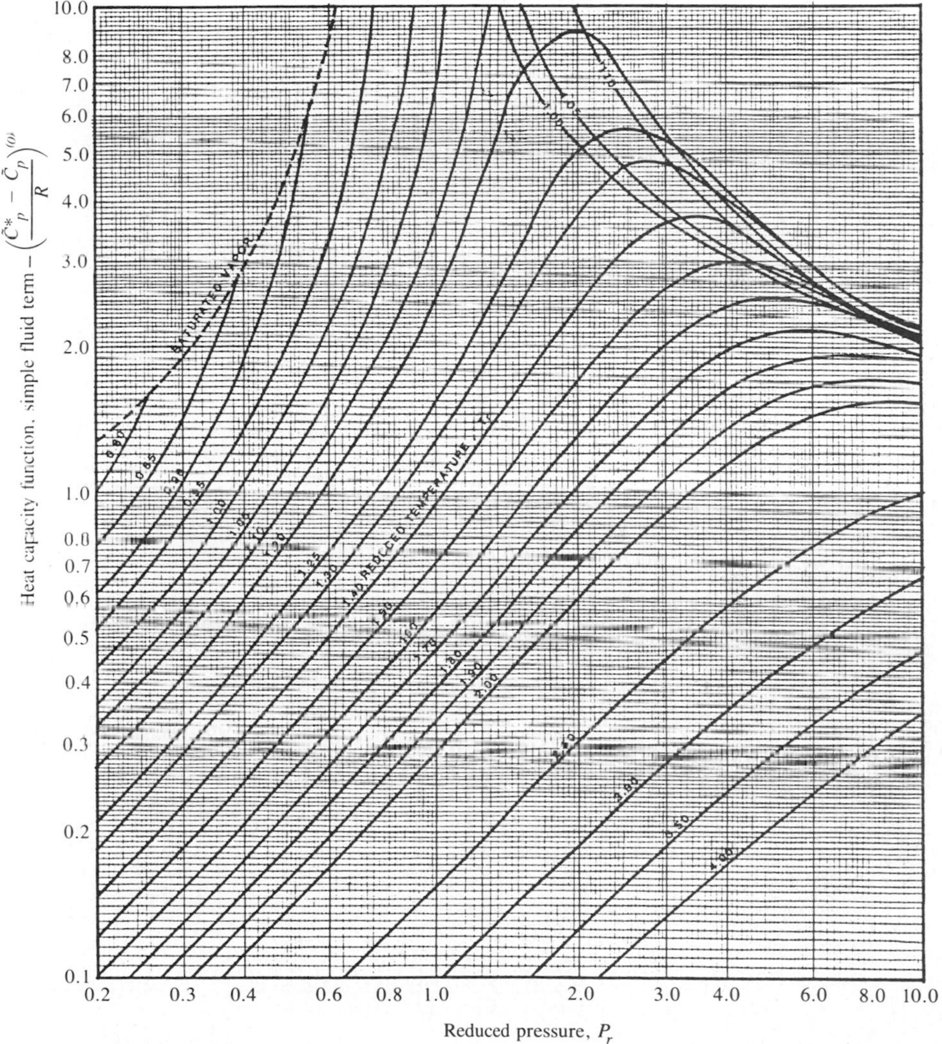

Figure 4.16 Effect of pressure on constant-pressure-gas heat capacity, simple fluid term. *(Adapted from "Technical Data Book—Petroleum Refining," 4th ed., copyright 1982 by American Petroleum Institute. Used with permission.)*

Using the departure function, the actual enthalpy can be calculated as

$$\Delta \overline{H} = \Delta \overline{H}^* - \frac{RT_c}{M} \frac{\Delta \overline{H}^* - \Delta \overline{H}}{RT_c} \tag{4.72}$$

For defined mixtures of normal fluids the simple molar average mixing rule may be used to calculate pseudocritical properties T_{pc}, P_{pc}, ω for use in calculating the overall departure function. The proper value of the ideal-gas heat capacity for use in this case is the weight average of the defined mixture. Methods for undefined mixtures have been proposed by Lee and Kesler (1975) and Huang and Daubert (1974).

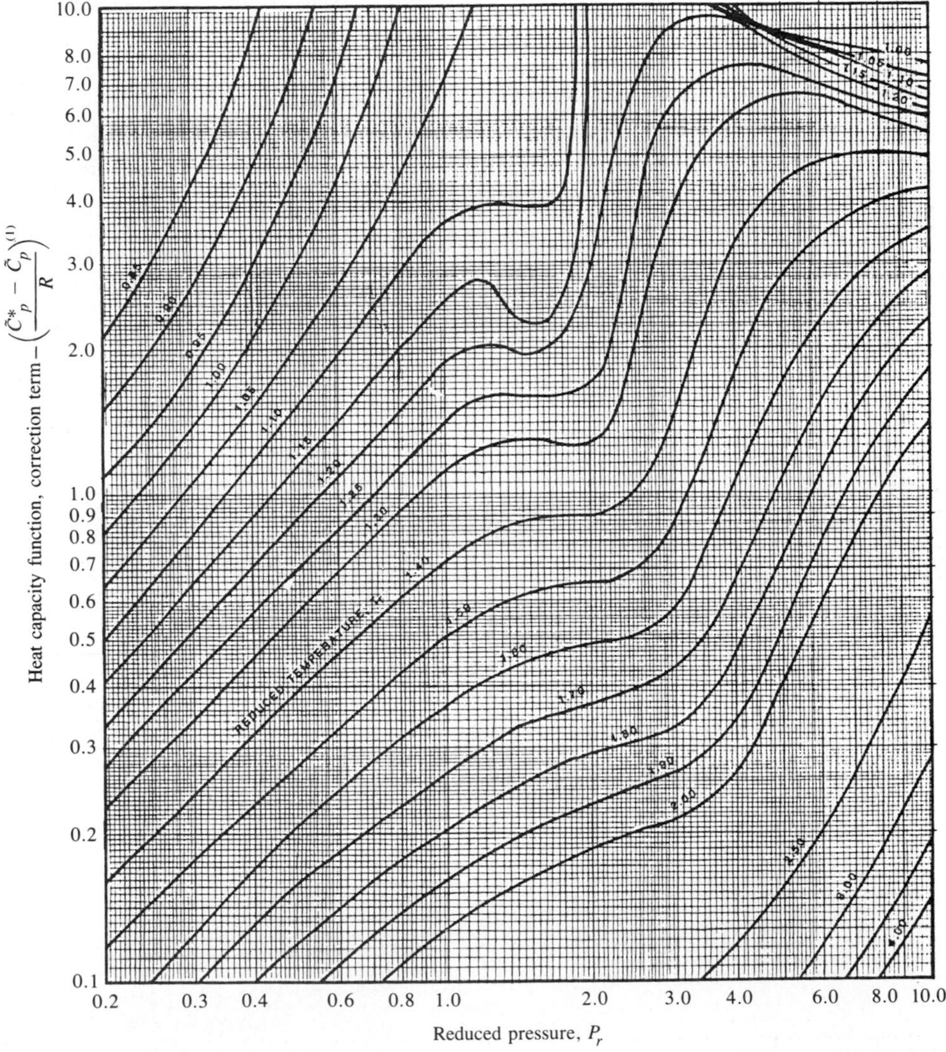

Figure 4.17 Effect of pressure on constant-pressure-gas heat capacity, correction term. *(Adapted from "Technical Data Book—Petroleum Refining," 4th ed., copyright 1982 by American Petroleum Institute. Used with permission.)*

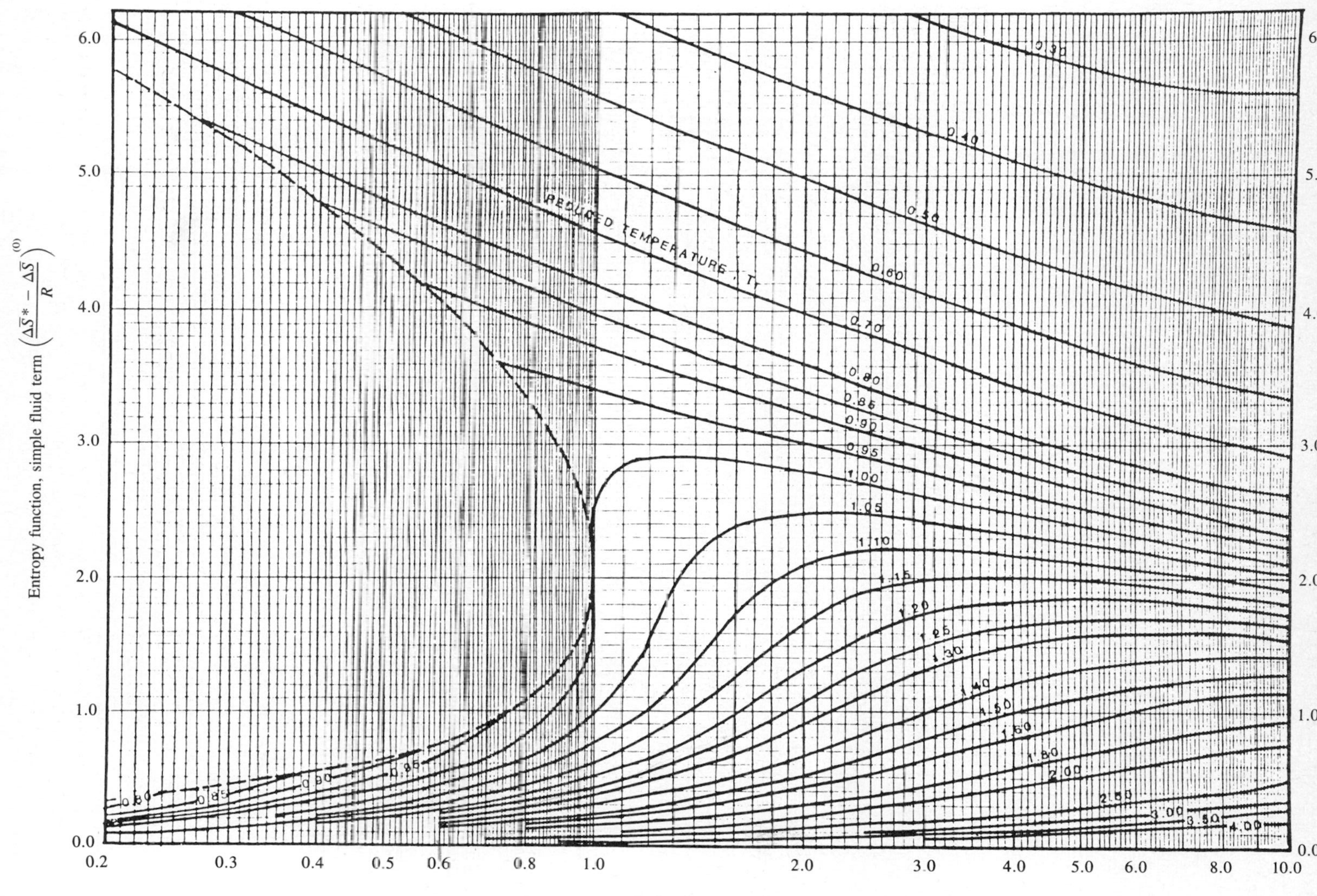

Figure 4.18 Effect of pressure on entropy, simple fluid term. *(Adapted from "Technical Data Book—Petroleum Refining," 4th ed., copyright 1982 by American Petroleum Institute. Used with permission.)*

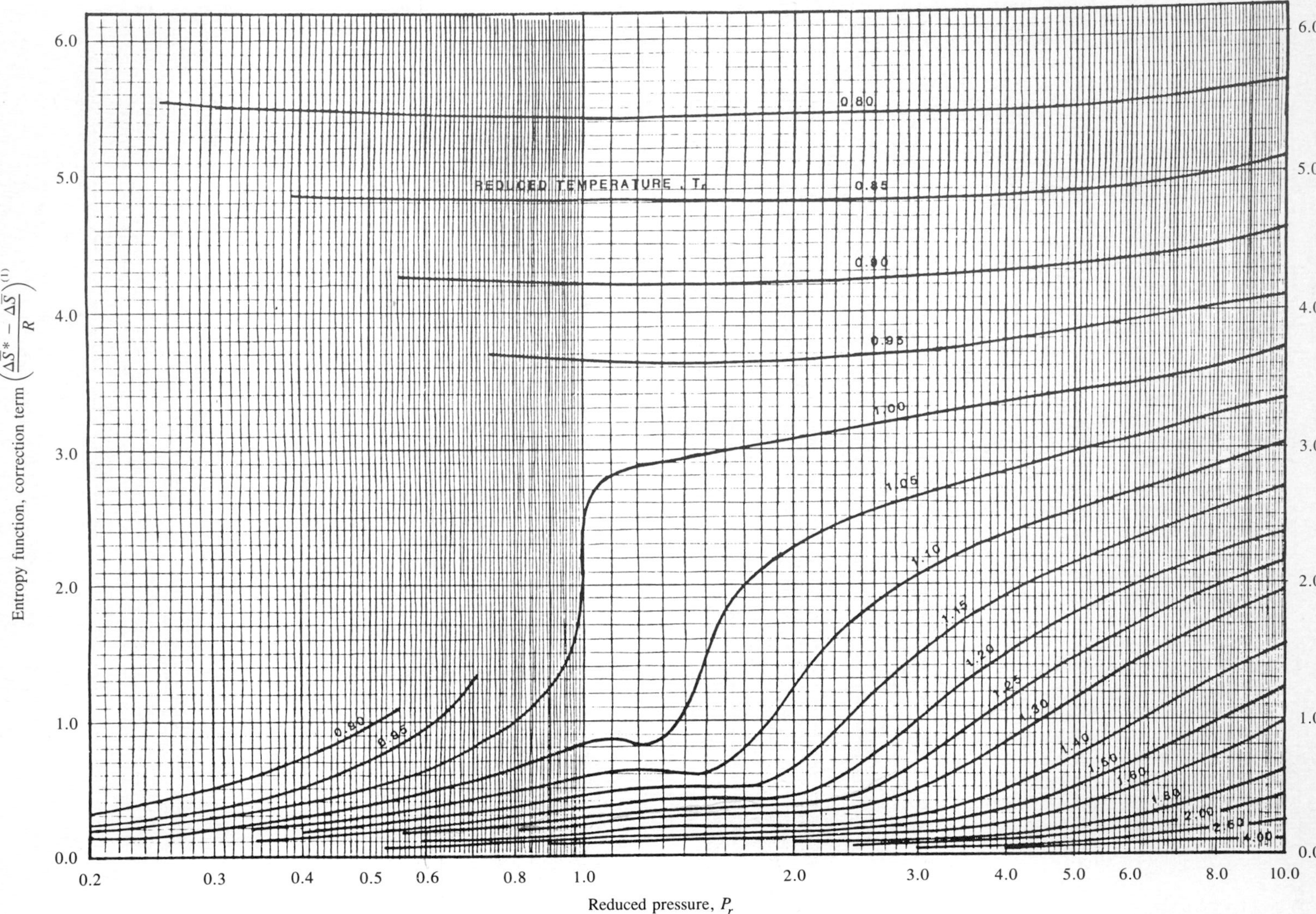

Figure 4.19 Effect of pressure on entropy, correction term. *(Adapted from "Technical Data Book—Petroleum Refining," 4th ed., copyright 1982 by American Petroleum Institute. Used with permission.)*

Heat capacity In the same manner as for enthalpy, the following equations can be derived for isobaric heat capacity.

Departure: $$\frac{\bar{C}_P^* - \bar{C}_P}{R} = \left(\frac{\bar{C}_P^* - \bar{C}_P}{R}\right)^{(0)} + \omega\left(\frac{\bar{C}_P^* - \bar{C}_P}{R}\right)^{(1)} \tag{4.73}$$

Property: $$C_P = C_P^* - \frac{R}{M}\frac{\bar{C}_P^* - \bar{C}_P}{R} \tag{4.74}$$

Figures 4.16 and 4.17 give the departures which are also available from the same source in tabular form. Errors in heat capacity rarely exceed 0.2 kJ/(kg · °C), except in the critical region where errors of 1.5 kJ/(kg · °C) or more may occur. Extrapolations to zero reduced pressure are again possible.

Similar expressions for isochoric heat capacity and for heat capacity ratio are available in the "Technical Data Book—Petroleum Refining."

Entropy Equations derived in the same manner as for enthalpy, taking into account the additional necessary variable that ideal-gas properties are dependent on pressure, follow:

Departure: $$\frac{\Delta\bar{S}^* - \Delta\bar{S}}{R} = \left(\frac{\Delta\bar{S}^* - \Delta\bar{S}}{R}\right)^{(0)} + \omega\left(\frac{\Delta\bar{S}^* - \Delta\bar{S}}{R}\right)^{(1)} + \ln P \tag{4.75}$$

where P = total pressure, kPa.

Property: $$S = S^* - \frac{R}{M}\frac{\Delta\bar{S}^* - \Delta\bar{S}}{R} \tag{4.76}$$

where S = total entropy referred to basis entropy of 1 kJ/(kg · K) for ideal gas at 0 K and 1 kPa

S^* = ideal-gas entropy

Figures 4.18 and 4.19 give the departures and tabular values that are available. As experimental data are normally not available for entropy, the method was checked as to its consistency with compressibility factor, enthalpy, and fugacity data with success. Extrapolations to reduced pressures of zero are again possible.

Example 4.19 Repeat Example 4.18, using the Pitzer technique.

$$T_r = 1.50 \qquad P_r = 1.13 \qquad \omega = 0.3448$$

From Fig. 4.14,

$$\left(\frac{\Delta\bar{H}^* - \Delta\bar{H}}{RT_c}\right)^{(0)} = 0.57$$

From Fig. 4.15,

$$\left(\frac{\Delta\overline{H}^* - \Delta\overline{H}}{RT_c}\right)^{(1)} \approx 0.05$$

Using Eq. (4.71),

$$\frac{\Delta\overline{H}^* - \Delta\overline{H}}{RT_c} = 0.57 + 0.3448(0.05) \approx 0.59$$

$$\Delta\overline{H}^* - \Delta\overline{H} = (0.59)(8.3143)(647.3)$$

$$= 3175 \text{ kJ/kmol}$$

If $\Delta H^* = 0$, then

$$\Delta\overline{H} = -3175 \text{ kJ/kmol}$$

The answer is essentially identical to the answer obtained by using the Lydersen method.

Exercise 4.R Calculate the enthalpy and entropy departures for ethane gas at 600°C and 6 MPa pressure from a base of 600°C and 0.1 MPa pressure by using Pitzer's techniques.

Free energy and internal energy Specific formulations for free energy and internal energy have not been derived. For the former, the use of fugacities, discussed in Chap. 7, may be used to calculate free energy, or the defining isothermal equation of $G = H - TS$ may be used. In the latter case, the defining equation $H = E + PV = E + ZRT$ or $E = H - ZRT$ can be easily applied.

Analytical representation of Pitzer-type method As discussed in Chap. 2 for compressibility factors, a modified Benedict-Webb-Rubin equation can be used to represent the departures by using the concept of a reference fluid. This method proposed by Lee and Kesler (1975) is presented for enthalpy, isobaric heat capacity, and entropy calculations below. Errors in using these methods are essentially equal to those for the graphical Pitzer method. Fugacity calculations by this method will be discussed in Chap. 7.

1. *Computer method for the enthalpy of pure-hydrocarbon liquids and real gases.* The following generalized correlation is recommended for calculating the pressure effect on the enthalpy of pure-hydrocarbon liquids and gases by a digital computer.

   The equation to be used is

$$\left(\frac{\overline{H}^\circ - \overline{H}}{RT_c}\right) = \left(\frac{\overline{H}^\circ - \overline{H}}{RT_{c,0}}\right)^{(0)} + \frac{\omega}{\omega^{(h)}}\left[\left(\frac{\overline{H}^\circ - \overline{H}}{RT_{c,h}}\right)^{(h)} - \left(\frac{\overline{H}^\circ - \overline{H}}{RT_{c,0}}\right)^{(0)}\right] \quad (4.77)$$

where $\dfrac{\overline{H}^\circ - \overline{H}}{RT_c}$ = dimensionless effect of pressure on enthalpy of fluid of interest

$\left(\dfrac{\overline{H}^\circ - \overline{H}}{RT_{c,0}}\right)^{(0)}$ = effect of pressure on enthalpy for simple fluid, to be calculated from Eq. (4.75)

$\left(\dfrac{\overline{H}^\circ - \overline{H}}{RT_{c,h}}\right)^{(h)}$ = effect of pressure on enthalpy for heavy reference fluid (n-octane), to be calculated from Eq. (4.75)

T_c = critical temperature of fluid for which enthalpy is desired, K

ω = acentric factor of fluid for which pressure effect on enthalpy is being sought

$\omega^{(h)}$ = acentric factor of heavy reference fluid = 0.3978

The dimensionless effects of pressure on the enthalpies of the simple fluid and the heavy reference fluid are to be calculated from the following equation:

$$\left(\frac{\overline{H}^\circ - \overline{H}}{RT_{c,i}}\right)^{(i)} = -T_r\left[z^{(i)} - 1 - \frac{b_2 + (2b_3/T_r) + (3b_4/T_r^2)}{T_r V_r} - \frac{c_2 - 3c_3/T_r}{2T_r V_r^2} + \frac{d_2}{5T_R V_r^5} + 3E\right] \tag{4.78}$$

where

$$E = \frac{c_4}{2T_r^3\gamma}\left[\beta + 1 - \left(\beta + 1 + \frac{\gamma}{V_r^2}\right)e^{-\gamma/V_r^2}\right]$$

$$\left(\frac{\overline{H}^\circ - \overline{H}}{RT_{c,i}}\right)^{(i)} = \begin{cases} \left(\dfrac{\overline{H}^\circ - \overline{H}}{RT_{c,0}}\right)^{(0)} & \text{when equation applied to simple fluid} \\ \left(\dfrac{\overline{H}^\circ - \overline{H}}{RT_{c,h}}\right)^{(h)} & \text{when equation applied to heavy reference fluid} \end{cases}$$

T_r = reduced temperature, T/T_c

T_c = critical temperature, K

T = temperature, K

$V_r = P_c V/RT_c$

$z^{(i)} = z^{(0)}$ or $z^{(h)}$, depending on whether equation is applied to simple fluid or heavy reference fluid

$z^{(0)}$ = compressibility factor of simple fluid

$z^{(h)}$ = compressibility factor of heavy reference fluid

b_2, b_3, b_4, c_2, c_3, c_4, d_2, γ, and β are two sets of constants, one set for the simple fluid and another for the heavy reference fluid. The values of the constants are given in Chap. 2.

With the term for dimensionless pressure effect on enthalpy from Eq. (4.76), the total enthalpy for the fluid of interest may be found from the equation

$$H = H^\circ - \frac{RT_c}{M}\frac{\overline{H}^\circ - \overline{H}}{RT_c} \tag{4.79}$$

where H = total enthalpy, in kJ/kg, referred to basis enthalpy of 0 kJ/kg for ideal gas at 0 K; $\overline{H}$ is analogous molar quantity
H° = ideal gas enthalpy, kJ/kg; $\overline{H}^\circ$ is analogous molar quantity
R = gas constant = 8.3140 kJ/(kmol·K)
M = molecular weight

2. *Computer method for the heat capacity of pure real gases*. The following generalized method is recommended for computer calculations of the pressure effect on the isobaric heat capacity of pure-hydrocarbon gases.

The equation to be used is

$$\frac{\overline{C}_P^\circ - \overline{C}_P}{R} = \left(\frac{\overline{C}_P^\circ - \overline{C}_P}{R}\right)^{(0)} + \frac{\omega}{\omega^{(h)}}\left|\left(\frac{\overline{C}_P^\circ - \overline{C}_P}{R}\right)^{(h)} - \left(\frac{\overline{C}_P^\circ - \overline{C}_P}{R}\right)^{(0)}\right| \tag{4.80}$$

where $\dfrac{\overline{C}_P^\circ - \overline{C}_P}{R}$ = dimensionless effect of pressure on isobaric heat capacity

$\left(\dfrac{\overline{C}_P^\circ - \overline{C}_P}{R}\right)^{(0)}$ = effect of pressure on isobaric heat capacity for simple fluid, to be calculated from Eq. (4.78)

$\left(\dfrac{\overline{C}_P^\circ - \overline{C}_P}{R}\right)^{(h)}$ = effect of pressure on isobaric heat capacity for heavy reference fluid (n-octane), to be calculated from Eq. (4.78)

ω = acentric factor of gas for which pressure effect on isobaric heat capacity is sought

$\omega^{(h)}$ = acentric factor of heavy reference fluid = 0.3978

The dimensionless effect of pressure on the isobaric heat capacity of the simple fluid and the heavy reference fluid is to be calculated from the following equation:

$$\left(\frac{\overline{C}_P^\circ - \overline{C}_P}{R}\right)^{(i)} = 1 + T_r\frac{(\partial P_r/\partial T_r)^2_{V_r}}{(\partial P_r/\partial V_r)_{T_r}} + \left(\frac{\Delta\overline{C}_V}{R}\right)^{(i)} \tag{4.81}$$

where

$$\left(\frac{\partial P_r}{\partial T_r}\right)_{V_r} = \frac{1}{V_r}\left\{1 + \frac{b_1 + (b_3/T_r^2) + (2b_4/T_r^3)}{V_r} + \frac{c_1 - 2c_3/T_r^3}{V_r^2} + \frac{d_1}{V_r^5} - \frac{2c_4}{T_v^3 V_r^2}\left[\left(\beta + \frac{\gamma}{V_r^2}\right)e^{-\gamma/V_r^2}\right]\right\}$$

$$\left(\frac{\partial P_r}{\partial V_r}\right)_{T_r} = -\frac{T_r}{V_r^2}\left(1 + \frac{2B}{V_r} + \frac{3C}{V_r^2} + \frac{6D}{V_r^5} + \frac{c_4}{T_r^3 V_r^2} \times \left\{3\beta + \left[5 - 2\left(\beta + \frac{\gamma}{V_r^2}\right)\right]\frac{\gamma}{V_r^2}\right\} e^{-\gamma/V_r^2}\right)$$

$$\left(\frac{\Delta \bar{C}_V}{R}\right)^{(i)} = \begin{cases} \left(\dfrac{\bar{C}_V{}^\circ - \bar{C}_V}{R}\right)^{(0)} & \text{when equation applied to simple fluid} \\ \left(\dfrac{\bar{C}_V^\circ - \bar{C}_V}{R}\right)^{(h)} & \text{when equation applied to heavy reference fluid} \end{cases}$$

T_r = reduced temperature, T/T_c
T = temperature, K
T_c = critical temperature of fluid whose heat capacity is desired, K
$V_r = P_c V/RT_c$
$B = b_1 - b_2/T_r - b_3/T_r^2 - b_4/T_r^3$
$C = c_1 - c_2/T_r + c_3/T_r^3$
$D = d_1 + d_2/T_r$

Note that b_1, b_2, b_3, b_4, c_1, c_2, c_3, d_1, d_2, c_4, γ, and β are two sets of constants, one for the simple fluid and another for the heavy reference fluid. The values of the constants are given in Chap. 2.

With the dimensionless effect of pressure on the isobaric heat capacity from Eq. (4.80), the total heat capacity may be calculated from the equation

$$C_P = C_P^\circ - \frac{R}{M}\left(\frac{\bar{C}_P^\circ - \bar{C}_P}{R}\right) \tag{4.82}$$

where C_P = isobaric heat capacity of real gas, kJ/(kg·C); $\bar{C}_P$ is analogous molar quantity
C_P° = isobaric heat capacity of ideal gas, kJ/(kg·C); $\bar{C}_P$ is analogous molar quantity
R = gas constant = 8.3140 kJ/(kmol·K)
M = molecular weight

Table 4.2 Free-energy functions, standard heat of formation, and ideal-gas enthalpies

		$(G_T{}^\circ - H_0{}^\circ)/T$, kJ/(kmol · K)						
		298.15 K	400 K	500 K	600 K	800 K	1000 K	1500 K
Methane	IG	152.67	162.67	170.62	177.44	189.20	199.45	221.29
Ethane	IG	189.28	201.67	212.21	221.84	239.37	255.31	290.20
Propane	IG	220.75	236.48	250.41	263.47	287.73	309.99	359.15
n-Butane	IG	245.31	265.94	284.34	301.54	333.46	376.56	427.19
Isobutane	IG	235.22	254.68	272.34	289.11	320.49	349.53	413.63
n-Pentane	IG	268.36	294.09	316.90	338.07	377.82	413.80	
Isopentane	IG	269.83	293.72	315.39	355.98	374.47	410.45	
Neopentane	IG	228.15	252.42	274.68	295.81	335.56	371.96	
n-Hexane	IG	292.46	323.05	350.12	375.30	422.17	465.26	
Isohexane	IG	292.55	321.12	347.10	371.96	417.98	460.66	
Ethylene	IG	184.1	195.1	204.0	212.2	226.8	239.8	267.7
Propylene	IG	221.6	236.0	248.3	259.7	280.6	299.6	340.9
1-Butene	IG	248.0	266.4	282.6	297.8	325.8	351.3	407.2
Isobutylene	IG	236.4	254.9	271.1	286.4	314.6	340.3	396.2
H_2S	IG	−172.21	−182.06	−189.64	−195.93	−206.15	−214.41	−230.43
SO_2	IG	212.81	223.47	231.93	239.13	251.13	260.99	280.15
SO_3	IG	217.25	229.34	239.35	248.14	263.23	275.97	301.31
O_2	IG	176.0	184.6	191.1	196.6	205.3	212.2	225.2
H_2	IG	102.2	110.7	117.0	122.2	130.5	137.0	149.0
H_2O	IG	155.6	165.3	172.8	179.0	188.9	196.8	211.9
N_2	IG	162.5	171.0	177.5	182.9	191.3	198.0	210.5
NH_3	IG	159.0	169.0	176.9	183.5	194.6	203.8	222.4
CO	IG	168.6	177.1	183.6	189.0	197.5	204.2	216.8
CO_2	IG	182.4	191.9	199.5	206.1	217.3	226.5	244.8
C(mon)	IG	136.1	142.5	147.3	151.3	157.4	162.1	170.7

Source: Abstracted from "Selected Values of Properties of Hydrocarbons and Related Compounds," Thermodynamics Research Center, Texas A & M University, College Station, Tex. (extant June 1983).

3. *Computer method for the entropy of pure-hydrocarbon liquids and real gases.* The following generalized correlation is recommended for calculating the entropy of pure-hydrocarbon liquids and gases by a digital computer.

Three other procedures, those for density, fugacity, and enthalpy, are required to use this method. The results are combined by using the following equation:

$$\frac{\bar{S}^\circ - \bar{S}}{R} = \frac{\bar{H}^\circ - \bar{H}}{RT} + \ln\frac{f}{P} + \ln P \tag{4.83}$$

where $\dfrac{\bar{S}^\circ - \bar{S}}{R}$ = dimensionless effect of pressure on entropy

$\dfrac{\bar{H}^\circ - \bar{H}}{RT}$ = dimensionless effect of pressure on enthalpy from Eq. (4.76)

$\ln(f/P)$ = dimensionless fugacity term from Chap. 7

P = pressure, kPa

$-\Delta H_{f0}$, (kJ/kmol) (10^{-3})	$H_T^\circ - H_0^\circ$, [kJ/(kmol · K)] (10^{-3})						
	298.15 K	400 K	500 K	600 K	800 K	1000 K	1500 K
66.505	10.016	13.887	18.238	23.192	34.819	48.493	89.286
68.20	11.874	17.874	25.058	33.430	53.220	76.358	144.139
81.67	14.740	23.276	33.623	45.689	74.182	107.194	203.217
97.15	19.276	30.648	44.350	60.250	97.613	140.666	265.224
105.56	17.933	29.200	42.928	58.911	96.483	139.829	264.805
115.02	24.184	38.074	54.810	74.433	120.499	174.054	
119.50	22.008	35.857	52.718	72.383	118.826	172.799	
135.19	23.179	38.032	55.522	75.722	123.846	179.912	
130.04	28.702	45.187	65.187	88.492	143.093	206.271	
135.39	26.317	42.970	63.137	86.734	142.256	206.271	
−60.79	10.570	15.534	21.420	28.180	43.869	61.785	113.4
−35.45	13.550	20.888	29.620	39.734	63.418	90.794	169.8
−20.76	17.213	27.142	39.139	52.953	85.269	122.44	229.6
−4.10	17.087	27.301	39.407	53.372	85.771	122.94	230.2
74.1	9.955	13.494	17.118	20.908	29.014	37.799	62.065
294.3	10.578	14.847	19.384	24.205	34.472	45.297	73.683
390.0	11.704	17.255	23.337	29.903	44.054	59.080	98.597
0	8.684	11.711	14.772	17.936	24.528	31.400	49.309
0	8.471	11.432	14.355	17.286	23.178	29.161	44.777
239.0	9.907	13.359	16.831	20.408	27.906	35.899	58.022
0	8.673	11.645	14.588	17.573	23.726	30.148	47.097
38.97	10.051	13.831	17.874	22.244	31.910	42.701	73.661
113.9	8.676	11.653	14.610	17.622	23.859	30.375	47.542
393.3	9.369	13.374	17.679	22.282	32.187	42.784	71.106
0	6.540	8.662	10.743	12.824	16.986	21.146	31.549

The total entropy is found by using the dimensionless pressure effect on entropy with the equation

$$S = S^\circ - \frac{R}{M}\left(\frac{\bar{S}^\circ - \bar{S}}{R}\right) \tag{4.84}$$

where S = total entropy referred to basis entropy of 1 kJ/(kg·K) for ideal gas at 0 K and 1 kPa, kJ/(kg·K); $\bar{S}$ is the analogous molar quantity

S° = ideal gas entropy, kJ/(kg·K); $\bar{S}^\circ$ is analogous molar quantity

R = gas constant = 8.3140 kJ/(kmol·K)

M = molecular weight

Subroutines for implementing the procedures for enthalpy departures, isobaric and isochoric heat capacity departures, and entropy departures are given in App. E.

4.5.4 Free-Energy Functions and Tabular Data

Free-energy data have been tabulated (1) as values at an arbitrary temperature, usually 25°C, which can then be converted to other temperatures with known heat capacities, (2) as tables of heats of formation and absolute entropies at a selected base temperature which can be combined to calculate free energies, after which heat capacities can be used to convert the values to any temperature, and (3) as free-energy–enthalpy differences at many temperatures, the free-energy functions, in their standard states, i.e., relative to the elements and their conditions.

The standard free energy of formation of any compound at any temperature can be expressed as

$$\left(\frac{\Delta G_f^\circ}{T}\right)_T = \left(\frac{G_T^\circ - H_0^\circ}{T} + \frac{\Delta H_{f0}^\circ}{T}\right)_{\text{comp}} - \sum\left(\frac{G_T^\circ - H_0^\circ}{T}\right)_{\text{elem}} \tag{4.85}$$

where G_T° = molar free energy in ideal-gas state at temperature T

H_0° = molar enthalpy of compound or element at 0 K; H_T° at temperature T

ΔH_{f0}° = standard molar heat of formation at 0 K (zero for elements)

Since

$$\Delta H_T = \sum_{\text{Pr}}\left[(H_T^\circ - H_0^\circ) + H_{f0}^\circ\right] - \sum_{\text{Re}}\left[(H_T^\circ - H_0^\circ) + \Delta H_{f0}^\circ\right] \tag{4.86}$$

$$\frac{\Delta G^\circ}{T} = \sum_{\text{Pr}}\left[\frac{G_T^\circ - H_0^\circ}{T} + \frac{\Delta H_{f0}^\circ}{T}\right] - \sum_{\text{Re}}\left[\frac{G_T^\circ - H_0^\circ}{T} + \frac{\Delta H_{f0}^\circ}{T}\right] \tag{4.87}$$

All equations refer to all compounds and elements in their standard states.

Table 4.2 lists representative values necessary for use of these equations which will become quite useful for equilibrium calculations in Chap. 9.

Exercise 4.S Determine the free-energy change for the reaction of ethylene and hydrogen to produce ethane at 1000 K by using free-energy functions.

4.6 PRACTICAL THERMODYNAMIC CYCLES

In Sec. 4.3 ideal cycles which are the basis of practical processes were discussed in some detail and provided a limit in work-producing ability for any cyclical process. All practical cycles have a lower efficiency. This section will attempt to discuss idealized cycles of two types important to chemical engineers: open cycles, where the working fluid passes through the process (such as the gasoline and diesel engine, gas turbine, and jet engine), and the closed cycle, where the working fluid is contained withing the cycle (such as certain turbines, steam engines, and vapor compression cycles used for refrigeration). The case of air liquefaction which introduces the Joule-Thompson coefficient is also discussed.

4.6.1 The Otto Cycle—The Gasoline Engine

An air-standard Otto cycle consists of six steps: intake of air, compression, ignition, power, expansion, and exhaust. The first and sixth steps are normally neglected as they occur at constant pressure. The other four steps are normally taken as the cycle operating on air which is assumed to be an ideal gas. The cycle shown on a *PV* diagram is shown in Fig. 4.20.

AB = adiabatic compression

BC = ignition with air addition at constant volume

CD = adiabatic power expansion

DA = expansion with air rejection at constant volume

The compression (AB) and expansion (CD) steps are taken to be adiabatic, while the heat addition (BC) Q_1 and heat removal (DA) Q_2 are assumed to be at constant volume.

Thus, $Q_1 = C_V(T_C - T_B)$ and $Q_2 = C_V(T_A - T_D)$. Since the change in internal energy is zero over the cycle, then $Q_1 + Q_2 = W_{AB} + W_{CD}$. Therefore, the

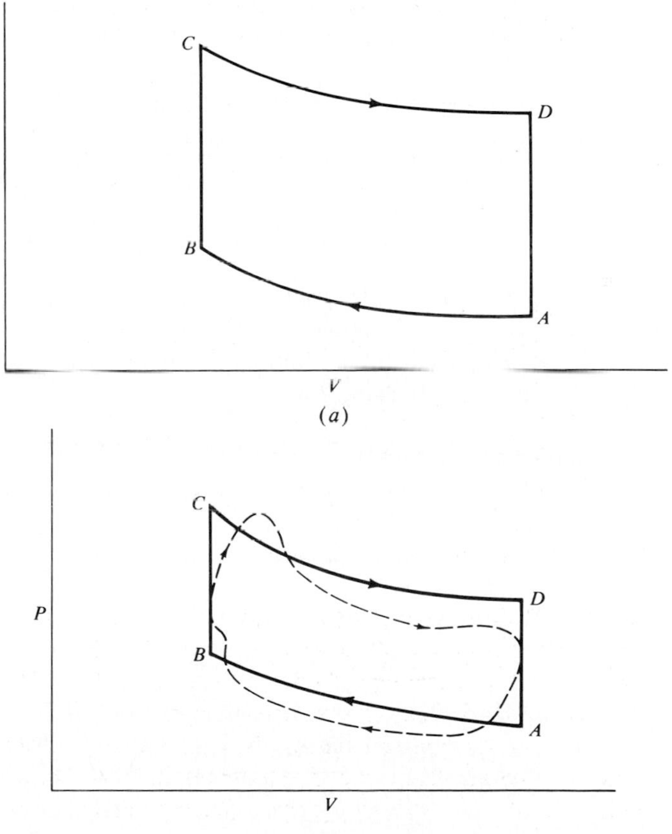

Figure 4.20 The Otto cycle. (*a*) Ideal; (*b*) actual.

efficiency of the cycle is

$$\eta_0 = \frac{W_{\text{cycles}}}{Q_1} = \frac{Q_1 + Q_2}{Q_1}$$

Substituting from above,

$$\eta_0 = 1 - \frac{T_D - T_A}{T_C - T_B} \tag{4.88}$$

Usually internal-combustion engines are compared by using the compression ratio r, that is, the ratio of the volume at A to the volume at B, $r = V_A/V_B$.

Using the relationships for adiabatic processes developed in Chap. 2,

$$T_B = T_A\left(\frac{V_A}{V_B}\right)^{k-1} = T_A r^{k-1}$$

$$T_C = T_B\frac{P_C}{P_B} = T_A r^{k-1}\frac{P_C}{P_B}$$

$$T_D = T_C\left(\frac{V_B}{V_A}\right)^{k-1} = T_C\frac{1}{r^{k-1}} = T_A\frac{P_C}{P_B}$$

If these are substituted into Eq. (4.88) for the efficiency of the Otto cycle,

$$\eta_0 = 1 - \frac{1}{r^{k-1}} \tag{4.89}$$

Thus, the efficiency only depends on the compression ratio. The net shaft work done in one cycle can then be calculated by

$$W_s = \eta_0 n C_V(T_C - T_B) = \eta_0 n\frac{R}{k-1}(T_C - T_B) \tag{4.90}$$

Since this is an idealized analysis, an actual spark-ignition engine will not work quite the same, as shown by the dashed line in Fig. 4.20. However, the approximation is valid for preliminary calculations. A detailed analysis of such engines is given by Hougen, Watson, and Ragatz (1959).

Example 4.20 Consider an air-standard Otto cycle operating on 5 kg of air with inlet conditions of 0.08 MPa and 37°C. A compression ratio of 10 is used and 500 kJ of heat is added during ignition. Determine Q, W, P, and T for each step of the process and the overall efficiency of the process.

COMPRESSION STEP $A \rightarrow B$

$$T_A = 310\text{ K} \qquad P_A = 80\text{ kPa} \qquad Q_{AB} = 0 \qquad \text{(adiabatic)}$$

$$T_B = T_A r^{k-1} = (310)(10)^{0.4} = 778.7\text{ K}$$

$$V_A = \frac{RT_A}{P_A} = \frac{(8.3143)(310)}{80} = 32.22\text{ m}^3/\text{kmol}$$

$$r = \frac{V_A}{V_B}$$

Therefore

$$V_B = \frac{32.22}{10} = 3.222 \text{ m}^3/\text{kmol}$$

$$P_B = \frac{RT_B}{V_B} = \frac{(8.3143)(778.7)}{3.222} = 2009.4 \text{ kPa}$$

$$W_{AB} = \frac{-nRT_A}{k-1}\left[\left(\frac{P_B}{P_A}\right)^{(k-1)/k} - 1\right]$$

$$= \frac{\frac{5}{29}(8.3143)(310)}{0.4}\left[\left(\frac{2009.4}{80}\right)^{0.2857} - 1\right]$$

$$= -1679.5 \text{ kJ}$$

IGNITION STEP $B \rightarrow C$

$$V_B = V_C = 3.222 \text{ m}^3/\text{kmol} \qquad Q_{BC} = 500 \text{ kJ} \qquad W_{BC} = 0$$

Since $Q_{BC} = nC_V(T_C - T_B)$,

$$T_C - T_B = \frac{500}{\frac{5}{29}(20.93)} = 138.6$$

$$T_C = 917.3 \text{ K}$$

$$P_C = \frac{RT_C}{V_C} = \frac{(8.3143)(917.3)}{3.222} = 2367.1 \text{ kPa}$$

POWER STEP $C \rightarrow D$

$$Q_{CD} = 0 \text{ (adiabatic)}$$

$$T_D = T_C\frac{1}{r^{k-1}} = 917.3\left(\frac{1}{10^{0.4}}\right) = 365.2 \text{ K}$$

$$V_D = V_A = 32.22 \text{ m}^3/\text{kmol}$$

$$\frac{P_C V_C}{T_C} = \frac{P_D V_D}{T_D}$$

Thus

$$P_D = \left(\frac{T_D}{T_C}\right)\left(\frac{V_C}{V_D}\right)P_C = \left(\frac{365.2}{917.3}\right)\left(\frac{3.222}{32.22}\right)(2367.1)$$

$$= 94.2 \text{ kPa}$$

$$W_{CD} = \frac{-nRT_C}{k-1}\left[\left(\frac{P_D}{P_C}\right)^{(k-1)/k} - 1\right] = +1978.7 \text{ kJ}$$

EXPANSION STEP $D \rightarrow A$

$$W_{DA} = 0$$

$$Q_{DA} = nC_V(T_A - T_D) = \frac{5}{29}(20.93)(310 - 365.2)$$

$$= -199.2 \text{ kJ}$$

CHECKING

$$Q_{BC} + Q_{DA} = W_{AB} + W_{CD}$$
$$500 - 199.2 = -1679.5 + 1978.7$$
$$300.8 = 299.2 \qquad \text{(within calculational tolerance)}$$

EFFICIENCY FROM EQ. (4.89)

$$\eta_0 = 1 - \frac{1}{r^{k-1}} = 1 - \frac{1}{10^{0.4}} = 0.602$$

or

$$\eta_0 = \frac{Q_{BC} + Q_{DA}}{Q_{BC}} = \frac{300.8}{500} = 0.602$$

Exercise 4.T Prove Eq. (4.89). For an eight-cylinder engine with total displacement of 5.0×10^{-3} m^3, determine what compression ratio would be required to deliver 1 kJ of work/cycle. Assume a temperature rise on ignition of 2500°C.

4.6.2 The Compression-Ignition Cycle—The Diesel Engine

A compression-ignition engine takes air into an engine and compresses it to ignition conditions, after which the fuel is sprayed into the combustion chamber by a pump. This differs from the standard internal-combustion engine in that the carburetor is replaced by an injection pump and the ignition is direct. Full air is always available as there is no throttle. These engines comprise two types: the *diesel cycle*, where isentropic compression and expansion are combined with fuel injection at constant pressure and combustion at constant volume, and the *dual cycle*, where part of the combustion occurs at constant volume before the constant-pressure portion. The dual cycle is usually more efficient.

The air-standard diesel cycle This ideal cycle is shown on a PV diagram in Fig. 4.21 with the actual cycle shown as a dashed line for comparison.

The compression (AB) and expansion (CD) steps are adiabatic, while the injection and combustion (BC) is at constant pressure with absorption of heat Q_1, and the rejection (DA) is at constant volume with production of heat Q_2. The efficiency of such a cycle is

$$\eta_D = \frac{W_{\text{net}}}{Q_1} = \frac{W_{AB} + W_{BC} + W_{CD}}{Q_1}$$

Thus

$$\eta_D = \frac{n(E_C - E_D) + P_B(V_C - V_B) - n(E_B - E_A)}{n(H_C - H_B)}$$

or

$$\eta_D = \frac{nC_V(T_C - T_D) + nR(T_C - T_B) - nC_V(T_B - T_A)}{nC_P(T_C - T_B)}$$

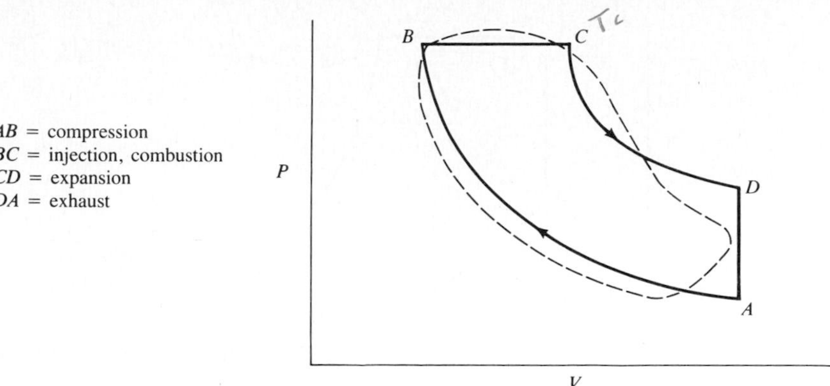

Figure 4.21 Air-standard diesel cycle.

Reducing the equation,

$$\eta_D = 1 - \frac{1}{k}\frac{T_D - T_A}{T_C - T_B} \tag{4.91}$$

The compression ratio $r_C = V_A/V_B$, unlike the Otto cycle, differs from the expansion ratio $r_e = V_D/V_C$. Defining the cutoff or load ratio $r_l = V_C/V_B$, then $r_C = r_l r_e$. Temperature relations for the adiabatic processes can be used to determine that $T_B = T_A r_C^{k-1}$, $T_C = T_A r_C^{k-1} r_l$, and $T_D = T_A r_l^k$. Substituting these expressions in the efficiency definition,

$$\eta_D = 1 - \frac{1}{r_C^{k-1}} \frac{r_l^k - 1}{k(r_l - 1)} \tag{4.92}$$

Then the net shaft work per cycle can be calculated as

$$W_S = n\eta_D C_P (T_C - T_B) = n\eta_D R \frac{k}{k-1}(T_C - T_B) \tag{4.93}$$

where

$$n = \frac{P_{\text{atm}}(V_A - V_B)}{RT_{\text{atm}}}$$

Equation (4.92) shows that at a given compression ratio the efficiency of a diesel cycle is different from that of the Otto cycle only by the parenthetical term which is always greater than unity. Thus, in general $\eta_0 > \eta_D$ for the same compression ratio. However, diesel engines are normally operated at compression ratios of 13 to 17 rather than the ratio of 6 to 8 for the gasoline engine to alleviate this problem.

The dual cycle The dual cycle is shown on a PV diagram in Fig. 4.22. Similar relations for efficiency can be derived. In this case, BB' and $B'C$ are both parts of the fuel-injection step. The other steps are analogous to the diesel cycle.

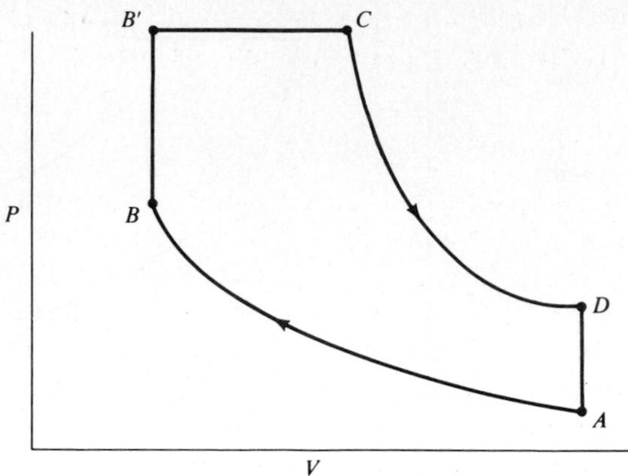

Figure 4.22 Dual cycle.

Example 4.21 Consider an air-standard diesel cycle operating on 5 kg of air with inlet conditions of 0.08 MPa and 37°C. For the same efficiency as the Otto cycle in Example 4.20, determine the necessary compression ratio if T_C and T_D are as calculated in the previous example.

SOLUTION

$$T_A = 310 \text{ K} \qquad T_C = 917.3 \text{ K} \qquad T_D = 365.2 \text{ K}$$

$$n = 0.602$$

From Eq. (4.91),

$$\eta_D = 1 - \frac{1}{k}\frac{T_D - T_A}{T_C - T_B}$$

$$0.602 = 1 - \frac{1}{1.4}\frac{365.2 - 310}{917.3 - T_B}$$

Therefore

$$T_B = 818.2 \text{ K}$$

$$r_C = \frac{V_A}{V_B}$$

Therefore

$$\frac{T_B}{T_A} = r_C^{k-1} \qquad \text{(adiabatic)}$$

$$\frac{818.2}{310} = r_C^{0.4}$$

$$r_C = 734 \qquad \text{(not practical)}$$

Discuss how you might operate this cycle practically to obtain this efficiency.

Exercise 4.U Compare the efficiency of an air-standard Otto cycle, an air-standard diesel cycle, and a dual cycle if each cycle operates at the same compression ratio and the isobaric expansion portion of the fuel-injection step is one-third the length of the total expansion.

4.6.3 Joule and Brayton Cycles—The Gas Turbine

A gas turbine cycle consists of the isentropic compression of air in a compressor (AB), constant-pressure heating (BC) in a combustion chamber, isentropic expansion through a turbine (CD), and constant-pressure cooling of the effluent (DA) as shown in Fig. 4.23. If such a cycle is operated in a closed loop, it is termed a *Joule cycle*; if operated as an open system, it is termed a *Brayton cycle*. A plot of the ideal cycle is also shown in Fig. 4.23 on PV coordinates, with an actual cycle shown by dashed lines.

The efficiency of such a cycle is defined as

$$\eta_J = \frac{W_{\text{net}}}{\Delta H_C} \tag{4.94}$$

where W_{net} = net shaft work
ΔH_C = heat of combustion of fuel = Q

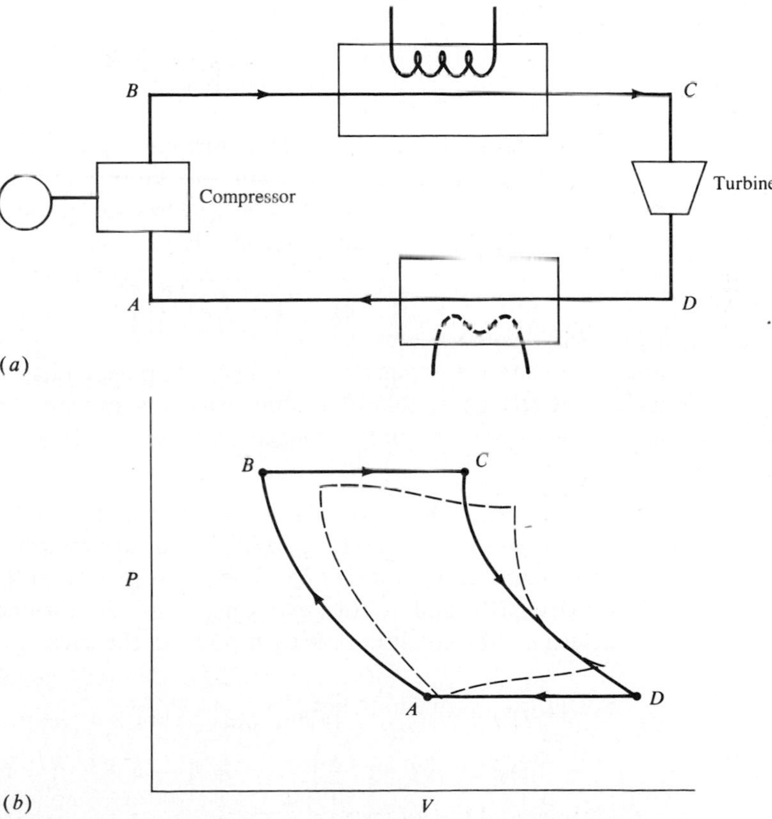

Figure 4.23 Gas turbine cycle. (a) Schematic diagram; (b) PV diagram.

While only strictly correct for the closed Joule cycle, this ratio can be used for the open Brayton cycle with little error as the changes in properties of the air stream resulting from fuel injection are minimal. For the Brayton cycle,

$$W_{\text{net}} = W_{\text{turbine}} + W_{\text{compressor}}$$
$$= (H_C - H_D) - (H_B - H_A)$$
$$Q = H_C - H_B$$

Therefore,

$$\eta_J = \frac{(H_C - H_D) - (H_B - H_A)}{H_C - H_B}$$

If heat capacities can be assumed to be equal,

$$\eta_J = \frac{(T_C - T_D) - (T_B - T_A)}{T_C - T_B} = 1 - \frac{T_D - T_A}{T_C - T_B} \tag{4.95}$$

Since the pressure ratios are equal for the compression and expansion steps,

$$\frac{P_B}{P_A} = \frac{P_C}{P_D} \qquad \text{and} \qquad \frac{T_B}{T_A} = \frac{T_C}{T_D} = \left(\frac{P_B}{P_A}\right)^{(k-1)/k}$$

Thus, substituting,

$$\eta_J = 1 - \left(\frac{P_B}{P_A}\right)^{(1-k)/k} \tag{4.96}$$

Therefore, ideally the efficiency of the turbine depends only on the pressure ratio and not on the temperature. This will only be true for the standard cycle. If the expanded gas is not discharged but is used to preheat the air discharged from the compressor, derivation of the efficiency shows

$$\eta_{J_R} = 1 - \frac{T_A}{T_C}\left(\frac{P_B}{P_A}\right)^{(k-1)/k}$$

and is temperature-dependent as well as being dependent on the pressure ratio in an inverse relation to that of a nonregenerative process. Actual turbine operation and efficiencies are given by Hougen, Watson, and Ragatz (1959).

Example 4.22 Determine the ideal thermal efficiency of a gas turbine operating in air at 25°C and 0.1 MPa with a turbine inlet temperature of 900°K and a compression ratio of 4.0. The pressure drop in the combustion chamber is 0.01 MPa and in the exit line from the turbine is 0.005 MPa. Also determine the conditions at each point in the cycle.

SOLUTION From Eq. (4.96),

$$\text{Ideal } \eta_J = 1 - \left(\frac{P_B}{P_A}\right)^{(1-k)/k} = 1 - 4^{-0.4/1.4} = 1 - 0.673 = 0.327$$

$$\text{Actual } \eta_J = \frac{(H_C - H_D) - (H_B - H_A)}{H_C - H_B}$$

At point A:

$$T_A = 298 \text{ K} \qquad P_A = 0.1 \text{ MPa}$$

At point B:

$$P_B = 0.4 \text{ MPa} \qquad T_B = T_A\left(\frac{P_B}{P_A}\right)^{(k-1)/k}$$

If adiabatic,

$$T_B = (298)(4)^{0.2857} = 442.8 \text{ K}$$

At point C:

$$P_C = 0.4 - 0.01 = 0.39 \text{ MPa} \qquad T_C = 900 \text{ K}$$

At point D:

$$P_D = 0.1 + 0.005 = 0.105 \text{ MPa}$$

If adiabatic,

$$T_D = T_C\left(\frac{P_D}{P_C}\right)^{(k-1)/k}$$

$$= 900\left(\frac{0.105}{0.39}\right)^{0.2857} = 618.6 \text{ K}$$

To determine the actual efficiency, the efficiency of each part of the cycle would need to be known, the actual temperatures calculated, and the enthalpy of air at each of these conditions used.

Exercise 4.V Draw the Joule-Brayton cycle on a TS diagram and compare the cycle with similar diagrams drawn for the Otto and diesel cycles. Repeat on HS diagrams.

4.6.4 Jet Engines—The Turbojet

The turbojet engine is a modified gas turbine engine where the compressor-turbine combination does not provide net mechanical work but only provides useful output as kinetic energy and pressure of the exhaust which causes thrust by expansion through a nozzle. Diagrammatically the turbojet is shown in Fig. 4.24 together with the idealized process on a PV diagram.

The dashed line shows the direction of the process if an afterburner is not utilized. The internal or thermal efficiency of the cycle is defined as

$$\eta_T = \frac{\Delta E_k}{Q_{\text{fuel}}} = \frac{\left(u_G^2 - u_A^2\right)/2}{Q_{\text{fuel}}} \tag{4.97}$$

where ΔE_k = change in kinetic energy
Q_{fuel} = heating value of fuel

The propulsion efficiency is the ratio of the propulsion work to the increase in

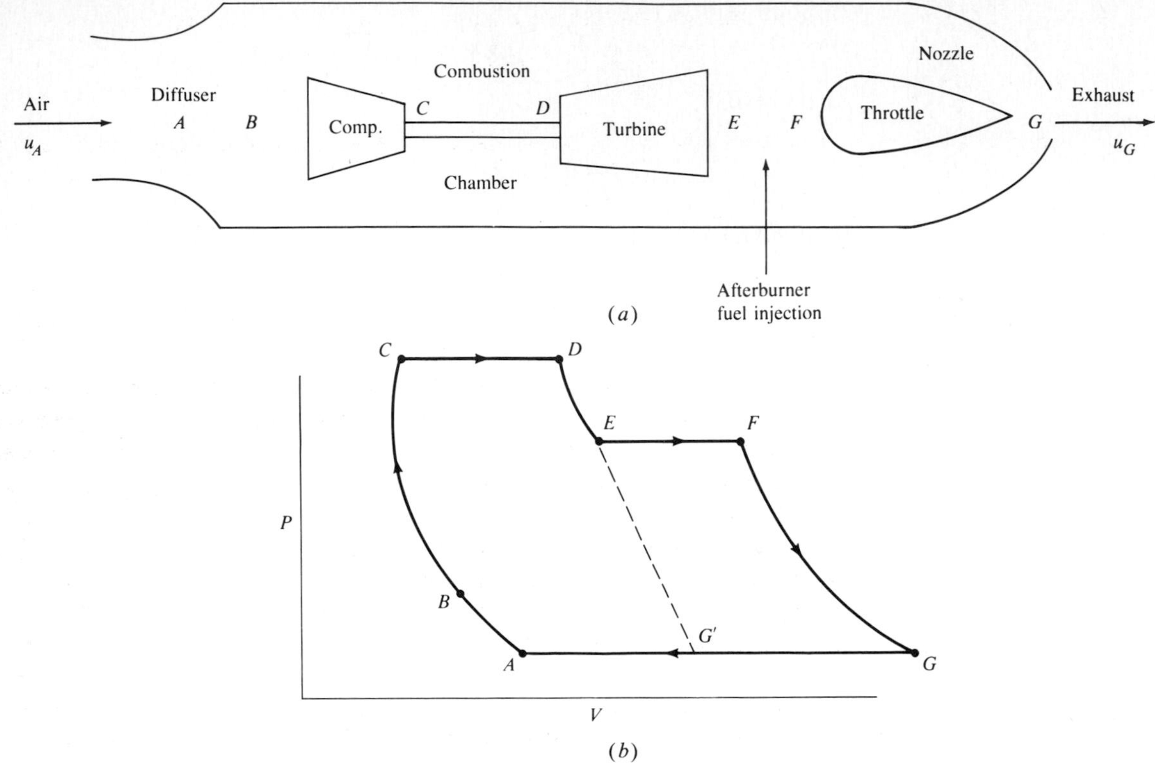

Figure 4.24 Turbojet engine. (*a*) Schematic diagram; (*b*) *PV* diagram.

kinetic energy:

$$n_P = \frac{W}{\Delta E_k} = \frac{(u_G - u_A)u_A}{(u_G^2 - u_A^2)/2} \tag{4.98}$$

The overall efficiency is

$$\eta = n_T n_P = \frac{(u_G - u_A)u_A}{Q_{\text{fuel}}} \tag{4.99}$$

4.6.5 The Rankine Cycle—The Steam Engine

The Rankine cycle is a reversible, ideal cycle which closely approximates the cycle of operations in a steam power plant as shown in Fig. 4.25.

The process consists of a constant-volume pumping of liquid (AB), production of saturated liquid at constant pressure (BC), vaporization at constant pressure (CD), superheating at constant pressure (DD'), isentropic expansion through a turbine ($D'E$), and condensation of the vapor in a condenser (EA). The process is shown on PV and TS diagrams in Fig. 4.25.

The steps and the associated heat effects are:

AB: Water pumped from P_A to P_B with an increase in enthalpy from H_A to H_B with a negligible temperature rise. The water is normally at a temperature

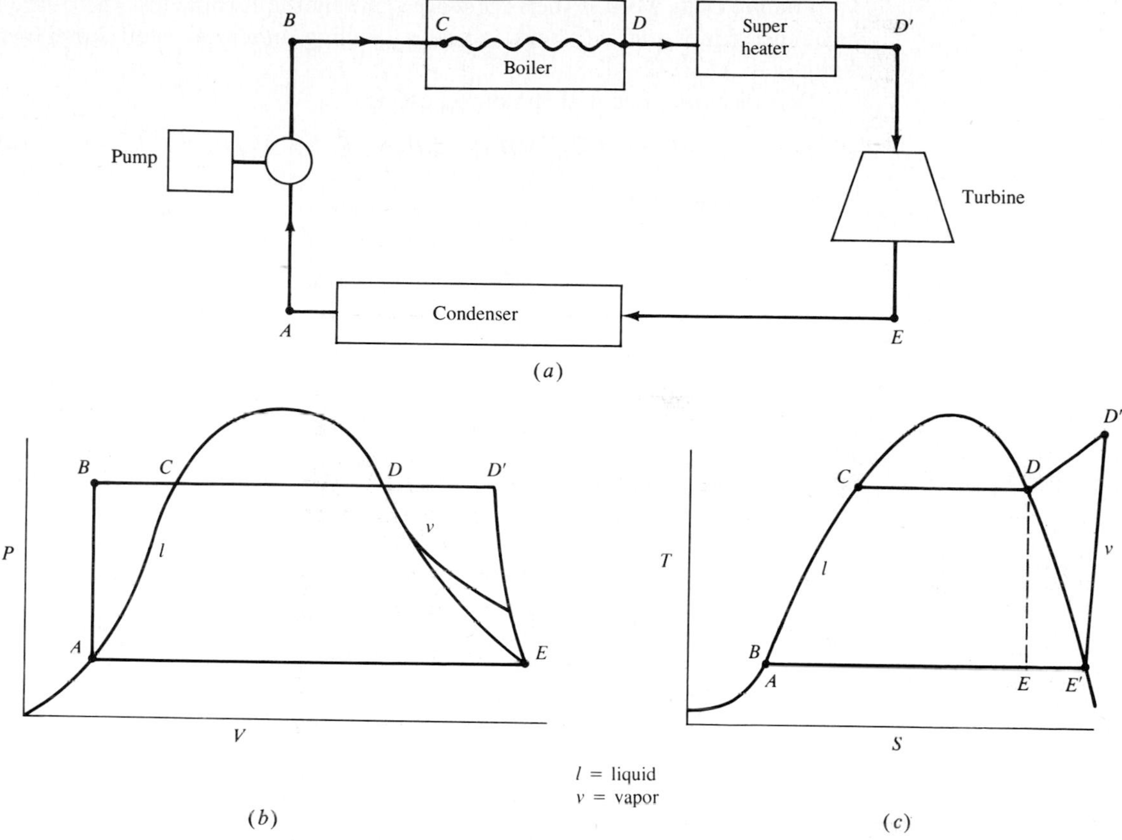

Figure 4.25 The Rankine cycle. (*a*) Schematic diagram; (*b*) *PV* and *TS* diagrams.

slightly lower than the saturation temperature at P_B. The entropy remains essentially constant. The work done $-W_{AB} = \int \overline{V}\, dP$. Since the specific volume of the liquid will not change significantly, $-W_{AB} = \overline{V}(P_B - P_A)$. This work will be equal to its increase in enthalpy, $-W_{AB} = H_B - H_A$.

BCD: Water is heated to saturation at C and then vaporizes isothermally to form saturated vapor at D. Thus

$$Q_{BCD} = H_D - H_B = \int_{S_B}^{S_D} T\, dS$$

$\overline{DD'}$: If a superheater is used, the saturated vapor will be increased in temperature with the addition of heat $Q_{DD'} = H_{D'} - H_D$.

DE or $D'E'$: Isentropic expansion through the turbine takes place.

$$W_{DE} = H_D - H_E \quad \text{or} \quad W_{D'E'} = H_{D'} - H'_E$$

E or $E'A$: Condensation of the vapor rejects heat.

$$Q_{EA} = H_A - H_E \quad \text{or} \quad Q_{E'A} = H_A - H_{E'}$$

The net cycle work is then the work done by the turbine less the pump work and must equal the net heat input, neglecting potential- and kinetic-energy changes as $\Delta E = 0$.

Consider the cycle without superheating:

$$W_{net} = (H_D - H_E) - (H_B - H_A) = Q_{BCD} + Q_{EA} \tag{4.100}$$

Thus, the efficiency is

$$\eta_R = \frac{W_{net}}{Q_{BCD}} = \frac{Q_{BCD} + Q_{EA}}{Q_{BCD}}$$

$$= \frac{(H_D - H_E) - (H_B - H_A)}{H_D - H_B} \tag{4.101}$$

Since $H_B - H_A$ is very small, it is often neglected in the calculations. Therefore

$$\eta_R = \frac{H_D - H_E}{H_D - H_A} \tag{4.102}$$

If superheat is taken into account, a similar derivation is

$$W_{net} = (H_{D'} - H_E) - (H_B - H_A) = Q_{BCD} + Q_{DD'} + Q_{EA}$$

$$\eta_{R,S} = \frac{W_{net}}{Q_{BCD} + Q_{DD'}} = \frac{(H_{D'} - H_E) - (H_B - H_A)}{(H_D - H_B) + (H_{D'} - H_D)}$$

If $H_B - H_A$ is again assumed to be negligible,

$$\eta_{R,S} = \frac{H_{D'} - H_E}{H_{D'} - H_B} \tag{4.103}$$

The efficiency of the Rankine cycle compared to the Carnot cycle can easily be seen on a *TS* diagram in Fig. 4.26. The efficiency of a Carnot cycle is

$$\eta = \frac{\text{I}}{\text{I} + \text{IV}}$$

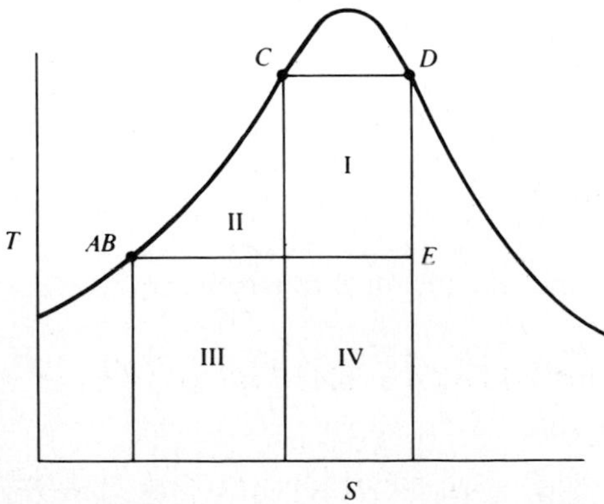

Figure 4.26 Comparison of Rankine and Carnot cycles.

while the efficiency of the Rankine cycle is

$$\eta_R = \frac{\text{I} + \text{II}}{\text{I} + \text{II} + \text{III} + \text{IV}}$$

where I to IV are areas. It is quite obvious that the Rankine cycle efficiency will always be lower than that of the Carnot cycle. Analysis of the Rankine cycle by area analysis on a *TS* diagram gives insight into the problem.

Many modifications of the Rankine cycle exist to improve its efficiency. These include (1) increasing the inlet pressure, (2) increasing the superheat temperature, and (3) decreasing the exhaust pressure. Reheating steam after partial expansion through the turbine is also used to reduce moisture content, although it usually lowers the thermodynamic efficiency. Analysis of each of these modifications on a *TS* diagram will illustrate their utility.

Example 4.23 Assuming an ideal Rankinc cycle where $H_C - H_B = H_C - H_A$ and all heat is added and removed from the working fluid at constant pressure, calculate the thermal efficiency of a cycle which isentropically expands steam at 20 MPa to 0.06 MPa if (*a*) the high-pressure steam is saturated and (*b*) the high-pressure steam is superheated by 300°C. Follow the nomenclature of Sec. 4.6.5 and use App. B.

(*a*)

$$P_D = 20\ \text{MPa} \qquad P_E = 0.06\ \text{MPa}$$
$$T_D = 365.8°\text{C} \qquad T_E^s = 85.94°\text{C}$$
$$H_D = 2409.7\ \text{kJ/kg}$$
$$S_D = 4.9269\ \text{kJ/(kg} \cdot \text{K)}$$

$$h_f = 359.86 \qquad h_g = 2653.5$$
$$S_f = 1.1453 \qquad S_g = 7.5320$$

$$S_E = S_D = 4.9269 = x(1.1453) + (1 - x)(7.5320)$$

where x = fraction liquid. Thus

$$x = 0.4079$$
$$1 - x - 0.5921$$

Therefore

$$H_E = (0.4079)(359.86) + (0.5921)(2653.5) = 1717.9$$

Since by Eq. (4.102)

$$\eta_R = \frac{H_D - H_E}{H_D - H_A}$$

and H_A = saturated liquid at 0.06 MPa = 359.9, then

$$\eta_R = \frac{2409.7 - 1717.9}{2409.7 - 359.9} = 0.3375$$

(*b*) $P'_D = 20$ MPa → Same conditions as in (*a*)
$T'_D = 365.8 + 300 = 665.8°C$ → Same conditions as in (*a*)
$H'_D = 3717.9$ kJ/kg
$S'_D = 6.7039$ kJ/(kg · K)

$\bar{h}'_f = 359.86 \qquad h'_g = 2653.5$
$S'_f = 1.1453 \qquad S'_g = 7.5320$

$$S'_E = S' = 6.7039 = x(1.1453) + (1 - x)(7.5320)$$
$$x = 0.1297$$
$$1 - x = 0.8703$$

Thus

$$H_E = (0.1297)(359.86) + (0.8703)(2653.5) = 2356.0$$

Since $H_A = H_B$, using Eq. (4.103) gives

$$\eta_{R,S} = \frac{H_{D'} - H_E}{H_{D'} - H_B} = \frac{3717.9 - 2356.0}{3717.9 - 359.9} = 0.4056$$

Exercise 4.W Starting with a standard Rankine cycle, analyze the following modifications by drawing illustrative diagrams on *TS* diagrams for each of the situations below and showing the effects on the cycle's efficiency.

1. Increase the inlet pressure by 33 percent.
2. Decrease the exhaust pressure by 33 percent.
3. Increase the superheat temperature by 300 K.

Apply each case to the problem of Example 4.23*b*.

4.6.6 Refrigeration Cycles

The basic principles of the reverse Carnot or refrigeration cycle were discussed in Sec. 4.3.3. This section will basically consider the standard vapor-compression and heat pump cycles and comment on other cycles in common use.

Vapor-compression cycle The vapor-compression cycle is shown diagrammatically in Fig. 4.27.

The closed cycle uses a mechanical compressor to isentropically compress to high pressure and temperature (AB) vapor leaving the evaporator after heat absorption (DA) at low pressure and temperature. The heat is then restricted by condensation of the vapor in the condenser (BC), after which the vapor is isenthalpically expanded through a valve (CD) to low temperature and pressure to complete the cycle. The cycle is sketched on *PV* and *TS* coordinates in Fig. 4.27.

Since $\Delta E = 0$ for the cycle, then the net work done on the system equals the heat added to the system. As defined earlier, the coefficient of performance (COP) of the cycle equals the heat input divided by the net work. Therefore

$$\text{COP} = \frac{Q_{DA}}{W_{\text{net}}} = \frac{Q_{DA}}{-Q_{DA} - Q_{BC}} \tag{4.104}$$

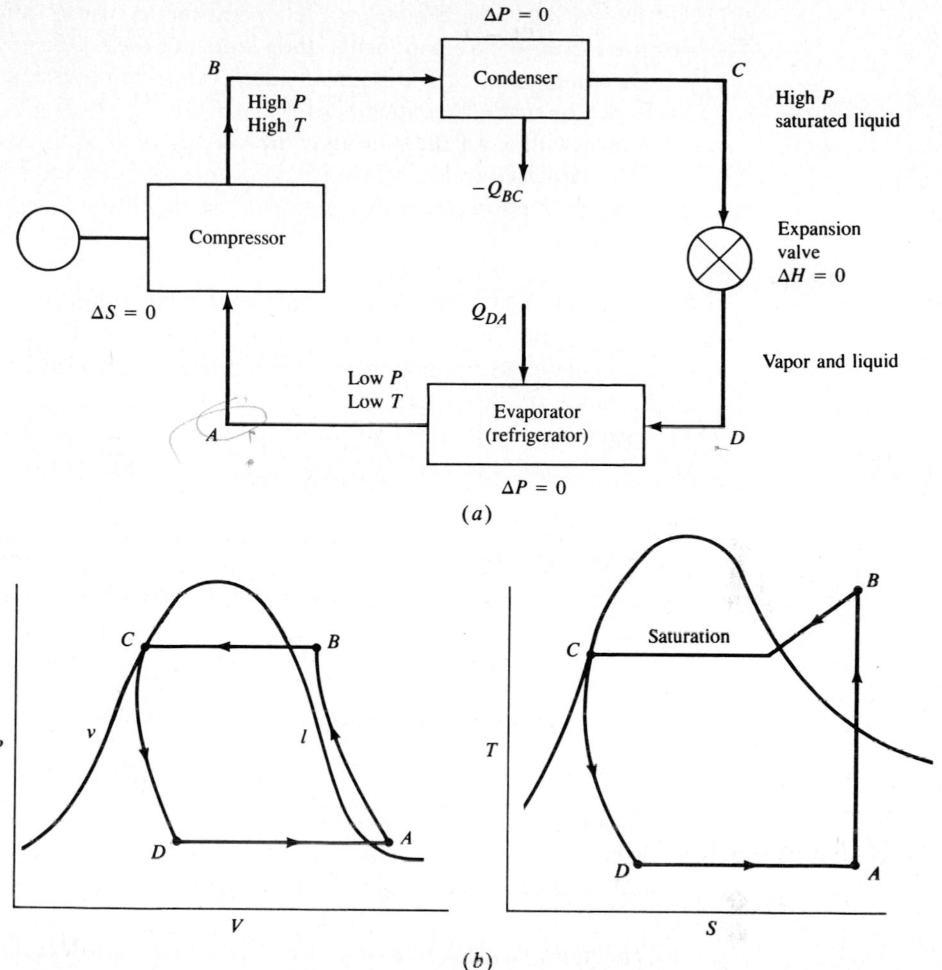

Figure 4.27 Vapor-compression cycle. (*a*) Schematic diagram; (*b*) *PV* and *TS* diagrams.

Considering each step,

$$Q_{DA} = H_A - H_D \qquad Q_{AB} = 0$$

$$Q_{BC} = H_C - H_B \qquad Q_{CD} = 0$$

therefore,

$$\text{COP} = \frac{H_A - H_D}{-(H_A - H_D) - (H_C - H_B)} = \frac{H_A - H_D}{H_B - H_A - H_C + H_D}$$

Since $H_C = H_D$,

$$\text{COP} = \frac{H_A - H_D}{H_B - H_A} \tag{4.105}$$

This cycle assumes reversible operation as usual, and actual cycles will require larger work components, thus reducing the COP.

A comparison of COP for an ideal vapor-compression cycle and a Carnot cycle can be made by remembering that $COP_{Carnot} = T_2/(T_1 - T_2)$. It can easily be shown with a *TS* diagram that the Carnot cycle will have a higher efficiency.

The circulation rate of a refrigeration cycle is an important parameter. This rate depends upon $H_A - H_D$ and will be in units of *kg/time* per amount of refrigeration.

Example 4.24 Consider an ideal vapor-compression cycle operating on steam leaving the condenser as saturated liquid at 30°C and evaporating at 5°C. Calculate the circulation rate for a refrigeration load of 1 MJ/h.

SOLUTION

$$\frac{\text{kg steam}}{\text{h}} = \frac{10^6}{H_A - H_D}$$

From App. B,

$$H_A = \text{saturated vapor at } 5°\text{C} = 2510.6$$

$$H_D = H_C = \text{saturated liquid at } 30°\text{C} = 125.78$$

Thus,

$$\frac{\text{kg steam}}{\text{h}} = \frac{10^6}{2510.6 - 125.78} = 419$$

Exercise 4.X Make a comparison of the COP for an ideal vapor-compression cycle with that of a Carnot cycle, by using a *TS* diagram. Calculate the COP in both cases for the conditions of Example 4.24.

Refrigerants Refrigerants must be selected to maximize the work of expansion and secure the required pressure range for operation between any two temperature levels. The working fluids should have (1) a low boiling point to allow both liquid and vapor to exist at low temperature and pressure, (2) a high critical temperature so that liquid can be formed under the operating conditions, (3) a high latent heat of vaporization and vapor heat capacity to keep the required flow rate low, and (4) a low freezing point to prevent plugging. Practical, nonthermodynamic considerations include inertness of the material, flammability characteristics, toxicity, stability, irritability, and cost.

Modern refrigerants include primarily Freons for household and moderate commercial refrigerating tasks and ammonia for large-volume commercial refrigerators. Sulfur dioxide has been used as a refrigerant but is no longer used for fairly obvious reasons. Thermodynamic properties of Freon 12, ammonia, and sulfur dioxide are given in Chap. 5.

Heat pumps As noted earlier heat pumps have come into vogue as a way of obtaining heating in winter and cooling in summer. The analysis of the heat pump for cooling is identical to that of the vapor-compression cycle, with provisions for reversing the cycle in winter when heating is required. The coefficient of performance when *heating* is the ratio of heat rejected at the higher temperature level to the net mechanical work of compression:

$$\text{COP} = \frac{-Q_{BC}}{W_{\text{net}}} = \frac{H_B - H_C}{H_B - H_A} \tag{4.106}$$

Example 4.25 Consider your home to be heated by a heat pump in winter and air-conditioned in summer by the same device. The average winter temperatures are 5°C outside and 20°C inside. The average summer temperatures are 30°C outside and 26°C inside. Assume the temperature difference between the fluid in the cycle and the surroundings is 5°C in all cases. For an ideal cycle, determine the work required in both cases as a fraction of the heat input.

SOLUTION For the heat pump, assume the ideal case of a Carnot cycle. In winter, working temperatures are $T_{\text{out}} = 5 - 5 = 0°\text{C}$ and $T_{\text{in}} = 20 + 5 = 25°\text{C}$. Q_{BC} is the heat extracted from the outside. Therefore

$$\frac{W_{\text{cycle}}}{Q_{BC}} = \frac{T_{\text{in}} - T_{\text{out}}}{T_{\text{in}}} = \frac{298 - 273}{298}$$

$$W_{\text{cycle}} = 0.084 Q_{BC}$$

In summer, $T_{\text{out}} = 30 + 5 = 35°\text{C}$ and $T_{\text{in}} = 26 - 5 = 21°\text{C}$. Q_{DA} is the heat absorption from the room. Thus

$$\frac{W_{\text{cycle}}}{Q_{DA}} = \frac{T_{\text{out}} - T_{\text{in}}}{T_{\text{in}}} = \frac{308 - 294}{294}$$

$$W_{\text{cycle}} = 0.048 Q_{DA}$$

What does this analysis say about the applicability of heat pumps?

Exercise 4.Y Consider a combination heat pump–air conditioner working between the temperatures as given in Example 4.25. The summer cooling load is 200,000 kJ/h, while the winter heating load is 100,000 kJ/h. Compare the power requirements and comment on your answer.

Absorption refrigeration Absorption refrigeration is used to absorb heat at low temperature and pressure and reject heat at an intermediate temperature and high pressure following an increase in temperature from some high-temperature source such as a gas burner. No mechanical energy is used except for low-level liquid pumping. The refrigerant is compressed from the low pressure to the high pressure through absorption at a low temperature. The refrigerant is desorbed at high pressure by heat addition from the high-temperature source. Ammonia is normally used as the refrigerant and water is used as the absorbent. Diagrammati-

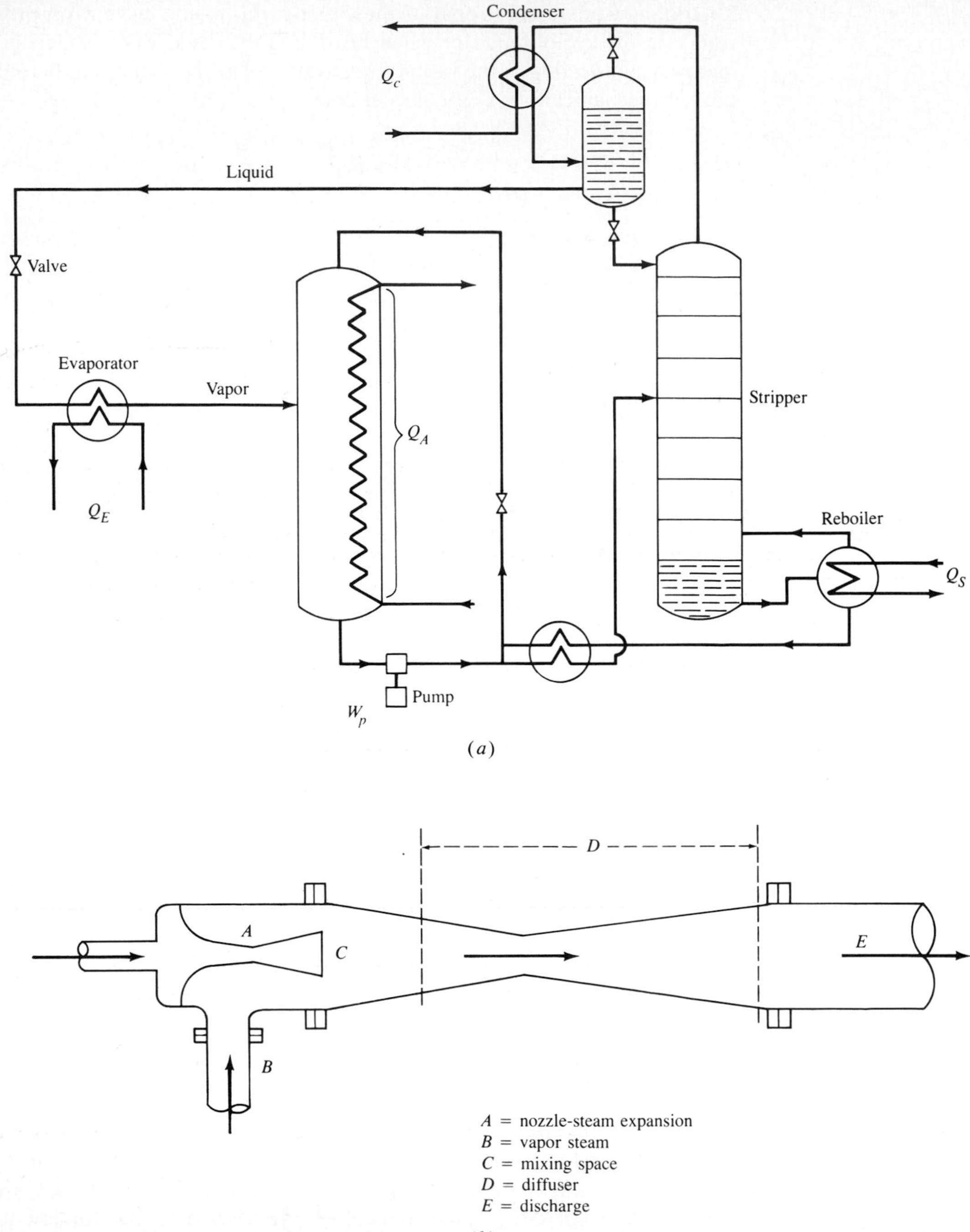

Figure 4.28 Absorption and ejection refrigeration. (*a*) Absorption system; (*b*) ejector.

cally the system is shown in Fig. 4.28. For the overall process, an energy balance yields

$$Q_E + Q_S = Q_C + Q_A + W_P \tag{4.107}$$

where Q_E = heat absorbed in evaporation
Q_S = heat added in stripper reboiler
Q_C = heat removed by condenser
Q_A = heat removed by absorber
W_P = work done on system by pump

Enthalpy balances on each part of the system are necessary for the solution of this equation. Enthalpy-concentration data are available for common systems such as ammonia and water.

Steam-jet or vapor-ejection refrigeration This system consists of a steam (vapor) jet inserted into a water (liquid) stream, as shown diagrammatically in Fig. 4.28*b*. The rapid evaporation of liquid as the vapor condenses chills the liquid and maintains a low pressure. The chilled water can then be used for condenser operation or air conditioning and is economical and simple if steam is available. This method is often used to maintain pressures in evaporators.

Example 4.26 A drawing of a vapor-recompression evaporator is shown.

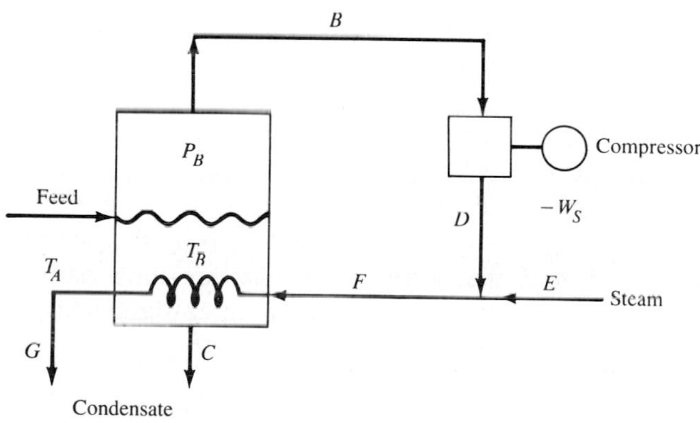

Determine the amount of work required for compression if the following conditions hold.

$M_A = 500$ kg/h	$C_P = 3.2$ kJ/(kg · K)	$T_A = 20°C$
$M_C = 300$ kg/h	$C_P = 2.8$ kJ/(kg · K)	$T_C = T_B = 80°C$
$M_E = 50$ kg/h	$T_E = 120°C$ (saturated)	$T_D = 120°C$

Condensate is saturated liquid at 120°C.

SOLUTION An energy balance on the system assuming no heat losses and a base temperature of 0°C is

$$-W_S + M_A H_A + M_E H_E = (M_B + M_E) H_G + M_C H_C$$

$$-W_S + (500)(3.2)(20 - 0) + (50)(2706.3)$$
$$= (200 + 50)(503.7) + (300)(2.8)(80 - 0)$$

$$-W_S = 125{,}925 + 67{,}200 - 135{,}315 - 32{,}000$$
$$= 25{,}810 \text{ kJ/h}$$

Important

Exercise 4.Z Water at 0.2 MPa and 27°C is expanded through a valve into an insulated chamber at 0.8 kPa pressure. Determine the temperature to which the water is chilled and the amount of water vaporized in the process.

4.6.7 Liquefaction

Liquefaction of gases by special techniques is necessary where the critical temperature of the gas is below the temperature of any available cooling water. These liquefied gases include important materials such as hydrogen, helium, oxygen, and nitrogen. In the liquefied form, they can be more readily transported and stored. Also, such low-temperature materials are necessary for certain types of processing. A special use is to allow air to be separated into its various components by liquefaction followed by distillation.

Liquefaction processes merely compress and precool the gases by heat exchange and self-cooling. Two types of processes are used, free expansion and isentropic expansion.

Free expansion A typical system for free expansion of gases is shown diagrammatically in Fig. 4.29. The throttling valve operates insenthalpically and the separator is adiabatic.

The amount of gas liquefied may be easily determined by a series of enthalpy balances over the entire system, the throttling valve, and separator, and the heat exchange system. Let x = fraction of gas liquefied.

System: $$H_C = xH_E + (1 - x)H_G \tag{4.108}$$

Heat exchanger: $$H_C + (1 - x)H_F = H_D + (1 - x)H_G \tag{4.109}$$

Separator: Since $\Delta H_{\text{throttling valve}} = 0$,

$$H_D = xH_E + (1 - x)H_F \tag{4.110}$$

Usually T_C, P_C, T_E, P_E, are T_G and known and x and T_D are unknown. Since only two independent equations exist, these can then be solved. For example,

$$x = \frac{H_G - H_C}{H_G - H_E}$$

from Eq. (4.108). H_D and thence T_D can be determined by using the second or third equation. Trial and error is usually required. Experimental data are normally used although generalized correlations can be used to estimate enthalpies.

The Linde air-liquefaction system shown in Fig. 4.30 is a practical example of a free-expansion process used commercially to liquefy air prior to the distillative separation of nitrogen, oxygen, and argon from the air.

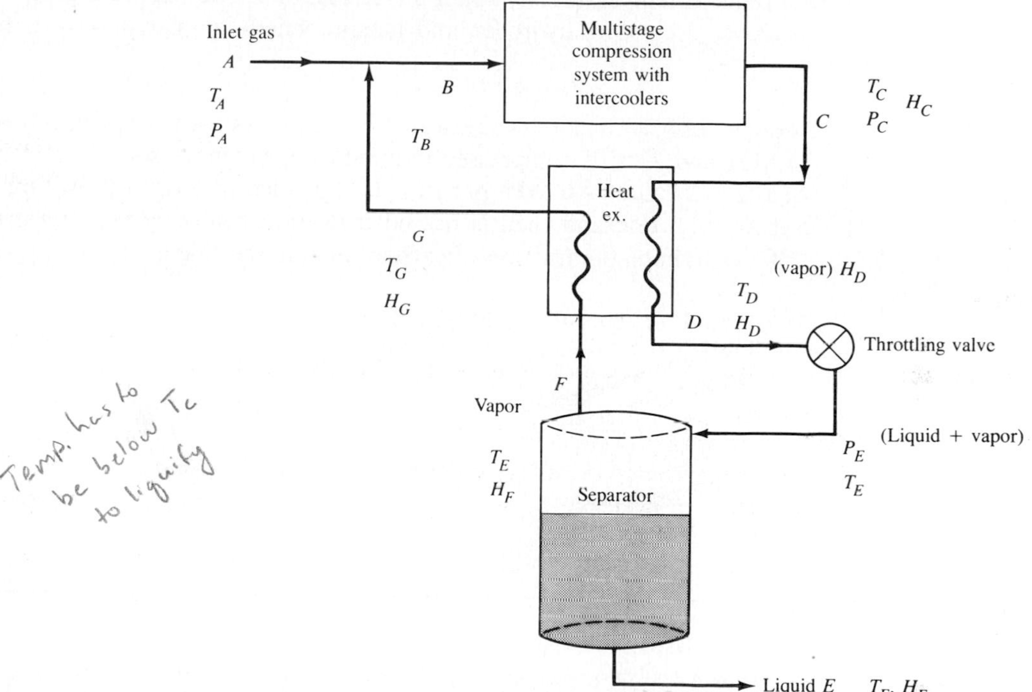

Figure 4.29 Free expansion of gases, schematic diagram.

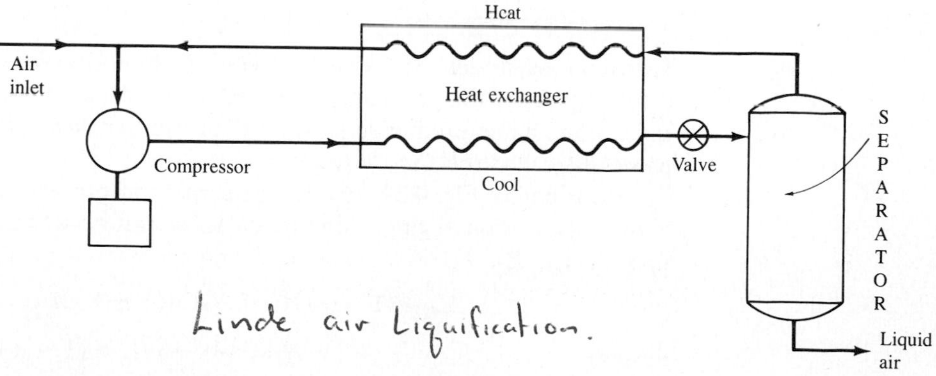

Figure 4.30 Air liquefaction, schematic diagram.

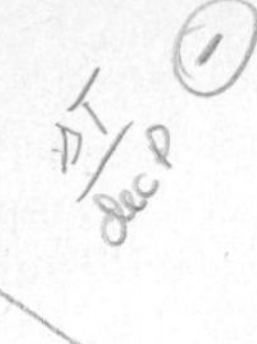

Through the throttling valve the temperature drops, as the gas is nonideal. This process is known as the *Joule-Thompson effect*, which states that the change in temperature with decrease in pressure is negative at constant enthalpy; that is, $(\partial H/\partial P)_H = +$. If this is not true, liquefaction cannot occur by free expansion. Obviously, the process cannot work for an ideal gas as $\Delta H \approx C_P \Delta T$ and if $\Delta H = 0$, ΔT must be zero. In addition, the process will not work for real gases above a reduced temperature of about 6 (see Fig. 4.7). The only real gases where such behavior occurs are hydrogen and helium, where another process is necessary.

Example 4.27 To make dry ice, a free expansion is used. Carbon dioxide at 0.1 MPa and 27°C is compressed in a multistage compressor, with interstage cooling to 27°C, to 4.0 MPa pressure, is expanded through a throttling valve back to 0.1 MPa, and then is passed through a heat exchanger, leaving at 15°C. Determine the fraction of carbon dioxide solidified.

SOLUTION Following Fig. 4.29 from Fig. 5.17.

$$H_C(300\text{ K}, 4.0\text{ MPa}) = 150\text{ Btu/lb}$$

$$H_E(\text{sat. solid at } 0.1\text{ MPa}) = -115\text{ Btu/lb}$$

$$H_G(288\text{ K}, 0.1\text{ MPa}) \approx 168\text{ Btu/lb}$$

Thus

$$\text{Fraction of solid} = \frac{H_G - H_C}{H_G - H_E} = \frac{168 - 150}{168 - (-115)} = \frac{18}{283} = 0.064$$

Exercise 4.AA If ethylene is to be liquefied in a free-expansion system, calculate the fraction of ethylene liquefied, the temperature of the liquid ethylene, and the heat requirement for the heat exchanger if the ethylene is compressed to 10 MPa and 27°C, the separator operates at 1 MPa, and the gaseous ethylene leaves the heat exchanger at 25°C.

Isentropic expansion To expand gases that cannot cool by expansion through a throttling valve, the gases are expanded adiabatically, hence isentropically, in an expansion engine producing power W for use in other parts of the process. This process is called the *Claude process*.

Considering Fig. 4.29 for the free-expansion process and replacing the valve with an expansion engine, enthalpy balances can be set up.

Heat exchanger:

$$H_C + (1 - x)H_F = H_D + (1 - x)H_G \tag{4.111}$$

System:

$$H_C = W + xH_E + (1 - x)H_G \tag{4.112}$$

Separator and engine:

$$H_D = W + xH_E + (1 - x)H_F \tag{4.113}$$

An entropy balance can also be written over the engine.

$$S_D = xS_E + (1 - x)S_F \tag{4.114}$$

Since two enthalpy balances and the entropy balance are independent, they can be solved for three unknowns, usually x, T_D, and W.

If the heat exchanger enthalpy balance and the entropy balance are combined and x is eliminated,

$$\frac{(H_C - H_D) - (H_G - H_F)}{H_G - H_F} - \frac{S_D - S_F}{S_F - S_E} = 0 \tag{4.115}$$

The value of T_D can be determined by trial-and-error solution of this equation. x can then be determined from either of these equations. W can be determined from either of the remaining enthalpy balances.

Example 4.28 Steam is expanded through a throttling value in a Joule-Thompson expansion from 0.2 to 0.1 MPa and 110°C. What is the fraction of vapor in the inlet stream? Repeat the problem for an isentropic expansion.

JOULE-THOMPSON VALVE: ISENTHALPIC

Steam 0.2 MPa → ⊗ → Steam 0.1 MPa 110°C

$\Delta H = 0$

$$H = 2696.5$$
$$h_g = 2706.7 \qquad H = 2696.5 \text{ kJ/kg} \qquad \text{(App. B)}$$
$$h_f = 504.7$$

x = fraction of vapor in inlet stream

Therefore

$$2696.5 = x(2706.7) + (1 - x)(504.7)$$

$$x = 0.995$$
$$T = 120.2°\text{C}$$

ISENTROPIC EXPANSION ENGINE

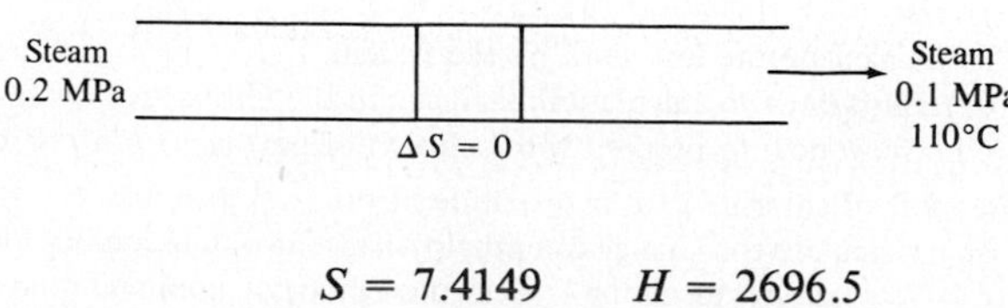

$$S = 7.4149 \qquad H = 2696.5$$
$$S_g = 7.1271 \qquad S = 7.4149$$

Therefore steam is superheated vapor:

$$x = 1.000$$
$$T = 179°C$$

Exercise 4.BB Repeat Exercise 4.Z with an isentropic engine substituted for the throttling valve.

4.7 POSTSCRIPT

This chapter attempted to give a working acknowledge of common calculations that utilize the second law of thermodynamics. More-advanced treatments of the subject are given in the reference list of Chap. 1 and should be consulted for more detailed treatment of the subject.

4.8 REFERENCES

Denbigh, K.: "The Principles of Chemical Equilibrium," Cambridge, New York, 1981.

Hill, T. L.: "An Introduction to Statistical Thermodynamics," Addison-Wesley, Reading, Mass., 1960.

Hougen, O. A., K. M. Watson, and R. A. Ragatz: "Chemical Process Principles, Part II, Thermodynamics," 2d ed., Wiley, New York, 1959.

Huang, P. K., and T. E. Daubert: *Ind. Eng. Chem. Process Des.* Dev., **13**:359 (1974).

Lee, B. I., and M. G. Kesler: *AICHE J.*, **21**:510 (1975).

Lewis, G. N., and M. Randall: "Thermodynamics and the Free Energy of Chemical Substances," McGraw-Hill, New York, 1923; 2d ed. with K. S. Pitzer and L. Brewer, 1961.

Sussman, M. V.: "Availability (Exergy) Analysis," Mulliken House, Lexington, Mass., 1980.

4.9 PROBLEMS

1 During an irreversible process taking place in a piston-cylinder device, the system receives 200 Btu of heat from a reservoir at 800°R and the internal-energy change is 50 Btu. The system is then restored to its original state by a reversible process where the only heat transfer is between the heat reservoir at 800°R and the system. The total change in entropy after both processes is +0.03 Btu/°R. Calculate the heat transfer with respect to the system and the work done by the system during the reversible process.

2 A gas of constant-pressure heat capacity of 45 kJ/(kmol · K) is expanded from 0.5 MPa and 100°C to 0.1 MPA and 20°C with an external coolant at 10°C (the surroundings).

(*a*) If the gas is ideal, calculate the change in entropy of the system, of the surroundings, and of the universe.

(*b*) Calculate the lost work by the system.

(*c*) Show how to calculate the change in the Gibbs free energy of the system.

(*d*) Show how to proceed with solution of part (*a*) for a nonideal gas.

3 One mole of chlorine is to be expanded from 91.3 atm and 417 K to 76.1 atm and 843 K.

(*a*) Calculate the change in enthalpy and entropy assuming ideal gas.

(*b*) Repeat (*a*), using the Lydersen method for nonideal fluids.

(*c*) Compare and comment on your answers.

4 Determine the changes in enthalpy, entropy, internal energy, and Gibbs free energy for the following processes:

(*a*) 5 kg of water vaporized at 100°C and 1 atm

(*b*) 1 kmol of saturated liquid ammonia vaporized at a constant pressure of 1 atm

5 For water at 1 atm total pressure, calculate the difference in entropy between water vapor at 100°C and 1 atm and ice at 0°C and 1 atm. Assume mean heat capacities.

6 Calculate the entropy of mixing of five different ideal gases each at 1 atm and 25°C to form a mixture at the same pressure and temperature containing gases *A*, *B*, *C*, *D*, and *E* in molar amounts of 0.1, 0.2, 0.3, 0.25, and 0.15, respectively. How would the results be affected if the total pressure was 10 atm? If the temperature was 1000 K?

7 A steam turbine operates adiabatically and receives 2000 lbm of steam/h at 150 lb/in^2 abs and 700°F. After doing work by expansion it discharges steam at 14.696 lb/in^2 abs and 300°F. Determine the horsepower developed by the turbine and the efficiency of the turbine. Compare the turbine with one that operates reversibly and adiabatically from the same initial conditions to the same final pressure.

8 A heat exchanger for cooling a hot-hydrocarbon liquid used 10,000 lb/h of cooling water entering the exchanger at 70°F. The hot oil flows at 5000 lb/h entering at 300°F and leaving at 150°F with a mean heat capacity of 0.6 Btu/(lb · °F). The exchanger can be assumed to have no heat loss.

(*a*) Calculate the change in entropy of the oil.

(*b*) Calculate the total change in entropy as a result of the process.

(*c*) Determine how much work could possibly have been obtained if the cooling of the hot oil were carried out by using the heat to operate a reversible Carnot engine with a 70°F sink temperature.

9 Calculate the absolute entropy of methanol at 700 K given the following data:

Phase transition at 157.4 K, ΔH = 154.3 cal/g mol
Melting point at 175.22 K, ΔH_{fus} = 757.0 cal/g mol
Heat of vaporization at 298.16 K = 8943.7 cal/g mol
Specific heat data – T(K), C_P, cal/(g mol · °C)

T	C_P	T	C_P
18.8	1.109	181.09	16.60
24.43	1.959	235.84	17.41
30.72	2.829	273.58	18.30
37.64	3.962	292.01	19.11
69.95	8.001	298.16	10.8
118.79	11.64	400	12.7
153.98	14.12	500	14.5
164.14	11.29	600	16.3
167.75	11.68	700	17.8

10 If 1 kg of liquid water is heated at a constant pressure of 1 atm from 20°C to vapor at 100°C:

(*a*) What is the maximum percent of the heat added that can be converted to work with surroundings at 20°C? What happens to the rest of the heat?

(*b*) Determine the entropy change in the system, the surroundings, and the universe.

(*c*) If the heat came from a furnace at 260°C, what is the entropy change from the heating process and the lost work in the process?

11 If 1 kg of saturated steam at 200°C is doubled in volume in a free expansion, determine its change in entropy, Gibbs free energy, and Helmholtz free energy by using data from the steam tables.

12 A power plant has a maximum operating temperature of 1400°F and river cooling water at 50°F. What is the maximum efficiency of the power plant?

13 A heat pump is to maintain an inside temperature of 70°F with heat flux losses of 1000 Btu/(h · °F difference between the inside temperature and the outside temperature).

(*a*) If the average winter temperature is 40°F, determine the minimum required work.

(*b*) If the average heat flux gain in the summer is equal to the heat flux loss in the winter, what is the maximum allowable outside temperature for the same work requirement? What is its coefficient of performance?

14 Consider the level of temperature for an ideal heat engine and an ideal refrigerator. Compare an increase in the hot-reservoir temperature with a decrease in the cold-reservoir temperature and each effect on the efficiencies if the magnitudes of the temperature differences are equal.

15 If 100,000 lb/h of combustion products is cooled from 700 to 500°F by preheating 90,000 lb/h of air from 80 to 200°F in a heat exchanger:

(*a*) Calculate the availability of energy in the products before and after cooling.

(*b*) If the combustion products transferred the heat reversibly through heat engines, how hot would the air be heated and how much power would the engines develop? (Setup only.)

16 Air is compressed reversibly in a steady-state flow apparatus from 14 lb/in^2 abs to 100 lb/in^2 gauge. Air enters at 80°F. Calculate the work of compression, the heat transfer, and the entropy change per pound of air for an isothermal, adiabatic, and polytropic ($k = 1.25$) process. Sketch each process on PV and TS diagram.

17 Freon 12 (see data in Chap. 5) is stored at 80°F and 60 lb/in^2 abs. To fill pressure cylinders holding 100 lbm of Freon 12, it is compressed adiabatically to 150 lb/in^2 abs and 180°F and condensed to saturated liquid at 150 lb/in^2 abs. Heat loss in filling causes the final cylinder to be at 150 lb/in^2 abs and 80°F. Determine the volume of the cylinder, the work requirement and heat requirement per cylinder filled, and the changes in entropy of the overall system, the surroundings at 80°F, and the universe.

18 Saturated-liquid ethylene at 1 atm is available: It is to be used at 35 atm and 0°C.

(*a*) What is the availability of the feed ethylene?

(*b*) Compare the minimum work required for the process with the work required and work lost for three practical methods:

(*i*) Constant-pressure heating to 0°C followed by reversible isothermal compression

(*ii*) Vaporization and superheating followed by reversible adiabatic compression

(*iii*) Reversible adiabatic pumping to the final pressure followed by heating to the final pressure

19 What pressure would be necessary to increase the vapor pressure of carbon dioxide at 0°F from 305 to 330 lb/in^2 abs?

20 Consider the change in enthalpy and entropy for methane between the ideal gas at 150°F and atmospheric pressure and the gas at 150°F and 100 atm pressure. Calculate the values assuming:

(*a*) Ideal gas

(*b*) van der Waals equation applicable

(*c*) Redlich-Kwong-Soave equation applicable

(*d*) Lydersen corresponding states applicable

(*e*) Pitzer corresponding states applicable

21 If 1 mol of chlorine is expanded from 91.3 atm and T_c to P_c and 834 K, calculate the change in enthalpy and entropy assuming ideal-gas behavior and the theorem of corresponding states.

Compare your answers.

22 Propane is expanded in an adiabatic turbine from 150°C and 27 atm to 100°C and 3.5 atm. Using the Pitzer corresponding-states technique, determine the entropy change per pound of propane as it flows through the turbine, the work done per pound of propane, and the isentropic efficiency of the turbine.

23 An advertisement says that a steam turbine will produce 3500 hp by using steam at 500°F and 100 lb/in^2 gauge with an exit pressure of 2 lb/in^2 gauge in the amount of 30,000 lb/h. Evaluate the advertisement.

24 An air-standard Otto cycle operates with a compression ratio of 8, inlet conditions of 60°F and 14.7 lb/in^2 abs, and a heat addition of 1200 Btu/lb air. The fuel and air in stoichiometric proportions gives off 1200 Btu/lb of mixture on combustion. Determine the thermal efficiency of the cycle and the maximum temperature and pressure in the cycle.

25 An air-standard diesel cycle operates at a compression ratio of 14, inlet conditions of 60°F and 14.7 lb/in^2 abs, and a maximum temperature of 2000 K. Determine the thermal efficiency of the cycle.

26 An air-standard Brayton cycle operates at a pressure ratio of 4 across the compressor with inlet air entering the compressor at 14.7 lb/in^2 abs and 60°F. The maximum cycle temperature is 2000°R, and 1200 lb/min of air flows through the cycle. Determine the work of the compressor, the work of the turbine, and the thermal efficiency of the cycle.

27 A turbojet engine operates with a compression ratio of 6 which raises the turbine inlet temperature to 2000°R and an afterburner which raises the temperature to 2400°R. If the plane is flying at 30,000 ft at a temperature of −30°F and a pressure of 0.5 atm, determine the overall efficiency of the engine. Assume the ratio of the cross-sectional area of the inlet diffuser to the exhaust nozzle is 2.5, the heating value of the fuel is 250 Btu/lb, and properties of air approximate the mixture. Repeat if the afterburner is not used. Repeat if the plane flew at sea level and 70°F.

28 The following conditions exist in an ideal steam power plant cycle:

Exit of boiler: 2.5 MPa, 320°C
Turbine: 80 percent efficiency, exit pressure 0.014 MPa
Condenser: exit temperature 38°C
Pump: 70 percent efficiency, exit pressure 3.5 MPa
Other conditions normal

Determine the efficiency of the cycle, the temperature and pressure at each point in the cycle, and the work and heat for each piece of equipment.

29 Repeat Prob. 28 if a superheater which raises the boiler exit steam by 100°C at constant pressure is added to the system.

30 Consider an open-cycle steam power plant where the turbine exhausts to the atmosphere. For the boiler and pump exit conditions of Prob. 28, an 80 percent efficient turbine, and water available at 38°C, determine the efficiency of the cycle.

31 What would be the coefficient of performance of an ideal refrigeration cycle with condensation at 100°F and evaporation at 0°F for the working fluids ammonia, sulfur dioxide, and Freon 12? (See properties in Chap. 5.)

32 A refrigeration cycle using Freon 12 operates as follows:

Compressor inlet: 20°F, 25 lb/in^2 abs
Compressor outlet: 170°F, 175 lb/in^3 abs
Compressor power input: 2.5 hp
Condenser outlet: 100°F, 165 lb/in^2 abs
Evaporator outlet: 20°F, 25 lb/in^2 abs

Determine the work and heat requirements for each unit, the tons of refrigeration provided, the overall coefficient of performance, and the lost work in the process.

33 Compare heating your house with electric resistance heating or a heat pump for an average outside temperature of 30°F and inside temperature of 70°F.

34 Consider an ideal ammonia-water absorption refrigeration system with evaporator operating at 40°F and condenser and absorber operating at 90°F. The mixture leaves the condenser at 180 lb/in^2 abs and the mixture entering the absorber contains 5 mol % ammonia. Determine Q_E, Q_S, Q_A, and the conditions at each point.

35 Determine the amount of steam required per pound of water evaporated at 40°F by using a steam jet of saturated steam at 100 lb/in^2 abs and condensation at 1 lb/in^2 abs.

36 What is the quality of ammonia exiting a throttling valve at 30 lb/in^2 abs if saturated liquid ammonia at 80°F enters the valve?

37 Assume air is to be liquefied in a Linde air-liquefaction system. The air enters the Joule-Thompson valve at 80°F and 100 atm and expands to 1 atm. The air flows at 50 ft^3/min at 60°F and 1 atm. Assume no heat leaks, zero temperature differences at the warm end of the exchanger, and adiabatic compression. Calculate the liquid production per hour and the fraction of the air liquefied. Repeat if a heat leak of 1.2 Btu/lb entering air and a 10°F temperature difference exists at the warm end of the exchanger.

38 A Linde type of process is used to solidify carbon dioxide by operating the exit of the expansion valve below the triple point. If CO_2 enters the compressor at 1 atm and 70°F and leaves at 800 lb/in^2 abs and 90°F, the valve throttles to 1 atm, and the heat exchanger returns vapor at 70°F, determine the fraction of the CO_2 solidified, the temperature of the solid CO_2 and the work requirement for adiabatic compression.

39 Ethylene is to be liquefied in an isentropic-expansion process. The ethylene leaves the compressor at 5 MPa and 100°C, leaves the valve at 1 MPa, and returns to the compressor at 25°C. Determine the fraction condensed, T, P, H, and S at each point, and the heat transferred in the exchanger. (Data are given in Chap. 5.)

40 Repeat Prob. 39, replacing the isentropic engine with an isenthalpic valve.

CHAPTER

FIVE

RELATIONSHIPS AMONG THERMODYNAMIC PROPERTIES—GRAPHICAL REPRESENTATION OF PROPERTIES AND PROCESSES

To utilize thermodynamic properties and principles to most advantage, it is necessary to understand the many relationships among properties and to be capable of calculating values of properties given the values of other properties. This chapter will begin with a general discussion of the fundamental properties and a review of their defining equations. The differential relationships between properties will be discussed, and the methods for manipulation of the properties by differential techniques and by the method of jacobians will be analyzed in detail. Methods for determining pressure and temperature effects on properties will be advanced.

Using the relationships derived in this chapter, experimental data, and generalized equations of state, methods for the construction of thermodynamic diagrams will be outlined. Finally, thermodynamic diagrams will be given for common materials, especially those which do not easily lend themselves to calculation.

5.1 THERMODYNAMIC PROPERTIES

Thermodynamic properties can be divided into several categories: absolute reference properties, fundamental and derived energy properties, point or state properties, and path properties.

The absolute reference properties are temperature, pressure, volume, composition, and entropy. Temperature, pressure, and composition are intensive properties, while volume and entropy are extensive properties. They are all state properties and do not depend on the path of any process.

The fundamental energy properties are internal energy, heat, and work. All are extensive properties. Internal energy is a point function, while heat and work are a function of the path of the process under consideration. The primary derived energy properties are enthalpy, Gibbs free energy, and Helmholtz free energy. All are extensive and state properties.

Other properties derived from the reference and energy properties are heat capacities, the compressibility coefficient, the coefficient of thermal expansion, and the Joule-Thompson coefficient, which are all state properties and extensive.

For any system where only pressure-volume work is of importance, if any two of the reference properties of temperature, pressure, volume, and entropy of a pure substance or fixed-composition, single-phase mixture are specified, the relative values of all other point or state properties are also fixed. Thus, for example, pressure and temperature will fix volume, entropy, internal energy, etc., in such a case. For any specific fixed temperature and pressure, then any third property would represent a surface in space for each phase. The accumulation of such surfaces forms a complicated contour in space. For example, if volume were chosen as the third property, lines of constant values of all other properties such as entropy and internal energy could be placed on the surface. If composition is not fixed, a four-dimensional representation would be necessary. Obviously, such pictorial representation is not practical, and mathematical manipulation of the variables will be much more useful than graphical representation.

The mathematical representation of this principle, called the *state principle*, is

$$(\text{Any property})_{\text{phase composition}} = f(T, P \text{ or } T, V \text{ or } P, V \text{ or } S, T, \text{etc.})$$

The differential relations between the energy properties as derived earlier for systems of constant composition and mass, operating reversibly, and with only pressure as an external force are

$$dE = dQ - dW = T\,dS - P\,dV$$

$$dH = dE + d(PV) = dE + P\,dV + V\,dP$$

$$dG = dH - d(TS) = dH - T\,dS - S\,dT$$

$$dA = dE - d(TS) = dE - T\,dS - S\,dT$$

Eliminating all properties except reference properties on the right, the energy properties can be expressed in terms of the four basic reference properties:

$$dE = T\,dS - P\,dV \tag{5.1}$$

$$dH = T\,dS + V\,dP \tag{5.2}$$

$$dG = -S\,dT + V\,dP \tag{5.3}$$

$$dA = -S\,dT - P\,dV \tag{5.4}$$

These relationships will be the basis for the derivations that will follow in this chapter when coupled with the state principle. Normally pressure and temperature will be used as the two independent properties since they are more easily measured, although volume is sometimes used as one property. Entropy is seldom used as it cannot be measured.

Example 5.1 Using the differential relations among properties, derive the differential equations for the change of enthalpy with entropy at constant pressure and the change of internal energy with volume at constant entropy.

SOLUTION From Eq. (5.2),

$$dH = T\,dS + V\,dP$$

At constant pressure, $dH = T\,dS$. Therefore

$$\left(\frac{\partial H}{\partial S}\right)_P = T$$

From Eq. (5.1),

$$dE = T\,dS - P\,dV$$

At constant entropy, $dE = -P\,dV$. Therefore

$$\left(\frac{\partial E}{\partial V}\right)_S = -P$$

Exercise 5.A In the same manner as Example 5.1, derive equations for the change of Helmholtz free energy with temperature at constant volume and the change of Gibbs free energy with temperature at constant pressure.

5.2 MATHEMATICAL RELATIONSHIPS AMONG BASIC PROPERTIES—THE MAXWELL RELATIONS

Certain relationships between total and partial derivatives and properties of derivatives and integrals must be reviewed to understand the derivations of relationships between properties.

An exact differential equation is an equation that relates any dependent variable which is a single-valued continuous function to any number of independent variables. For example,

$$v = f(x, y, z)$$

$$dv = \left(\frac{\partial v}{\partial x}\right)_{y,z} dx + \left(\frac{\partial v}{\partial y}\right)_{x,z} dy + \left(\frac{\partial v}{\partial x}\right)_{x,y} dz \tag{5.5}$$

The integral of dv between any two conditions only depends on the value of v at the two end points and not on the path. For example, if

$$dv = \left(\frac{\partial v}{\partial x}\right)_y dx + \left(\frac{\partial y}{\partial y}\right)_x dy$$

then

$$\int_1^2 dv = v_2 - v_1 = f(x, y)$$

If a variable v is continuous, then a mathematical requirement exists between any pair of independent variables. For example, if

$$v = f(x, y, z)$$

then

$$\frac{\partial}{\partial y}\left(\frac{\partial v}{\partial x}\right)_y = \frac{\partial}{\partial x}\left(\frac{\partial v}{\partial y}\right)_x$$

or

$$\frac{\partial^2 v}{\partial y\, \partial x} = \frac{\partial^2 v}{\partial x\, \partial y} \tag{5.6}$$

Similar relationships for x and z and y and z can be similarly derived.

Consider a two-independent-variable exact differential equation

$$dv = R\, dx + Q\, dy \tag{5.7}$$

where $R = f(x)$ and $Q = f(y)$. Exactness requires that $R = (\partial v/\partial x)_y$ and $Q = (\partial v/\partial y)_x$. Thus

$$\left(\frac{\partial R}{\partial y}\right)_x = \frac{\partial^2 v}{\partial x\, \partial y} \qquad \text{and} \qquad \left(\frac{\partial Q}{\partial x}\right)_y = \frac{\partial^2 v}{\partial y\, \partial x}$$

Equating,

$$\left(\frac{\partial R}{\partial y}\right)_x = \left(\frac{\partial Q}{\partial x}\right)_y \tag{5.8}$$

If Eq. (5.7) is followed and v is a constant, then $R\, dx + Q\, dy = 0$, which leads to

$$\left(\frac{\partial v}{\partial x}\right)_y = -\frac{(\partial y/\partial x)_v}{(\partial y/\partial v)_x} \tag{5.9}$$

If Eq. (5.7) is divided by dy and a variable w is held constant,

$$\left(\frac{\partial v}{\partial y}\right)_w = \left(\frac{\partial v}{\partial x}\right)_y\left(\frac{\partial x}{\partial y}\right)_w + \left(\frac{\partial v}{\partial y}\right)_x \tag{5.10}$$

The numerator and denominator of a partial derivative can be divided by the same differential with the same variable as a constant:

$$\left(\frac{\partial v}{\partial x}\right)_y = \frac{(\partial v/\partial w)_y}{(\partial x/\partial w)_y} = \left(\frac{\partial v}{\partial w}\right)_y\left(\frac{\partial w}{\partial x}\right)_y \tag{5.11}$$

Starting with Eq. (5.7), where $v = f(x, y)$, and also assuming that $v = f(s, t)$ and $x = f(s, t)$, Eqs. (5.9) and (5.10) can be applied to both v and x. The resulting equation for $(\partial v/\partial s)_y$ can be divided by the equation for $(\partial x/\partial s)_y$ to yield

$$\left(\frac{\partial v}{\partial x}\right)_y = \frac{(\partial v/\partial s)_t(\partial y/\partial t)_s - (\partial v/\partial t)_s(\partial y/\partial s)_t}{(\partial x/\partial s)_t(\partial y/\partial t)_s - (\partial x/\partial t)_s(\partial y/\partial s)_t} \tag{5.12}$$

This equation will be useful in deriving the Bridgeman table of thermodynamic relations.

Each of Eqs. (5.6) to (5.12) will be of use in deriving thermodynamic relations.

5.2.1 Maxwell Relations

The four Maxwell relations are important thermodynamic partial derivative identities which allow the substitution of measurable properties for unmeasurable properties in thermodynamic relationships. They are derived by utilizing the mathematical equations (5.7) and (5.8) in conjunction with the energy property relationships (5.1) to (5.4).

Consider internal energy:

$$dE = T\,dS - P\,dV \tag{5.1}$$

Comparing with Eqs. (5.7) and (5.8),

$$dv = R\,dx + Q\,dy$$

$$\left(\frac{\partial R}{\partial y}\right)_x = \left(\frac{\partial Q}{\partial x}\right)_y$$

Thus $R = T$, $Q = -P$, $x = S$, and $y = V$.

Thus from (5.8),

$$\left(\frac{\partial T}{\partial V}\right)_S = -\left(\frac{\partial P}{\partial S}\right)_V \tag{5.13}$$

This is the first Maxwell relation.

Using a similar approach, three additional relations can be developed from Eqs. (5.2) to (5.4).

From (5.2): $$\left(\frac{\partial T}{\partial P}\right)_S = \left(\frac{\partial V}{\partial S}\right)_P \tag{5.14}$$

From (5.3): $$-\left(\frac{\partial S}{\partial P}\right)_T = \left(\frac{\partial V}{\partial T}\right)_P \tag{5.15}$$

From (5.4): $$\left(\frac{\partial S}{\partial V}\right)_T = \left(\frac{\partial P}{\partial T}\right)_V \tag{5.16}$$

These relationships become quite important since, for example, the property $(\partial S/\partial P)_T$ cannot be experimentally measured, although its identity $(\partial V/\partial T)_P$ can be measured relatively easily for substitution in thermodynamic calculations.

Heat capacities expressions can be derived from basic equations as

From Eq. (5.2): $$dH = T\,dS + V\,dP$$

$$C_P = \left(\frac{\partial H}{\partial T}\right)_P = T\left(\frac{\partial S}{\partial T}\right)_P + V\left(\frac{\partial P}{\partial T}\right)_P^{\!\!0}$$

or

$$\left(\frac{\partial S}{\partial T}\right)_P = \frac{C_P}{T} \tag{5.17}$$

From Eq. (5.1): $$dE = T\,dS - P\,dV$$

$$C_V = \left(\frac{\partial E}{\partial T}\right)_V = T\left(\frac{\partial S}{\partial T}\right)_V - P\left(\frac{\partial V}{\partial T}\right)_V^{\!\!\!\!\!\!\!\!/\;0}$$

or

$$\left(\frac{\partial S}{\partial T}\right)_V = \frac{C_V}{T} \tag{5.18}$$

These two identities supplement the Maxwell relations.

In addition to these relations, an expression for the heat capacity ratio, $k = C_P/C_V$, can be derived from these relations:

$$\frac{C_P}{C_V} = \frac{(\partial S/\partial T)_P}{(\partial S/\partial T)_V} = -\left(\frac{\partial P}{\partial V}\right)_S\left(\frac{\partial T}{\partial P}\right)_V\left(\frac{\partial V}{\partial T}\right)_P \tag{5.19}$$

5.2.2 Bridgeman Tables

Bridgeman (1926) has developed a series of expressions for setting up equations for any partial derivative in terms of measurable properties C_P, $(\partial V/\partial P)_T$, and $(\partial V/\partial T)_P$. P, V, T, S, E, H, G, and A are covered. Table 5.1 was developed by using Eq. (5.12), where s is temperature, t is pressure, and v is the property desired leading to Eq. (5.20).

$$\frac{(\partial v)_y}{(\partial x)_y} = \frac{(\partial v/\partial T)_P(\partial y/\partial P)_T - (\partial v/\partial P)_T(\partial y/\partial T)_P}{(\partial x/\partial T)_P(\partial y/\partial P)_T - (\partial x/\partial P)_T(\partial y/\partial T)_P} \tag{5.20}$$

This is a simple method to derive various relationships among properties.

Example 5.2 Derive an expression for $C_P - C_V$ in terms of PVT properties.

SOLUTION Using Eqs. (5.17) and (5.18),

$$C_P - C_V = T\left(\frac{\partial S}{\partial T}\right)_P - T\left(\frac{\partial S}{\partial T}\right)_V$$

Since from Eq. (5.9)

$$\left(\frac{\partial v}{\partial x}\right)_y = -\frac{(\partial y/\partial x)_v}{(\partial y/\partial v)_x}$$

therefore

$$\left(\frac{\partial S}{\partial T}\right)_P = -\frac{(\partial P/\partial T)_S}{(\partial P/\partial S)_T}$$

$$\left(\frac{\partial S}{\partial T}\right)_V = -\frac{(\partial V/\partial T)_S}{(\partial V/\partial S)_T}$$

$$C_P - C_V = T\left[\frac{(\partial V/\partial T)_S}{(\partial V/\partial S)_T} - \frac{(\partial P/\partial T)_S}{(\partial P/\partial S)_T}\right]$$

Table 5.1 Condensed summary of thermodynamic relationships

Pressure constant and pressure variable

$$(\partial V)_P = -(\partial P)_V = \left(\frac{\partial V}{\partial T}\right)_P$$

$$(\partial T)_P = -(\partial P)_T = 1$$

$$(\partial S)_P = -(\partial P)_S = \frac{C_P}{T}$$

$$(\partial E)_P = -(\partial P)_E = C_P - P\left(\frac{\partial V}{\partial T}\right)_P$$

$$(\partial H)_P = -(\partial P)_H = C_P$$

$$(\partial A)_P = -(\partial P)_A = -\left[S + P\left(\frac{\partial V}{\partial T}\right)_P\right]$$

$$(\partial G)_P = -(\partial P)_G = -S$$

Temperature constant and temperature variable

$$(\partial V)_T = -(\partial T)_V = -\left(\frac{\partial V}{\partial P}\right)_T$$

$$(\partial S)_T = -(\partial T)_S = \left(\frac{\partial V}{\partial T}\right)_P$$

$$(\partial E)_T = -(\partial T)_E = T\left(\frac{\partial V}{\partial T}\right)_P + P\left(\frac{\partial V}{\partial P}\right)_T$$

$$(\partial H)_T = -(\partial T)_H = -V + T\left(\frac{\partial V}{\partial T}\right)_P$$

$$(\partial A)_T = -(\partial T)_A = P\left(\frac{\partial V}{\partial P}\right)_T$$

$$(\partial G)_T = -(\partial T)_G = -V$$

Volume constant and volume variable

$$(\partial S)_V = -(\partial V)_S = \frac{1}{T}\left[C_P\left(\frac{\partial V}{\partial P}\right)_T + T\left(\frac{\partial V}{\partial T}\right)_P^2\right]$$

$$(\partial E)_V = -(\partial V)_E = C_P\left(\frac{\partial V}{\partial P}\right)_T + T\left(\frac{\partial V}{\partial T}\right)_P^2$$

$$(\partial H)_V = -(\partial V)_H = C_P\left(\frac{\partial V}{\partial P}\right)_T + T\left(\frac{\partial V}{\partial T}\right)_P^2 - V\left(\frac{\partial V}{\partial T}\right)_P$$

$$(\partial A)_V = -(\partial V)_A = -S\left(\frac{\partial V}{\partial P}\right)_T$$

$$(\partial G)_V = -(\partial V)_G = -\left[V\left(\frac{\partial V}{\partial T}\right)_P + S\left(\frac{\partial V}{\partial P}\right)_T\right]$$

Entropy constant and entropy variable

$$(\partial E)_S = -(\partial S)_E = \frac{P}{T}\left[C_P\left(\frac{\partial V}{\partial P}\right)_T + T\left(\frac{\partial V}{\partial T}\right)_P^2\right]$$

Table 5.1 Continued

Entropy constant and entropy variable
$(\partial H)_S = -(\partial S)_H = -\dfrac{VC_P}{T}$
$(\partial A)_S = -(\partial S)_A = \dfrac{1}{T}\left\{P\left[C_P\left(\dfrac{\partial V}{\partial P}\right)_T + T\left(\dfrac{\partial V}{\partial T}\right)_P^2\right] + ST\left(\dfrac{\partial V}{\partial T}\right)_P\right\}$
$(\partial G)_S = -(\partial S)_G = -\dfrac{1}{T}\left[VC_P - ST\left(\dfrac{\partial V}{\partial T}\right)_P\right]$
Internal energy constant and internal energy variable
$(\partial H)_E = -(\partial E)_H = -V\left[C_P - P\left(\dfrac{\partial V}{\partial T}\right)_P\right] - P\left[C_P\left(\dfrac{\partial V}{\partial P}\right)_T + T\left(\dfrac{\partial V}{\partial T}\right)_P^2\right]$
$(\partial A)_E = -(\partial E)_A = P\left[C_P\left(\dfrac{\partial V}{\partial P}\right)_T + T\left(\dfrac{\partial V}{\partial T}\right)_P^2\right] + S\left[T\left(\dfrac{\partial V}{\partial T}\right)_P + P\left(\dfrac{\partial V}{\partial P}\right)_T\right]$
$(\partial G)_E = -(\partial E)_G = -V\left[C_P - P\left(\dfrac{\partial V}{\partial T}\right)_P\right] + S\left[T\left(\dfrac{\partial V}{\partial T}\right)_P + P\left(\dfrac{\partial V}{\partial P}\right)_T\right]$
Enthalpy constant and enthalpy variable
$(\partial A)_H = -(\partial H)_A = -\left[S + P\left(\dfrac{\partial V}{\partial T}\right)_P\right]\left[V - T\left(\dfrac{\partial V}{\partial T}\right)_P\right] + pC_P\left(\dfrac{\partial V}{\partial P}\right)_T$
$(\partial G)_H = -(\partial H)_G = -V(C_P + S) + TS\left(\dfrac{\partial V}{\partial T}\right)_P$
Free-energy constant and free-energy variable
$(\partial A)_G = -(\partial G)_A = -S\left[V + P\left(\dfrac{\partial V}{\partial P}\right)_T\right] - PV\left(\dfrac{\partial V}{\partial T}\right)_P$

Source: Bridgman (1926).

From Maxwell relations,

$$\left(\frac{\partial V}{\partial S}\right)_T = \left(\frac{\partial T}{\partial P}\right)_V \quad \text{and} \quad \left(\frac{\partial P}{\partial S}\right)_T = -\left(\frac{\partial T}{\partial V}\right)_P$$

$$C_P - C_V = T\left[\frac{(\partial V/\partial T)_S}{(\partial T/\partial P)_V} + \frac{(\partial P/\partial T)_S}{(\partial T/\partial V)_P}\right]$$

This expression is in terms of PVT properties although it would be difficult to gather data at constant entropy. Thus, additional substitutions are necessary, as will be shown later.

Example 5.3 Verify the expression for $(\partial S)_V$ in Table 5.1:

$$(\partial S)_V = -(\partial V)_S = \frac{1}{T}\left[C_P\left(\frac{\partial V}{\partial P}\right)_T + T\left(\frac{\partial V}{\partial T}\right)_P^2\right]$$

SOLUTION Using the notation of Eq. (5.20),

$$(\partial S)_V = \left(\frac{\partial S}{\partial T}\right)_P \left(\frac{\partial V}{\partial P}\right)_T - \left(\frac{\partial S}{\partial P}\right)_T \left(\frac{\partial V}{\partial T}\right)_P$$

Substituting Eqs. (5.17) and (5.18),

$$(\partial S)_V = \frac{C_P}{T}\left(\frac{\partial V}{\partial P}\right)_T - \left(\frac{\partial V}{\partial T}\right)_P \left(\frac{\partial V}{\partial T}\right)_P$$

$$= \frac{1}{T}\left[C_P\left(\frac{\partial V}{\partial P}\right)_T + T\left(\frac{\partial V}{\partial T}\right)_P^2\right]$$

Exercise 5.B Using a Maxwell relation, verify the accuracy of the data below for steam at 300 lb/in^2 abs and 500°F where $V = 1.765$ ft^3/lb and $S = 1.5701$ Btu/(lb · °R).

$t = 500$°F		$V = 1.7675$ ft^3/lb	
V	S	t, °F	P, lb/in^2 abs
2.151	1.5949	425.82	270
2.063	1.5897	450.03	280
1.9809	1.5846	474.76	290
1.9047	1.5796	500.00	300
1.8338	1.5748	525.80	310
1.7675	1.5701	551.94	320
1.7054	1.5655	578.42	330
1.6472	1.5611	605.21	340
1.5925	1.5567		
1.5410	1.5524		

Exercise 5.C Starting with Eqs. (5.3) and (5.4) and using Eq. (5.9), show that

$$\left(\frac{\partial G}{\partial P}\right)_A = \frac{SP(\partial V/\partial P)_T}{S + P(\partial V/\partial T)_P} + V$$

Exercise 5.D Using the Bridgman Table 5.1, develop the expression in Exercise 5.C.

5.3 ADDITIONAL RELATIONSHIPS AMONG PROPERTIES—JACOBIANS

To derive the relationships among properties, much manipulation of partial differentials is necessary. Thus, the jacobian notation is introduced to simplify the derivation of such relationships. A jacobian is simply the determinant of a series of partial derivatives.

Consider

$$\frac{\partial(x, y)}{\partial(T, P)} = \begin{vmatrix} \left(\frac{\partial x}{\partial T}\right)_P & \left(\frac{\partial y}{\partial T}\right)_P \\ \left(\frac{\partial x}{\partial P}\right)_T & \left(\frac{\partial y}{\partial P}\right)_T \end{vmatrix}$$

$$= \left(\frac{\partial x}{\partial T}\right)_P \left(\frac{\partial y}{\partial P}\right)_T - \left(\frac{\partial x}{\partial P}\right)_T \left(\frac{\partial y}{\partial T}\right)_P \tag{5.21}$$

If T and P are constants, the jacobian $J(x, y) = \partial(x, y)/\partial(T, P)$. The notation $J(x, y)$ will be shortened to $[x, y]$ as $J(T, P) = 1$.

By reference to Eq. (5.21), it is apparent that jacobians have the following properties:

$$\left(\frac{\partial X}{\partial Y}\right)_Z = \frac{[X, Z]}{[Y, Z]} \tag{5.22}$$

$$[X, Y] = -[Y, X] \tag{5.23}$$

$$[X, Y] = X\,dY \tag{5.24}$$

$$[X, X] = 0 \qquad [Y, Y] = 0 \tag{5.25}$$

Consider the exact differential of Eq. (5.7): $dv = R\,dx + Q\,dy$. In jacobian notation this expression can be written

$$[v, z] = R[x, z] + Q[y, z] \tag{5.26}$$

where z is another thermodynamic property different from v. For example, from Eq. (5.2), $dH = T\,dS + V\,dP$. In jacobian notation,

$$[H, z] = T[S, z] + V(P, z)$$

where z can be any property except enthalpy.

Using jacobians, many relationships can be derived including the reduction of any Maxwell relations to the equivalent expression

$$[T, S] = [P, V] \tag{5.27}$$

For example, from Eq. (5.4),

$$\left(\frac{\partial S}{\partial V}\right)_T = \left(\frac{\partial P}{\partial T}\right)_V$$

Using Eq. (5.22),

$$\frac{[S, T]}{[V, T]} = \frac{[P, V]}{[T, V]}$$

Using Eq. (5.23),

$$[V, T] = -[T, V]$$

therefore

$$[S, T] = -[P, V]$$

Again using Eq. (5.23),

$$[T,S] = [P,V] \tag{5.27}$$

The equations for heat capacities, heat capacity ratios, the coefficient of thermal expansion, and the isothermal coefficient of compressibility can also be written in this form:

$$C_P = T\left(\frac{\partial S}{\partial T}\right)_P = T\frac{[S,P]}{[T,P]} \tag{5.28}$$

$$C_V = T\left(\frac{\partial S}{\partial T}\right)_V = T\frac{[S,V]}{[T,V]} \tag{5.29}$$

$$\beta = \frac{1}{V}\left(\frac{\partial V}{\partial T}\right)_P = \frac{1}{V}\frac{[V,P]}{[T,P]} \tag{5.30}$$

where β = coefficient of thermal expansion

$$\kappa = -\frac{1}{V}\left(\frac{\partial V}{\partial P}\right)_T = -\frac{1}{V}\frac{[V,T]}{[P,T]} \tag{5.31}$$

where κ = isothermal coefficient of compressibility.

Thermodynamic functions can all be reduced to functions of measurable properties. Internal energy, enthalpy, Gibbs free energy, and Helmholtz free energy can all be eliminated in preference to pressure, volume, temperature, and entropy by using differential relations. Either the standard differential equation form of Eq. (5.7) or the jacobian form of Eq. (5.26) can be used, the latter being the simpler method. Entropy can be eliminated by use of Eqs. (5.28) and (5.29) for C_P and C_V, respectively, and by use of the Maxwell identity, Eq. (5.27).

For example, from Eq. (5.1), $dE = T\,dS - P\,dV$. Thus

$$[E,T] = T[S,T] - P[V,T]$$

Substitute Eq. (5.27):

$$[E,T] = T[V,P] - P[V,T]$$

Substitute Eq. (5.30):

$$[E,T] = -T\beta V[P,T] - P[V,T]$$

therefore

$$[dE = -T\beta V\,dP - P\,dV]_T$$

The same types of manipulations can be carried out for any property to derive an equation applicable to the value of the property under any one or set of constant reference properties—pressure, temperature, volume, composition, or entropy.

Relationships which allow calculation of pressure and temperature effects by using the principles of both Secs. 5.2 and 5.3 will be derived in Sec. 5.4 and will be applied to ideal gases and other equations of state.

Example 5.4 Using jacobians, show that

$$\left(\frac{\partial H}{\partial T}\right)_V = C_V + V\left(\frac{\partial P}{\partial T}\right)_V$$

SOLUTION From Eq. (5.2), $dH = T\,dS + V\,dP$. Using the form of Eq. (5.26), therefore,

$$[H,V] = T[S,V] + V[P,V]$$

Using the form of Eq. (5.22),

$$\left(\frac{\partial H}{\partial T}\right)_V = \frac{[H,V]}{[T,V]} = \frac{T[S,V] + V[P,V]}{[T,V]}$$

Substituting Eq. (5.29), where

$$C_V = T\frac{[S,V]}{[T,V]}$$

$$\left(\frac{\partial H}{\partial T}\right)_V = C_V + V\left(\frac{\partial P}{\partial T}\right)_V$$

Exercise 5.E Show that each of the four Maxwell relations reduces to Eq. (5.27).

Exercise 5.F By use of jacobians, show that:

1. $\left(\frac{\partial H}{\partial V}\right)_T = -\frac{1}{\kappa_T} + T\left(\frac{\partial P}{\partial T}\right)_V$
2. $\left(\frac{\partial H}{\partial P}\right)_T = \bar{V}(1 - \beta_T T)$
3. $\left(\frac{\partial T}{\partial P}\right)_S = \frac{\bar{V}\beta_T T}{C_P}$
4. $\left(\frac{\partial P}{\partial T}\right)_V = \frac{\beta_T}{\kappa_T}$
5. $-\left(\frac{\partial P}{\partial V}\right)_T = \left(\frac{\partial P}{\partial T}\right)_V\left(\frac{\partial T}{\partial V}\right)_P$

Evaluate the right-hand side of each for an ideal gas.

5.4 CHANGES IN PROPERTIES—TEMPERATURE AND PRESSURE EFFECTS

Any state property such as entropy, internal energy, enthalpy, Gibbs free energy, or Helmholtz free energy may be assumed to be a function of two of the reference properties, normally temperature and pressure. For any general property Q an exact differential equation can be written to show this relationship:

$$Q = Q(P,T)$$

$$dQ = \left(\frac{\partial Q}{\partial T}\right)_P dT + \left(\frac{\partial Q}{\partial P}\right)_T dP$$

Therefore

$$\Delta Q = \int_{T_1}^{T_2} \left(\frac{\partial Q}{\partial T} \right)_P dT + \int_{P_1}^{P_2} \left(\frac{\partial Q}{\partial P} \right)_T dP \tag{5.32}$$

Each of these changes can be evaluated separately and the sum calculated to determine the total change. Both partial derivatives must be expressed in terms of measurable properties by use of Eqs. (5.1) to (5.4) and jacobians. Integration then requires the partial derivative to be substituted in terms of an equation of state—ideal-gas, analytical, or generalized. Such derivations can easily be carried out for the common properties.

5.4.1 Entropy

If temperature and pressure are considered,

$$dS = \left(\frac{\partial S}{\partial P} \right)_T dP + \left(\frac{\partial S}{\partial T} \right)_P dT$$

From Eq. (5.15),

$$\left(\frac{\partial S}{\partial P} \right)_T = -\left(\frac{\partial V}{\partial T} \right)_P$$

From Eq. (5.17),

$$\left(\frac{\partial S}{\partial T} \right)_P = \frac{C_P}{T}$$

Substituting,

$$dS = -\left(\frac{\partial V}{\partial T} \right)_P dP + \frac{C_P}{T} dT \tag{5.33}$$

$$\Delta S = \int_{P_0}^{P} -\left(\frac{\partial V}{\partial T} \right)_P dP + \int_{T_0}^{T} \frac{C_P}{T} dT \tag{5.34}$$

Similarly, if volume and temperature are considered,

$$dS = \left(\frac{\partial S}{\partial V} \right)_T dV + \left(\frac{\partial S}{\partial T} \right)_V dT$$

Substituting from Eqs. (5.16) and (5.18),

$$dS = \left(\frac{\partial P}{\partial T} \right)_V dV + \frac{C_V}{T} dT$$

or

$$dS = -\frac{(\partial V/\partial T)_P}{(\partial V/\partial P)_T} dV + \frac{C_V}{T} dT \tag{5.35}$$

5.4.2 Internal Energy

Considering pressure and temperature as variables,

$$dE = \left(\frac{\partial E}{\partial P} \right)_T dP + \left(\frac{\partial E}{\partial T} \right)_P dT$$

Since $dE = T\,dS - P\,dV$,

$$\left(\frac{\partial E}{\partial P}\right)_T = \frac{T[S,T] - P[V,T]}{[P,T]} = \frac{T[V,P] - P[V,T]}{-[T,P]}$$

$$= -T\left(\frac{\partial V}{\partial T}\right)_P - P\left(\frac{\partial V}{\partial P}\right)_T$$

$$= -T\beta V - P(-V\kappa)$$

Also,

$$\left(\frac{\partial E}{\partial T}\right)_P = C_P - P\left(\frac{\partial V}{\partial T}\right)_P$$

$$= C_P - \beta VP$$

Therefore

$$dE = (-T\beta V + PV\kappa)\,dP + (C_P - \beta VP)\,dT \tag{5.36}$$

If the system is at constant volume this expression can be easily integrated.

Another method of arriving at an expression for dE is to consider defining Eq. (5.1):

$$dE = T\,dS - P\,dV$$

Substitute Eq. (5.33) for dS:

$$dE = -T\left(\frac{\partial V}{\partial T}\right)_P dP + C_P\,dT - P\,dV \tag{5.37}$$

Therefore

$$\Delta E = \int_{P_0}^{P} -T\left(\frac{\partial V}{\partial T}\right)_P dP + \int_{T_0}^{T} C_P\,dT - \int_{V_0}^{V} P\,dV$$

Similarly, substituting Eq. (5.35) for dS,

$$dE = -\left[P + T\frac{(\partial V/\partial T)_P}{(\partial V/\partial P)_T}\right] dV + C_V\,dT \tag{5.38}$$

5.4.3 Enthalpy

Considering pressure and temperature as variables,

$$dH = \left(\frac{\partial H}{\partial P}\right)_T dP + \left(\frac{\partial H}{\partial T}\right)_P dT$$

$$\left(\frac{\partial H}{\partial T}\right)_P = C_p$$

$$\left(\frac{\partial H}{\partial P}\right)_T = \frac{[H,T]}{[P,T]} = \frac{T[S,T] + V[P,T]}{[P,T]}$$

$$= T\frac{[S,T]}{[P,T]} + V = -T\frac{[V,P]}{[T,P]} + V$$

$$= V - T\left(\frac{\partial V}{\partial T}\right)_P$$

Therefore

$$dH = \left[V - T\left(\frac{\partial V}{\partial T}\right)_P\right] dP + C_P\, dT \tag{5.39}$$

$$\Delta H = \int_{P_0}^{P}\left[V - T\left(\frac{\partial V}{\partial T}\right)_P\right] dP + \int_{T_0}^{T} C_P\, dT \tag{5.40}$$

As in the case of internal energy, Eq. (5.39) could also be derived by substituting Eq. (5.33) for dS and the definition of C_P into the defining equation (5.2): $dH = T\,dS + V\,dP$. If volume and temperature are taken as the variables, Eq. (5.41) can be easily derived by using similar methods.

$$dH = V\,dP + C_V\,dT - T\left[\frac{(\partial V/\partial T)_P}{(\partial V/\partial P)_T}\right] dV \tag{5.41}$$

5.4.4 Free Energy

Similar equations can be derived for both free energies. Another expression of importance for the temperature effect on Gibbs free energy can be derived.

Since

$$G \equiv H - TS \tag{4.36}$$

at constant temperature

$$\Delta G = \Delta H - T\Delta S$$

$$\frac{\Delta G}{T} = \frac{\Delta H}{T} - \Delta S$$

Differentiating at constant pressure with respect to temperature yields

$$\left[\frac{\partial(\Delta G/T)}{\partial T}\right]_P = \frac{1}{T}\left(\frac{\partial \Delta H}{\partial T}\right)_P - \frac{\Delta H}{T^2} - \left(\frac{\partial \Delta S}{\partial T}\right)_P$$

Since, from Eq. (5.17),

$$\left(\frac{\partial H}{\partial T}\right)_P = T\left(\frac{\partial S}{\partial T}\right)_P$$

$$\left[\frac{\partial(\Delta G/T)}{\partial T}\right]_P = -\frac{\Delta H}{T^2} \tag{5.42}$$

This equation will be quite useful when considering chemical reactions in Chap. 9.

5.4.5 Heat Capacities

As defined earlier,

$$C_P = \left(\frac{\partial H}{\partial T}\right)_P \quad \text{and} \quad C_V = \left(\frac{\partial E}{\partial T}\right)_V$$

The effects of pressure and volume on each of these variables are sometimes important. If the principle of Eq. (5.8) is applied to Eq. (5.33), Eq. (5.43) results:

$$\left(\frac{\partial R}{\partial y}\right)_x = \left(\frac{\partial Q}{\partial x}\right)_y \tag{5.8}$$

where $R = (\partial v/\partial x)_y$ and $Q = (\partial v/\partial y)_x$,

$$dS = -\left(\frac{\partial V}{\partial T}\right)_P dP + \frac{C_P}{T}\,dT \tag{5.33}$$

thus

$$\left(\frac{\partial C_P}{\partial P}\right)_T = -T\left(\frac{\partial^2 V}{\partial T^2}\right)_P \tag{5.43}$$

If Eq. (5.11) is applied to Eq. (5.33), Eq. (5.44) results:

$$\left(\frac{\partial C_P}{\partial V}\right)_T = -T\left(\frac{\partial P}{\partial V}\right)_T\left(\frac{\partial^2 V}{\partial T^2}\right)_P \tag{5.44}$$

Similarly, Eqs. (5.45) and (5.46) can be derived from Eq. (5.35):

$$\left(\frac{\partial C_V}{\partial V}\right)_T = T\left(\frac{\partial^2 P}{\partial T^2}\right)_V \tag{5.45}$$

$$\left(\frac{\partial C_V}{\partial P}\right)_T = T\left(\frac{\partial V}{\partial P}\right)_T\left(\frac{\partial^2 P}{\partial T^2}\right)_V \tag{5.46}$$

5.4.6 Applications to Ideal Gases

The generalized equations for each of the thermodynamic properties, such as Eqs. (5.34) for ΔS and (5.40) for ΔH, require an equation of state for actual calculation of the property change. The simplest equation is that of the ideal gas which will be illustrated for these two properties. Similar equations can be developed for ΔE, ΔG, and ΔA.

For the ideal gas, $PV = RT$.

Eq. (5.34):
$$\Delta S = \int_{P_0}^{P} -\left(\frac{\partial V}{\partial T}\right)_P dP + \int_{T_0}^{T}\frac{C_P}{T}\,dT$$

Eq. (5.40)
$$\Delta H = \int_{P_0}^{P}\left(V - T\left(\frac{\partial V}{\partial T}\right)_P\right) dP + \int_{T_0}^{T} C_P\,dT$$

$$\left(\frac{\partial V}{\partial T}\right)_P = \frac{R}{P} = \frac{V}{T} \tag{5.47}$$

Therefore

$$\Delta S^* = -\int_{P_0}^{P} R\frac{dP}{P} + \int_{T_0}^{T}\frac{C_P}{T}\,dT$$

$$= R\ln\frac{P_0}{P} + \int_{T_0}^{T}\frac{C_P}{T}\,dT \tag{5.48}$$

and

$$\Delta H^* = \int_{P_0}^{P}\left[V - T\left(\frac{V}{T}\right)\right] dP + \int_{T_0}^{T} C_P\,dT$$

$$= \int_{T_0}^{T} C_P\,dT \tag{5.49}$$

For derivation of other expressions, the other first derivatives are necessary.

$$\left(\frac{\partial P}{\partial T}\right)_V = \frac{R}{V} = \frac{P}{T} \tag{5.50}$$

$$\left(\frac{\partial V}{\partial P}\right)_T = -\frac{V}{P} = -\frac{V^2}{RT} = -\frac{RT}{P^2} \tag{5.51}$$

Other derivatives of importance for ideal-gas calculations can easily be derived and are listed for reference.

$$\left(\frac{\partial E^*}{\partial P}\right)_T = 0 \qquad \left(\frac{\partial E^*}{\partial V}\right)_T = 0 \tag{5.52}$$

$$\left(\frac{\partial H^*}{\partial P}\right)_T = 0 \qquad \left(\frac{\partial H^*}{\partial V}\right)_T = 0 \tag{5.53}$$

$$\left(\frac{\partial C_P^*}{\partial P}\right)_T = 0 \qquad \left(\frac{\partial C_P^*}{\partial V}\right)_T = 0 \tag{5.54}$$

$$\left(\frac{\partial C_V^*}{\partial P}\right)_T = 0 \qquad \left(\frac{\partial C_V^*}{\partial V}\right)_T = 0 \tag{5.55}$$

$$\left(\frac{\partial T}{\partial P}\right)_{H^*} = 0 \tag{5.56}$$

$$\left(\frac{\partial T}{\partial P}\right)_{E^*} = 0 \tag{5.57}$$

$$\left(\frac{\partial T}{\partial P}\right)_{S^*} = \frac{TR}{PC_P} \tag{5.58}$$

$$\left(\frac{\partial T}{\partial V}\right)_{S^*} = -\frac{TR}{VC_V} \tag{5.59}$$

5.4.7 Application to Other Equations of State

Based on the principles shown in this chapter, expressions for ΔH and ΔS have been derived for the simplest practical equation of state, the van der Waals equation, in Sec. 4.5.1. For comparison purposes, the van der Waals equation,

$$P = \frac{RT}{\overline{V} - b} - \frac{a}{\overline{V}^2}$$

yielded the following equations:

$$\overline{\Delta S} = R \ln \frac{\overline{V} - b}{\overline{V}' - b} + \int_{T'}^{T} \frac{C_P}{T}\, dT \tag{5.60}$$

where V' and T' are the reference state taken at very low pressure where ideality may be assumed,

$$\overline{\Delta H} = (P\overline{V} - P'\overline{V}') + \left(\frac{a}{\overline{V}'} - \frac{a}{\overline{V}}\right) + \int_{T'}^{T} C_P\, dT \tag{5.61}$$

Departure functions, i.e., the difference between $\Delta\overline{Q}^*$ and $\Delta\overline{Q}$, were also given for each:

$$\Delta\overline{H}^* - \Delta\overline{H} = RT - P\overline{V} + \frac{a}{\overline{V}} \tag{5.62}$$

$$\Delta\overline{S}^* - \Delta\overline{S} = R\ln\frac{P'}{P} - R\ln\frac{\overline{V} - b}{\overline{V}' - b} \tag{5.63}$$

Similar expressions for $\Delta\overline{H}$ and $\Delta\overline{S}$ may be derived for any given equation of state including the Soave-Redlich-Kwong equation which is recommended for many calculations. The expression for $\Delta\overline{H} - \Delta\overline{H}^*$ is given in Chap. 4.

The use of the generalized theorem of corresponding states can also be used to evaluate the derivatives in equations such as (5.34) and (5.40), which lead to the departure functions discussed in Chap. 4 on which the Lydersen and Pitzer charts are based.

Example 5.5 Repeat Example 4.18 if water follows the van der Waals equation of state.

$$a = 0.548 \times 10^6\ (\text{Pa} \cdot \text{m}^6)/\text{kmol}^2 \qquad b = 0.0306\ \text{m}^3/\text{kmol}$$

SOLUTION Using Eq. (5.62),

$$\Delta\overline{H}^* - \Delta\overline{H} = RT - P\overline{V} + \frac{a}{\overline{V}}$$

$$T = 700^\circ\text{C}$$

$$P = 25\ \text{MPa}$$

Using Eq. (2.13),

$$P = \frac{RT}{\overline{V} - b} - \frac{a}{\overline{V}^2}$$

$$25 \times 10^6 = \frac{(8314.3)(973.1)}{\overline{V} - 0.0306} - \frac{0.548 \times 10^6}{\overline{V}^2}$$

If ideal,

$$\overline{V}^* = \frac{RT}{P} = \frac{(8314.3)(973.1)}{25 \times 10^6} = 0.323\ \text{m}^3$$

By trial and error, if $\overline{V} = 0.25$,

$$25 = \frac{8.0906}{0.2194} - \frac{0.548}{(0.25)^2} = 36.88 - 8.77 \neq 28.11$$

If $\overline{V} = 0.27$,

$$25 = 33.80 - 7.52 \neq 26.28$$

If $\overline{V} = 0.29$,

$$25 = 31.19 - 6.52 \approx 24.67$$

Interpolating, $\overline{V} \approx 0.286$ m³/kmol. Thus

$$\Delta\overline{H}^* - \Delta\overline{H} = 8.0906 \times 10^6 - (25 \times 10^6)(0.286) + \frac{0.548 \times 10^6}{0.286}$$

$$= 2.85 \times 10^6 \text{ J/kmol} = 158 \text{ kJ/kg}$$

From Example 4.18, the value from the steam tables is 150.7 and the value from the Lydersen technique is 176 kJ/kg. Thus, the van der Waals equation is more accurate in this range.

Exercise 5.G Show the derivation of Eq. (5.41).

Exercise 5.H Derive comparable equations to Eqs. (5.62) and (5.63) for a gas that follows the Redlich-Kwong-Soave equation of state.

Exercise 5.I Calculate the entropy departure for Example 5.5. Derive an expression for the internal-energy departure.

5.5 THERMODYNAMIC DIAGRAMS—UTILITY AND CONSTRUCTION

Thermodynamic diagrams are of great importance in hand calculations of thermodynamic properties and in the thermodynamic analysis of processes. Experimental data are obviously the most accurate method of constructing such diagrams, although often sufficient data are not available to cover the entire range of properties and conditions. In this case, the experimental data are often supplemented and extended by proper utilization of an equation of state. Alternatively, an entire thermodynamic diagram can be constructed by using either an analytical or a generalized equation of state. Very few materials except water have sufficient experimental data for complete construction of thermodynamic diagrams.

This section will first discuss the various types of diagrams which are used. Methods for the construction of diagrams from experimental data, analytical equations of state, and the generalized equations of state will follow.

5.5.1 Types of Diagrams

Various types of diagrams are important for different applications. The *PV*, *PH*, *HT*, *TS*, and *HS* diagrams of ammonia will be used for illustrative purposes.

***PV* diagram** The pressure-volume diagram is the basic diagram used to plot *PVT* data and was shown diagrammatically in Fig. 2.3. It is useful for obtaining values

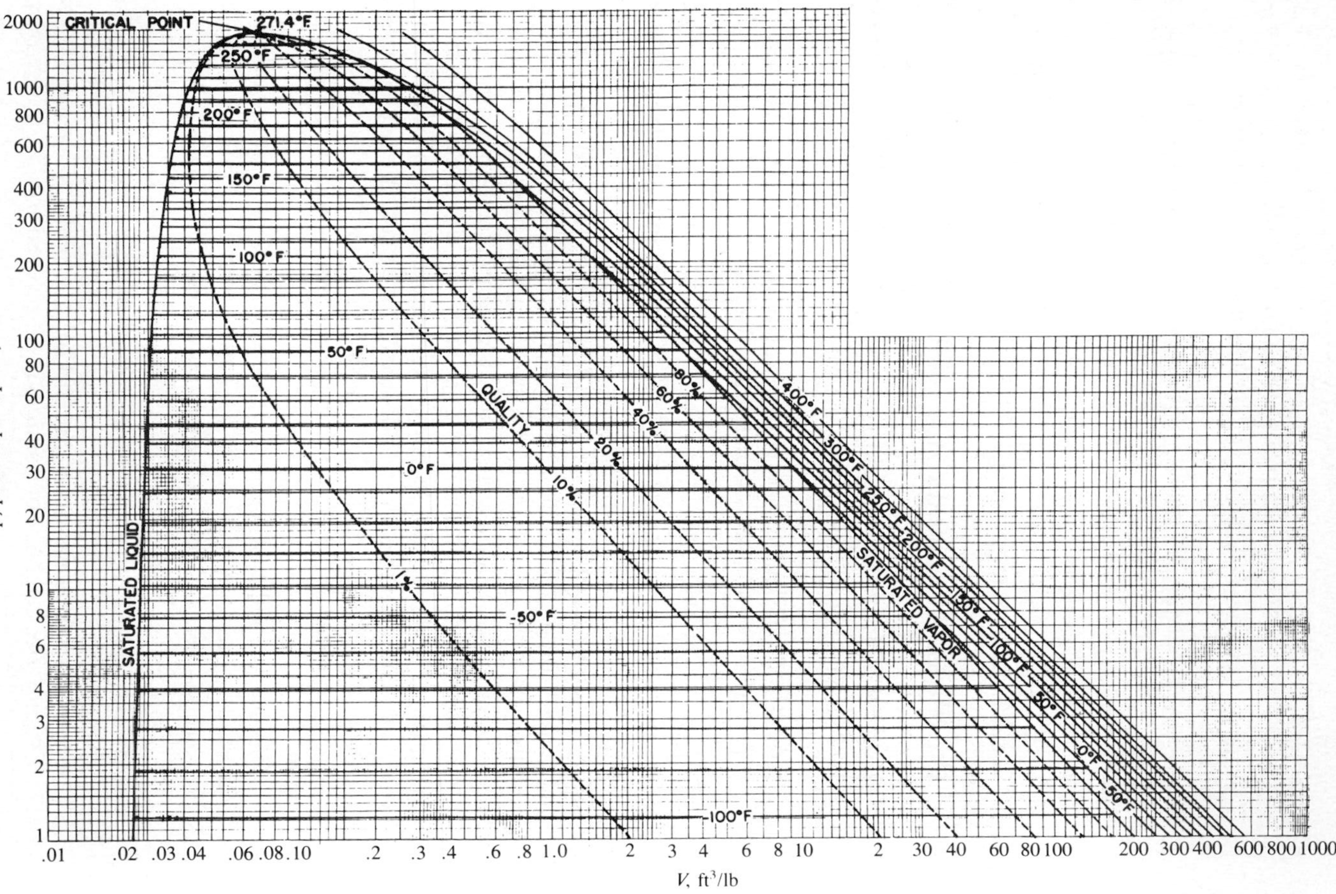

Figure 5.1 Pressure-volume diagram for ammonia, constructed from generalized corresponding-states methods.

of the three reference properties for work calculations. Figure 5.1 is a representative chart for ammonia. The two-phase envelope is shown with the critical point, with horizontal lines of constant temperature and lines of "quality" fraction or percent of vapor in the two-phase mixture within the envelope. Isotherms for superheated vapor are also usually given in the vapor region. This basic diagram is used as a starting point for the construction of other diagrams.

PH **diagrams** Pressure-enthalpy diagrams are useful in the analysis of any processes where constant-pressure and/or constant-enthalpy processes occur. Refrigeration processes have both types of processes, with evaporators and condensers operating at constant pressure and throttling values operating at constant enthalpy. A general plot is shown in Fig. 5.2. The two-phase envelope is given with the critical point. Within the envelope horizontal lines of constant pressure *and* temperature are given. Temperature lines are almost vertical in the liquid-phase region, become horizontal in the two-phase region, and tail off in the vapor-phase region. Lines of constant entropy are often given and cross from the vapor region into the two-phase region continuously. Lines of constant volume (dashed) exhibit an inflection point as they cross from the vapor- to the two-phase region. Lines of quality are often plotted in the two-phase region. A typical example for ammonia is given in Fig. 5.3.

HT **diagrams** Enthalpy-temperature diagrams are useful in throttling-process calculations and in constant-pressure-flow-process analysis. A typical diagram for ammonia is shown in Fig. 5.4. This diagram shows similar data to the *PH* diagram and is often used for the same purposes. As the lines of constant volume

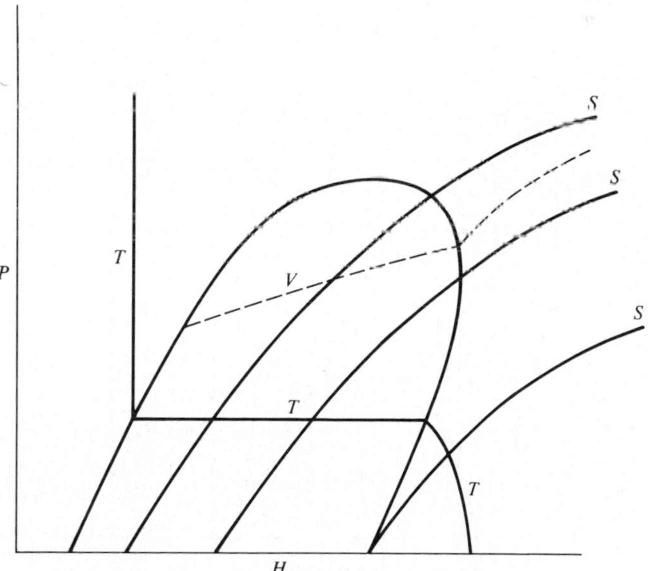

Figure 5.2 Schematic pressure-enthalpy diagram.

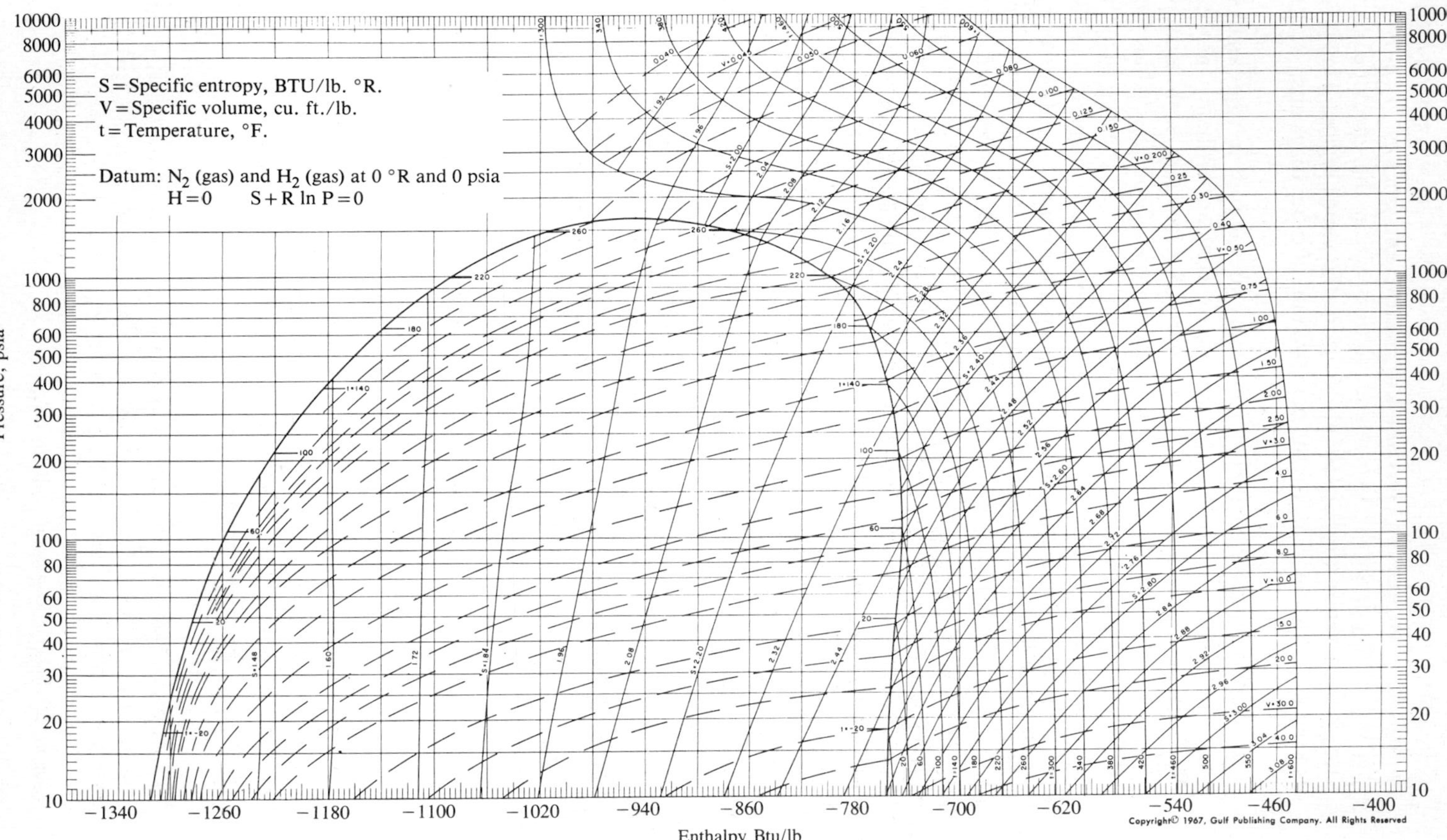

Figure 5.3 Pressure-enthalpy diagram for ammonia. *(From Lawrence N. Canjar and Francis S. Manning, "Thermodynamic Properties and Reduced Correlations for Gases," Copyright © 1967 by Gulf, Houston. All rights reserved. Used with permission.)*

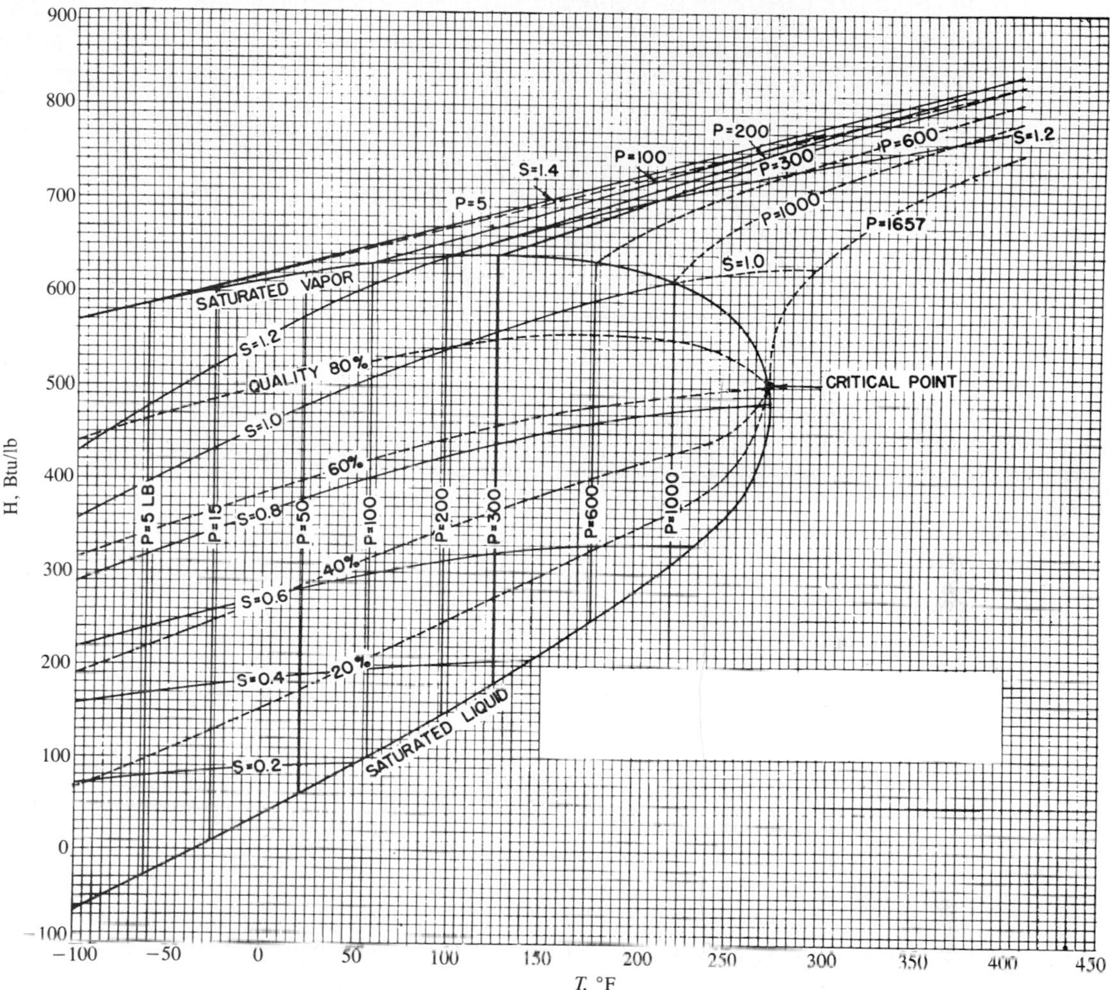

Figure 5.4 Enthalpy temperature diagram for ammonia, constructed from generalized corresponding-states methods.

exhibited an inflection as they passed the saturated-vapor line in the *PH* diagram, such lines plotted on an *HT* diagram would have the same behavior. The vertical distance between the saturated-vapor and saturated-liquid lines is the heat of vaporization at that temperature and pressure.

TS **diagrams** Temperature-entropy diagrams are of value in the analysis of any process which operates isentropically such as a compressor. Figure 5.5 shows a general diagram for a pure substance. Within the two-phase envelope are horizontal lines of constant pressure, lines of constant enthalpy, and lines of constant quality. Constant-enthalpy and constant-pressure lines are also shown in the single-phase regions. The diagram shows pressure lines including the critical

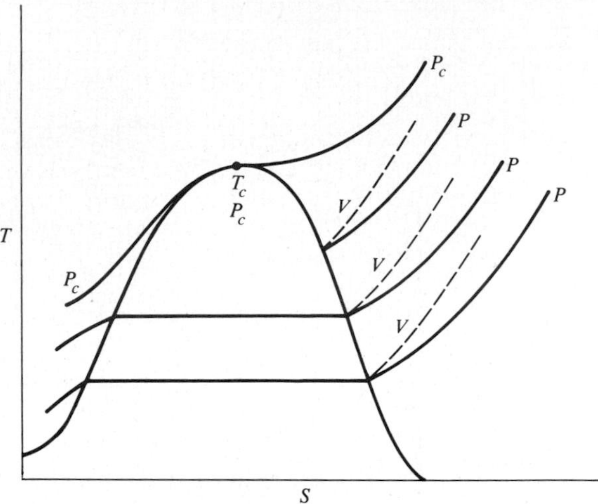

Figure 5.5 Schematic temperature-entropy diagram.

isobar. Lines of constant volume are also often given as noted by dashed lines. The horizontal distance across the two-phase envelope is the latent heat of vaporization divided by the temperature for that constant pressure. A typical diagram for ammonia is given in Fig. 5.6.

***HS* (Mollier) diagrams** The enthalpy-entropy diagram, usually called the *Mollier diagram*, is useful for any energy analysis in flow processes. It is especially used for calculating temperature changes in isentropic and isenthalpic processes. Figure 5.7 is a representative example for ammonia. Within the two-phase envelope, lines of constant temperature and pressure exist, while in the vapor region, these lines will separate with the pressure line rising continuously and the temperature line inflecting toward the horizontal. Lines of constant quality are also usually drawn in the two-phase region.

5.5.2 Construction of Diagrams

Thermodynamic diagrams can be constructed by utilizing experimental *PVT* data or various analytical or generalized equations of state to estimate *PVT* data as long as the critical temperature and critical pressure are known. To estimate enthalpies, the ideal-gas, zero-pressure heat capacity of the gas must be known as a function of temperature.

Construction from *PVT* data *PV* diagrams with lines of constant temperature can be constructed directly from the experimental data. Enthalpy and entropy can be evaluated for constructing *PH*, *HT*, *TS*, and *HS* diagrams by direct application of Eqs. (5.34) and (5.40) to calculate ΔS and ΔH, respectively. Obviously both enthalpy and entropy plots must have a base temperature and pressure for the ideal-gas properties. In the SI system the base is normally taken at 0 K, at which

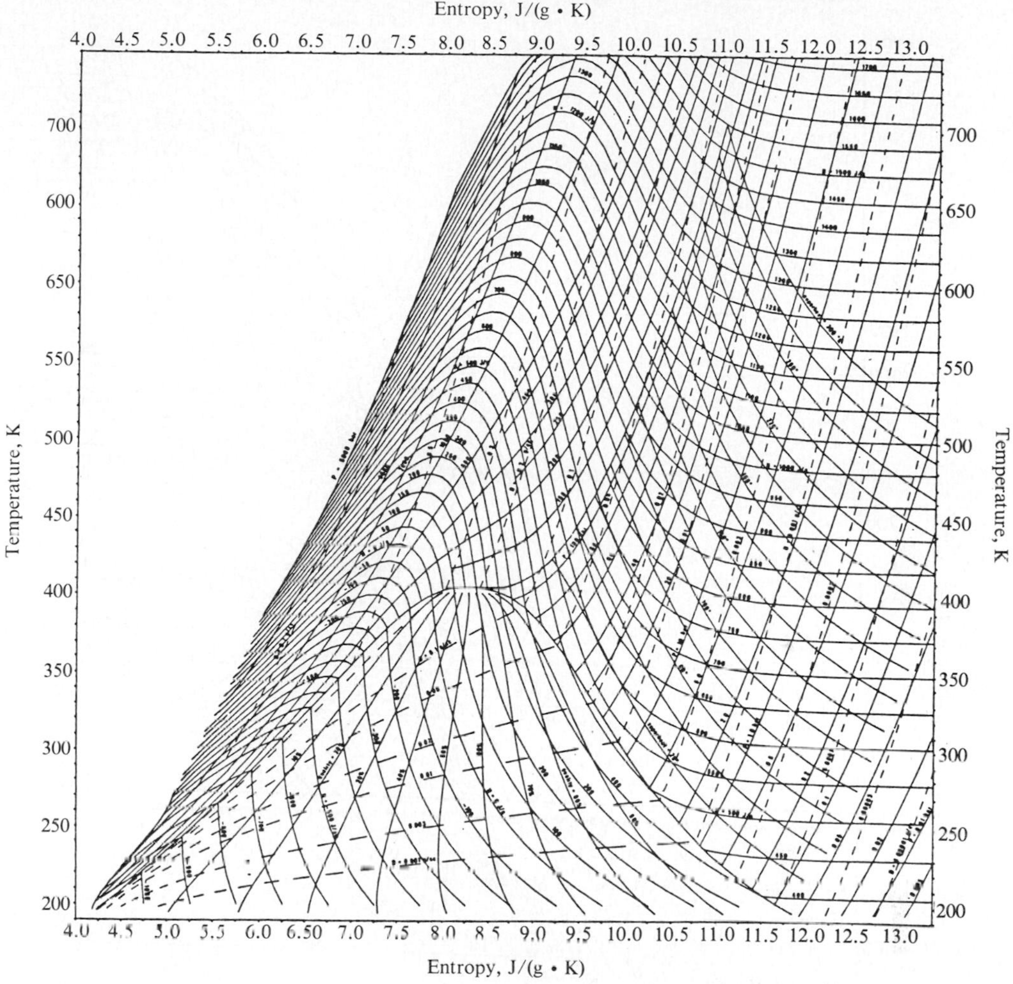

Figure 5.6 Temperature-entropy diagram for ammonia. *(From Haar and Gallagher, "Thermodynamic Properties of Ammonia," J. Phys. Chem. Ref. Data, 7: 635–792, copyright 1978 by American Institute of Physics. Used with permission.)*

point the enthalpy is taken as zero and the entropy is taken as 1 kJ/(kg · K) for the ideal gas at 0 K and 1 kPa total pressure. In the English system of units different bases are taken, some of the most common being −200, 0, and 32°F and a pressure of 1 lb/in^2 abs for entropy.

Construction from analytical equations of state Equations such as the van der Waals or Soave-Redlich-Kwong equation can be used to estimate *PVT* data from critical data, after which the methods become the same as those used for experimental data. Alternatively, the equation of state can be substituted into the defining equation for ΔH or ΔS as shown earlier, and derived equations as given,

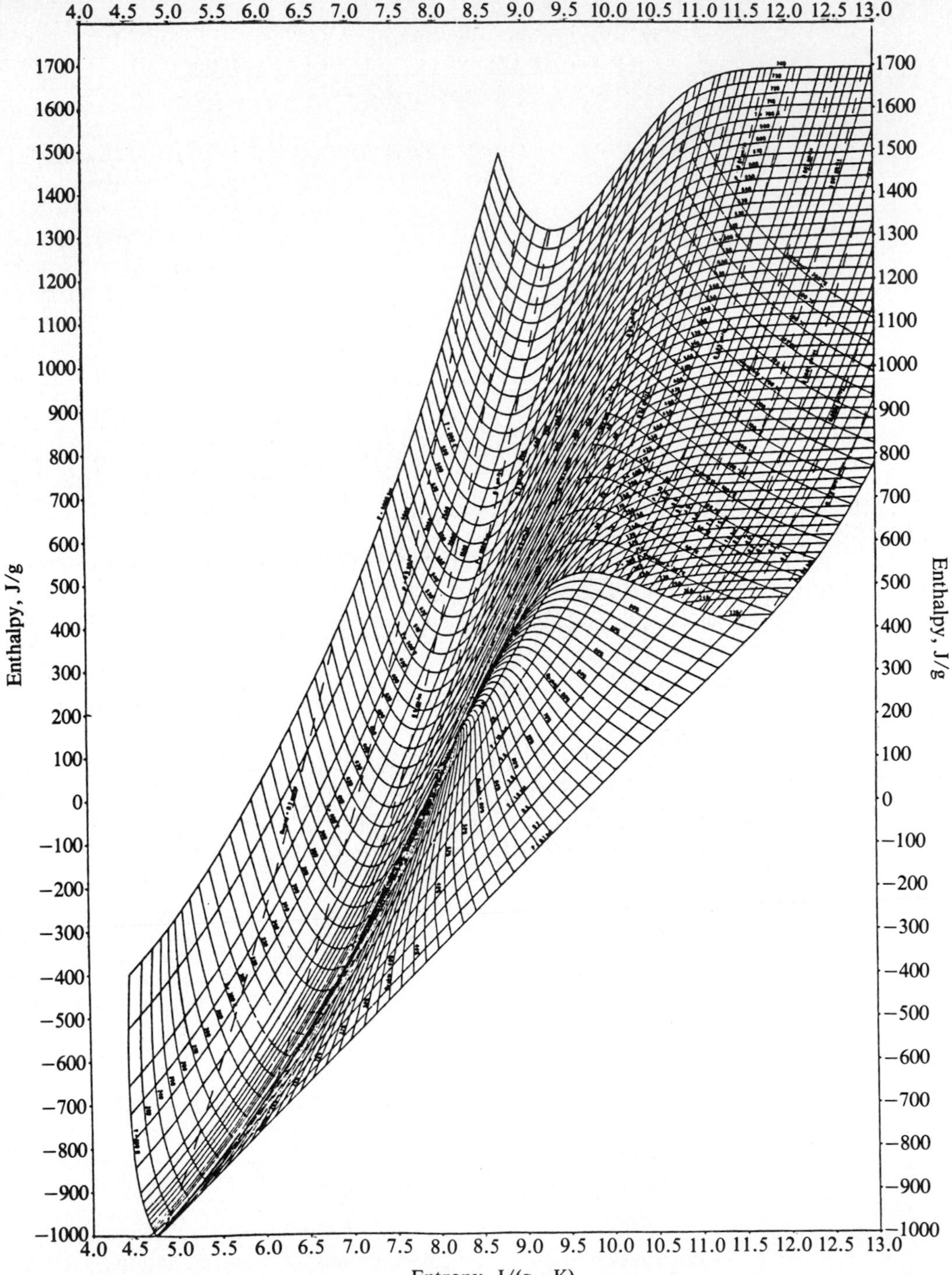

Figure 5.7 Enthalpy-entropy (Mollier) diagram for ammonia. *(From Haar and Gallagher, "Thermodynamic Properties of Ammonia," J. Phys. Chem. Ref. Data, 7: 635–792, copyright 1978 by American Institute of Physics. Used with permission.)*

for example, by Eqs. (5.60) and (5.61) for ΔS and ΔH, respectively, for a van der Waals gas can be utilized. Such expressions are often used for the computerized generation of thermodynamic diagrams.

Construction from generalized equations of state The Lydersen or Pitzer type of correlation can be used to generate thermodynamic diagrams. This section will discuss in detail methods for constructing *HT* and *TS* diagrams. *PH* and *HS* diagrams can be readily constructed by using the same general procedures.

The entire procedure by which these diagrams are constructed is by first calculating ideal-gas properties by the standard methods, followed by use of enthalpy and entropy departures as discussed in Sec. 4.5.3 to correct for pressure. Either the figures or alternate computer methods discussed in Chap. 4 may be used to evaluate these departures.

As noted before, an expression for ideal-gas heat capacity as a function of temperature is required. In addition, if latent heats are experimentally known, they can be used as an alternate to prediction by the corresponding-states technique.

Enthalpy-temperature diagrams To begin the preparation of the diagram, the range of temperature and pressure to be covered and the base must be selected. Since often the properties are not important below a certain temperature, this temperature is often taken as the temperature base rather than 0 K. At this selected temperature the saturated-liquid condition is chosen, fixing the pressure. Thus, a common reference state is saturated liquid at its pressure at the chosen temperature.

Consider the diagram of Fig. 5.8. This discussion will show how to calculate enthalpies at the base temperature T_0, any temperature below the critical T, and at the critical, T_c. Obviously the network of lines can be as fine or as coarse as accuracy in use demands.

A: The enthalpy at point A (T_0, P_0, the saturation pressure of the compound at T_0) is taken equal to zero: $H_A = 0$.

B: The enthalpy at point B can be determined as $H_B = H_A + \lambda_{P_0, T_0}$, with λ determined from experimental data or from the generalized charts (e.g., Figs. 4.7 or 4.14) by reading the value of $H^* - H$ from each of the saturated-vapor and saturated-liquid curves at $T_{r0} = T_0/T_c$ and $P_{r0} = P_0/P_c$,

$$\lambda_{P_0, T_0} = (H^* - H)_{\text{sat liq}} - (H^* - H)_{\text{sat vap}}$$

C: The enthalpy at C is determined by reading $H^* - H$ from the generalized chart at T_{r0} and P_{r0} for the vapor and adding it to H_B.

$$H_C^* = H_B + (H_C^* - H_B)$$

$D \to H$: An infinite set of such points exists. Calculations are identical for each.

D: $$H_D^* = H_C^* + \int_{T_0}^{T} C_P^* \, dT$$

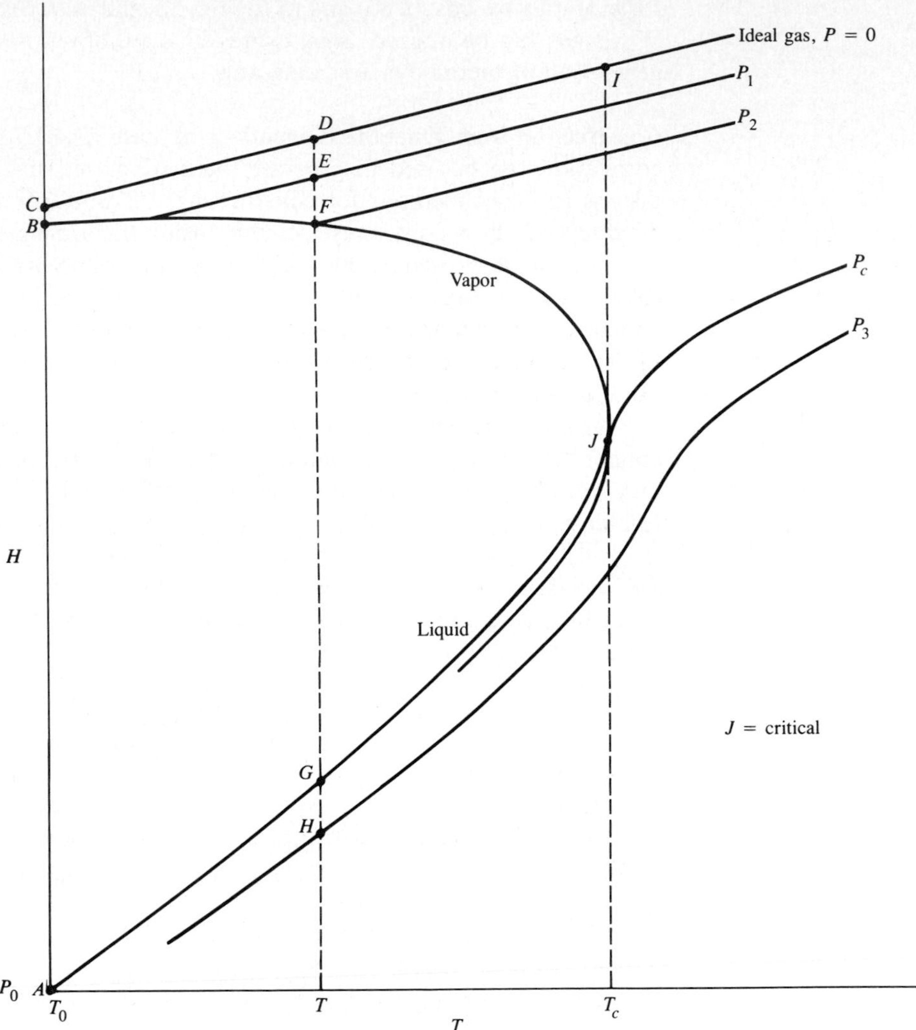

Figure 5.8 Generalized enthalpy-temperature diagram construction.

This is easily calculated by using the expression for ideal-gas heat capacity.

E: For any arbitrary pressure above the dew point, for example, P_1, calculate $P_{r1} = P_2/P_c$ and $T_r = T/T_c$. From the generalized chart, calculate $H_D^* - H_E$.

$$H_E = H_D^* - (H_D^* - H_E)$$

F: For the dew point pressure at T, P_2, calculate $P_{r2} = P_2/P_c$ and determine $H_D^* - H_F$ from the vapor line of the two-phase generalized envelope at P_{r2} and T_r.

$$H_F = H_D^* - (H_D^* - H_F)$$

G: For the same pressure and temperature as F, determine $H_D^* - H_G$ from the liquid line of the two-phase generalized envelope at P_{r2} and T_r.

$$H_G = H_D^* - (H_D^* - H_G)$$

H: For any arbitrary pressure above the bubble point pressure, for example, P_3, calculate $P_{r3} = P_3/P_c$. From the generalized chart determine $H_D^* - H_H$ at P_{r3} and T_r.

$$H_H = H_D^* - (H_D^* - H_H)$$

Note that there can be any number of points E and H.

$I \rightarrow J$: This set of points corresponds to the critical temperature T_c with $T_{rc} = 1$.

I: For the ideal gas,

$$H_I^* = H_C^* + \int_{T_0}^{T_c} C_P^* \, dT$$

J: For any pressure above zero including the critical pressure, P_1, P_2, P_c, P_3, calculate $P_{rn} = P_n/P_c$. Determine $H_I^* - H_J$ at any $P_{r,n}$ and $T_r = 1$. Thus

$$H_J = H_I^* - (H_I^* - H_J)$$

This procedure will yield a total enthalpy-temperature diagram.

Temperature-entropy diagrams Temperature-entropy diagrams can be developed in the same manner as enthalpy-temperature diagrams except in this case there is the pressure effect on ideal-gas entropy. Thus, as shown previously, $H_D^* = H_E^* = H_F^*$. For the case of entropy, $S_D^* \neq S_E^* \neq S_F^*$, and an extra calculation enters into the process of diagram construction.

Following the same base and example as for the PT diagram and Fig. 5.9, the entropy S_A, saturated liquid at T_0 and P_0 is either assumed to be zero or calculated from tabulated values. In either case, S_A is now known.

B: S_B is calculated either by using an experimental value of heat of vaporization at T_0 and P_0 and dividing by T_0 and adding to S_A,

$$S_B = S_A + \frac{\lambda_{T_0, P_0}}{T_0}$$

or by taking the difference between the generalized vapor- and liquid-departure curves (e.g., Fig. 4.10 or 4.18) at T_{r0} and P_{r0} and adding to S_A. That is,

$$S_B = S_A + (S_C^* - S_A)_{\text{sat liq}} - (S_C^* - S_A)_{\text{sat vap}}$$

C: S_C^* is calculated by adding the departure of vapor at T_{r0} and P_{r0} to S_B.

$$S_C^* = S_B + (S_C^* - S_B)$$

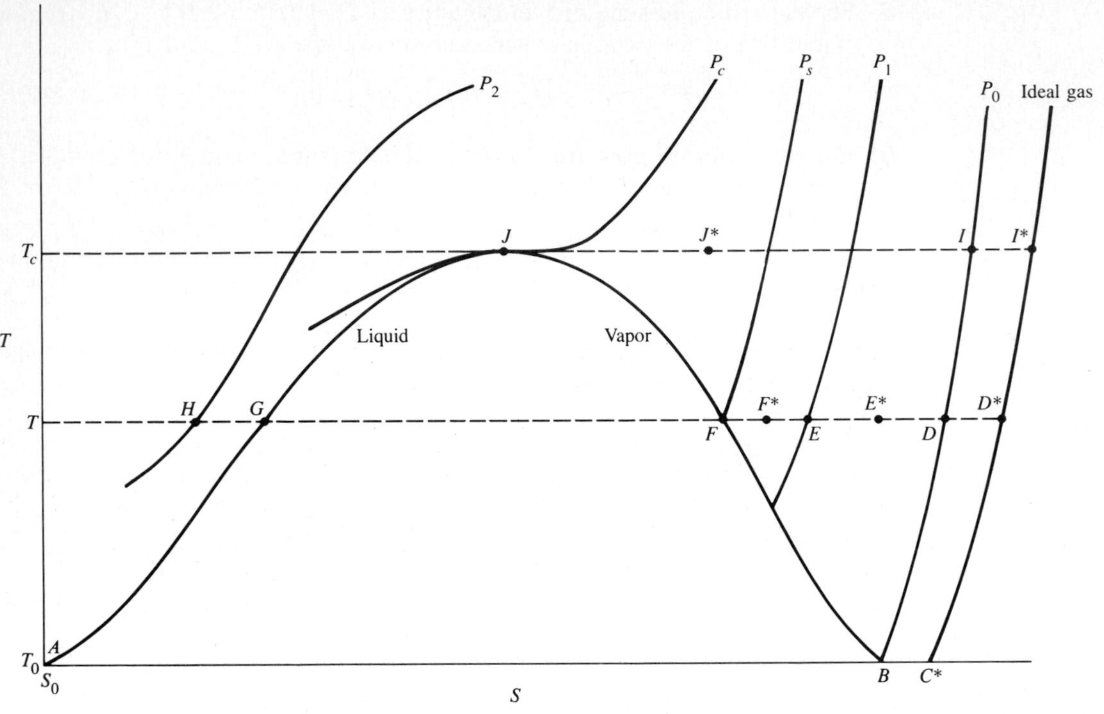

Figure 5.9 Generalized temperature-entropy diagram construction.

D^*: The ideal-gas value is changed for temperature in the usual way.

$$S_D^* = S_C^* + \int_{T_0}^{T} \frac{C_P^*}{T}\, dT$$

D:
$$S_D = S_D^* - \left(S_D^* - S_D\right)_{T_{r0},\,P_{r0}}$$

E^*:
$$S_E^* = S_D^* - R \ln \frac{P_E}{P_0}$$

E:
$$S_E = S_E^* - \left(S_E^* - S_E\right)_{T_r,\,P_r}$$

F^*:
$$S_F^* = S_D^* - R \ln \frac{P_F}{P_0}$$

F:
$$S_F = S_F^* - \left(S_F^* - S_F\right)_{\text{sat vap}} \qquad \text{at } T_r,\, P_{r,s}$$

G:
$$S_G = S_F - \left(S_F^* - S_F\right)_{\text{sat liq}} + \left(S_F^* - S_F\right)_{\text{sat vap}}$$

H^*:
$$S_H^* = S_D^* - R \ln \frac{P_H}{P_0}$$

H:
$$S_H = S_H^* - \left(S_H^* - S_H\right)_{T_r,\,P_{r2}}$$

$$I^*: \qquad S_I^* = S_C^* + \int_{T_0}^{T_c} \frac{C_P^*}{T}\, dT$$

$$I: \qquad S_I = S_I^* - \left(S_I^* - S_I\right)_{T_r=1,\, P_{r0}}$$

$$J^*: \qquad S_J^* = S_I^* - R \ln \frac{P_c}{P_0}$$

$$J: \qquad S_J = S_J^* - \left(S_J^* - S_J\right)_{T_r=1,\, P_r=1}$$

This procedure can be used for constructing an entire TS diagram.

Other diagrams Using the calculations made for the TS and HT diagrams, a Mollier (HS) diagram can be directly plotted. The PH diagram can be plotted directly from the calculations for the HT diagram.

Often charts are made more complex by placing lines of constant enthalpy on temperature-entropy diagrams and lines of constant entropy on enthalpy-temperature or pressure-enthalpy diagrams. As shown earlier, lines of constant volume and quality are also often plotted. Care must be taken when such lines are placed on diagrams as they often become so complex that they are almost unreadable. It often is more desirable to have a series of plots than one plot with a maze of lines.

5.6 REPRESENTATIVE THERMODYNAMIC DIAGRAMS AND DATA

Many compilations of thermodynamic diagrams and data exist in the literature. Some of the most useful general collections are listed below:

"Selected Values of Properties of Hydrocarbons and Related Compounds," API Research Project 44 and TRC Data Project, Thermodynamics Research Center, Texas A&M University, College Station, Tex., loose-leaf, extant 1983. Primarily hydrocarbons, nonhydrocarbon gases, and common low-molecular-weight organic materials. Data only.

"Technical Data Book—Petroleum Refining," 4th ed., American Petroleum Institute, Washington, extant 1982. Primarily hydrocarbons and nonhydrocarbon gases. Data and diagrams.

"JANAF Thermochemical Tables," 2d ed., NSRDS-NBS37, 1971. Supplements in *J. Phys. Chem. Ref. Data*, 3:311 (1974); 4:1 (1975); 7:793 (1978). Mainly inorganics and gases. Data only.

"Thermodynamic Properties and Reduced Correlations for Gases," Lawrence Canjar and Francis Manning, Gulf, Houston, 1967. Light hydrocarbons and nonhydrocarbon gases. Data and diagrams.

"Physical Properties of Hydrocarbons," vols. 1 and 2, Robert Gallant, Gulf, Houston, 1968 and 1970. Light hydrocarbons, gases, and miscellaneous organics, many predicted properties. Diagrams.

"Tables of Physical and Thermodynamic Properties of Pure Compounds," T. E. Daubert and R. P. Danner, American Institute of Chemical Engineers, New York, extant 1984. Important evaluated physical properties of most important chemical compounds—top 193 in 1984, adding to a total of 1000 by 1987.

The most widely used thermodynamic data are those for water. Since water properties must be known very accurately for some applications, most of the compilations have been based on experimental PVT data and extended and extrapolated with great care. Keenan et al. (1978) is one of the best sources of

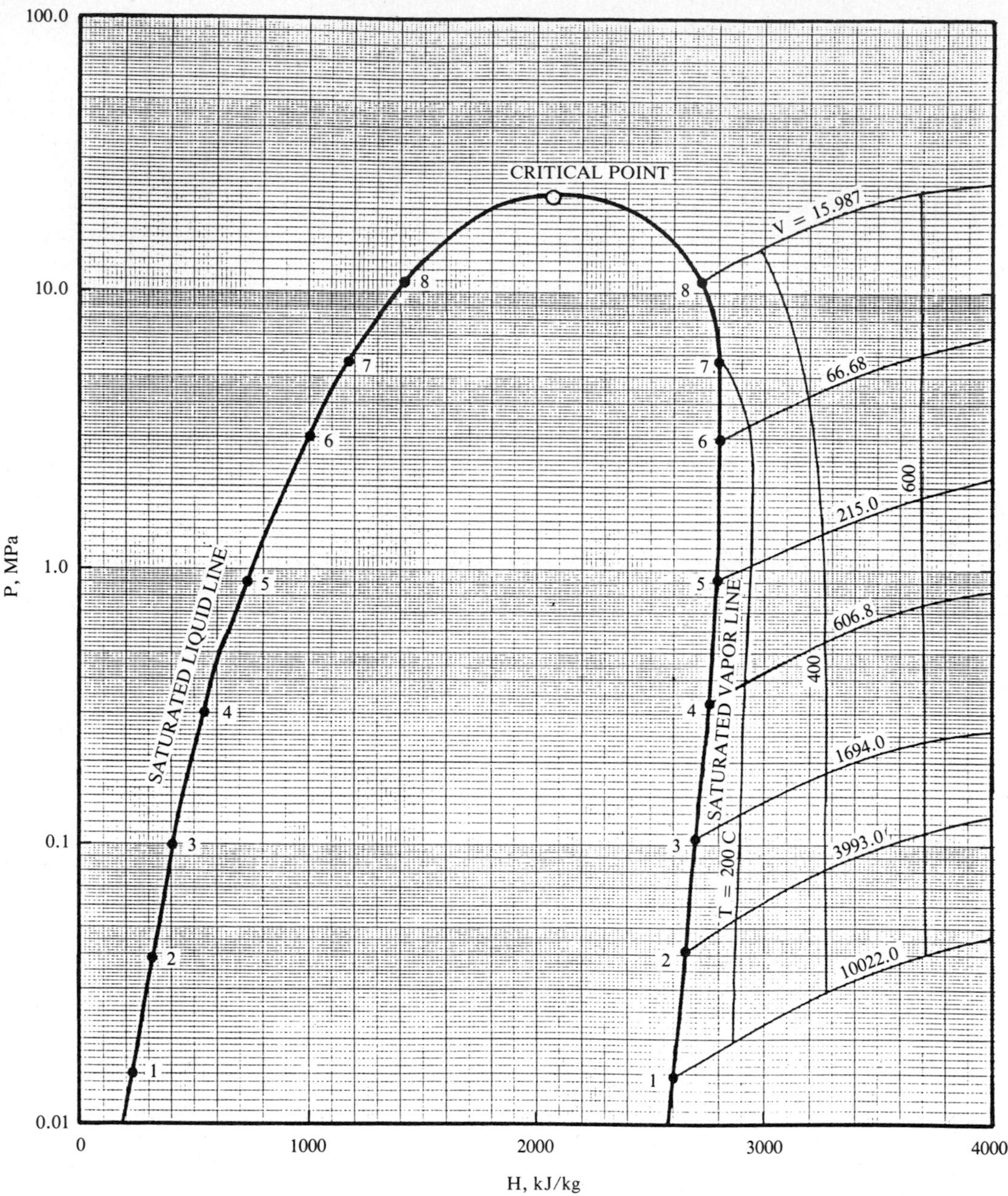

Figure 5.10 Pressure-enthalpy diagram for water.

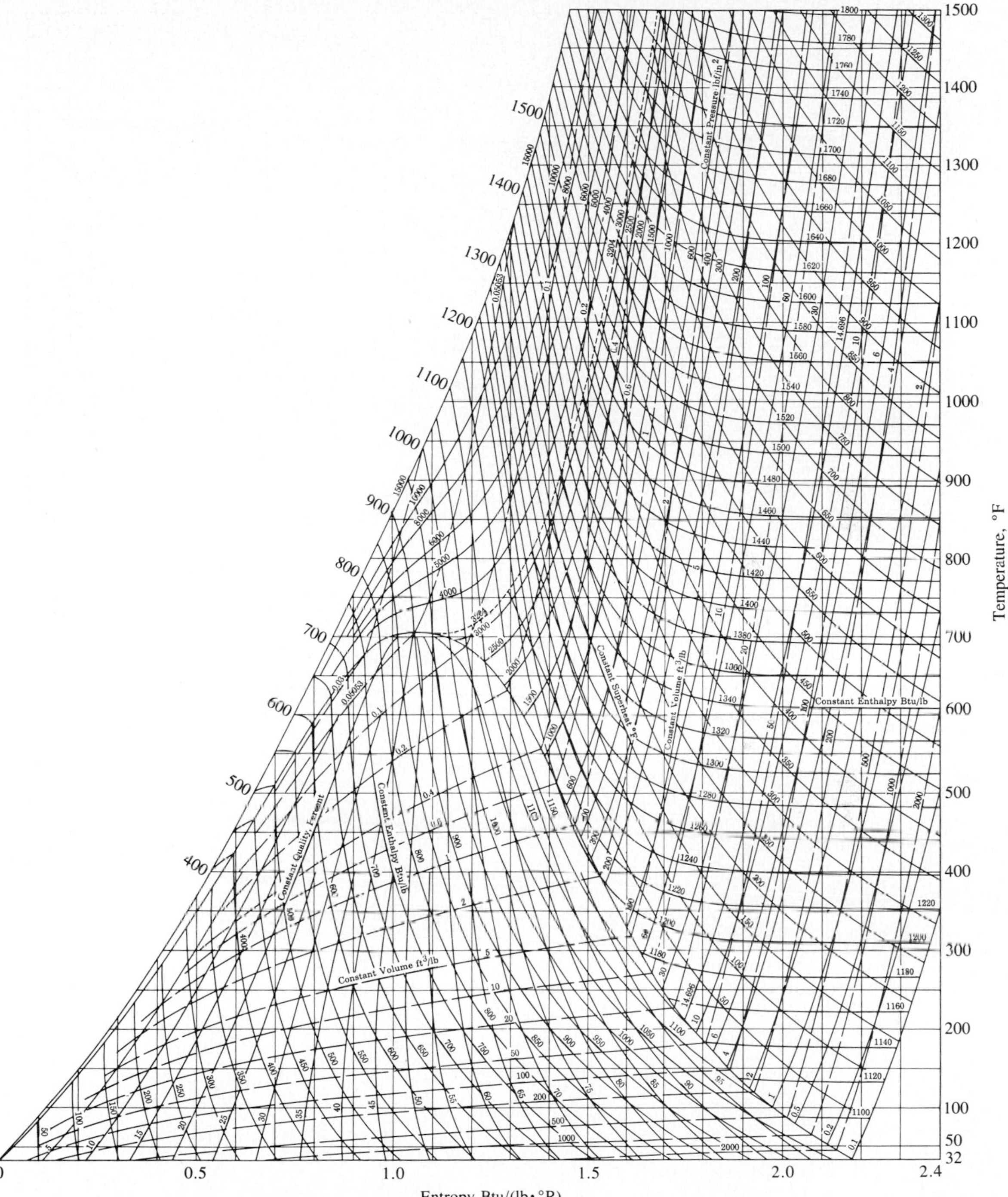

Figure 5.11 Temperature-entropy diagram for water. *(From Keenan, Keyes, Hill, and Moore, "Steam Tables—SI Units," copyright 1978 by Wiley. Used with permission.)*

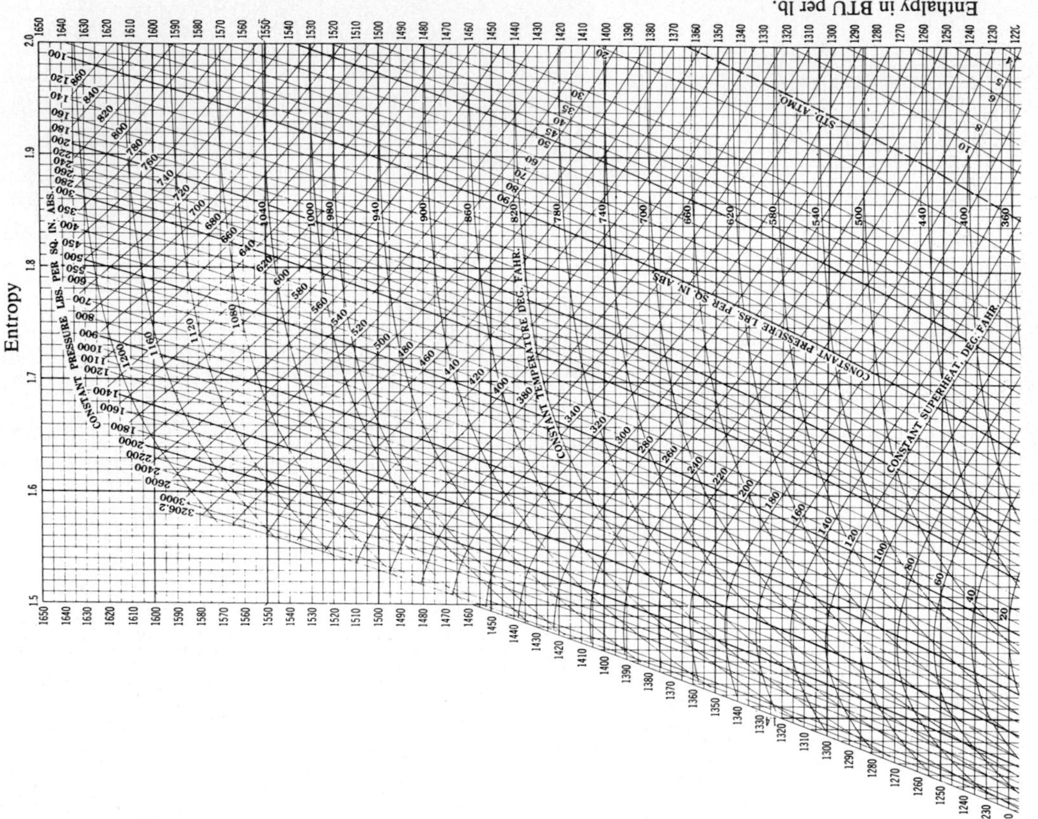

Entropy
Enthalpy in BTU per lb.
CONSTANT PRESSURE LBS. PER SQ. IN. ABS.
CONSTANT TEMPERATURE DEG. FAHR.
CONSTANT PRESSURE LBS. PER SQ. IN. ABS.
CONSTANT SUPERHEAT, DEG. FAHR.
STD. ATMO.

Figure 5.12

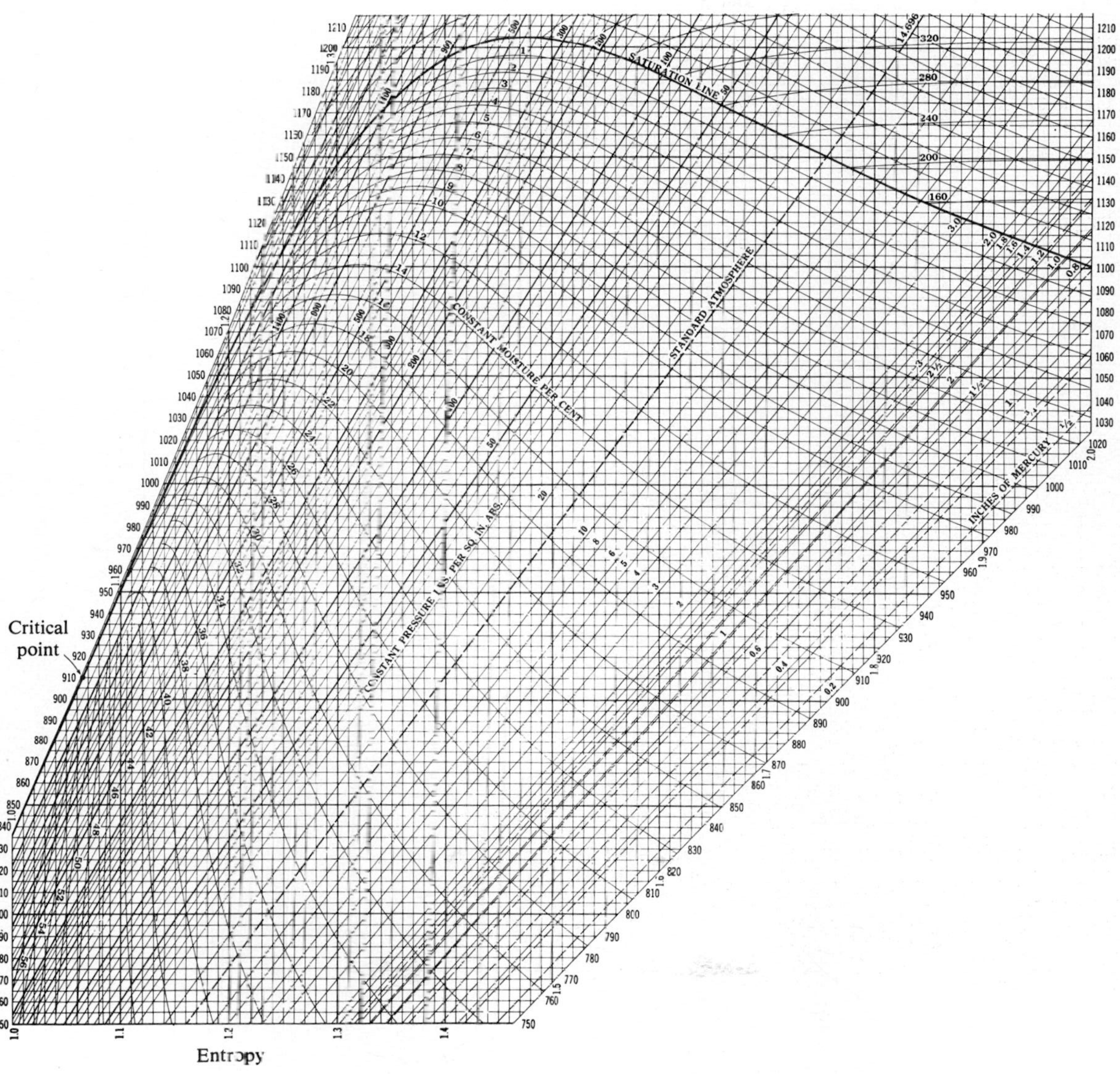

Figure 5.12 Enthalpy-entropy (Mollier) diagram for water. *(From "Steam Tables," copyright 1967 by Combustion Engineering, Inc. Used with permission.)*

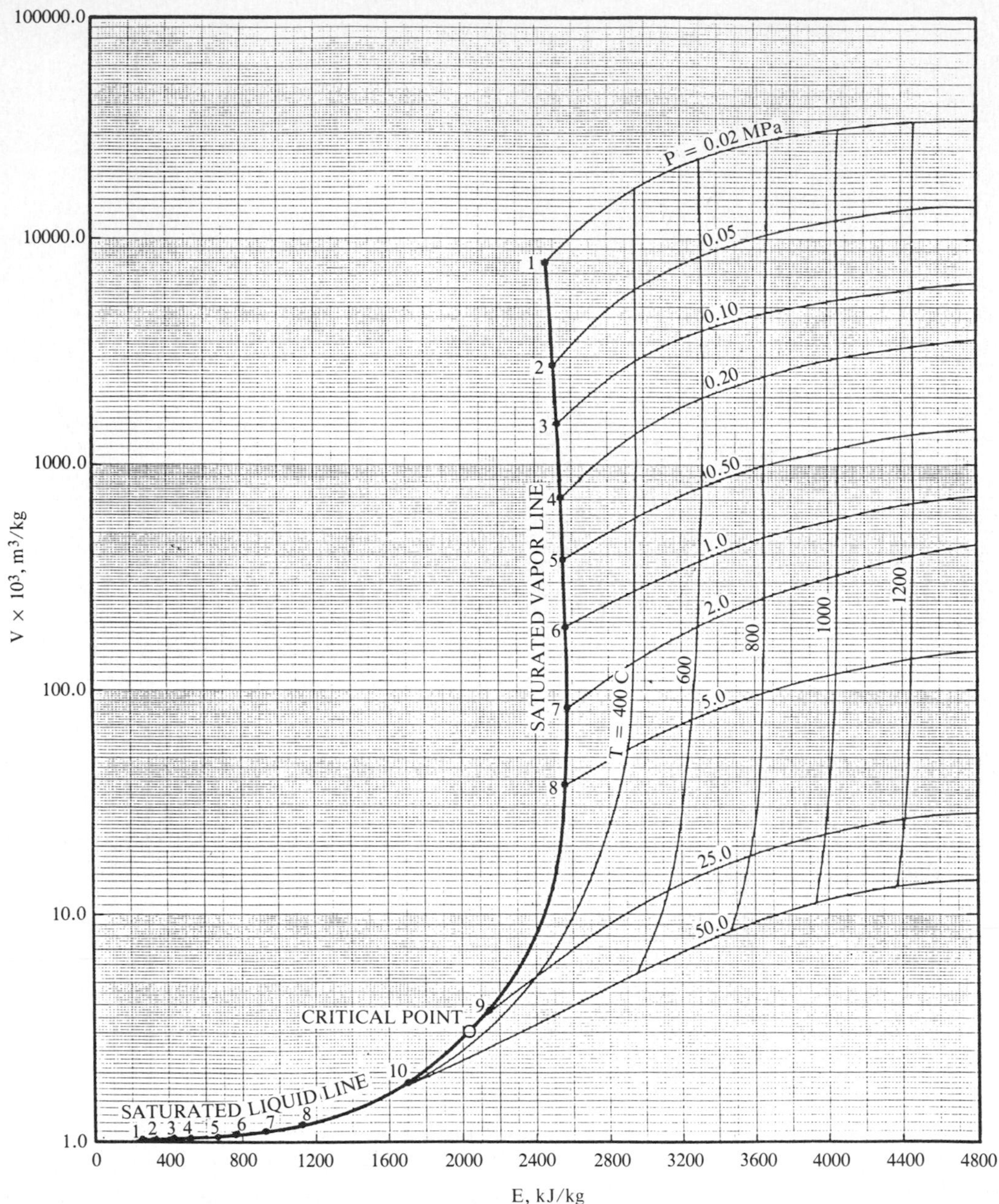

Figure 5.13 Internal energy-volume diagram for water.

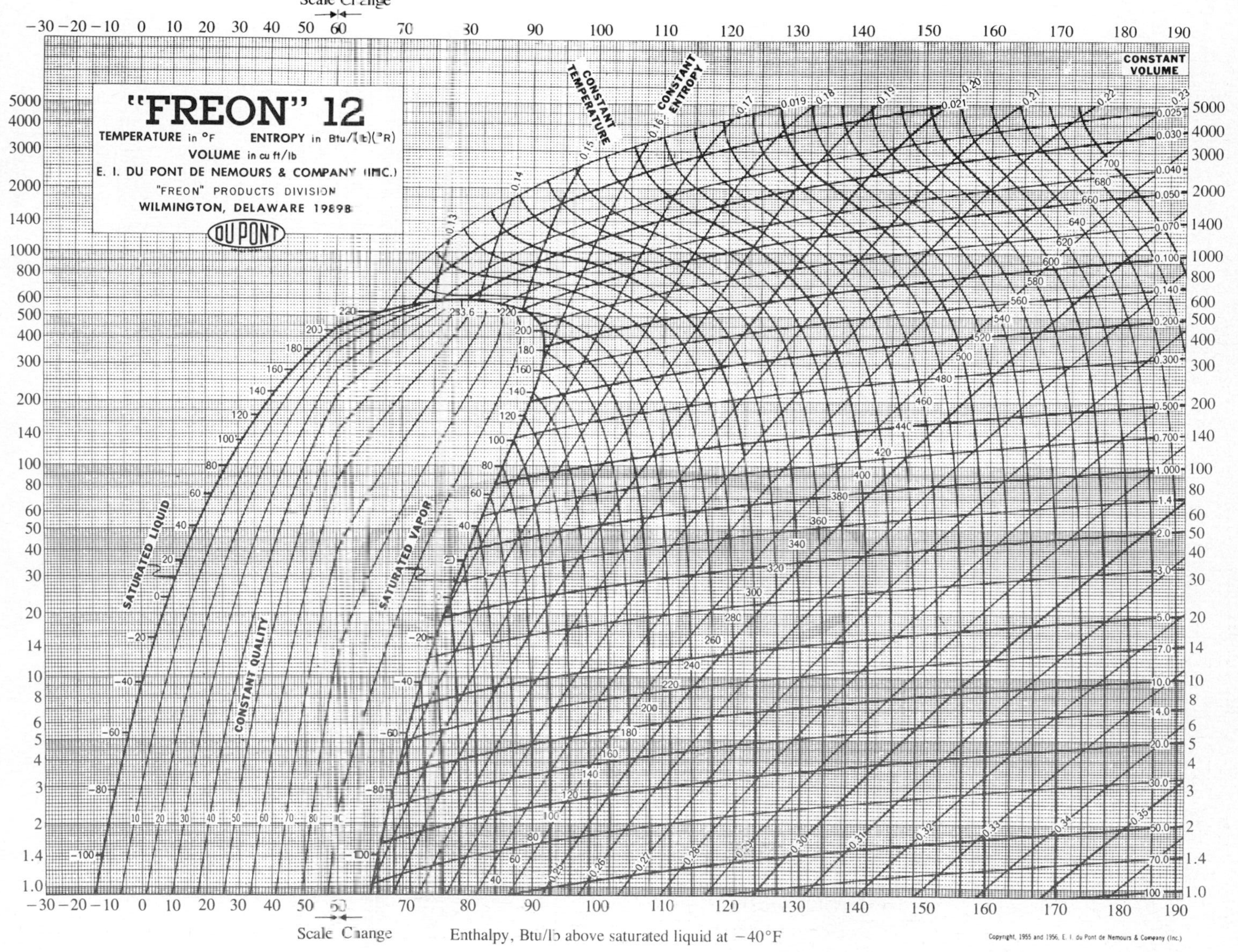

Figure 5.14 Pressure-enthalpy diagram for Freon 12. *(Adapted from "Thermodynamic Properties of Freon 12," T-12, copyright 1956 by E. I. du Pont de Nemours & Co. Used with permission.)*

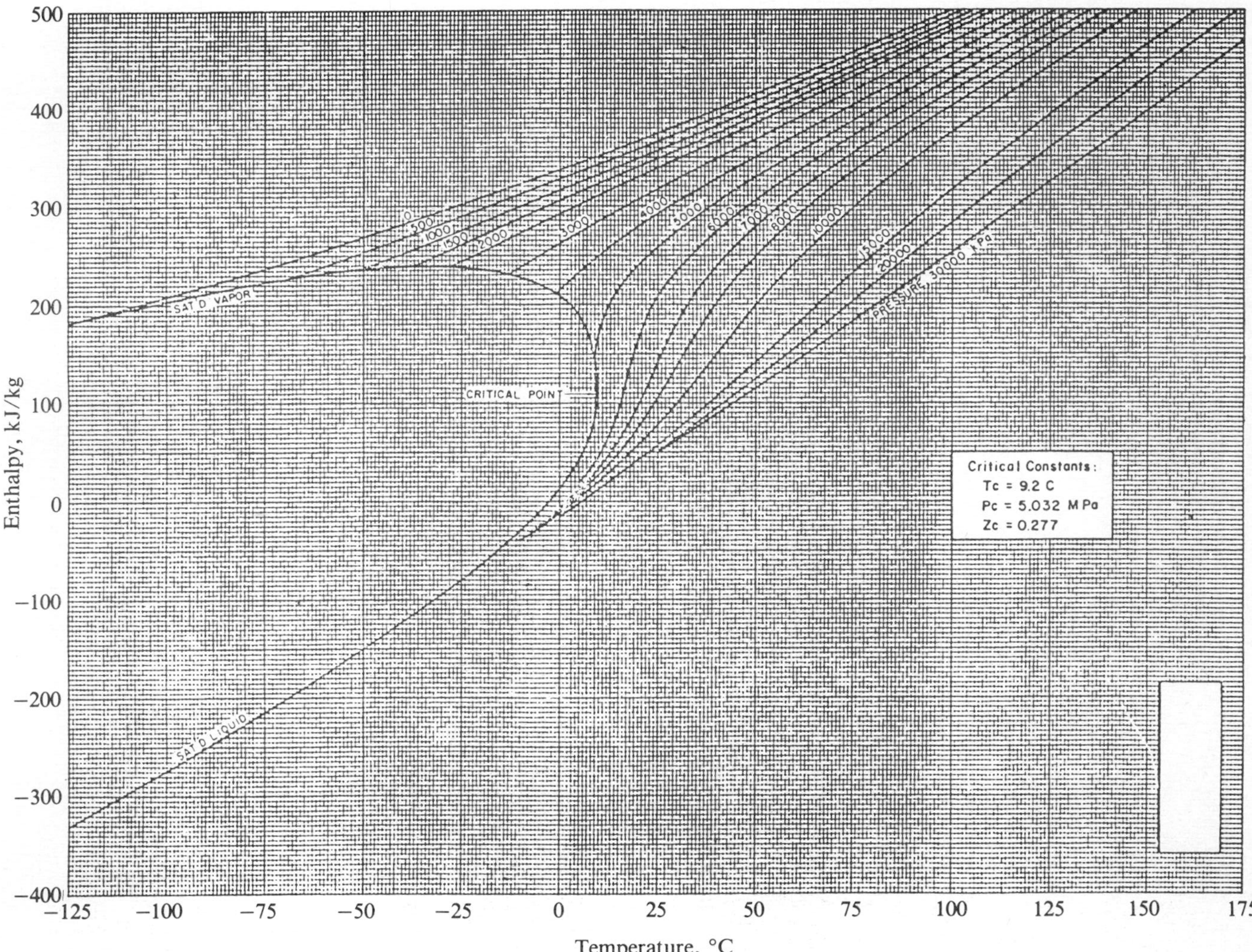

Figure 5.15 Enthalpy-temperature-pressure diagram for ethylene. *(Adapted from "Technical Data Book—Petroleum Refining," 4th ed., copyright 1982 by American Petroleum Institute. Used with permission.)*

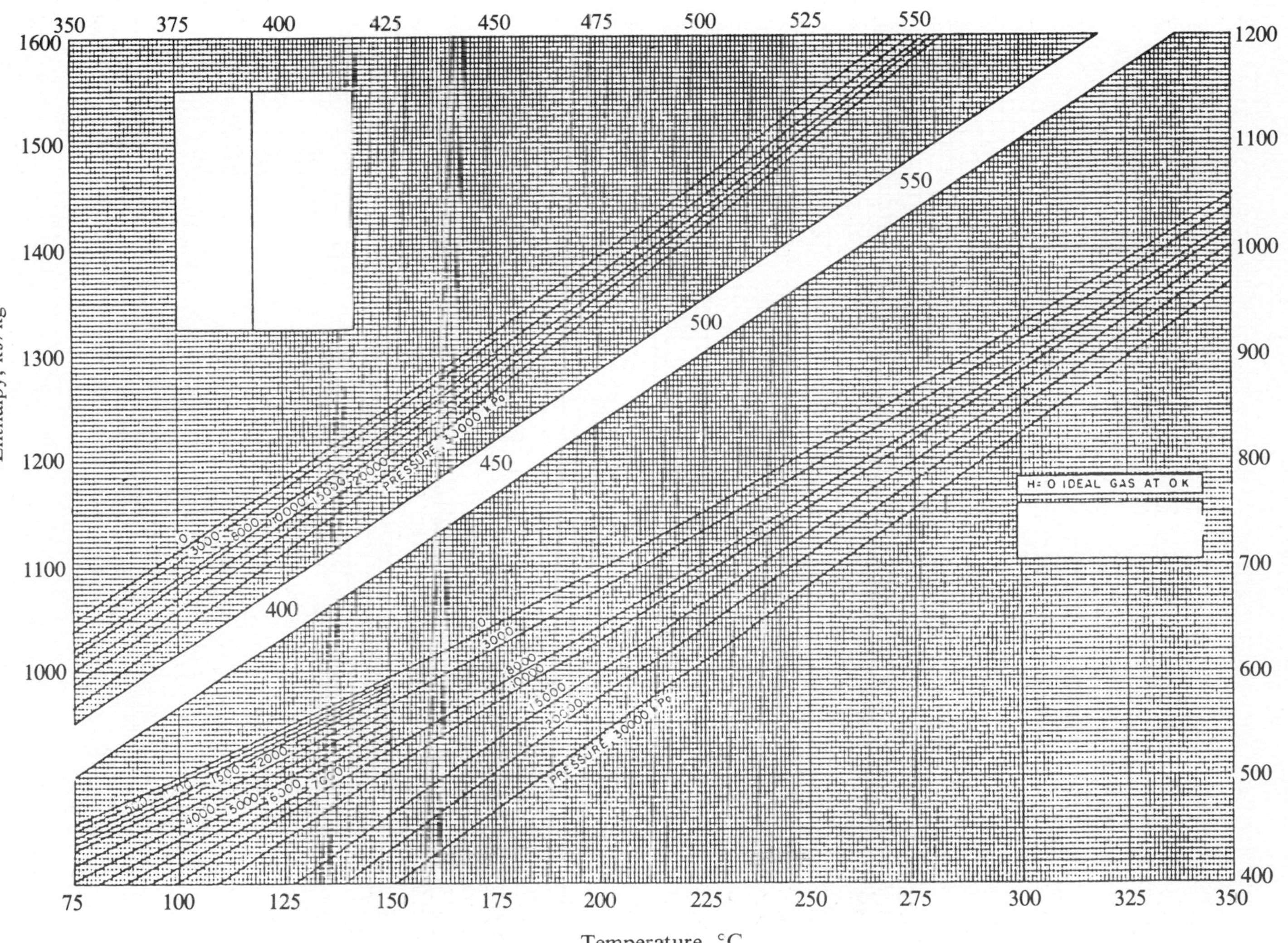

Temperature, °C

Figure 5.15 Continued

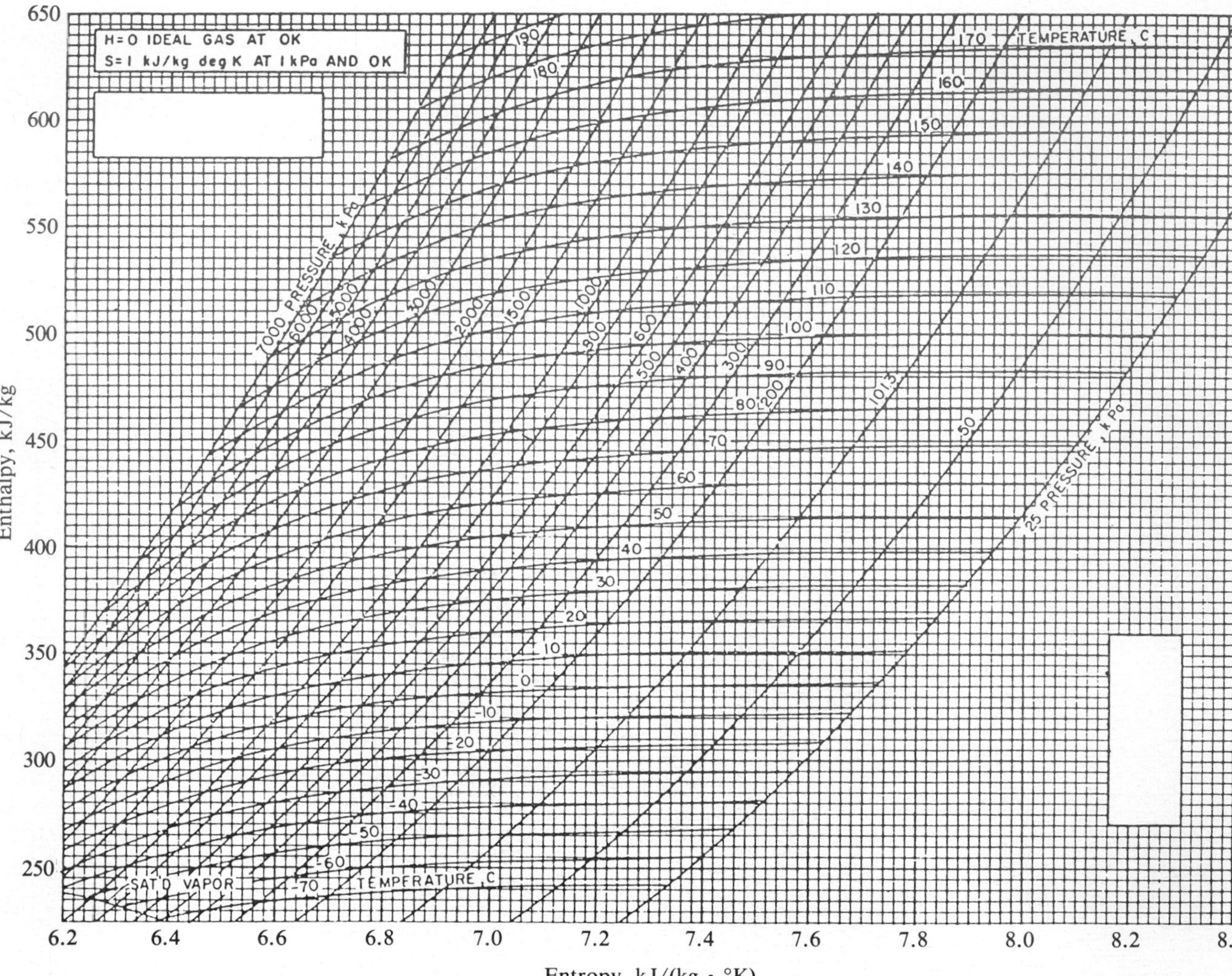

Figure 5.16 Enthalpy-entropy (Mollier) diagram for ethylene. *(Adapted from "Technical Data Book—Petroleum Refining, 4th ed., copyright 1982 by American Petroleum Institute. Used with permission.)*

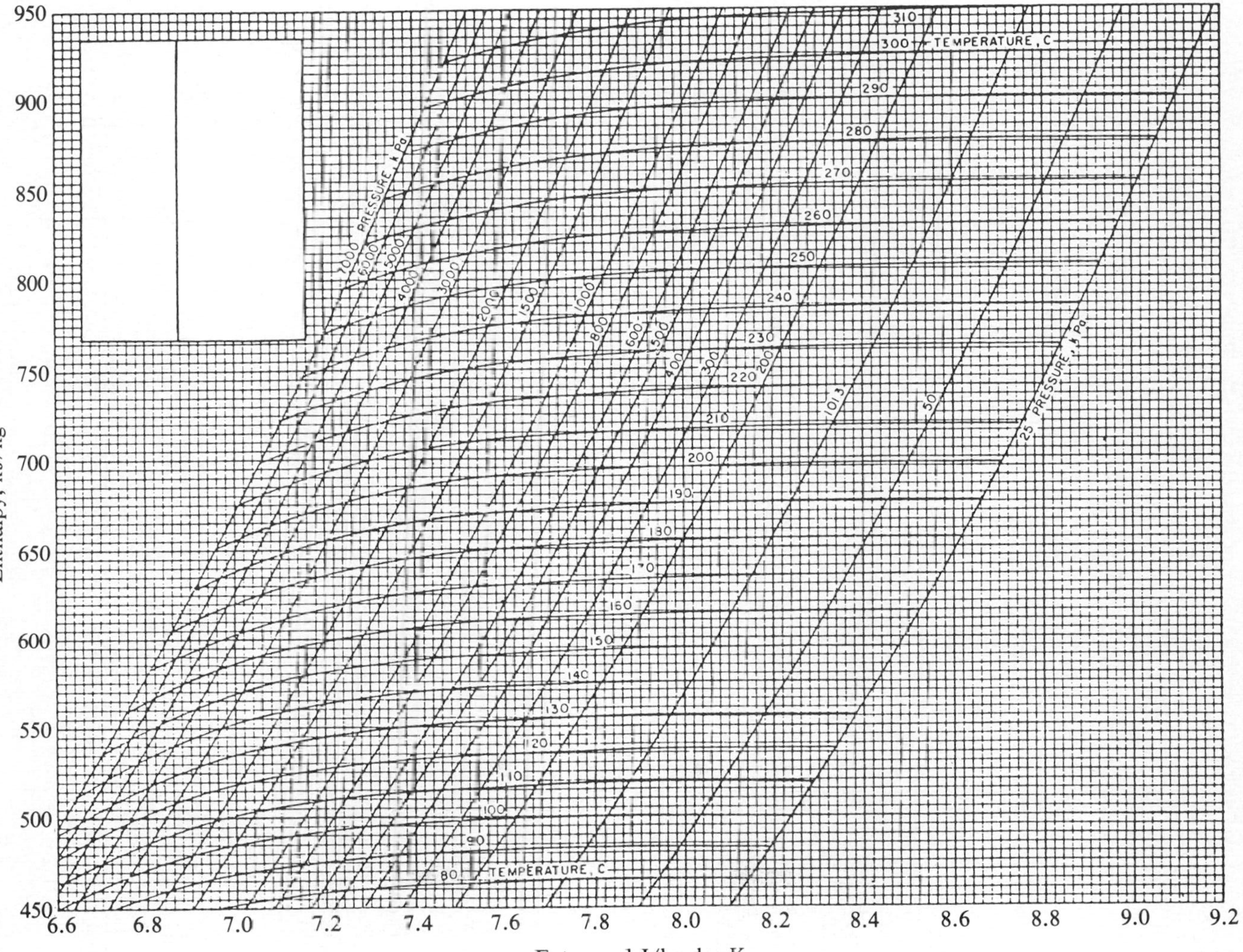

Figure 5.16 Continued

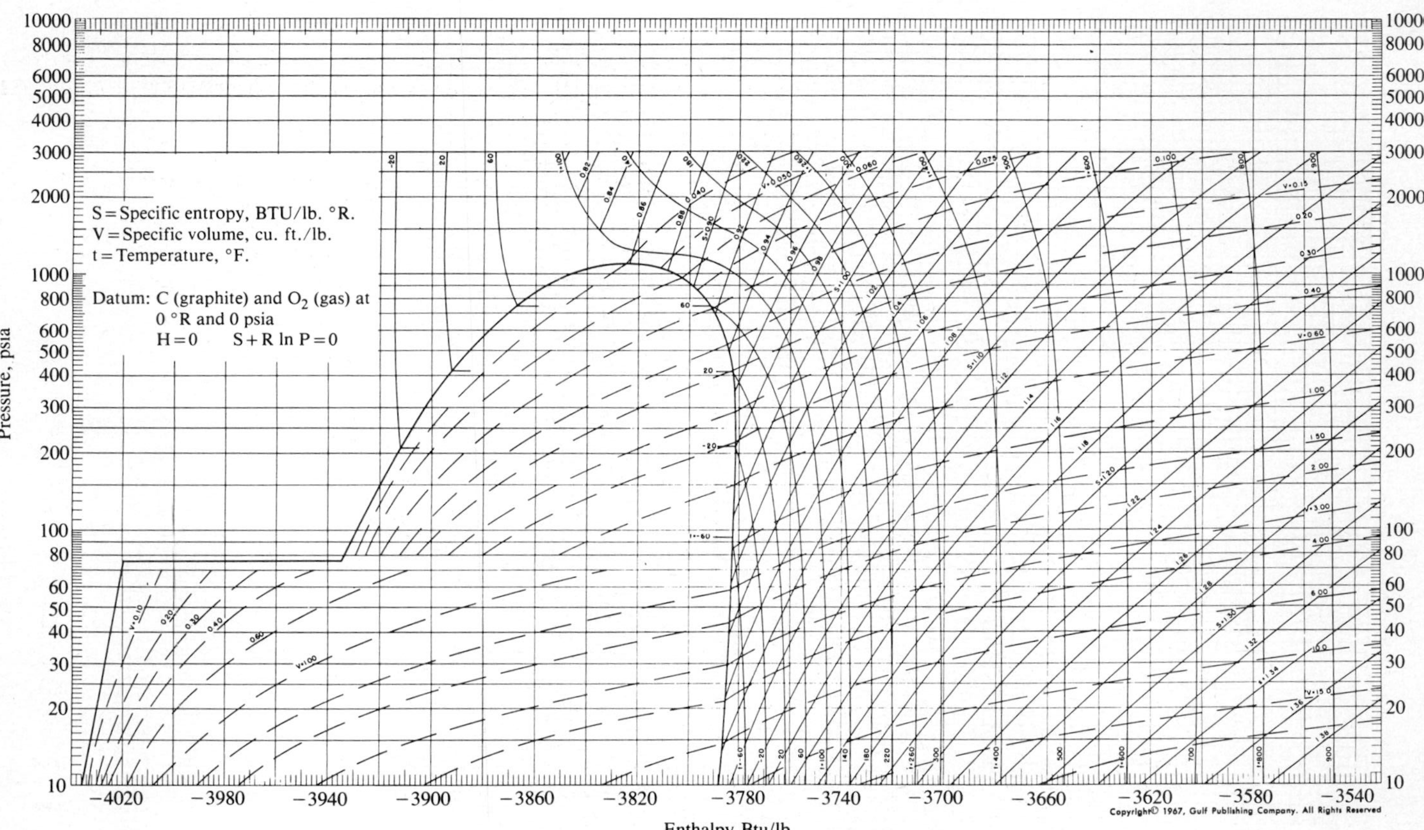

Figure 5.17 Pressure-enthalpy diagram for carbon dioxide. *(From Lawrence N. Canjar and Francis S. Manning, "Thermodynamic Properties and Reduced Correlations for Gases," copyright © 1967 by Gulf, Houston. All rights reserved. Used with permission.)*

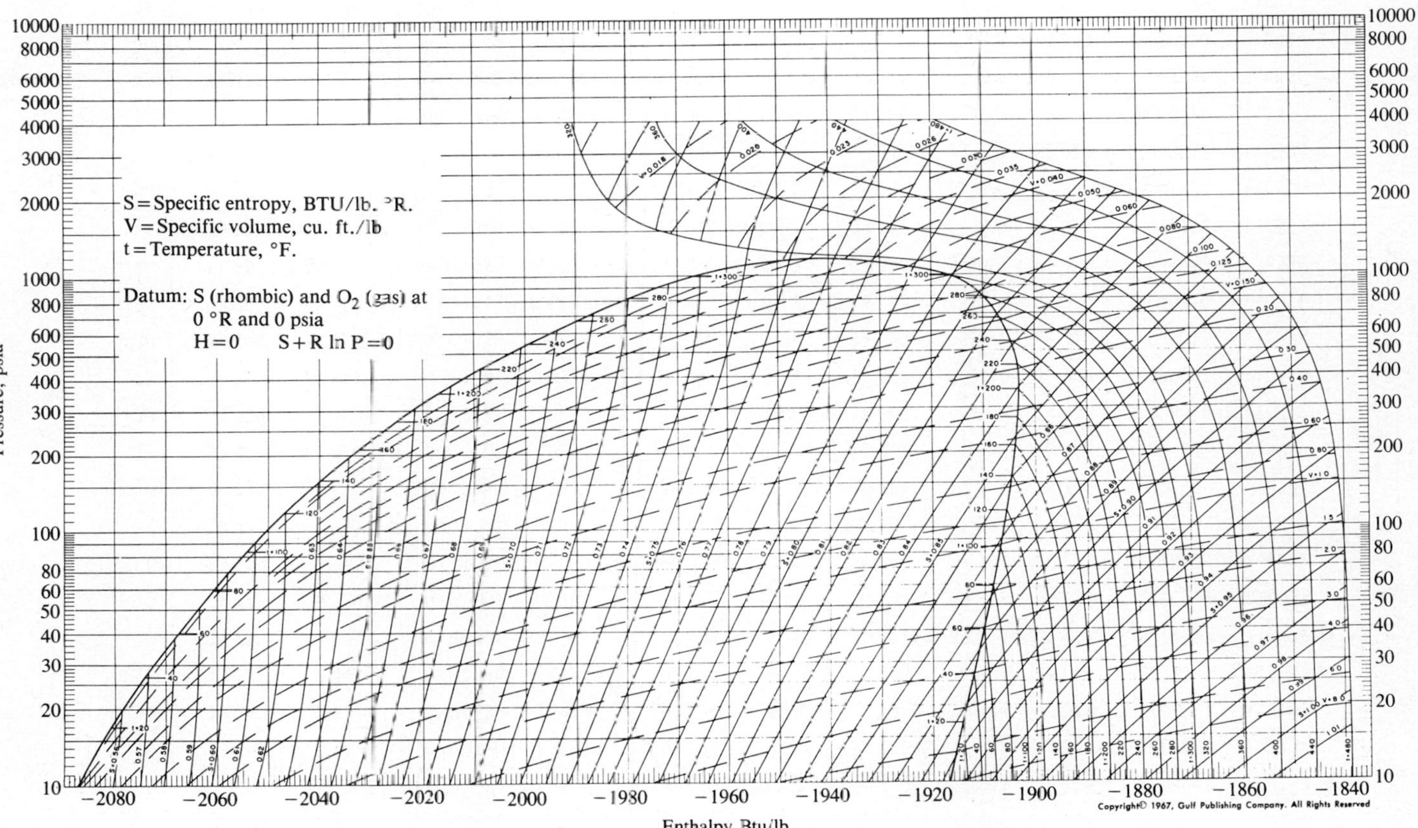

Figure 5.18 Pressure-enthalpy diagram for sulfur dioxide. *(From Lawrence N. Canjar and Francis S. Manning, "Thermodynamic Properties and Reduced Correlations for Gases," copyright © 1967 by Gulf, Houston. All rights reserved. Used with permission.)*

Table 5.2 Freon 12—saturation properties

Temp., °F	Pressure, lb/in²		Volume, ft³/lb		Density, lb/ft³		Enthalpy, Btu/lb			Entropy, Btu/(lb · °R)	
	abs	gauge	Liquid v_f	Vapor v_g	Liquid $1/v_f$	Vapor $1/v_g$	Liquid h_f	Latent h_{fg}	Vapor h_g	Liquid s_f	Vapor s_g
−40	9.3076	10.9709*	0.010564	3.8750	94.661	0.25806	0	72.913	72.913	0	0.17373
−35	10.586	8.367*	0.010618	3.4373	94.178	0.29093	1.0546	72.409	73.464	0.002492	0.17299
−30	11.999	5.490*	0.010674	3.0585	93.690	0.32696	2.1120	71.903	74.015	0.004961	0.17229
−25	13.556	2.320*	0.010730	2.7295	93.197	0.36636	3.1724	71.391	74.563	0.007407	0.17164
−20	15.267	0.571	0.010788	2.4429	92.699	0.40934	4.2357	70.874	75.110	0.009831	0.17102
−15	17.141	2.445	0.010846	2.1924	92.197	0.45612	5.3020	70.352	75.654	0.012234	0.17043
−10	19.189	4.493	0.010906	1.9727	91.689	0.50693	6.3716	69.824	76.196	0.014617	0.16989
−5	21.422	6.726	0.010968	1.7794	91.177	0.56199	7.4444	69.291	76.735	0.016979	0.16937
0	23.849	9.153	0.011030	1.6089	90.659	0.62156	8.5207	68.750	77.271	0.019323	0.16888
5	26.483	11.787	0.011094	1.4580	90.135	0.68588	9.6005	68.204	77.805	0.021647	0.16842
10	29.335	14.639	0.011160	1.3241	89.606	0.75523	10.684	67.651	78.335	0.023954	0.16798
15	32.415	17.719	0.011227	1.2050	89.070	0.82986	11.771	67.090	78.861	0.026243	0.16758
20	35.736	21.040	0.011296	1.0988	88.529	0.91006	12.863	66.522	79.385	0.028515	0.16719
25	39.310	24.614	0.011366	1.0039	87.981	0.99613	13.958	65.946	79.904	0.030772	0.16683
30	43.148	28.452	0.011438	0.91880	87.426	1.0884	15.058	65.361	80.419	0.033013	0.16648
35	47.263	32.567	0.011512	0.84237	86.865	1.1871	16.163	64.767	80.930	0.035240	0.16616
40	51.667	36.971	0.011588	0.77357	86.296	1.2927	17.273	64.163	81.436	0.037453	0.16586
45	56.373	41.677	0.011666	0.71149	85.720	1.4055	18.387	63.550	81.937	0.039652	0.16557
50	61.394	46.698	0.011746	0.65537	85.136	1.5258	19.507	62.926	82.433	0.041839	0.16530
55	66.743	52.047	0.011828	0.60453	84.544	1.6542	20.634	62.290	82.924	0.044015	0.16504
60	72.433	57.737	0.011913	0.55839	83.944	1.7909	21.766	61.643	83.409	0.046180	0.16479
65	78.477	63.781	0.012000	0.51642	83.335	1.9364	22.905	60.982	83.887	0.048336	0.16456
70	84.888	70.192	0.012089	0.47818	82.717	2.0913	24.050	60.309	84.359	0.050482	0.16434
75	91.682	76.986	0.012182	0.44327	82.089	2.2560	25.204	59.621	84.825	0.052620	0.16412
80	98.870	84.174	0.012277	0.41135	81.450	2.4310	26.365	58.917	85.282	0.054751	0.16392
85	106.47	91.77	0.012376	0.38212	80.802	2.6170	27.534	58.198	85.732	0.056877	0.16372
90	114.49	99.79	0.012478	0.35529	80.142	2.8146	28.713	57.461	86.174	0.058997	0.16353

95	122.95	108.25	0.012583	0.33063	79.470	3.0245	29.901	56.705	86.606	0.061113	0.16334
100	131.86	117.16	0.012693	0.30794	78.785	3.2474	31.100	55.929	87.029	0.063227	0.16315
105	141.25	126.55	0.012806	0.28701	78.088	3.4842	32.310	55.132	87.442	0.065339	0.16297
110	151.11	136.41	0.012924	0.26769	77.376	3.7357	33.531	54.313	87.844	0.067451	0.16279
115	161.47	146.77	0.013047	0.24982	76.649	4.0029	34.765	53.468	88.233	0.069564	0.16260
120	172.35	157.65	0.013174	0.23326	75.906	4.2870	36.013	52.597	88.610	0.071680	0.16241
125	183.76	169.06	0.013308	0.21791	75.145	4.5891	37.275	51.698	88.973	0.073800	0.16222
130	195.71	181.01	0.013447	0.20364	74.367	4.9107	38.553	50.768	89.321	0.075927	0.16202
135	208.22	193.52	0.013593	0.19036	73.568	5.2532	39.848	49.805	89.653	0.078061	0.16181
140	221.32	206.62	0.013745	0.17799	72.748	5.6184	41.162	48.805	89.967	0.080205	0.16159
145	235.00	220.30	0.013907	0.16644	71.904	6.0083	42.495	47.766	90.261	0.082361	0.16135
150	249.31	234.61	0.014078	0.15564	71.035	6.4252	43.850	46.684	90.534	0.084531	0.16110
155	264.24	249.54	0.014258	0.14552	70.137	6.8717	45.229	45.554	90.783	0.086719	0.16083
160	279.82	265.12	0.014449	0.13604	69.209	7.3509	46.633	44.373	91.006	0.088927	0.16053
165	296.07	281.37	0.014653	0.12712	68.245	7.8665	48.065	43.134	91.199	0.091159	0.16021
170	313.00	298.30	0.014871	0.11873	67.244	8.4228	49.529	41.830	91.359	0.093418	0.15985
175	330.64	315.94	0.015106	0.11080	66.198	9.0252	51.026	40.455	91.481	0.095709	0.15945
180	349.00	334.30	0.015360	0.10330	65.102	9.6802	52.562	38.999	91.561	0.098039	0.15900
185	368.11	353.41	0.015637	0.096190	63.949	10.396	54.141	37.449	91.590	0.10041	0.15850
190	387.98	373.28	0.015942	0.089418	62.728	11.183	55.769	35.792	91.561	0.10284	0.15793
195	408.63	393.93	0.016280	0.082946	61.426	12.056	57.453	34.009	91.462	0.10532	0.15727
200	430.09	415.39	0.016659	0.076728	60.026	13.033	59.203	32.075	91.278	0.10789	0.15651
205	452.38	437.68	0.017094	0.070714	58.502	14.141	61.032	29.955	90.987	0.11055	0.15561
210	475.52	460.82	0.017601	0.064843	56.816	15.422	62.959	27.599	90.558	0.11332	0.15453
215	499.53	484.83	0.018212	0.059030	54.908	16.941	65.014	24.925	89.939	0.11626	0.15320
220	524.43	509.73	0.018986	0.053140	52.670	18.818	67.246	21.790	89.036	0.11943	0.15149
225	550.26	535.56	0.020053	0.046900	49.868	21.322	69.763	17.888	87.651	0.12298	0.14911
230	577.03	562.33	0.021854	0.039435	45.758	25.358	72.893	12.229	85.122	0.12739	0.14512
233.6 (critical)	596.9	582.2	0.02870	0.02870	34.84	34.84	78.86	0	78.86	0.1359	0.1359

* Inches of mercury below 1 atm.

Source: Freon 12, Freon Products Division, E. I. du Pont de Nemours & Co. (Inc.) (with permission).

Table 5.3 Freon 12—Superheated vapor properties at various absolute pressures

V = volume, ft^3/lb; H = enthalpy, Btu/lb; S = entropy, Btu/(lb · °R) (saturation properties in parentheses)

1.0 lb/in^2			Temp.
V (30.896)	H (65.229)	(0.18977)	°F (−109.24 °F)
...	...	...	−110
31.730	66.303	0.19279	−100
32.631	67.482	0.19602	−90
33.531	68.679	0.19922	−80
34.429	69.892	0.20237	−70
35.327	71.123	0.20549	−60
36.223	72.370	0.20857	−50
37.119	73.633	0.21162	−40
38.014	74.913	0.21463	−30
38.908	76.208	0.21761	−20
39.802	77.519	0.22056	−10
40.695	78.845	0.22347	0
41.587	80.186	0.22636	10
42.480	81.542	0.22922	20
43.372	82.912	0.23204	30
44.263	84.297	0.23484	40
45.154	85.695	0.23761	50
46.045	87.107	0.24036	60
46.936	88.533	0.24307	70
47.826	89.971	0.24576	80
48.716	91.423	0.24843	90
49.606	92.887	0.25107	100
50.496	94.364	0.25368	110
51.386	95.853	0.25628	120
52.275	97.354	0.25884	130
53.165	98.867	0.26139	140
54.054	100.391	0.26391	150
54.943	101.926	0.26640	160
55.832	103.472	0.26888	170
56.721	105.030	0.27133	180

5.0 lb/in^2			Temp.
V (6.9069)	H (70.432)	S (0.17759)	°F (−62.35 °F)
...	...	...	−70
6.9509	70.729	0.17834	−60
7.1378	72.003	0.18149	−50
7.3239	73.291	0.18459	-40
7.5092	74.593	0.18766	−30
7.6938	75.909	0.19069	−20
7.8777	77.239	0.19368	−10
8.0611	78.582	0.19663	0
8.2441	79.939	0.19955	10
8.4265	81.309	0.20244	20
8.6086	82.693	0.20529	30
8.7903	84.090	0.20812	40
8.9717	85.500	0.21091	50
9.1528	86.922	0.21367	60
9.3336	88.358	0.21641	70
9.5142	89.806	0.21912	80
9.6945	91.266	0.22180	90
9.8747	92.738	0.22445	100
10.055	94.222	0.22708	110
10.234	95.717	0.22968	120
10.414	97.224	0.23226	130
10.594	98.743	0.23481	140
10.773	100.272	0.23734	150
10.952	101.812	0.23985	160
11.131	103.363	0.24233	170
11.311	104.925	0.24479	180
11.489	106.497	0.24723	190
11.668	108.079	0.24964	200
11.847	109.670	0.25204	210
12.026	111.272	0.25441	220

14.696 lb/in^2			Temp.
V (2.5315)	H (74.933)	S (0.17121)	°F (−21.62 °F)
2.5424	75.155	0.17172	−20
2.6097	76.535	0.17482	−10
2.6764	77.924	0.17788	0
2.7425	79.322	0.18089	10
2.8082	80.730	0.18385	20
2.8734	82.149	0.18678	30
2.9382	83.577	0.18967	40
3.0026	85.016	0.19252	50
3.0667	86.466	0.19534	60
3.1306	87.926	0.19812	70
3.1942	89.396	0.20087	80
3.2575	90.877	0.20359	90
3.3207	92.369	0.20628	100
3.3836	93.871	0.20894	110
3.4464	95.383	0.21157	120
3.5090	96.906	0.21417	130
3.5715	98.439	0.21675	140
3.6338	99.981	0.21930	150
3.6960	101.534	0.22183	160
3.7581	103.097	0.22433	170
3.8201	104.669	0.22681	180
3.8821	106.251	0.22926	190
3.9439	107.842	0.23169	200
4.0056	109.443	0.23410	210
4.0673	111.053	0.23648	220
4.1289	112.671	0.23885	230
4.1905	114.299	0.24119	240
4.2520	115.935	0.24351	250
4.3134	117.580	0.24581	260
4.3748	119.233	0.24810	270

25 lb/in^2			Temp.
V (1.5392)	H (77.510)	S (0.16867)	°F (2.23 °F)
1.5714	78.635	0.17108	10
1.6125	80.088	0.17414	20
1.6531	81.547	0.17715	30
1.6932	83.012	0.18012	40
1.7329	84.485	0.18304	50
1.7723	85.965	0.18591	60
1.8114	87.453	0.18875	70

50 lb/in^2			Temp.
V (0.79824)	H (81.249)	S (0.16597)	°F (38.15 °F)
0.80248	81.540	0.16655	40
0.82502	83.109	0.16966	50
0.84713	84.676	0.17271	60
0.86886	86.243	0.17569	70
0.89025	87.811	0.17862	80
0.91134	89.380	0.18151	90
0.93216	90.953	0.18434	100

100 lb/in^2			Temp.
V (0.40674)	H (85.351)	S (0.16389)	°F (80.76 °F)
—	—	—	80
0.41876	86.964	0.16685	90
0.43138	88.694	0.16996	100
0.44365	90.410	0.17300	110
0.45562	92.116	0.17597	120
0.46733	93.814	0.17888	130

1.8502	88.950	0.19155	80	0.95275	92.529	0.18713	110	0.49009	97.197	0.18452	150
1.8888	90.455	0.19431	90	0.97313	94.110	0.18988	120	0.50118	98.884	0.18726	160
1.9271	91.968	0.19704	100	0.99332	95.695	0.19259	130	0.51212	100.571	0.18996	170
1.9653	93.490	0.19973	110	1.0133	97.286	0.19527	140	0.52291	102.257	0.19262	180
2.0032	95.021	0.20240	120	1.0332	98.382	0.19791	150	0.53358	103.944	0.19524	190
2.0410	96.561	0.20503	130	1.0529	100.485	0.20051	160	0.54413	105.633	0.19782	200
2.0786	98.110	0.20763	140	1.0725	102.093	0.20309	170	0.55457	107.324	0.20036	210
2.1161	99.667	0.21021	150	1.0920	103.708	0.20563	180	0.56492	109.018	0.20287	220
2.1535	101.234	0.21276	160	1.1114	105.330	0.20815	190	0.57519	110.714	0.20535	230
2.1908	102.809	0.21528	170	1.1307	106.958	0.21064	200	0.58538	112.414	0.20780	240
2.2279	104.393	0.21778	180	1.1499	108.593	0.21310	210	0.59549	114.119	0.21022	250
2.2650	105.986	0.22025	190	1.1690	110.235	0.21553	220	0.60554	115.828	0.21261	260
2.3019	107.588	0.22269	200	1.1880	111.883	0.21794	230	0.61553	117.540	0.21497	270
2.3388	109.198	0.22512	210	1.2070	113.539	0.22032	240	0.62546	119.258	0.21731	280
2.3756	110.817	0.22752	220	1.2259	115.202	0.22268	250	0.63534	120.980	0.21962	290
2.4124	112.444	0.22989	230	1.2447	116.871	0.22502	260	0.64518	112.707	0.22191	300
2.4491	114.080	0.23225	240	1.2636	118.547	0.22733	270	0.65497	124.439	0.22417	310
2.4857	115.723	0.23458	250	1.2823	120.231	0.22962	280	0.66472	126.176	0.22641	320
2.5223	117.375	0.23689	260	1.3010	121.921	0.23189	290	0.67444	127.917	0.22863	330
2.5588	119.035	0.23918	270	1.3197	123.618	0.23414	300	0.68411	129.665	0.23083	340
2.5953	120.703	0.24145	280	1.3383	125.321	0.23637	310	0.69376	131.417	0.23301	350
2.6317	122.378	0.24370	290	1.3569	127.032	0.23857	320	0.70338	133.174	0.23517	360
2.6681	124.061	0.24593	300	1.3754	128.749	0.24076	330	0.71296	134.937	0.23730	370
				150 lb/in²			Temp.	200 lb/in²			Temp.
				V	H	S	°F	V	H	S	°F
				(0.26974)	(87,800)	(0.16281)	(109.45°F)	(0.19891)	(89.439)	(0.16195)	(131.74°F)
				0.27029	87.906	0.16299	110	0.20579	91.137	0.16480	140
				0.28007	89.800	0.16629	120	0.21370	93.141	0.16811	150
				0.28943	91.662	0.16947	130	0.22121	95.100	0.17130	160
				0.29845	93.498	0.17256	140	0.22842	97.024	0.17438	170
				0.30718	95.313	0.17556	150	0.23535	98.921	0.17737	180
				0.31566	97.112	0.17849	160	0.24207	100.795	0.18027	190
				0.32392	98.899	0.18135	170	0.24860	102.652	0.18311	200
				0.33200	100.675	0.18415	180	0.25496	104.494	0.18588	210
				0.33992	102.444	0.18689	190	0.26117	106.325	0.18860	220
				0.34769	104.206	0.18958	200	0.26726	108.147	0.19126	230
				0.35533	105.965	0.19223	210	0.27323	109.962	0.19387	240
				0.36285	107.720	0.19483	220	0.27911	111.771	0.19644	250
				0.37028	109.474	0.19739	230	0.28489	113.576	0.19896	260
				0.37761	111.226	0.19992	240	0.29060	115.378	0.20145	270
				0.38486	112.979	0.20240	250	0.29623	117.178	0.20390	280
				0.39203	114.732	0.20485	260	0.30180	118.977	0.20632	290
				0.39913	116.486	0.20728	270	0.30730	120.775	0.20870	300
				0.40617	118.242	0.20967	280	0.31275	122.574	0.21105	310
				0.41315	120.000	0.21203	290	0.31815	124.373	0.21337	320
				0.42008	121.761	0.21436	300	0.32350	126.173	0.21567	330
				0.42695	123.524	0.21666	310	0.32881	127.974	0.21793	340
				0.43379	125.290	0.21894	320	0.33408	129.778	0.22017	350
				0.44058	127.060	0.22120	330	0.33932	131.583	0.22239	360
				0.44733	128.833	0.22343	340	0.34452	133.392	0.22458	370
				0.45405	130.609	0.22564	350	0.34969	135.202	0.22675	380
				0.46074	132.390	0.22782	360	0.35483	137.016	0.22890	390
				0.46739	134.174	0.22999	370	0.35994	138.832	0.23102	400
				0.47402	135.963	0.23213	380	0.36503	140.652	0.23313	410
				0.48062	137.755	0.23425	390	0.37010	142.475	0.23521	420
				0.48719	139.552	0.23635	400	0.37514	144.302	0.23728	430

such data. Some typical plots for the properties of water are given in Figs. 5.10 to 5.13, which represent *PH*, *TS*, *HS*, and *VE* plots, respectively. The last plot has not previously been discussed but is of value in calculations for constant-volume systems. Typical tabular data for water are given in App. B.

To complement the plots for ammonia shown earlier, some typical diagrams are given for Freon 12 (Fig. 5.14), ethylene (Figs. 5.15 and 5.16), CO_2 (Fig. 5.17), and SO_2 (Fig. 5.18). Tabular data for Freon 12 are given in Tables 5.2 and 5.3.

5.7 REFERENCES

Bridgman, P. W.: "Condensed Collection of Thermodynamic Formulas," Harvard, Cambridge, Mass., 1926.

Keenan, P. W., F. G. Keyes, P. G. Hill, and J. G. Moore: "Steam Tables, SI Units," Wiley, New York, 1978.

5.8 PROBLEMS

1 Starting with the differential relations among properties [Eqs. (5.1) to (5.4)], derive the relation

$$\left(\frac{\partial S}{\partial V}\right)_T = \left(\frac{\partial P}{\partial T}\right)_A$$

2 Derive the relations below:

$$(a)\ C_P - C_V = -\left(\frac{\partial P}{\partial T}\right)_V\left[P\left(\frac{\partial V}{\partial P}\right)_T + \left(\frac{\partial E}{\partial P}\right)_T\right]$$

$$(b)\ \left(\frac{\partial H}{\partial P}\right)_V - V = P\left(\frac{\partial V}{\partial P}\right)_T + \left(\frac{\partial E}{\partial P}\right)_T$$

3 Using Table 5.1, derive the equivalent of the following in terms of P, V, T.

$$(a)\ \left(\frac{\partial H}{\partial P}\right)_S \quad (b)\ \left(\frac{\partial H}{\partial S}\right)_P \quad (c)\ \left(\frac{\partial E}{\partial S}\right)_V \quad (d)\ \left(\frac{\partial E}{\partial V}\right)_S$$

$$(e)\ \left(\frac{\partial E}{\partial P}\right)_T \quad (f)\ \left(\frac{\partial E}{\partial S}\right)_T \quad (g)\ \left(\frac{\partial H}{\partial S}\right)_T \quad (h)\ \left(\frac{\partial G}{\partial T}\right)_S$$

4 For a certain substance at 25°C and 1 atm, $C_P = 138$ kJ/(kmol · K) and $V = 0.090$ m³/kmol. Also

$$\left(\frac{\partial V}{\partial T}\right)_P = 9.0(10^{-8})\ \text{m}^3/\text{K} \qquad \text{and} \qquad \left(\frac{\partial V}{\partial P}\right)_T = 9.0(10^{-9})\ \text{m}^3/\text{atm}$$

Calculate:

$$(a)\ \left(\frac{\partial E}{\partial P}\right)_T \quad (b)\ \left(\frac{\partial H}{\partial P}\right)_T \quad (c)\ \text{Joule-Thomson coefficient}$$

5 For a substance at 25°C and pressures to 10,000 atm, $\overline{V} = a + bP + cP^2$ cm³/mol. Calculate the isothermal increase in Gibbs free energy as pressure increases from 1000 to 10,000 atm.

6 Determine the derivatives below for an ideal gas, a real gas where $PV = ZRT$, and a van der Waals gas.

$$(a)\ \left(\frac{\partial V}{\partial T}\right)_P \quad (b)\ \left(\frac{\partial P}{\partial V}\right)_T \quad (c)\ \left(\frac{\partial P}{\partial T}\right)_V \quad (d)\ \left(\frac{\partial^2 V}{\partial T^2}\right)_P \quad (e)\ \left(\frac{\partial^2 P}{\partial T^2}\right)_V$$

7 Using jacobians, show that the following expressions are true. Evaluate each for an ideal gas.

$$(a)\ \left(\frac{\partial E}{\partial V}\right)_T = T\left(\frac{\partial P}{\partial T}\right) - P$$

$$(b)\ \left(\frac{\partial E}{\partial H}\right)_T = \frac{T - P(\partial T/\partial P)_V}{T - V(\partial T/\partial V)_P}$$

$$(c)\ \left(\frac{\partial T}{\partial V}\right)_S = -\frac{T(\partial P/\partial T)_V}{C_V}$$

$$(d)\ C_P = \frac{T(\partial V/\partial T)_P}{(\partial T/\partial P)_S}$$

$$(e)\ \left(\frac{\partial E}{\partial V}\right)_T = -\frac{C_V}{(\partial V/\partial T)_P} + \frac{(\partial E/\partial T)_P}{(\partial V/\partial T)_P}$$

$$(f)\ \left(\frac{\partial A}{\partial P}\right)_S = P\frac{C_V(\partial V/\partial T)_P}{C_P(\partial P/\partial T)_V} + \frac{S}{C_P}\left[\left(\frac{\partial V}{\partial P}\right)_T + P\right]$$

$$(g)\ \left(\frac{\partial T}{\partial V}\right)_P = -\left(\frac{\partial P}{\partial V}\right)_T\left(\frac{\partial T}{\partial P}\right)_V$$

8 A pure gas flows at a low rate through a well-insulated horizontal pipe at high pressure and is throttled to a slightly lower pressure. The equation of state $P(V - C) = RT$, where C is a positive constant, applies. Kinetic-energy changes are negligible. Prove whether or not the gas temperature rises or falls by throttling.

9 Using data of Figs. 5.15 and 5.16, calculate the changes in enthalpy and entropy and the final temperature when ethylene changes from 7000 kPa and 70°C to 101.3 kPa by (*a*) a throttling process and (*b*) a reversible adiabatic compression process.

10 Validate Eqs. (5.52) to (5.59).

11 Derive a general expression for the internal-energy departure in terms of P, V, T, β, and K and derivatives thereof only.

12 Prove that

$$\left[\frac{\partial(\overline{V}/T)}{\partial T}\right]_P = -\frac{1}{T^2}\left(\frac{\partial \overline{H}}{\partial P}\right)_T$$

What type of data would be necessary to prove the relation for a pure compound?

13 Calculate the change in enthalpy and entropy of carbon dioxide if it is compressed isothermally at 282°C from 1 to 60 atm by using:

(*a*) The ideal-gas law
(*b*) The van der Waals equation [$a = 3.6(10^6)(\text{cm}^6 \cdot \text{atm})/\text{mol}$; $b = 42.8\ \text{cm}^3/\text{mol}$]
(*c*) The Redlich-Kwong-Soave equation of state
(*d*) The Lydersen corresponding-states method
(*e*) The Pitzer corresponding-states method

14 Calculate the entropy, enthalpy, and specific volume of steam at 100 atm and 370°C by using the Lydersen approach and compare the results with the steam table.

15 Propane is heated from 70°F and 150 lb/in^2 abs to 500°F and 800 lb/in^2 abs. What is the change in internal energy, enthalpy, entropy, Gibbs free energy, and volume, assuming a 32°F and 1-atm ideal reference state. Use the Pitzer corresponding-states technique. What are Q and W?

16 Using Fig. 5.17 for carbon dioxide, calculate ΔG for isothermal expansion at 90°F from 1000 to 25 lb/in^2 abs.

17 Using Fig. 5.6 for ammonia, estimate C_P at 600 lb/in^2 abs and 180°F.

18 Given the data below for a pure substance, outline how you would proceed to develop both saturation and superheat tables analogous to the steam tables.

Vapor-pressure-temperature data
Critical properties
Saturated-liquid densities
An equation of state $P = f(T, V)$
Isochoric heat capacity at one specific volume

19 Freon 12 is to be expanded irreversibly and isothermally at 180°F from 300 to 35 lb/in² abs with the addition of 20 Btu/lb of heat. Calculate the work of the process by using (*a*) Fig. 5.14, (*b*) Table 5.2, (*c*) Lydersen, (*d*) ideal gas. Compare the answers.

20 Using predictive methods from Lee-Kesler-Pitzer, draw a Mollier diagram for propane with a 32°F saturated-liquid base for zero entropy and enthalpy. Show the saturation curve, isobars at $P_r = 0.25$, 0.50, 1, and 1.5, and isotherms at $T_r = 0.85$, 0.95, 1, and 1.1. Ideal-gas heat capacity data are given in Chap. 3. Vapor pressure data and heat of vaporization data are given below.

T, °F	$P°$, lb/in² abs	λ_v, Btu/lb
− 50	12.6	185
0	35	172
50	90	157
100	186	136
125	255	122
150	335	106
175	450	83
200	~ 575	44

CHAPTER

SIX

ESTIMATION OF AUXILIARY PHYSICAL PROPERTIES—PROPERTIES OF MIXTURES

To carry out thermodynamic calculations, basic properties of the fluid or fluid mixtures must be known or be able to be estimated. Such properties include liquid and vapor densities, molecular weight, normal boiling point, vapor pressure, critical temperature, critical pressure, critical volume or critical-compressibility factor, and acentric factor. If experimental data are available, these are obviously preferable and should be utilized. Otherwise, methods for estimation must be available. This chapter will give a brief presentation of commonly accepted methods of estimating such properties for use as precursors to thermodynamic analyses. For further information the three general works listed below (referred to in the text as TDB, TDM, and RPS, respectively) should be consulted.[1]

In addition, methods given to this point in this book for thermodynamic-property prediction have been advanced primarily for pure defined mixtures and undefined mixtures such as petroleum fractions or coal liquids. This chapter will

[1] "Technical Data Book—Petroleum Refining," 4th ed., T. E. Daubert and R. P. Danner, eds., American Petroleum Institute, Washington, extant 1984. A complete set of methods for predicting properties of hydrocarbons and their mixtures rigorously evaluated for accuracy. (TDB)

"Technical Data Manual," T. E. Daubert and R. P. Danner, eds., American Institute of Chemical Engineers, New York, extant 1984. A set of methods for predicting properties of nonhydrocarbons and their mixtures, developed with extensive evaluations using experimental data. (TDM)

R. C. Reid, J. M. Prausnitz, and T. K. Sherwood: "The Properties of Gases and Liquids," 3d ed., McGraw-Hill, New York, 1977. A general treatment of estimation methods for all types of compounds, some of which are evaluated for accuracy. (RPS)

also attempt to lead the reader to the preferred mixing rules for estimating composite properties of defined mixtures as well as to suggest methods for applying the techniques discussed to undefined mixtures.

6.1 DENSITY

Liquid densities are easily measured, and normally at least one experimental value is available. Data as a function of temperature are available in several sources, one of the most complete being those of the API Research Project 44 and TRC Data Project discussed in Chap. 5. Many nomographs are available for the relative density of pure compounds as a function of temperature including a complete set for most common hydrocarbons in the TDB. The Chemical Rubber "Chemistry and Physics Handbook," Lange's "Handbook of Chemistry," and the "Chemical Engineer's Handbook," are also good sources.

A liquid-density value at one temperature and pressure may be converted to another temperature and pressure by a simple correlation attributed to Lu (1959). The critical temperature and pressure of the compound are required, and correction factors are correlated as a function of reduced temperature and pressure:

$$\rho_{T,P} = \rho_{\text{ref}} \frac{C(T_r, P_r)}{C(T_{r,\text{ref}}, P_{r,\text{ref}})} \tag{6.1}$$

Figure 6.1 is used to determine the correction factors.

For defined mixtures, an estimate of the liquid density can be calculated by dividing the average molecular weight by the corresponding average molal volumes of the components of the mixture. This assumes ideality in mixing, which is sufficient for most purposes. Lu (1959) gives a more rigorous procedure.

For undefined mixtures of hydrocarbons, methods are given in the TDB for estimation of their liquid densities knowing the relative density at 15°C and the mean average boiling point of the mixture.

The saturated-liquid densities of hydrocarbons and nonhydrocarbons can most accurately be predicted by the modified Rackett method of Spencer and Danner (1972):

$$V_s = \left(\frac{RT_c}{P_c}\right) Z_{\text{RA}}^{1.0+(1.0-T_r)^{2/7}} \tag{6.2}$$

This equation is generally valid between the triple point and the critical point. Values of Z_{RA} are available in the Spencer and Danner (1972) article and in the TDB for common hydrocarbons and selected nonhydrocarbon liquids. Values for selected organic compounds are given in the TDM. Less than 1 percent average error was recorded when the equation was tested for a large experimental data set. If Z_{RA} is not available, Z_c may be used in its place; however average errors will increase to about 3 percent for hydrocarbons and organic nonhydrocarbons. An alternate method is that of Gunn and Yamada (1971) which requires knowledge of a reference-liquid molar volume, Z_c, T_c, and ω for use. The accuracy is slightly lower.

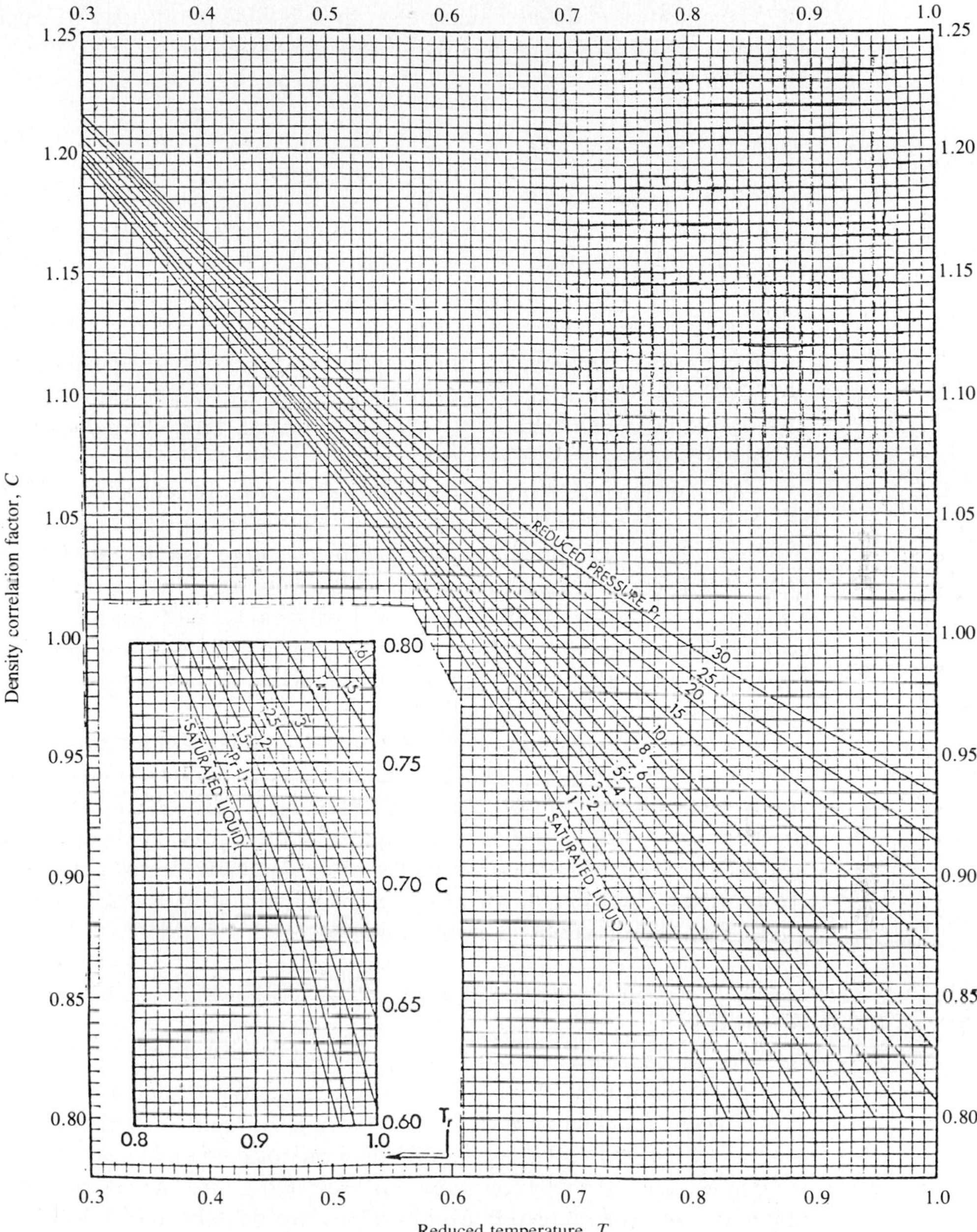

Figure 6.1 Effect of pressure and temperature on densities of pure and defined mixtures of hydrocarbons. *(Adapted from "Technical Data Book—Petroleum Refining," 4th ed., copyright 1982 by American Petroleum Institute. Used with permission.)*

The densities of defined mixtures at their bubble points may be calculated by using the modified Rackett method in the following form:

$$V_{\text{BP}} = V_{\bar{c}} Z_{\overline{\text{RA}}}^{1+(1-T_r)^{2/7}} \tag{6.3}$$

where

$$V_{\bar{c}} = \sum_{i=1}^{n} x_i V_{ci} \qquad T_r = \frac{T}{T_{mc}}$$

$$Z_{\overline{\text{RA}}} = \sum_{i=1}^{n} x_i Z_{\text{RA},i}$$

Errors for binary hydrocarbon mixtures average only 2 percent except that errors up to 20 percent can occur near the critical. If the pseudocritical temperature T_{pc} is used in place of T_{mc} (to be discussed later) to calculate T_r, errors average 7 percent.

Vapor densities for pure compounds are normally estimated by using either an analytical or a generalized equation of state as discussed in Chap. 2. Critical temperature, critical pressure, and a critical-compressibility factor or acentric factor are necessary input parameters. Generally the Pitzer type of approach is preferred for hydrocarbons, while the Lydersen type of approach gives better estimates for nonhydrocarbons. Errors in calculated compressibility factors for hydrocarbons average less than 1 percent, while errors for organic nonhydrocarbons average 5 percent.

For both defined and undefined gaseous hydrocarbon mixtures, vapor densities may be calculated with generalized equations of state by substituting T_{pc}, P_{pc}, and a molar average Z_c or ω in the correlations. Use of more-complex mixing rules does not improve the accuracy appreciably. Errors for defined hydrocarbon mixtures in the prediction of compressibility factors average 4 percent, with most errors under 2 percent. Error analyses for undefined hydrocarbon mixtures and nonhydrocarbon mixtures are not available.

6.2 MOLECULAR WEIGHT

The molecular weight of a pure compound is easily calculated from its structure. The molecular weight of a defined mixture is normally calculated as the molar average of the molecular weights of the components of the mixture. Prediction techniques are normally required for estimating the molecular weight of an undefined mixture as experimental values are seldom available and are often inaccurate even if available.

Methods for the estimation of undefined-mixture molecular weights are limited to hydrocarbons. Winn (1957) presented a nomographic method, while Kesler and Lee (1976) and Riazi (1979) derived empirical equations for such estimations as a function of the normal boiling point and relative density of the mixture. The Kesler and Lee and the Riazi methods are essentially equivalent in accuracy, with average errors of 10 to 12 percent over the range of molecular weights of 80 to 700. Below a molecular weight of 300, errors are from 3 to 5 percent.

Because of its simplicity the Riazi correlation is given as a method for the prediction of molecular weight (MW):

$$\mathrm{MW} = 1.6607(10^{-4})T_b^{2.1962}(\text{rel den})^{-1.0164} \tag{6.4}$$

This correlation predicts hydrocarbon molecular weights in the 80 to 300 molecular-weight range within an average error of 2.6 percent.

6.3 BOILING POINT AND VAPOR PRESSURE

Boiling points and vapor pressure are intimately related by the Clapeyron equation discussed in Chap. 3, Sec. 3.4:

$$\frac{dp}{dT} = \frac{\lambda}{T(\bar{V}_G - \bar{V}_L)} \tag{6.5}$$

This equation can be simplified to the Clausius-Clapeyron equation if the heat of vaporization is independent of temperature over the temperature range:

$$\ln\frac{P_2}{P_1} = -\frac{\lambda}{R}\left(\frac{1}{T_2} - \frac{1}{T_1}\right) \tag{6.6}$$

Given one point of vapor pressure, a vapor pressure can be calculated for another temperature, or vice versa. Often the normal boiling point is known, thus $P = 1$ atm and this point can be used to calculate additional points. If the heat of vaporization is not known, it can be estimated from the methods of Chap. 3.

Much vapor pressure data have been correlated by using the Antoine type of equation, a modification of the Clapeyron equation:

$$\ln P = A - \frac{B}{T + C} \tag{6.7}$$

where A, B, C are constants. Antoine constants have been determined for many compounds by regression of experimental data. Reliable compilations of pure-component vapor pressures are given by the API 44 and TRC projects. Antoine constants should be used only in the range for which they were derived, normally for pressures less than 2 atm.

The recent compilation of physical property data for the most important industrial chemicals (AIChE, 1984) gives correlation equations for vapor pressure for several hundred compounds derived from critical evaluation of experimental data and usually valid between the triple point and the critical point. The correlating equation is

$$\ln P = A + \frac{B}{T} + C\ln T + DT^E \tag{6.8}$$

where A, B, C, D, E are constants. Proper curvatures are well-defined by this equation.

For pure hydrocarbons, Lee and Kesler (1975) have proposed a corresponding-states technique for predicting vapor pressure if the critical temperature, critical pressure, and acentric factor are known:

$$\left[\ln P_r = (\ln P_r)^{(0)} + \omega(\ln P_r)^{(1)}\right]_{T_r} \tag{6.9}$$

where

$$P_r = \left(\frac{P}{P_c}\right)_{T_r} \qquad T_r = \frac{T}{T_c}$$

P = vapor pressure, kPa ω = acentric factor

P_c = critical pressure, kPa

$$(\ln P_r)^{(0)} = 5.92714 - \frac{6.09648}{T_r} - 1.28862 \ln T_r + 0.169347 T_r^6$$

$$(\ln P_r)^{(1)} = 15.2518 - \frac{15.6875}{T_r} - 13.4721 \ln T_r + 0.43577 T_r^6$$

If critical properties are known, this equation predicts pure-hydrocarbon vapor pressures within 3 to 4 percent, with larger errors if critical properties must be estimated. The equations are only valid if $0.30 < T_r < 1.0$. In addition, they cannot be used below the freezing point of the data. The method can also be used for narrow-boiling petroleum fractions, but no estimate of accuracy is available for these situations. Obviously, the equations can be reversed to determine the temperature at which a certain vapor pressure will exist, the boiling point, although such a procedure is trial and error.

If critical properties are not available or cannot be accurately estimated as is the case for many high-boiling hydrocarbons or narrow-boiling (less than about 30°C range) petroleum fractions, the method of Maxwell and Bonnell (1955) can be used for estimating vapor pressures, or conversely, a normal boiling point from a known vapor pressure. The equations necessary for such calculations are

$$\log P = \frac{3000.538X - 6.761560}{43X - 0.987672} - 0.8752041$$
$$X > 0.0022 \ (P < 0.267 \text{ kPa}) \tag{6.10}$$

$$\log P = \frac{2663.129X - 5.994296}{95.76X - 0.972546} - 0.8752041$$
$$0.0013 \le X \le 0.0022 \ (0.267 \text{ kPa} \le p \le 101.3 \text{ kPa}) \tag{6.11}$$

$$\log P = \frac{2770.085X - 6.412631}{36X - 0.989679} - 0.8752041 \tag{6.12}$$
$$X < 0.0013 \ (p > 101.3 \text{ kPa})$$

where

$$X = \frac{(T_b/T) - 0.00051606 T_b'}{748.1 - 0.38610 T_b'}$$

where T_b = normal boiling point, K

T_b' = corrected normal boiling point, K

$$T_b - T_b' = 1.39f\,(K - 12)\log\frac{P}{101.3}$$

$$f = \begin{cases} 1 & \text{if } T_b > 204°\text{C} \\ \dfrac{T_b - 659.7}{200} & \text{if } 93°\text{C} < T_b < 204°\text{C} \\ 0 & \text{if } T_b < 93°\text{C} \end{cases}$$

$$K = \frac{(1.8\text{MeABP})^{1/3}}{\text{rel den}}$$

MeABP = mean average boiling point, K

For vapor pressure greater than 0.13 kPa, experimental data are reproduced within an average error of 8 percent. Below this pressure, errors increase and average 30 percent between 10^{-7} and 0.13 kPa. This method is only applicable to hydrocarbons.

For nonhydrocarbon organic materials, the method of Reidel (1954*a*) is recommended. The normal boiling point must be known and the critical temperature and pressure must be known or predictable. Tables of parameters as a function of reduced temperature are required. The TDM gives the complete procedure. Above 15 kPa total pressure, accuracy is within 3 percent.

An alternate method for predicting the vapor pressure of nonhydrocarbon organics when the normal boiling point is unknown or if the pressure is below 15 kPa is the reference-substance correlative method of Othmer (1968). Two experimental points of data are required to derive the two empirical constants in the equation. A table of the constants for several hundred compounds is given in the TDM. The equation is

$$\log P_s = B \log P_{s,\text{water}} + A \tag{6.13}$$

where A and B are the constants and $P_{s,\text{water}}$ is the accurate vapor pressure of water at the given temperature.

For very heavy hydrocarbon materials, vapor pressures and the normal boiling point may be estimated from subatmospheric values for hydrocarbon-type materials by the SWAP correlations of Smith et al. (1976) and the subsequent work of Macknick et al. (1978).

No really accurate methods for the prediction of normal boiling points from other properties of a pure compound are generally available except by reverse application of vapor pressure expressions such as those of Lee and Kesler and of Maxwell and Bonnell. In addition, the SWAP correlation shown below can be useful for predicting normal boiling points for compounds of high molecular weight.

Original: $$T_b = 10.0 + 1.38855T_{10} - 2.051(10^{-4})T_{10}^2$$

Revised: $$T_b = 3.22 + 2.089T_{0.01} - 1.035(10^{-3})T_{0.01}^2 \tag{6.14}$$

where T_b = boiling point at 101.3 kPa (760 mmHg)
T_{10} = boiling point at 1.33 kPa (10 mmHg)
$T_{0.01}$ = boiling point at $1.33(10^{-3})$ kPa (0.01 mmHg)

6.4 CRITICAL TEMPERATURE

The utility of critical and pseudocritical temperatures is well-established and has been generally defined and discussed in Chap. 2, Sec. 2.3. Methods for their prediction are absolutely essential for thermodynamic calculations.

True critical properties predict the phase behavior of mixtures as to whether liquid, vapor, or mixed phases can exist and are used for equipment design calculations to determine what phases will exist in a certain, e.g., mass transfer or reactor process, step. Except where noted otherwise, critical properties of mixtures useful for thermodynamic-property prediction are usually pseudocritical (molar average) properties or mixture-correspondence properties according to the correlation requirements. In this book the pseudocritical properties are normally used unless explicitly stated otherwise.

For pure compounds the true critical and pseudocritical properties are equal. For true critical temperature of hydrocarbons, Spencer and Daubert (1973), determined that a modified Nokay (1959) method for estimation of T_c was the most accurate method of estimation, with an average error of 3.6°C in the C_1–C_{20} carbon atom range. For higher-molecular-weight materials the errors may be higher. The Nokay equation is given as Eq. (6.15) and the modified constants are given in Table 6.1

$$\log T_c = A + B \log(\text{rel den}) + C \log T_b \tag{6.15}$$

where T_c = critical temperature, K
rel den = relative density at 15°C, 0.1 MPa
T_b = normal boiling point, K

For pure nonhydrocarbons the method of Lydersen (1955) is recommended by the TDM, with an average deviation of 9°C in the C_1–C_{13} range. This group-contribution method is given as Eq. (6.16), with the group contributions in Table 6.2. Alternate methods and error estimate by homologous series are also

Table 6.1 Coefficients for Eq. (6.15)

Type compound	A	B	C
Paraffin	1.35940	0.43684	0.56224
Naphthene	0.65812	−0.07165	0.81196
Olefin	1.09534	0.27749	0.65563
Acetylene	0.74673	0.30381	0.79987
Diolefin	0.1384	−0.39618	0.99481
Aromatic	1.0615	0.22732	0.66929

available in the TDM.

$$T_c = T_b\left[0.567 + \Sigma\Delta_T - (\Sigma\Delta_T)^2\right]^{-1} \tag{6.16}$$

where $\Sigma\Delta_T$ = sum of critical-temperature-group contributions from Table 6.2

For mixtures of defined composition, Spencer, Daubert, and Danner (1973) determined that the method of Li (1972) was most accurate for estimating true critical temperatures for defined hydrocarbon-hydrocarbon and hydrocarbon-nonhydrocarbon systems excluding mixtures above 50% methane, hydrogen-containing systems, and mixtures rich in carbon dioxide, carbon monoxide, or nitrogen. Average errors approach 3°C. Li's combining rule is

$$T_{cm} = \sum_{i=1}^{n} \theta_i T_{ci} \tag{6.17}$$

where $\theta_i = \dfrac{x_i \overline{V}_{ci}}{\overline{V}_{pc}}$ = volumetric fraction of component i

$\overline{V}_{pc} = \sum_{i=1}^{n} x_i \overline{V}_{ci}$

x_i = mole fraction

$\overline{V}_{ci}$ = molar critical volume

$\overline{V}_{pc}$ = molar average critical volume

T_{cm} = true critical temperature of mixture, K

For mixtures of defined composition containing nonhydrocarbons, the mixing rules of Chueh and Prausnitz (1967*a*, *b*) were determined to be the most accurate with an average error of 11°C for the limited data set available for testing for the TDM. The use of the Li method gives an average error of 15°C. Neither method should be used for systems containing simple monatomic and diatomic molecules such as helium and nitrogen.

Methods are available for prediction of the true critical temperature of undefined mixtures such as natural-gas mixtures and petroleum fractions (TDB).

Pseudocritical temperatures of defined mixtures can be easily calculated from the properties of the pure components as a molar average. For undefined hydrocarbon mixtures, Winn (1957) and the TDB present nomographs for predicting pseudocritical temperature as a function of molal average boiling point and relative density at 15°C. Kesler and Lee (1976) have a similar correlation. Both correlations show an average error of about 1.5 percent in prediction when tested with pure compounds and defined mixtures in the C_5–C_{22} range. Riazi (1979) presents a simplified correlation which predicts within 1.3 percent for the same data set:

$$T_{pc} = 19.0623 T_b^{0.58848} (\text{rel den})^{0.3596} \tag{6.18}$$

where T_{pc} = pseudocritical temperature, K

T_b = normal boiling point, K

rel den = relative density

Table 6.2 Lydersen's critical-property increments

	Δ_T	Δ_P	Δ_V
Nonring increments			
$-CH_3$	0.020	0.227	55
$\overset{\vert}{-CH_2}$	0.020	0.227	55
$\overset{\vert}{\underset{\vert}{-CH}}$	0.012	0.210	51
$\overset{\vert}{\underset{\vert}{-C-}}$	0.00	0.210	41
$=CH_2$	0.018	0.198	45
$\overset{\vert}{\equiv CH}$	0.018	0.198	45
$\overset{\vert}{=C-}$	0.0	0.198	36
$=C=$	0.0	0.198	36
$\equiv CH$	0.005	0.153	(36)
$\equiv C-$	0.005	0.153	(36)
Ring increments			
$-CH_2-$	0.013	0.184	44.5
$\overset{\vert}{-CH}$	0.012	0.192	46
$\vert$			
$\overset{\vert}{\underset{\vert}{-C-}}$	(−0.007)	(0.154)	(31)
$\overset{\vert}{=CH}$	0.011	0.154	37
$\overset{\vert}{=C-}$	0.011	0.154	36
$=C=$	0.001	0.154	36

Table 6.2 Continued

	Δ_T	Δ_P	Δ_V
	Halogen increments		
—F	0.018	0.224	18
—Cl	0.017	0.320	49
—Br	0.010	(0.50)	(70)
—I	0.012	(0.83)	(95)
	Oxygen increments		
—OH (alcohols)	0.082	0.06	(18)
—OH (phenols)	0.031	(−0.02)	(3)
—O— (nonring)	0.021	0.16	20
—O— (ring)	(0.014)	(0.12)	(8)
| —C—O (nonring)	0.040	0.29	60
| —C═O (ring)	(0.033)	(0.2)	(50)
| HC═O (aldehyde)	0.048	0.33	73
—COOH (acid)	0.085	(0.4)	80
—COO— (ester)	0.047	0.47	80
═O (except for combinations above)	(0.02)	(0.12)	(11)
	Nitrogen increments		
—NH_2	0.031	0.095	28
| —NH (nonring)	0.031	0.135	(37)
| —NH (ring)	(0.024)	(0.09)	(27)
| —N— (nonring)	0.014	0.17	(42)
| —N— (ring)	(0.007)	(0.13)	(32)
—CN	(0.060)	(0.36)	(80)
—NO_2	(0.055)	(0.42)	(78)

Table 6.2 Continued

	Δ_T	Δ_P	Δ_V
	Sulfur increments		
—SH	0.015	0.27	55
—S— (nonring)	0.015	0.27	55
—S— (ring)	(0.008)	(0.24)	(45)
=S	(0.003)	(0.24)	(47)
	Miscellaneous		
—Si— (with vertical bonds above and below)	0.03	(0.54)	
—B— (with vertical bond below)	(0.03)		

Note: There are no increments for hydrogen. All bonds shown as free are connected with atoms other than hydrogen. Values in parentheses are based upon too few experimental values to be reliable. It has been suggested that the $\searrow\!\!\swarrow$ C—H ring increment common to two condensed saturated rings be given the value of $\Delta_T = 0.064$.

6.5 CRITICAL PRESSURE

Critical-pressure utility is similar to that of critical temperature as discussed in Sec. 6.4, and both true and pseudocritical properties are necessary for various types of calculations.

The true or pseudocritical pressure of pure hydrocarbons and monofunctional nonhydrocarbons have been shown by Spencer and Daubert (1973) to be best estimated by the method of Lydersen (1955):

$$P_c = \frac{0.1013\ \mathrm{MW}}{(\Sigma\Delta_P + 0.34)^2} \tag{6.19}$$

where MW = molecular weight
$\Sigma\Delta_P$ = sum of critical-pressure-group contributions from Table 6.2
P_c = critical pressure, MPa

When evaluated against C_1–C_{20} paraffins and C_3–C_{14} compounds for other hydrocarbon families, an average deviation from experimental data was 0.12 MPa and a maximum of 0.50 MPa. For nonhydrocarbon compounds, the TDM tested 139 C_1–C_{10} compounds ranging in critical pressure from 1.0 to 8.3 MPa, with an average deviation from experimental data of 0.2 MPa. Alternate methods and error estimates by homologous series are available in the TDM. The method of Ambrose (1978, 1979) has been recently shown to be slightly superior to that of Lydersen, although calculations are slightly more complex.

For calculating the true critical pressure of a defined mixture, the method of Kreglewski and Kay (1969) was shown by Spencer et al. (1973) to be superior for hydrocarbon mixtures excluding methane and inorganic gases, with errors for binary systems averaging 3.3 percent (0.13 MPa).

$$P_{cm} = P_{pc} + P_{pc}\left[5.808 + 4.93\sum_{i}^{n}(x_i\omega_i)\right]\frac{T_{cm} - T_{pc}}{T_{pc}} \tag{6.20}$$

where all properties are calculated as noted earlier. Recently the use of the Redlich-Kwong-Soave equation of state applied at the critical point has been shown to be an alternate accurate method.

For nonhydrocarbon systems, average deviations were 0.5 MPa for mixtures not containing inorganics whose critical loci often display large excess critical pressures. A complete procedure for use in these systems including values of liquid-molar volumes and acentric factors to be used is given in the TDM.

Pseudocritical pressures of defined mixtures are calculated as a molar average.

True critical pressures of undefined hydrocarbon mixtures can be estimated for petroleum fractions by the method of Edmister and Pollock (1948). Pseudocritical pressures of undefined hydrocarbon mixtures can be estimated by the method of Winn (1957), the TDB, and Kesler and Lee (1976) as a function of mean average boiling point and relative density, with an average deviation of 3 to 4 percent when tested with pure-hydrocarbon and defined mixture data.

Riazi (1979) has proposed a simple method for the prediction of pseudocritical pressure for petroleum fractions based on C_5–C_{22} range pure-hydrocarbon data, which predicts with an average deviation of about 3 percent.

$$P_{pc} = 5.5274 \times 10^9 T_b^{-2.3125}(\text{rel den})^{2.3201} \tag{6.21}$$

where P_{pc} = pseudocritical pressure , kPa
T_b = normal boiling point, K

6.6 CRITICAL VOLUME AND CRITICAL COMPRESSIBILITY FACTOR

Critical volume is not required by itself for corresponding-states calculations but is of importance in calculating critical-compressibility factors and critical densities from their definitions.

$$Z_c = \frac{P_C V_C}{nRT_c} \tag{6.22}$$

$$\rho_c = \frac{MP}{Z_c RT_c} \tag{6.23}$$

In addition, the parameter for pure components is important for various volumetric-average mixing rules to be discussed in Sec. 6.8.

Spencer and Daubert (1973) showed that for pure hydrocarbons, the method of Reidel (1954) is superior for critical-volume prediction. Average errors were 0.0874 dm^3/kg when tested on C_3–C_{18} hydrocarbons.

$$V_c = \frac{RT_c}{P_c[3.72 + 0.26(\alpha - 7.00)]} \tag{6.24}$$

$$\alpha = 5.811 + 4.919\omega \tag{6.25}$$

where V_c = critical volume, $dm^3/kmol$

For nonhydrocarbons the method of Lydersen (1955) is a reasonable method available for critical-volume prediction.

$$V_c = (40 + \Sigma\Delta_V)10^{-3} \tag{6.26}$$

where V_c = critical volume, $m^3/kmol$

$\Sigma\Delta_V$ = sum of critical-volume-group contributions from Table 6.2

The method of Fedors (1979) has been determined to be slightly more accurate than the method of Lydersen for all families of organic compounds containing C, H, O, S, N, or halogens in the C_1–C_7 range by the the TDM. The average deviation from experimental data is 0.007 $m^3/kmol$.

$$V_c = 0.0266 + \Sigma\Delta_V \tag{6.27}$$

where V_c = critical volume, $m^3/kmol$

$\Sigma\Delta_V$ = sum of critical-volume-group contributions from Table 6.2

The critical volume of a defined mixture can be estimated most accurately by the method of Chueh and Prausnitz (1967*a*), although errors can be rather large.

Other methods for predicting the critical-compressibility factor are available, e.g., from the saturated-liquid compressibility factor or saturated-liquid reduced temperature *and* the saturated-liquid reduced pressure as discussed by Hougen, Watson, and Ragatz. However, in general, Z_c should be calculated from its definition, Eq. (6.22).

6.7 ACENTRIC FACTOR

The acentric factor is an essential third parameter for the Pitzer type of generalized theorem of corresponding states as well as for certain analytical equations of state such as the Redlich-Kwong-Soave equation.

The acentric factor was defined in Chap. 2, Sec. 2.3, as

$$\omega = -\log\frac{P_S}{P_c} - 1.000 \tag{6.28}$$

where P_S = vapor pressure at T_r = 0.700. In general the best method for calculating acentric factors for pure components is from the defining equation by using accurate vapor pressure data. Passut and Danner (1973) have presented the most accurate set of acentric factors for hydrocarbons. The references on page 241 extended this list of factors. Certain of these values along with preferred values for selected nonhydrocarbons are available in App. A.

Often, accurate vapor pressure data at $T_r = 0.7$ are not available. Thus, a correlation for the acentric factor based on an easily determined, available property is desirable. Lee and Kesler (1975) proposed a correlation requiring the temperature, critical pressure, and normal boiling point which has been determined by Daubert and by Riazi (1979) to be the most accurate correlation for pure hydrocarbons in the boiling range of 0–350°C with an average deviation of only 3 percent:

$$\omega = \frac{\ln P_{br}^S - 5.92714 + 6.09648/T_{br} + 1.28862 \ln T_{br} - 0.169347 T_{br}^6}{15.2518 - 15.6875/T_{br} - 13.4721 \ln T_{br} + 0.43577 T_{br}^6} \tag{6.29}$$

where P_{br} = reduced vapor pressure at the boiling point
T_{br} = reduced normal boiling point

For undefined mixtures, the Lee and Kesler equation can be applied by using the mean-average boiling point for T_b. For heavy undefined mixtures Kesler and Lee (1976) propose a method applicable to materials with $T_{br} > 0.8$ which appears to predict reasonable values.

An alternate method for predicting acentric factors for both pure components and undefined mixtures is to predict the vapor pressure at a reduced temperature of 0.7 by using the Maxwell-Bonnell equations (6.10) to (6.12) and to simply calculate the acentric factor by using the definiting equation (6.28). Maslanik and Daubert (1982) have shown this method to be equally as good or better than any other predictor.

For defined mixtures the mixture acentric factor can be estimated by using the molar average mixing rule:

$$\omega = \sum_{i=1}^{n} x_i \omega_i \tag{6.30}$$

where x_i = mole fraction of component i
ω_i = acentric factor of component i

This approximation is satisfactory in most cases, with no alternate currently available.

6.8 COMBINING AND MIXING RULES

For mixtures of pure compounds, various rules have been noted for determining reference properties of the mixtures. For determining pseudocritical temperatures and pressures, the simple molar-average mixing rule of Kay (1936) is usually recommended. This rule is usually sufficient for pseudocritical temperature and for some pseudocritical pressures, especially for mixtures of similar compounds. However, when compounds are widely different in nature, properties will vary considerably. Although many authors also call mixture properties derived from other mixing rules pseudocriticals, to avoid confusion this book will term them *mixture correspondence properties*, as does the TDB. These more complex combining rules will be especially important for phase equilibria calculations to be discussed in Chap. 7.

For critical-pressure calculations, the simplest combining rule which assumes no interactions is that of Prausnitz and Gunn (1958):

$$P_{cm} = \frac{R\sum_i (x_i Z_{ci})}{\sum_i (x_i \bar{V}_{ci})} T_{pc} \tag{6.31}$$

This rule is useful, for example, in the method for calculating the density of hydrocarbon–nonpolar gas mixtures as given by the TDB.

When materials are dissimilar, interactions between molecules must be taken into account. Chueh and Prausnitz (1967*a*, *b*, *c*) have developed a set of combining rules that utilize a geometric-mean mixing rule and introduce an interaction parameter k_{ij} which takes into account each binary pair of parameters and has been determined to be very useful for vapor-liquid equilibrium calculations.

The Chueh and Prausnitz method defines a mixture correspondence temperature T_{mc}, defined in Eq. (6.32):

$$T_{mc} = \sum_{i=1}^{n} \sum_{j=1}^{n} \phi_i \phi_j T_{cij} \tag{6.32}$$

where

$$\phi_{i \text{ or } j} = \frac{x_i V_{ci}}{\Sigma x_i V_{ci}} \tag{6.33}$$

and

$$T_{cij} = \sqrt{T_{ci} T_{cj}}\,(1 - k_{ij}) \tag{6.34}$$

$$k_{ij} = 1.0 - \left[\frac{(V_{ci} V_{cj})^{1/3}}{(V_{ci}^{1/3} + V_{cj}^{1/3})/2} \right]^3 \tag{6.35}$$

where T_{mc} = mixture correspondence temperature. For the mixture-correspondence-temperature correlation, Eq. (6.32) can be expanded for any number of compounds, e.g., for binaries:

$$T_{mc} = \phi_1^2 T_{C_{11}} + 2\phi_1 \phi_2 T_{C_{12}} + \phi_2^2 T_{C_{22}} \tag{6.36}$$

This type of correlation has been shown to be useful in several estimation procedures, as, for example, for determining the effect of temperature and pressure on the densities of hydrocarbon mixtures of defined composition as discussed in the TDB.

Although for the procedure noted k_{ij} can be estimated from Eq. (6.35), usually values must be back-calculated from binary equilibrium data by use of an equation of state. It has been shown that k_{ij} for most practical purposes is independent of temperature, pressure, and composition.

While the generalized equations of state normally use molar-average mixing rules, the analytical equations of state normally require more complex rules, usually of the geometric-mean type. Reid, Prausnitz, and Sherwood (RPS, 1977) describe the mixing rules for the common equations of state as well as give a general background on the development of mixing rules.

As an example and because they will be of general use in Chap. 7, the mixing rules for the Redlich-Kwong-Soave equation of state,

$$P = \frac{RT}{\bar{V} - \bar{b}} - \frac{\overline{\alpha a}}{\bar{V}(V + \bar{b})} \tag{6.37}$$

are given below:

$$\overline{\alpha a} = \sum_{i=1}^{n} \sum_{j=1}^{n} x_i x_j \alpha_{ij} a_{ij} \tag{6.38}$$

$$\bar{b} = \sum_{j=1}^{n} x_j b_j \tag{6.39}$$

where $$\alpha_{ij} a_{ij} = (1 - k_{ij})(\alpha_i a_i \alpha_j a_j)^{1/2} \tag{6.40}$$

In this case k_{ij} must be determined be use of binary mixture data. This equation will be discussed in more detail in Chap. 7.

6.9 REFERENCES

AIChE Design Institute for Physical Property Data: "Data Compilation Tables of Properties of Pure Compounds," (edited by T. E. Daubert and R. P. Danner), AIChE, New York, extant 1984.

Ambrose, D.: National Phys. Lab. Report, *Chem.*, **92** (1978) and **98** (1979).

API Research Project 44: "Selected Values of Properties of Hydrocarbons and Related Compounds," Texas A & M, College Station, Tex., loose-leaf data sheets, extant 1982.

Chueh, P. L., and J. M. Prausnitz: *AIChE J.*, **13**:1099 (1967*a*).

______, and ______: *AIChE J.*, **13**:1107 (1967*b*).

______, and ______: *Ind. Eng. Chem. Fund.*, **6**:492 (1967*c*).

Edmister, W. C., and D. H. Pollock: *Chem. Eng. Prog.*, **44**:905 (1948).

Fedors, R. F.: *AIChE J.*

, **25**:202 (1979).

Gunn, R. D., and T. Yamada: *AIChE J.*, **17**:1341 (1971).

Kay, W. B.: *Ind. Eng. Chem.*, **28**:1014 (1936).

Kesler, M. G., and B. I. Lee: *Hydrocarbon Process.*, **55**(3):153 (1976).

Kreglewski, A., and W. B. Kay, *J. Phys Chem.*, **73**:33 (1969).

Lee, B. I., and M. G. Kesler: *AIChE J.*, **21**:510 (1975).

Li, C. C.: *Can. J. Chem. Eng.*, **49**:709 (1971); **50**:152 (1972).

Lu, B. C.: *Chem. Eng.*, **66**(9):137 (1959).

Lydersen, A. L.: "Estimation of Critical Properties of Organic Compounds by the Method of Group Contributions," *Univ. Wis. Eng. Exp. Stn. Rep.*, no. 3 (1955).

Macknick, A. B., J. Winnick, and J. M. Prausnitz: *AIChE J.*, **24**:731 (1978).

Maslanik, M. K., and T. E. Daubert: "Documentation of the Basis for the Selection of the Contents of Chapter 2, Characterization," in "Technical Data Book—Petroleum Refining," University Microfilms, Ann Arbor, Mich., 1982.

Maxwell, J. B., and L. S. Bonnell: "Vapor Pressure Charts for Petroleum Engineers," Esso Research and Engineering Company, 1955.

Nokay, R.: *Chem. Eng.*, **66**(4):147 (1959).

Othmer, D. F., and E. Yu: *Ind. Eng. Chem.*, **60**:22 (1968).

Passut, C. A., and R. P. Danner: *Ind. Eng. Chem. Process Des. Dev.*, **12**:365 (1973).

Prausnitz, J. M., and R. D. Gunn: *AIChE J.*, **4**:430, 494 (1958).

Reidel, L.: *Chem.-Ing.-Tech.*, **26**:83 (1954*a*).

_____: *Chem.-Ing.-Tech.*, **26**:679 (1954*b*).

Riazi, M. R.: Doctoral dissertation, Department of Chemical Engineering, The Pennsylvania State University, University Park, Pa., 1979.

Smith, G., J. Winnick, D. S. Abrams, and J. M. Prausnitz: *Can J. Chem. Eng.*, **54**:337 (1976).

Spencer, C. F., and R. P. Danner: *J. Chem. Eng. Data*, **17**:236 (1972).

_____, and T. E. Daubert: *AIChE J.*, **19**:482 (1973).

_____, _____, and R. P. Danner: *AIChE J.*, **19**:522 (1973).

Thermodynamics Research Center (TRC), "Selected Values of Properties of Chemical Compounds," Texas A & M, College Station, Tex., loose-leaf data sheets, extant 1982.

Winn, F. W.: *Pet. Refiner*, **36**(2):157 (1957).

CHAPTER

SEVEN

SOLUTION PROPERTIES AND PHYSICAL EQUILIBRIA

One of the most important applications of thermodynamics is the prediction of the equilibrium existing between phases where no chemical reactions occur. For example, calculations of vapor-liquid equilibria are essential for distillation and absorption equipment design, gas-solid and liquid-solid equilibria are essential for adsorber design, and liquid-liquid and liquid-solid equilibria must be estimated for extraction processes.

Certain additional concepts of thermodynamics are necessary to understand such processes and to calculate equilibria. Chemical potential, partial molar properties, the concepts of fugacity and activity, the difference between ideal and nonideal solutions, and the phase rule are the necessary fundamentals to be discussed in this chapter. Following these prerequisites, the application of the concepts to the prediction of vapor-liquid equilibria will be discussed thoroughly, while briefer treatments of the calculation of liquid-liquid, vapor-solid, and liquid-solid equilibria will be given.

7.1 PARTIAL MOLAR PROPERTIES AND CHEMICAL POTENTIAL

The state principle discussed in Chap. 5 specified that any thermodynamic property for a pure component can be specified as a function of any two reference properties, e.g., pressure and temperature. If a homogeneous mixture of two or more components exists, an additional reference property must be specified for each additional component; e.g., for a four-component mixture, temperature, pressure, and the moles or mole fraction of three of the components would usually be specified.

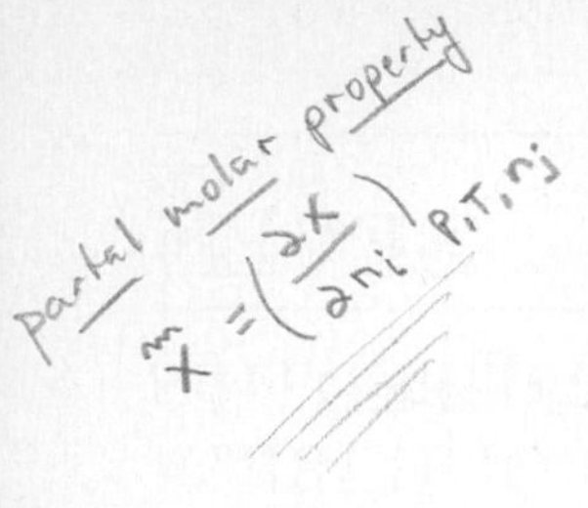

A partial molar property is a property of a compound when mixed with other components. The partial molar property $\tilde{X}$ of any component i is the differential change in the property with respect to the differential change in its amount at constant temperature, pressure, and composition. Volume can be substituted for pressure or temperature if desired. In general,

$$\tilde{X}_i = \left(\frac{\partial X}{\partial n_i}\right)_{P,T,n_j} \tag{7.1}$$

where n_i = component under discussion
n_j = all other components

For example, for volume and free energy,

$$\tilde{V}_i = \left(\frac{\partial V}{\partial n_i}\right)_{P,T,n_j}$$

$$\tilde{G}_i = \left(\frac{\partial G}{\partial n_i}\right)_{P,T,n_j}$$

Partial molar quantities can be summed to give an exact differential:

$$dX = \left(\frac{\partial X}{\partial P}\right)_{T,n_j} dP + \left(\frac{\partial X}{\partial T}\right)_{P,n_j} dT + \left(\frac{\partial X}{\partial n_1}\right)_{P,T,n_j} dn_1 + \left(\frac{\partial X}{\partial n_2}\right)_{P,T,n_j} dn_2 + \cdots \tag{7.2}$$

At constant pressure and temperature,

$$dX_{P,T} = \sum_i \left(\frac{\partial X}{\partial n_i}\right)_{P,T,n_j} dn_i \tag{7.3}$$

Thus any particular property may be calculated by summing the partial molar quantities of each component by using Eq. (7.3). The properties normally combined in this way are V, E, H, S, G, and A.

A partial molar property is the rate at which any extensive property of a solution changes with respect to the number of moles of any component i in the solution at constant temperature, pressure, and composition of the other components. Thus the property is intensive and depends on the composition of the solution. It is often determined from the slope of a plot of the property vs. the number of moles of component i in the solution at constant temperature and pressure.

Partial molar properties can be manipulated in the same way as total-energy properties by using the defining equations (5.2) to (5.4):

$$\tilde{H}_i = \tilde{E}_i + P\tilde{V}_i \tag{7.4}$$

$$\tilde{G}_i = \tilde{H}_i - T\tilde{S}_i \tag{7.5}$$

$$\tilde{A}_i = \tilde{E}_i - T\tilde{S}_i \tag{7.6}$$

A partial molar property is not the same as a molar property except for ideal solutions, which will be discussed later. This differentiation will be illustrated in the sections on ideal and nonideal solutions.

Example 7.1 Show graphically how you could determine the partial molar volumes of A and B in mixtures of A and B given molar volume-concentration data for the mixture at constant temperature and pressure.

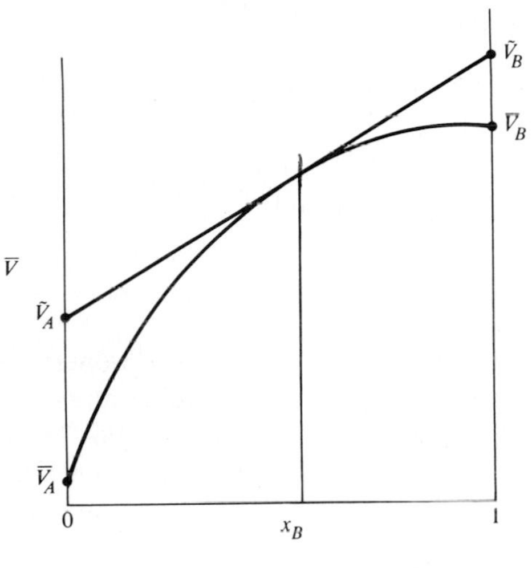

SOLUTION

$$\tilde{V}_i = \left(\frac{\partial V}{\partial n_i}\right)_{P,T,n_j}$$

where n = moles. Therefore

$$\tilde{V}_A = \left(\frac{\partial V}{\partial n_A}\right)_{P,T,n_B} \qquad \tilde{V}_B = \left(\frac{\partial V}{\partial n_B}\right)_{P,T,n_A}$$

Let x = mole fraction. Thus

$$\bar{V} = \tilde{V}_A x_A + \tilde{V}_B x_B$$

$$= (1 - x_B)\tilde{V}_A + x_B \tilde{V}_B$$

$$d\bar{V} = -(\tilde{V}_A - \tilde{V}_B)\, dx_B$$

Thus

$$\tilde{V}_A = \bar{V} - x_B \frac{\partial \bar{V}}{\partial x_B}$$

$$\tilde{V}_B = \bar{V} + (1 - x_B)\frac{\partial \bar{V}}{\partial x_B}$$

Plotting $\overline{V}$ against x_B, drawing a tangent at the desired point, and extrapolating to pure components yields $\tilde{V}_A$ and $\tilde{V}_B$.

Exercise 7.A Calculate the partial molar volumes of methanol and water at 0.5 mol fraction methanol given the following data at a temperature of 0°C and a pressure of 1 atm.

Mol fraction methanol	Molar volume, m^3/kmol
0	0.0181
0.114	0.0203
0.197	0.0219
0.249	0.0230
0.495	0.0283
0.692	0.0329
0.785	0.0352
0.892	0.0379
1	0.0407

Chemical potential μ is a fundamental-property relationship for single-component systems whether of fixed or variable mass or fixed or variable composition. The thermodynamic properties of a single-component system of fixed mass are defined by specifying mass and two independent variables chosen from temperature, pressure, volume, and entropy. Also, a mass or composition value must be considered for each component.

For an infinitesimal change in a system, the change in any property can be calculated in terms of partial molar properties in the general form of Eq. (7.2). For example, consider the change in enthalpy of a binary system if pressure and entropy are held constant.

$$dH = \left(\frac{\partial H}{\partial P}\right)_{S,n_1,n_2} dP + \left(\frac{\partial H}{\partial S}\right)_{P,n_1,n_2} dS + \left(\frac{\partial H}{\partial n_1}\right)_{S,P,n_2} dn_1 + \left(\frac{\partial H}{\partial n_2}\right)_{S,P,n_1} dn_2 \tag{7.7}$$

At constant composition, the third and fourth terms are zero. Since $dH = T\,dS + V\,dP$, then

$$\left(\frac{\partial H}{\partial P}\right)_{S,n_1,n_2} = V \quad \text{and} \quad \left(\frac{\partial H}{\partial S}\right)_{P,n_1,n_2} = T$$

Therefore

$$dH = T\,dS + V\,dP + (\mu_1)_{S,P}\,dn_1 + (\mu_2)_{S,P}\,dn_2$$

where

$$(\mu_1)_{S,P} = \left(\frac{\partial H}{\partial n_1}\right)_{S,P,n_2} \quad \text{and} \quad (\mu_2)_{S,P} = \left(\frac{\partial H}{\partial n_2}\right)_{S,P,n_1} \tag{7.8}$$

Pressure is intensive, while entropy is extensive and μ_1 and μ_2 are intensive. μ_1 and μ_2 are chemical potentials, intensive variables of composition. Obviously, Eq. (7.8) can be easily expanded to three or more component systems.

If the same derivation is carried out on E, G, and A, Eqs. (7.9) to (7.11) will result:

$$dE = T\,dS - P\,dV + (\mu_1)_{V,S}\,dn_1 + (\mu_2)_{V,S}\,dn_2 \tag{7.9}$$

$$dG = -S\,dT + V\,dP + (\mu_1)_{T,P}\,dn_1 + (\mu_2)_{T,P}\,dn_2 \tag{7.10}$$

$$dA = -S\,dT - P\,dV + (\mu_1)_{T,V}\,dn_1 + (\mu_2)_{T,V}\,dn_2 \tag{7.11}$$

The chemical potentials in the previous equations for a given component μ_i are equal. From the defining equations,

$$E = H - PV = G + TS - PV = A + TS \tag{7.12}$$

Taking the total differential,

$$\begin{aligned} dE &= dH - P\,dV - V\,dP \\ &= dG + T\,dS + S\,dT - P\,dV - V\,dP \\ &= dA + T\,dS + S\,dT \end{aligned} \tag{7.13}$$

Add $P\,dV - T\,dS$ to each term:

$$\begin{aligned} dE + P\,dV - T\,dS &= dH - T\,dS - V\,dP \\ &= dG + S\,dT - V\,dP \\ &= dA + P\,dV + S\,dT \end{aligned} \tag{7.14}$$

If the equalities of Eq. (7.14) are subtracted from Eqs. (7.8) to (7.11) and like coefficients are equated,

$$(\mu_1)_{S,P} = (\mu_1)_{V,S} = (\mu_1)_{T,P} = (\mu_1)_{T,V} = \mu_1 \tag{7.15}$$

Thus, from the definitions,

$$\begin{aligned} \mu_1 &= \left(\frac{\partial H}{\partial n_1}\right)_{S,P,n_2} = \left(\frac{\partial E}{\partial n_1}\right)_{V,S,n_2} = \left(\frac{\partial G}{\partial n_1}\right)_{T,P,n_2} \\ &= \left(\frac{\partial A}{\partial n_1}\right)_{T,V,n_2} \end{aligned} \tag{7.16}$$

or

$$\mu_1 = \tilde{G}_1 \tag{7.17}$$

Since partial molal quantities are normally all defined at constant temperature and pressure, only the chemical potential and partial molal Gibbs free energy are equal. The other state variable quantities are also chemical potentials when defined as in Eq. (7.16).

7.2 THE CONCEPT OF FUGACITY

The concept of fugacity was introduced by G. N. Lewis to allow the use of the form of the equation derived for the free-energy changes of an isothermal ideal gas for the real gas. By definition for any component of a mixture,

$$\left[d\tilde{G}_i = d\mu_i = RT\,d(\ln \bar{f}_i)\right]_T \tag{7.18}$$

$$dG = \begin{cases} -S\,dT + V\,dP & \text{for any pure component} \\ V\,dP & \text{at constant temperature} \\ \dfrac{RT}{P}\,dP = RT\,d(\ln P) & \text{for an ideal gas} \end{cases}$$

Thus, the form of the ideal gas has been adopted to the real gas with the fugacity f replacing the pressure P. In partial molar terms, $d\tilde{G}_i = \tilde{V}_i\,dP$, or

$$\left(\frac{\partial \tilde{G}_i}{\partial P}\right)_T = \left(\frac{\partial \mu_i}{\partial P}\right)_T = \tilde{V}_i \tag{7.19}$$

If Eqs. (7.18) and (7.19) are combined, an equation for the effect of pressure on fugacity at constant temperature results:

$$RT\,\partial(\ln \bar{f}_i) = \tilde{V}_i\,\partial P$$

or

$$\left[\frac{\partial(\ln \bar{f}_i)}{\partial P}\right]_T = \frac{\tilde{V}_i}{RT} \tag{7.20}$$

As discussed earlier, the fugacity of a pure compound in the ideal-gas state must be equal to the pressure. Formally this requires that at zero pressure the fugacity of any component divided by the pressure at the limit equals unity; i.e., as $P \to 0$, $f/P = 1$.

Integrating Eq. (7.18) for a pure component between any state f and the ideal-gas state f^*,

$$G^* - G = RT \ln \frac{f^*}{f} \tag{7.21}$$

If this equation is differentiated with respect to temperature at constant pressure,

$$\left(\frac{\partial G^*}{\partial T}\right)_P - \left(\frac{\partial G}{\partial T}\right)_P = R \ln \frac{f^*}{f} + RT\left[\frac{\partial(\ln f^*)}{\partial T}\right]_P - RT\left[\frac{\partial(\ln f)}{\partial T}\right]_P \tag{7.22}$$

Noting that $f^* = P$ and combining Eqs. (7.21) and (7.22),

$$\left(\frac{\partial G^*}{\partial T}\right)_P - \left(\frac{\partial G}{\partial T}\right)_P = \frac{G^* - G}{T} - RT\left[\frac{\partial(\ln f)}{\partial T}\right]_P \tag{7.23}$$

Combining Eq. (5.3), $G \equiv H - TS$, with Eq. (7.23) and noting that $(\partial G/\partial T)_P = -S$ from Eq. (5.3),

$$\frac{\partial(\ln f)}{\partial T} = \frac{H^* - H}{RT^2} \tag{7.24}$$

The same equation can then be written for any component i in a solution:

$$\left[\frac{\partial(\ln \bar{f}_i)}{\partial T}\right]_P = \frac{\tilde{H}_i^* - \tilde{H}_i}{RT^2} \tag{7.25}$$

Application of these equations to different types of solutions will be deferred to a later section.

Another derived concept on which many correlations are based is the fugacity coefficient ϕ, the ratio of the fugacity of a pure compound and its pressure:

$$\phi_i = \frac{f_i}{P_i} \tag{7.26}$$

Fugacities have units of pressure; thus fugacity coefficients are dimensionless. P_i is obviously equal to the total pressure of a system for a pure compound.

At constant temperature

$$dG_i = V_i \, dP \tag{7.27}$$

If Eq. (7.27) is combined with Eqs. (7.18) and (7.26) for a pure compound,

$$RT \, d\left[\ln(\phi_i P)\right] = V_i \, dP \tag{7.28}$$

This expression at constant temperature is equivalent to

$$d(\ln \phi_i) + d(\ln P) = \frac{V_i}{RT} \, dP$$

or

$$d(\ln \phi_i) + \frac{dP}{P} = \frac{V_i}{RT} \, dP$$

Since $Z_i = PV_i/RT$,

$$\left[d(\ln \phi_i) = (Z_i - 1)\frac{dP}{P}\right]_T \tag{7.29}$$

Since at zero pressure $\phi_i = 1$,

$$\left[\ln \phi_i = \int_0^P (Z_i - 1)\frac{dP}{P}\right]_T \tag{7.30}$$

Since the fugacity coefficient is dimensionless, it has been determined to be more easily predicted by generalized methods than has fugacity. Both analytical equations of state and corresponding-states techniques are used to predict fugacity coefficients, after which Eq. (7.26) is used to calculate for fugacity of any pure component.

Exercise 7.B Calculate the fugacity of nitrogen at 100°C and pressure of 10 and 50 atm from the following data:

P, atm	0	14	22	30	44	58	68
Z	1.00	0.95	0.92	0.89	0.84	0.79	0.76

7.2.1 Estimation of the Fugacity of Gases and Liquids by Analytical Equations of State

The fugacity of any fluid can be estimated by using defining equations (7.18) and (7.27):

$$d\bar{G}_i = RT \, d(\ln \bar{f}_i) \tag{7.18}$$

$$dG_i = V_i \, dP \tag{7.27}$$

Consider the simple van der Waals equation of state discussed in Chap. 2. Using

Eq. (7.27) on a molar basis for a pure component,

$$\int_{G_1}^{G_2} d\bar{G} = \int_{P_1}^{P_2} \bar{V}\, dP$$

$$\Delta G = \int_{V_1}^{V_2} \left[\frac{2a}{\bar{V}^2} - \frac{RT\bar{V}}{(\bar{V}-b)^2} \right] d\bar{V} \tag{7.31}$$

Integrating Eq. (7.18) between limits for a pure component,

$$\Delta G = RT \ln \frac{f_2}{f_1} \tag{7.32}$$

Combination of Eqs. (7.31) and (7.32) will yield an expression for the fugacity. If the lower integration limit is taken at very low pressure where $f = P$, then the fugacity at a higher pressure can be estimated from this relation.

Exercise 7.C Derive an expression for estimating the fugacity of any pure component from the van der Waals equation of state, starting with Eqs. (7.31) and (7.32).

Normally, fugacity coefficients rather than fugacities are calculated and converted to fugacities by the defining equation (7.26). Lin and Daubert (1978) have reviewed the various analytical methods available, including the Lee and Kesler (1975) Benedict-Webb-Rubin type of equations, the Redlich-Kwong equation, Soave's modified Redlich-Kwong equation, and the truncated virial equation of state. When these equations were evaluated for their ability to predict fugacity coefficients of pure-hydrocarbon vapors, the Soave and Lee-Kesler methods were determined to be superior over a wide temperature range, although the virial equation was somewhat superior below reduced temperatures of 0.9. For the fugacity coefficient of a component in a mixture, the Soave, Redlich-Kwong, and virial equations were essentially equivalent below 20 atm total pressure, while the Lee-Kesler equation predicted higher values. Since essentially no mixture data on fugacity coefficients are available, the accuracy of the methods for mixtures can only be assumed and cannot normally be verified.

The Lee-Kesler method for fugacity coefficients is equivalent to their method for enthalpy discussed in Chap. 3.

$$\ln \phi = (\ln \phi)^{(0)} + \frac{\omega}{\omega^{(h)}} \left[(\ln \phi)^{(h)} - (\ln \phi)^{(0)} \right] \tag{7.33}$$

where $\phi = f/P$

f = fugacity of pure hydrocarbon

P = pressure

$(\ln \phi)^{(0)}$ = simple fluid fugacity coefficient function, to be calculated from Eq. (7.34)

$(\ln \phi)^{(h)}$ = heavy reference fluid (n-octane) fugacity coefficient function to be calculated from Eq. (7.34)

ω = acentric factor of compound for which fugacity is desired

$\omega^{(h)}$ = acentric factor of heavy reference fluid = 0.3978

The fugacity coefficient functions $(\ln\phi)^{(0)}$ and $(\ln\phi)^{(h)}$ may be calculated from the following equation:

$$(\ln\phi)^{(i)} = z^{(i)} - 1 - \ln(z^{(i)}) + \frac{B}{V_r} + \frac{C}{2V_r^2} + \frac{D}{5V_r^5} + E \qquad (7.34)$$

where

$$B = b_1 - \frac{b_2}{T_r} - \frac{b_2}{T_r^2} - \frac{b_4}{T_r^3}$$

$$C = c_1 - \frac{c_2}{T_r} + \frac{c_3}{T_r^3}$$

$$D = d_1 + \frac{d_2}{T_r}$$

$$E = \frac{c_4}{2T_r^3\gamma}\left[\beta + 1 - \left(\beta + 1 + \frac{\gamma}{V_r^2}\right)e^{-\gamma/V_r^2}\right]$$

$$(\ln\phi)^{(i)} = \begin{cases} (\ln\phi)^{(0)} & \text{when applied to simple fluid} \\ (\ln\phi)^{(h)} & \text{when applied to heavy reference fluid} \end{cases}$$

$z^{(i)}$ = compressibility factor of either simple fluid ($z^{(0)}$) or heavy reference fluid ($z^{(h)}$), depending on which fugacity function is being calculated

T_r = reduced temperature, T/T_c

T = temperature

T_c = critical temperature

V_r = reduced volume = $\dfrac{p_c V}{RT_c}$

and b_1, b_2, b_3, b_4, c_1, c_2, c_3, c_4, d_1, d_2, γ, and β are two sets of constants, one for the simple fluid and the other for the heavy reference fluid, given in Chap. 3.

The Soave procedure includes the following general equation for pure components in either gaseous or liquid phase:

$$\ln\phi = (Z-1) - \ln(Z-B) - \frac{A}{B}\ln\frac{Z+B}{Z} \qquad (7.35)$$

where

$$Z = \frac{PV}{RT} \qquad A = \frac{aP}{R^2T^2} \qquad B = \frac{bP}{RT}$$

$$a = 0.42747\frac{R^2T_c^2}{P_c}\left[1 + (0.48508 + 1.55171\omega - 0.15613\omega^2)\left(1 - \sqrt{T_r}\right)\right]^2$$

$$b = 0.08664\frac{RT_c}{P_c}$$

For mixtures of nonpolar hydrocarbons, the Soave procedure gives the following equation:

$$\ln \bar{\phi}_i = \ln \frac{\bar{f}_i}{Py_i} = \frac{b_i}{b_m}(Z_m - 1) - \ln(Z_m - B_m) - \frac{A_m}{B_m}\left(\frac{2a_i^{0.5}}{a_m^{0.5}} - \frac{b_i}{b_m}\right)\ln\left(1 + \frac{B_m}{Z_m}\right) \tag{7.36}$$

where

$$Z_m = \frac{PV_m}{RT} \qquad A_m = \frac{a_m P}{R^2 T^2} \qquad B_m = \frac{b_m P}{RT}$$

$$a_m = \left(\sum_i y_i a_i^{0.5}\right)^2$$

$$b_m = \sum_i y_i b_i$$

Other procedures also quite useful for computer calculation of fugacity coefficients, including use of the virial equation, are reviewed by Prausnitz et al. (1980).

Example 7.2 Calculate the fugacity of isobutane at 154.5°C and 8620 kPa by using the Redlich-Kwong-Soave equation. Its molar volume is 0.154 m^3/kmol. From App. A, T_c = 135°C, P_c = 3648 kPa, and taking ω = 0.1756,

$$T_r = \frac{154.5 + 273.1}{135.0 + 273.1} = 1.05 \qquad P_r = \frac{8620}{3648} = 2.36$$

$$f = \phi P$$

From Eq. (7.35),

$$\ln \phi = (Z - 1) - \ln(Z - B) - \frac{A}{B}\ln\frac{Z + B}{Z}$$

$$Z = \frac{PV}{RT} = \frac{(8.620 \times 10^6)(0.154)}{(8.3143 \times 10^3)(427.6)} = 0.373$$

$$a = \frac{(0.42747)(8.3143 \times 10^3)^2(408.1)^2}{3.648 \times 10^6}$$

$$\times \left\{1 + \left[0.48508 + 1.55171(0.1756) - 0.15613(0.1756)^2\right](1 - \sqrt{1.05})\right\}^2$$

$$= 1{,}299{,}383$$

$$b = \frac{(0.08664)(8.3143 \times 10^3)(408.1)}{3.648 \times 10^6}$$

$$= 0.0806$$

$$A = \frac{(1{,}299{,}383)(8.620 \times 10^6)}{(8.3143 \times 10^3)^2(427.6)^2} = 0.8861$$

$$B = \frac{(0.0806)(8.620 \times 10^6)}{(8.3143 \times 10^3)(427.6)} = 0.1954$$

$$\ln \phi = (0.373 - 1) - \ln(0.373 - 0.1954) - \frac{0.8861}{0.1954} \ln \frac{0.373 + 0.1954}{0.373}$$

$$= -0.627 + 1.7282 - 1.9103 = -0.809$$

$$\phi = 0.445$$

$$f = (0.445)(8620) = 3838 \text{ kPa}$$

7.2.2 Estimation of the Fugacity of Gases and Liquids by Generalized Equations of State

The two generalized equations of state for estimation of fugacity coefficients and hence fugacity are those of Pitzer and Lydersen as discussed in Chap. 4 for enthalpy and entropy prediction.

The Lydersen method is based on Eq. (7.30):

$$\ln \phi_i = \int_0^P (Z_i - 1) \frac{dP}{P} \tag{7.30}$$

To evaluate the integral of this equation at the lower limit where $Z_i/P \to \infty$, the equation is rearranged such that the lower limit is the ideal case where $\phi_i = 1$ at a reference pressure P_0.

$$\ln \phi_i = -\int_{P_0}^{P} \frac{1 - Z_i}{P} \, dP \tag{7.37}$$

This equation can be evaluated.

Equation (7.37) is written in reduced form,

$$\ln \phi_i = -\int_{P_{0,r}}^{P_r} \frac{1 - Z_i}{P_r} \, dP_r \tag{7.38}$$

This equation was then graphically integrated with pure-component compressibility factor data where the lower integration limit can be evaluated from the slope $(1 - Z_{i,0})/P_{r,0}$ of any compressibility isotherm. The resulting reduced plot is given as Fig. 7.1 for a constant critical compressibility factor of 0.27, with an expanded plot for the low-pressure range in Fig. 7.2.

To correct for values of the critical compressibility factor other than 0.27, Fig. 7.3 gives values for correction factors D_a and D_b to be used when the critical compressibility factor is greater than 0.27 or less than 0.27, respectively, together with Eq. (7.39).

$$(\phi_i)_{Z_c} = (\phi_i)_{Z_c = 0.27} 10^{D(Z_c - 0.27)} \tag{7.39}$$

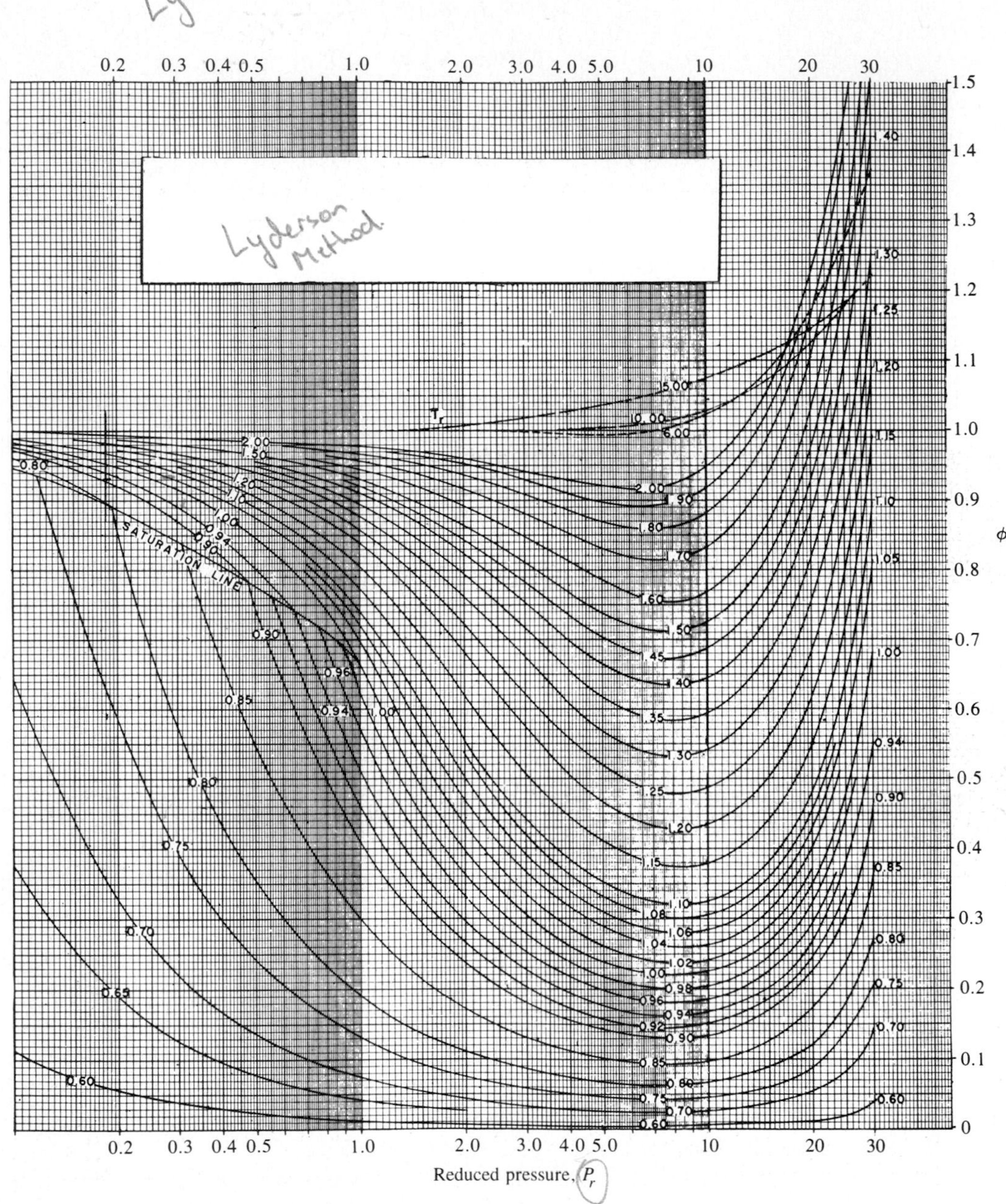

Figure 7.1 Generalized fugacity coefficients of pure gases and liquids at $Z_c = 0.27$. *(Adapted from Hougen, Watson, and Ragatz, "Chemical Process Principles Charts," 3d ed., copyright 1964 by Wiley. Used with permission.)*

Where $(\phi_i)_{Z_c}$ almost equals $(\phi_i)_{Z_c=0.27}$, Eq. (7.40) may be used:

$$(\phi_i)_{Z_c} = (\phi_i)_{Z_c=0.27}[1 + 2303D(Z_c - 0.27)] \tag{7.40}$$

Hougen, Watson, and Ragatz (1959) state that above a reduced pressure of 1.2, the same values of D as given for $P_r = 1.2$ may be used for liquids below a reduced temperature of 0.8 and for vapors above a reduced temperature of 1.2. Above a reduced pressure of 1.2, values are unreliable for $0.8 < T_r < 1.2$.

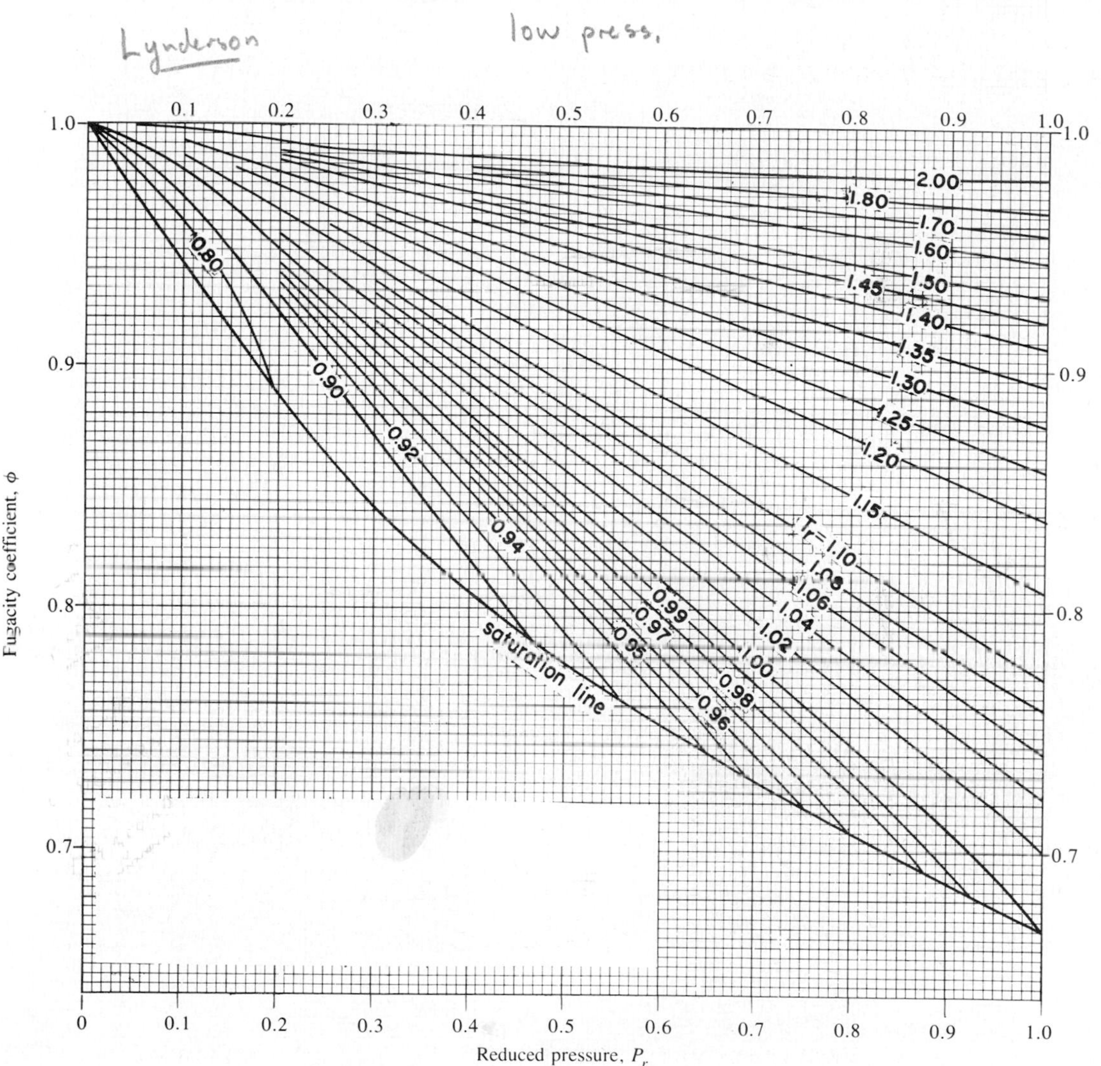

Figure 7.2 Generalized fugacity coefficients of pure gases at low pressure and $Z_c = 0.27$. *(Adapted from Hougen, Watson, and Ragatz, "Chemical Process Principles Charts," 3d ed., copyright 1964 by Wiley. Used with permission.)*

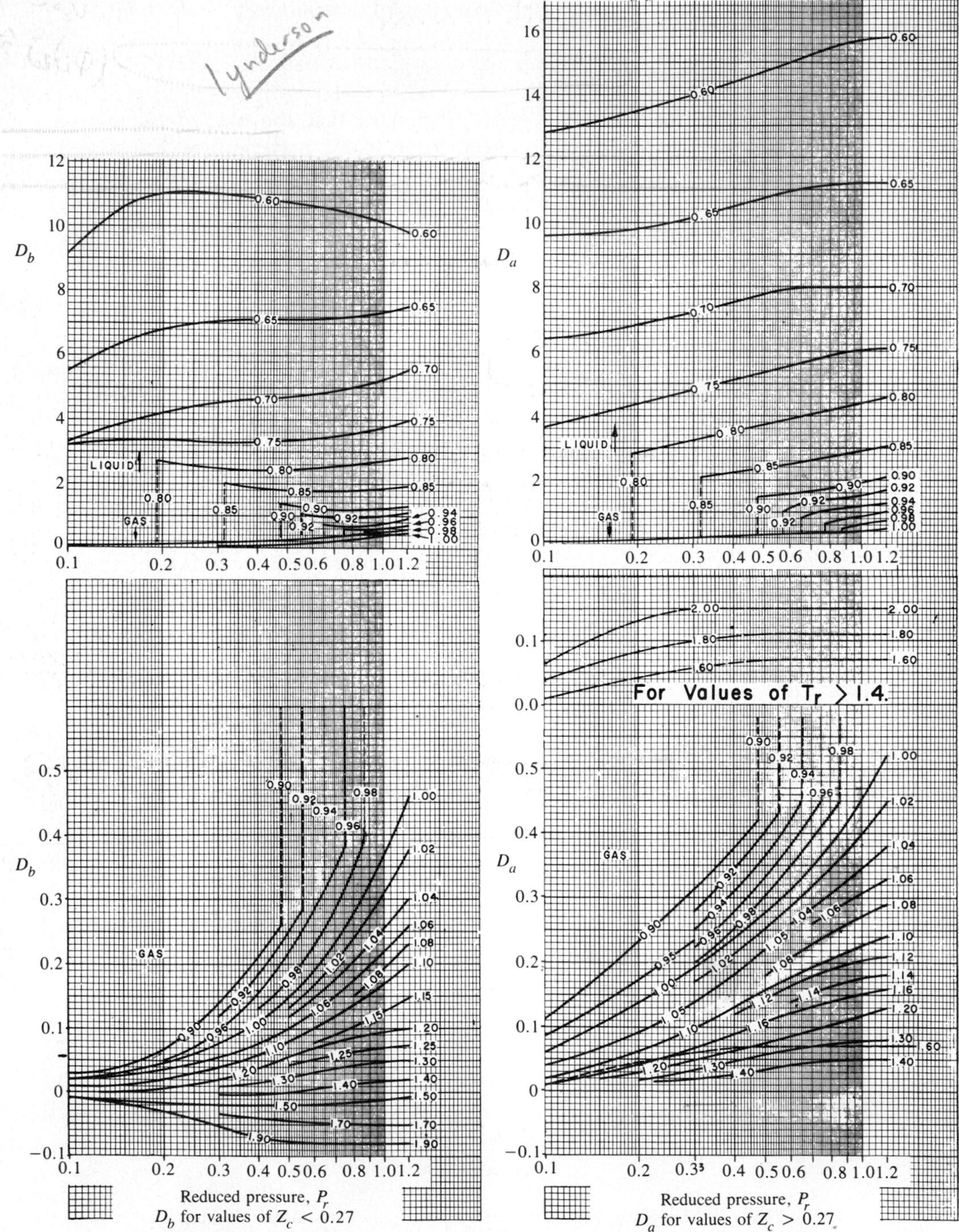

Figure 7.3 Generalized deviation of fugacity coefficients from values at $Z_c = 0.27$. *(Adapted from Hougen, Watson, and Ragatz, "Chemical Process Principles Charts," 3d ed., copyright 1964 by Wiley. Used with permission.)*

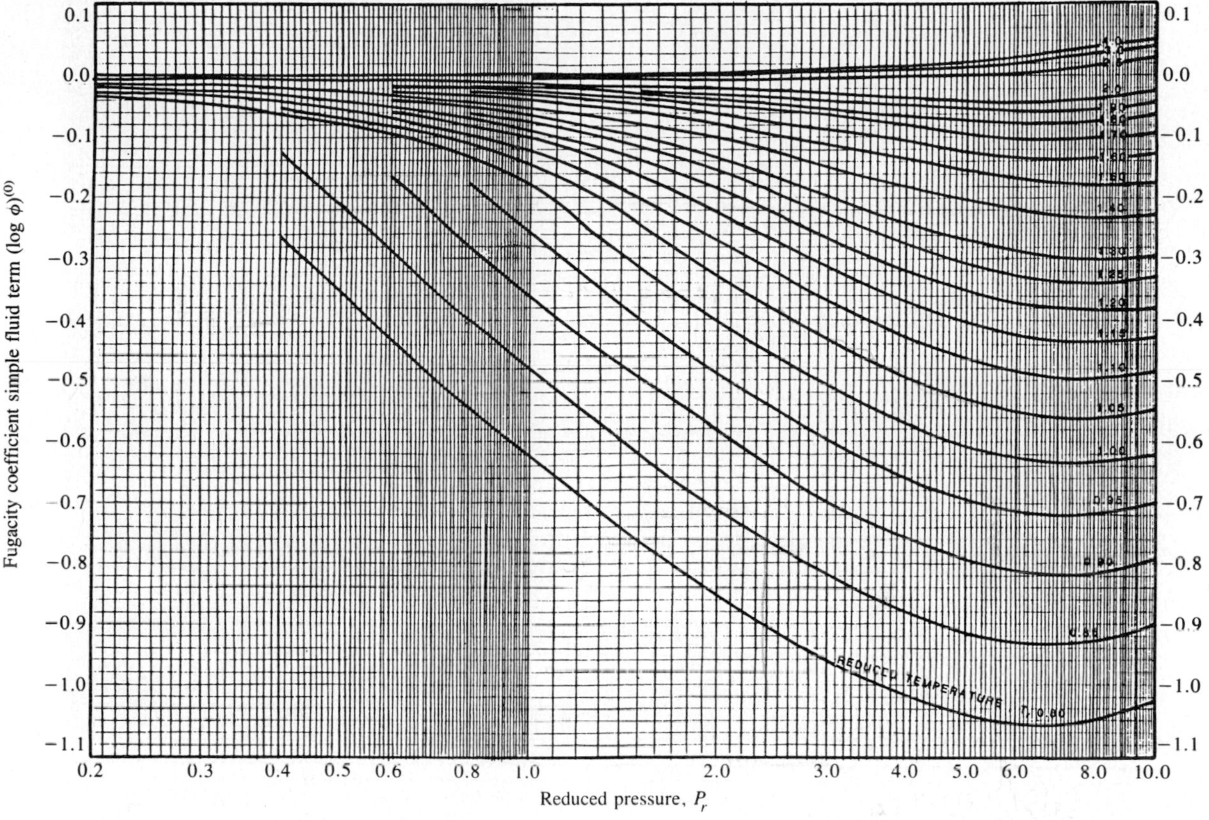

Figure 7.4 Generalized fugacity coefficients at $T_r > 0.80$, simple fluid term. *(Adapted from "Technical Data Book—Petroleum Refining," 4th ed., copyright 1982 by American Petroleum Institute. Used with permission.)*

Example 7.3 Repeat Example 7.2 by using the Lydersen method.

SOLUTION From Example 7.2

$$T_r = 1.05$$

$$P_r = 2.36$$

$$V_c = 0.263 \text{ m}^3/\text{kmol} \qquad \text{(from App. A)}$$

$$T_c = 135°\text{C}$$

$$P_c = 3648 \text{ kPa}$$

$$Z_c = \frac{P_c V_c}{RT_c} = \frac{(3.648 \times 10^6)(0.263)}{(8.3143 \times 10^3)(408.1)} = 0.283$$

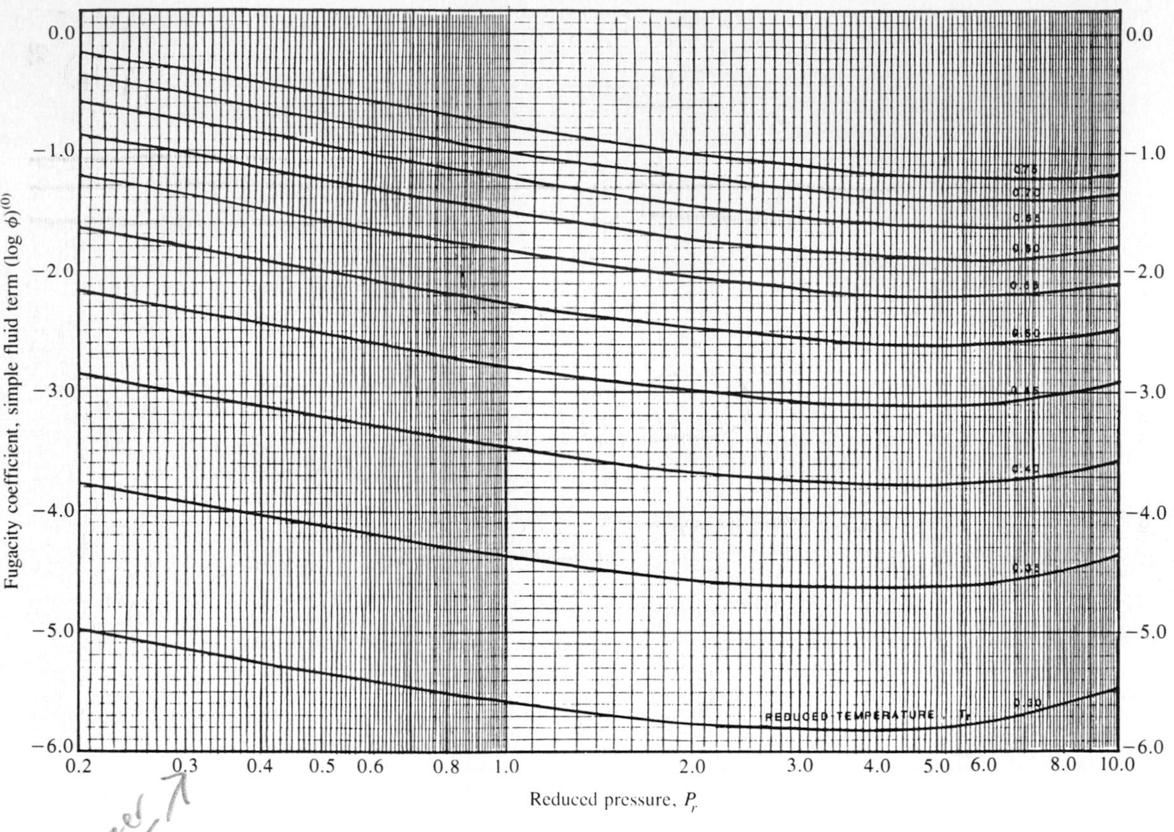

Figure 7.5 Generalized fugacity coefficients at $T_r < 0.80$, simple fluid term. *(Adapted from "Technical Data Book—Petroleum Refining," 4th ed., copyright 1982 by American Petroleum Institute. Used with permission.)*

Using Fig. 7.1,

$$\phi_{Z_c=0.27} = 0.44$$

From Eq. (7.40),

$$\phi_{Z_c} = \phi_{Z_c=0.27} 10^{D(Z_c-0.27)}$$

Using Fig. 7.3,

$$\phi_{Z_c} = (0.44)10^{0.35(0.283-0.27)}$$

$$= (0.44)(1.01) = 0.444$$

Therefore

$$f = \phi P = (0.444)(8620) = 3827 \text{ kPa}$$

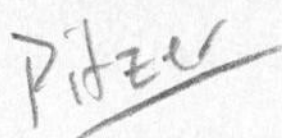

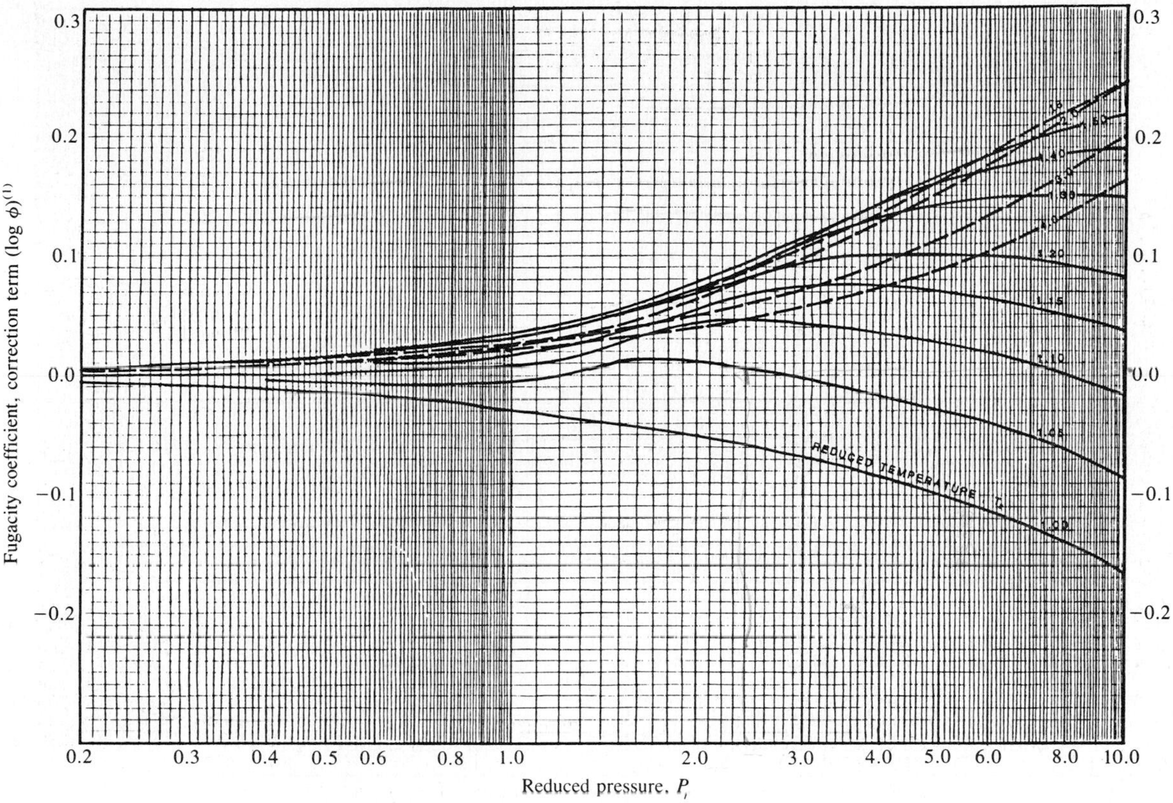

Figure 7.6 Generalized fugacity coefficients at $T_r > 1.00$, correction term. *(Adapted from "Technical Data Book—Petroleum Refining," 4th ed., copyright 1982 by American Petroleum Institute. Used with permission.)*

The Pitzer type of method follows the same general procedure as for enthalpy and entropy in Chap. 4 as shown by Eq. (7.41):

$$\log \phi = (\log \phi)^{(0)} + \omega(\log \phi)^{(1)} \tag{7.41}$$

where $(\log \phi)^{(0)}$ = fugacity coefficient of simple fluid shown in Figs. 7.4 and 7.5
$(\log \phi)^{(1)}$ = correction term for molecular acentricity shown in Figs. 7.6 and 7.7
ω = acentric factor

Application of this procedure should be limited to normal fluids, primarily nonpolar hydrocarbons.

A subroutine for calculating the fugacity coefficient using the Lee-Kesler analytical representation of the method is given in App. E.

Exercise 7.D Repeat Example 7.2 by using the Pitzer method.

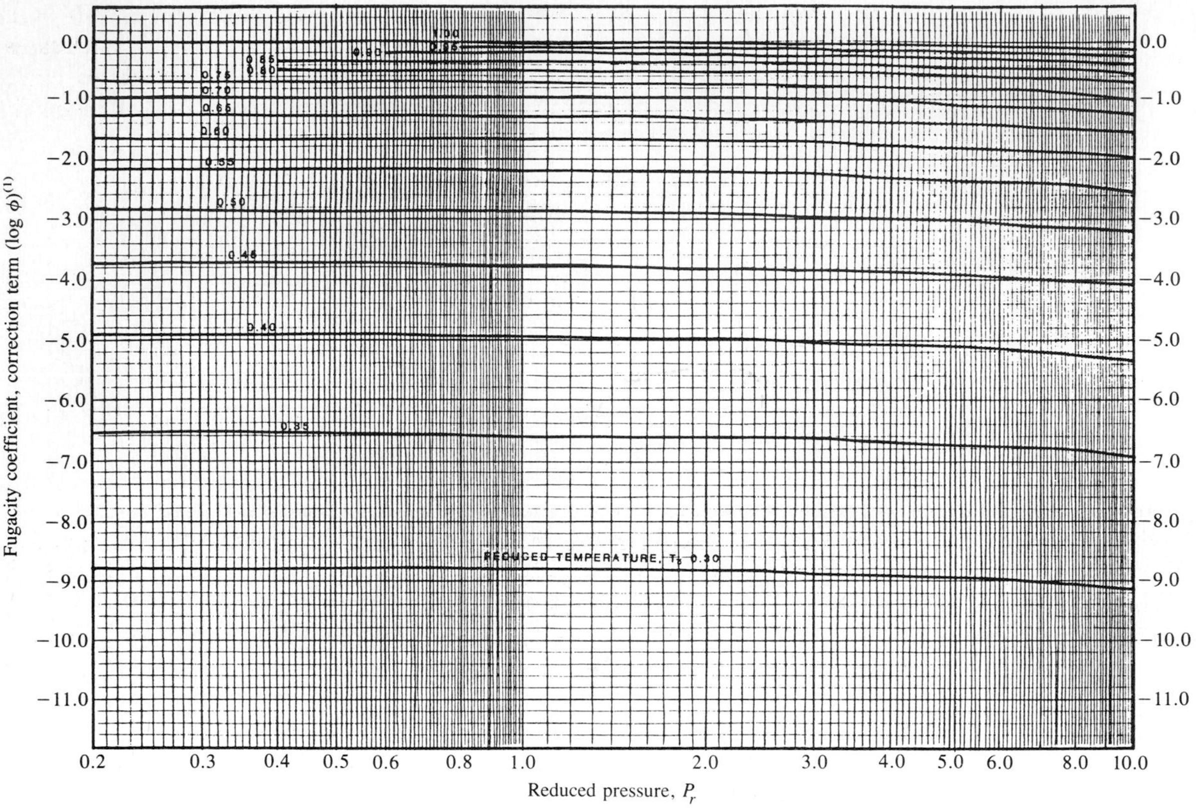

Figure 7.7 Generalized fugacity coefficients at $T_r < 1.00$, correction term. *(Adapted from "Technical Data Book—Petroleum Refining," 4th ed., copyright 1982 by American Petroleum Institute. Used with permission.)*

7.2.3 Liquid and Solid Fugacities

Liquid fugacities can be estimated from the corresponding-states correlations discussed in the previous section. In addition, the principle discussed in Chap. 4, that the free-energy change of the entire system at equilibrium is zero, also allows the estimation of liquid or solid fugacities from vapor fugacities.

$$\Delta G_{T,P,\text{eq.}} = RT \ln \frac{f_{\text{vap}}}{f_{\text{liq}}} = 0 \tag{7.42}$$

Therefore,

$$f_{\text{vap,eq.}} = f_{\text{liq,eq.}} \tag{7.43}$$

This principle would also apply to vapor-solid or liquid-solid equilibrium. In addition, so long as equilibrium is established, the number of phases is immaterial and the fugacities of each phase are equal. As noted earlier, at very low pressures vapors behave as ideal gases, and the fugacities of liquids and solids will be equal to their vapor pressures at the temperature of interest.

The effect of pressure on Gibbs free energy and hence on fugacity was discussed in Chap. 4, Sec. 4.4.2. Increased pressure on solids and liquids increases their vapor pressures or, more accurately, the fugacity of the vapors in equilibrium with the liquid or solid. Except at very high pressures the effect is small. However, it can be estimated by

$$\Delta G_T = RT\int_{f_0}^{f} d\ln f = \int_{P_0}^{P} \bar{V}\,dP$$

or

$$\left(\ln\frac{f}{f_0}\right)_T = \frac{1}{RT}\int_{P_0}^{P} \bar{V}\,dP \tag{7.44}$$

where f_0 = fugacity at the saturation vapor pressure P_0
f = fugacity at higher pressure P
$\bar{V}$ = liquid (or solid) molar volume

This is a modification of the familiar Poynting correction.

Example 7.4 Liquid chlorine at 25°C has a vapor pressure of 0.77 MPa and a fugacity of 0.70 MPa. Liquid chlorine has a molar volume of 5.1×10^{-2} m³/kmol. What is its fugacity at 10 MPa and 25°C?

SOLUTION Using Eq. (7.44),

$$\ln\frac{f}{f_0} = \frac{\bar{V}}{RT}\int_{P_0}^{P} dP$$

$$\ln\frac{f}{0.70} = \frac{(5.1)(10^{-2})(10 - 0.77)10^6}{(8.3143)(10^3)(298)}$$

$$= 0.190$$

$$\frac{f}{0.70} = 1.209$$

$$f = 0.846 \text{ MPa}$$

7.3 IDEAL AND NONIDEAL SOLUTIONS

7.3.1 Ideal Solutions

Solutions for which the fugacity of each component i is equal to the mole fraction x_i of i times the fugacity of the pure component, f_i, at the same temperature and pressure as the solution are termed *ideal*.

$$(\bar{f}_i = x_i f_i)_{P,T} \tag{7.45}$$

This general expression defining an ideal solution reduces to the more common but less general forms for gases and liquids, Dalton's and Raoult's laws, respectively.

In review, Dalton's law states that in a mixture of ideal gases each component exerts a pressure which depends on the total volume, temperature, and number of moles of component:

$$\bar{P}_i = n_i \frac{RT}{V} = n_i \frac{RT}{\bar{V} n_t}$$

$$= y_i \frac{RT}{\bar{V}} \tag{7.46}$$

where $\bar{P}_i$ = partial pressure of i
y_i = mole fraction of i in mixture
V = total volume of mixture
$\bar{V}$ = molar volume of mixture

Thus $$P_T = \sum_i \bar{P}_i \tag{7.47}$$

where P_T = total pressure.

Similarly Amagat's law, the law of additive volumes, would apply to ideal-gas mixtures which are held at constant total pressure, temperature and number of moles of component:

$$V_i = n_i \frac{RT}{P_T} = \frac{n_i}{n_t} V = y_i V \tag{7.48}$$

$$V = \sum_i V_i \tag{7.49}$$

Raoult's law for ideal-liquid solutions is a direct application of Eq. (7.45), where the forces between identical molecules or unlike molecules can be assumed to be equal. This law is usually limited to mixtures of nearly adjacent compounds in homologous series, e.g., methanol and ethanol, n-pentane and n-heptane at low pressures. The low-pressure criterion allows the pure-component fugacity f_i to be replaced by the pure-component vapor pressure P_i° such that

$$\bar{P}_i = x_i P_i^\circ \tag{7.50}$$

$$P_T = \sum_i x_i P_i^\circ \tag{7.51}$$

If a liquid solution is ideal, its vapor will be almost certainly ideal and Dalton's and Raoult's laws can be combined.

$$\bar{P}_i = x_i P_i^\circ = y_i P_T \tag{7.52}$$

This relation is of great value in predicting the vapor-liquid equilibria of ideal solutions and will be discussed in more detail.

Ideal solutions, in general, assume no chemical interactions, similar molecular sizes, and equal attractive and repulsive forces between like and unlike molecules. Thus, only mutual solubility results.

If Eq. (7.20) is applied to both the fugacity in solution $\bar{f}_i$ and the fugacity of the pure component f_i at the same temperature and pressure,

$$\left[\frac{\partial \ln(\bar{f}_i/f_i)}{\partial P}\right]_T = \frac{\tilde{V}_i - \bar{V}_i}{RT} \tag{7.53}$$

The left side of Eq. (7.53) reduces to zero at constant composition; thus

$$\tilde{V}_i = \overline{V}_i = 0 \tag{7.54}$$

leading to a total volume change on mixing at constant temperature of zero:

$$\Delta \overline{V}_M = \sum_i n_i(\tilde{V}_i - \overline{V}_i) = 0 \tag{7.55}$$

Thus, the molar volume of an ideal solution is

$$\overline{V} = \sum_i x_i \overline{V}_i \tag{7.56}$$

Applying Eq. (7.25) to both a component in solution and the pure fluid,

$$\frac{\partial \ln(\hat{f}_i/f_i)}{\partial T} = -\frac{\tilde{H}_i - \overline{H}_i}{RT^2} \tag{7.57}$$

For an ideal solution, the left side of Eq. (7.57) reduces to zero, and the change in enthalpy on mixing will be zero:

$$\Delta \overline{H}_M = \sum_i n_i(\tilde{H}_i - \overline{H}_i) = 0 \tag{7.59}$$

Thus the molar enthalpy of an ideal solution is

$$\overline{H} = \sum_i x_i \overline{H}_i \tag{7.59}$$

Similarly, since $\Delta H_M = \Delta E_M + P\Delta V_M$ and ΔH_M and ΔV_M are zero for ideal solution, ΔE_M must equal zero for an ideal solution and

$$\overline{E} = \sum_i x_i \overline{E}_i$$

To determine the change of Gibbs or Helmholtz free energy on mixing, consider the defining equations

$$\Delta G_M = \Delta H_M - T\Delta S_M$$
$$\Delta A_M = \Delta E_M - T\Delta S_M$$

Since ΔH_M and ΔE_M are zero for ideal solutions,

$$\Delta G_M = \Delta A_M = -T\Delta S_M \tag{7.60}$$

Thus, the ideal-solution entropy of mixing is required. As derived in Sec. 4.7.2,

$$\Delta S = \sum_i (-n_i R \ln x_i) \tag{7.61}$$

Thus, for any component i,

$$\Delta S_{i,M} = -Rn_i \ln x_i \tag{7.62}$$

Thus the entropy and free energies after mixing are not simply molar averages but are modified for the entropy of mixing:

$$S = \sum_i n_i S_i - \sum_i n_i R \ln x_i \tag{7.63}$$

$$G = \sum_i n_i H_i - T\sum_i n_i S_i + \sum_i n_i RT \ln x_i \tag{7.64}$$

Care should be taken in assuming ideal-solution behavior. For gases, such behavior is limited to low pressures, less than about 10 atm, but often includes chemically different species. For liquids, such behavior is limited to essentially adjacent members of homologous series at pressures usually not appreciably higher than 1 atm.

Example 7.5 The system acetone (A) = ethanol (E) at 0.1013 MPa total pressure is normal and miscible with no azeotropes. From the following point of equilibrium data, determine whether or not the system is ideal. $x_A = 0.40$, $y_A = 0.605$, $T = 66°C$.

SOLUTION If the system is ideal, Eq. (7.52) should hold:

$$\overline{P}_i = x_i P_i^\circ = y_i P_T$$

Vapor pressure data for acetone and ethanol yield

$$P_A^\circ = 0.1373 \text{ MPa} \qquad P_E^\circ = 0.0600 \text{ MPa}$$

For acetone,

$$\overline{P}_A = y_A P_T = x_A P_A^\circ$$

$$(0.605)(0.1013) = (0.40)(0.1373)$$

$$0.0613 \neq 0.0549$$

For ethanol,

$$\overline{P}_E = y_E P_T = x_E P_E^\circ$$

$$(0.395)(0.1013) = (0.60)(0.0600)$$

$$0.0400 \neq 0.0360$$

In both cases a positive deviation from Raoult's law is observed, indicating nonideal behavior.

Is there any region in which you would expect ideal behavior to occur?

Exercise 7.E A mixture of benzene, toluene, and *m*-xylene in the amount of 2.0, 61.0, and 37.0 mol %, respectively, exists at 0.1013 MPa pressure. Estimate the initial boiling point of the mixture if a liquid and the composition of the initial vapor. Calculate the Gibbs free energy of mixing if it exists as an ideal solution. Vapor pressure data are:

Benzene:	80.3°C = 101.3 kPa	110.4°C = 235.3 kPa
Toluene:	80.3°C = 39.2 kPa	110.4°C = 101.3 kPa
m-Xylene:	20°C = 0.80 kPa	100°C = 30.66 kPa

7.3.2 Nonideal Solutions

Many different nonideal-liquid solutions exist according to their behavior with respect to changes in volume or entropy or enthalpy on mixing and vary with structure and force differences between molecules.

Often the solute portion of a nonideal-liquid solution can be assumed to follow a simple law known as *Henry's law* at low concentrations:

$$\bar{P}_i = H_i x_i \tag{7.65}$$

where H = Henry's law constant. This law will reduce to Raoult's law, Eq. (7.50), for the ideal case where $H_i = P_i^\circ$ and is a more general form of Raoult's law. Essentially all liquids will follow Henry's law close to 0 mol fraction, but many will deviate from the law above 0.01–0.02 mol fraction, and almost all liquids will deviate above 0.10 mol fraction. Thus, this law is often termed the *solute law*, while Raoult's law is termed the *solvent law*. A combination of these laws is often used for the prediction of vapor-liquid equilibria, as will be discussed in Chap. 8. Henry's law constants arc usually obtained experimentally.

Hildebrand and Scott (1950) and Hildebrand, Prausnitz, and Scott (1970) describe several different types of nonideal solutions according to their behavior.

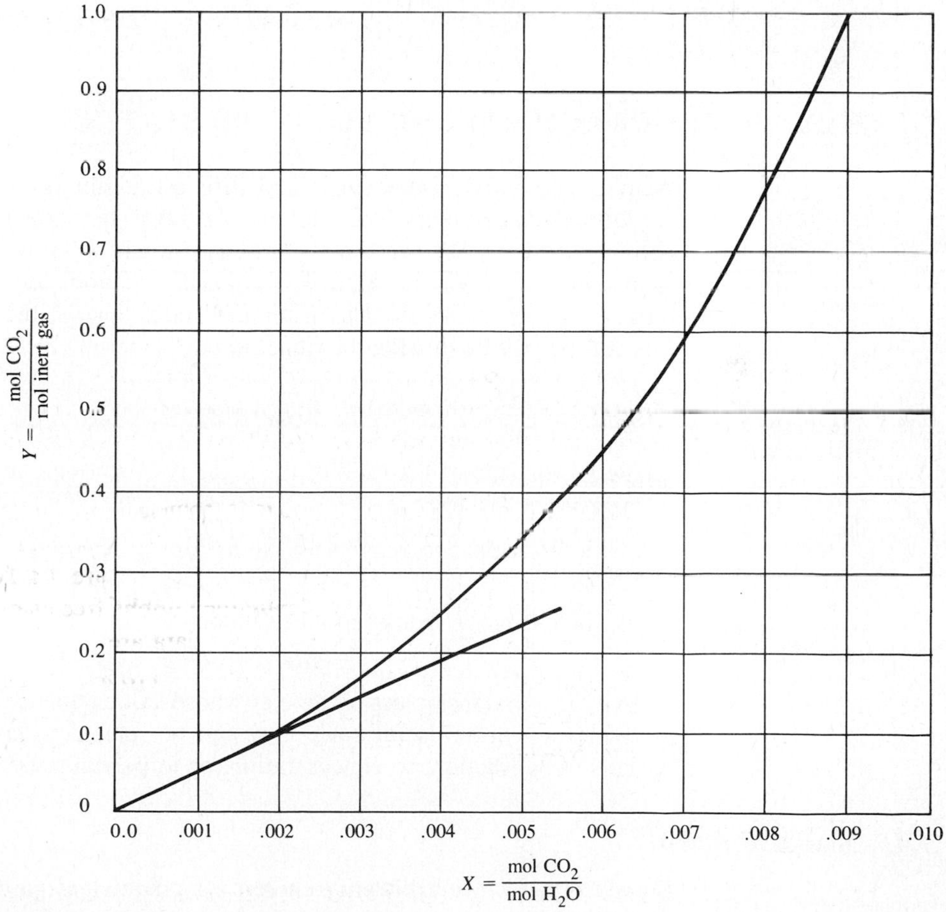

Figure 7.8 Equilibrium data for Example 7.6.

Such solutions will be discussed in context of the next section with the introduction of the term *activity*. Extensive discussion of conformal or thermodynamic nonideal-solution theory is beyond the scope of this book.

Example 7.6 Equilibrium data for the CO_2–water system at 2.53 MPa and 12.4°C are given in Fig. 7.8. Determine in what range Henry's law is followed and the value of the Henry's law constant.

SOLUTION Henry's law is followed where the equilibrium line is straight, i.e., up to $X = 0.0012$ and $Y = 0.06$. Using Eq. (7.65) for the liquid and Dalton's law for the vapor,

$$\bar{P}_{CO_2} = H_{CO_2} x_{CO_2} = y_{CO_2} P_T$$

$$y_{CO_2} = \frac{0.06}{1.06} = 0.0566 \qquad H_{CO_2}(0.0012) = (0.0566)(2.53)$$

$$x_{CO_2} = \frac{0.0012}{1.0012} \approx 0.0012 \qquad H_{CO_2} = 119.33 \text{ MPa}$$

7.4 THE CONCEPT OF ACTIVITY AND STANDARD STATES

Activity and activity coefficients and their estimation are important concepts for equilibrium calculations. The reference, base, or standard state on which values of activities and activity coefficients are based must be known in order to interpret such values. This section will define and show methods for testing thermodynamic data as well as define the usual standard states. Use of the concepts discussed in this section will be detailed in Chap. 8.

7.4.1 Activity

Activity is defined as the ratio of the fugacity of a compound at a state divided by its fugacity at a reference state:

$$\left[a_i \equiv \frac{\bar{f}_i}{f_i^\circ}\right]_T \tag{7.66}$$

Thus, the change in free energy can be defined as

$$\Delta G_{i,T} = RT \ln a_{i,T} \tag{7.67}$$

Activity is a dimensionless parameter whose value depends on the standard state. Activities are quite useful since they can be related to composition parameters such as mole fraction, to concentration, or to partial pressure.

7.4.2 Standard States

Standard states are arbitrarily chosen for advantages gained in use for calculations. Each component has a standard state which is always defined with respect to the system temperature, but the pressure of the standard state may vary according to the system under study. The standard state may not be able to be set at conditions that can be measured experimentally although it must be able to be calculated. For any compound the same standard state is used for any one set of calculations, although it may change for another set of calculations.

Common standard states used follow:

Gases One of two standard states may be used.

1. The pure gas as an ideal gas at 1 atm total pressure. Such a definition makes the fugacity unity in units of atmospheres, which then makes the activity of a compound merely equal to its fugacity. If the gas is also ideal, the activity is then the partial pressure of the gas. Much use will be made of this in Chap. 9.
2. The pure gas at the pressure of the system. If the total pressure is higher than the saturation pressure of the gas, the gas would condense and the state is only imaginary. For ideal-equilibrium systems, the activity of the compound equals its mole fraction, making calculations simpler.

Liquids Three standard states are used.

1. Pure liquid at 1 atm total pressure. If a liquid at the system temperature has a vapor pressure greater than 1 atm, this value becomes imaginary.
2. Pure liquid at the system pressure. If a liquid is above its saturation or critical temperature, this state is imaginary.
3. Pure liquid at its vapor pressure.

Solids The solid standard state is normally taken as the pure solid at 1 atm total pressure. Actual solid activities are not necessarily unity as often stated.

Nonelectrolyte solutions Often several other standard states relating to their activities at infinite dilution are used, especially for very slightly soluble materials.

1. The first is a hypothetical liquid state defined by referring the compound to its behavior at infinite dilution at a certain pressure, although the magnitude of the pressure is usually unimportant.

   As $x_i \to 0$, Henry's law will be followed and Eqs. (7.68) and (7.69) result:

$$\bar{f}_i = H_i x_i$$

   hence

$$\left(\frac{\partial \bar{f}_i}{\partial x_i}\right)_{x_i=0} = H_i \tag{7.68}$$

$$\frac{a_i'}{x_i} = \frac{f_i'}{f_i^{\circ\prime} x_i} = 1$$

hence

$$\left[\frac{\partial \bar{f}_i}{\partial x_i}\right] = f_i^{\circ\prime} \tag{7.69}$$

Equating,

$$f_i^{\circ\prime} = H_i' \tag{7.70}$$

2. Similarly a standard state related to unity molality (m) may be defined as the molality approaches zero. Here

$$f_i^{\circ\prime\prime} = H_i'' \tag{7.71}$$

where $H_i'' = \bar{f}_i/m_i$ and is a Henry's law constant defined in molality units rather than mole fraction.

Thus three different standard-state fugacities can be defined by equating:

$$\bar{f}_i = a_i f_i^{\circ} = a_i' f_i^{\circ\prime} = a_i'' f_i^{\circ\prime\prime} \tag{7.72}$$

Such activities can be determined by plotting fugacities of the component vs. mole fraction and noting the intercept at mole fraction equals 1, which gives $f_i{}^{\circ}$, $f_i^{\circ\prime}$ can be determined by drawing a tangent to the fugacity curve at 0 mol fraction and extending it to unit mol fraction. If fugacity is plotted against molality, $f_i^{\circ\prime\prime}$ can be obtained by drawing a tangent to the fugacity vs. molality curve at $m = 0$ and extending it to $m = 1.0$.

Electrolyte solutions For weak and strong electrolyte solutions, the standard state varies. Henry's law is approached at zero molality if complete ionization is assumed. For weak electrolytes various extrapolations to infinite dilution are usually used for the solute standard states for both ions and undissociated molecules. Lewis and Randall, Pitzer and Brewer (1961), and Denbigh (1981) discuss these methods in some detail.

7.4.3 Activity Coefficients

An activity coefficient is the correction factor which measures the departure of a solution from ideal behavior for a given standard state. It is defined in various ways with respect to the various composition parameters and is always the empirical ratio of activity to that composition parameter. With respect to the pure component, infinite-dilution mole fraction, and infinite-dilution molality formulations, the activity coefficient γ is defined as

$$\gamma_i = \frac{a_i}{x_i} = \frac{f'}{f_i^{\circ} x_i} \tag{7.73}$$

$$\gamma_i' = \frac{a_i'}{x_i} = \frac{\bar{f}_i}{f_i^{\circ\prime} x_i} \tag{7.74}$$

$$\gamma_i'' = \frac{a_i''}{m_i} = \frac{f_i}{f_i^{\circ\prime\prime} m_i} \tag{7.75}$$

Activity coefficients and activities can be calculated when fugacities are known and can be used to predict actual equilibria without ideal-solution assumptions such as Raoult's law; for example,

Ideal partial pressure (Raoult's law):

$$\bar{P}_i = x_i P_i^\circ$$

Actual partial pressure:

$$\bar{P}_i = a_i P_i^\circ = \gamma_i x_i P_i^\circ \tag{7.76}$$

When fugacities are unknown or not calculable by generalized techniques, as, for example, for very high boiling liquids and solids, the ratio of activities at different standard states can be evaluated from activity coefficients at infinite dilution, Eqs. (7.74) and (7.75), where $\gamma_i' = 1$ and $\gamma_i'' = 1$. From Eqs. (7.72) and (7.73) at infinite dilution and letting $\gamma_i = \gamma_i^\circ$, an expression for a_i can be derived:

$$a_i = \gamma_i^\circ a_i' = \frac{\gamma_i^\circ a_i'' \text{MW}}{1000} \tag{7.77}$$

where MW = molecular weight of the solvent.

The effect of temperature on activity coefficients can be shown by combining Eqs. (7.25) and (7.73):

$$\left[\frac{\partial(\ln \gamma_i)}{\partial T}\right]_{P, x_i} = \left(\frac{\bar{H}_i^\circ - \tilde{H}_i}{RT^2}\right)_{P, x_i} \tag{7.78}$$

where $\bar{H}_i^\circ$ and $\tilde{H}_i$ are for the phase being considered; their difference is the partial heat of mixing compound i from the standard state to the solution at the system pressure, composition, and phase considered.

The effect of pressure on activity coefficients is easily derived from Eqs. (7.20) and (7.73):

$$\left[\frac{\partial(\ln \gamma_i)}{\partial P}\right]_{T, x_i} = \left(\frac{\tilde{V}_i - \bar{V}_i^\circ}{RT}\right)_{T, x_i} \tag{7.79}$$

where $\tilde{V}_i$ and $\bar{V}_i^\circ$ are for the phase being considered. This effect is negligible at and below atmospheric pressure.

Example 7.7 Toluene and acetic acid vapor-liquid equilibrium data are available at 343 K. At these conditions the vapor pressures are 26.9 kPa for toluene and 18.1 kPa for acetic acid. At 0.486 liquid mol fraction for toluene, the partial pressures are 17.4 kPa for toluene and 11.8 kPa for acetic acid. Extrapolation of the data gives a Henry's law constant of 55 kPa for acetic acid.

Calculate the activities and activity coefficients of acetic acid for the given data point, with a standard-state fugacity established (a) by Henry's law, (b) as pure liquid at 343 K. Repeat (b) for toluene. Show whether the point of data given follows ideal behavior.

SOLUTION (*a*) From Eq. (7.66),

$$a_i = \frac{\bar{f}_i}{f_i^\circ}$$

From Eq. (7.73),

$$\gamma_i = \frac{a_i}{x_i}$$

For acetic acid with Henry's law standard state,

$$\bar{f}_i = \overline{P}_i \qquad \text{(low pressure)}$$

$$f_i^\circ = H_i$$

$$a_i = \frac{\overline{P}_i}{H_i} = \frac{\overline{P}_i}{55}$$

At $x_i = 0.514$, $\overline{P}_i = 11.8$. Therefore

$$a_i = \frac{11.8}{55} = 0.2145$$

$$\gamma_i = \frac{0.2145}{0.514} = 0.4174$$

(*b*) For acetic acid with pure-liquid standard state,

$$\bar{f}_i = \overline{P}_i \qquad f_i^\circ = P_i^\circ$$

$$a_i = \frac{\overline{P}_i}{P_i^\circ} = \frac{\overline{P}_i}{18.1}$$

Therefore

$$a_i = \frac{11.8}{18.1} = 0.652 \qquad \gamma_i = \frac{0.652}{0.514} = 1.27$$

For toluene with pure-liquid standard state,

$$\bar{f}_i = \overline{P}_i \qquad f_i^\circ = P_i^\circ$$

$$a_i = \frac{P_i}{P_i^\circ} = \frac{\overline{P}_i}{26.9}$$

Therefore

$$a_i = \frac{17.4}{26.9} = 0.647 \qquad \gamma_i = \frac{0.647}{0.486} = 1.331$$

If Raoult's law is followed,

$$\overline{P}_i = x_i P_i^\circ$$

For toluene,

$$17.4 \neq (0.486)(26.9)$$

$$17.4 \neq 13.07$$

This is obvious as $a_i \neq 1$.

The same result would be observed for acetic acid.

Exercise 7.F Calculate the activity coefficient for ammonia in a 10 mol % solution with water at 0°C. Use the conventional liquid standard state. Determine if the system follows Henry's law.

Mol % NH_3	5	10	15	50	100
Partial pressure of ammonia, kPa	1.79	3.58	6.20	133.4	429.3

7.4.4 The Gibbs-Duhem Equation

J. Willard Gibbs advanced a general relationship among the chemical potentials of the components of a solution in terms of partial molal properties. Although this equation can be similarly derived for any molar property, it is most used when applied to Gibbs free energy and its derived properties fugacity and activity. Only the derivation leading to these properties is given here.

Consider the total free energy of a solution at constant temperature and pressure and using defining Eq. (7.17):

$$G_{P,T} = \left(\sum_i n_i \tilde{G}_i\right)_{P,T} = \left(\sum_i n_i \mu_i\right)_{P,T} \tag{7.80}$$

Differentiating,

$$dG_{P,T} = \left(\sum_i n_i \, d\mu_i\right)_{P,T} + \left(\sum_i \mu_i \, dn_i\right)_{P,T} \tag{7.81}$$

Since all $d\mu_i$'s are zero at constant temperature and pressure,

$$dG_{P,T} = \left(\sum_i \mu_i \, dn_i\right)_{P,T} \tag{7.82}$$

Subtracting Eq. (7.82) from Eq. (7.81),

$$\left(\sum_i n_i \, d\mu_i\right)_{P,T} = 0 \tag{7.83}$$

In terms of mole fractions, since $x_i = n_i/n_T$,

$$\left(\sum_i x_i \, d\mu_i\right)_{P,T} = 0 \tag{7.84}$$

Since $[d\mu_i = RT\, d(\ln \bar{f}_i)]_T$, Eq. (7.84) can be rewritten as

$$\sum_i x_i \, d(\ln \bar{f}_i) = 0 \tag{7.85}$$

This is the usual form of the Gibbs-Duhem equation at constant temperature and pressure and is applicable to any number of components. For example, for a binary mixture of A and B,

$$\left[x_A \frac{\partial(\ln \bar{f}_A)}{\partial x_A}\right]_{P,T} = \left[-x_B \frac{\partial(\ln \bar{f}_B)}{\partial x_A}\right] \tag{7.86}$$

Since $a_i = \bar{f}_i/f_i^\circ$ and f_i° is constant at a fixed temperature and pressure,

$$\left[\sum_i x_i\,\partial(\ln a_i)\right]_{P,T} = 0 \tag{7.87}$$

Since $a_i = x_i\gamma_i$, it can be readily shown that

$$\left[\sum_i x_i\,\partial(\ln \gamma_i)\right]_{P,T} = 0 \tag{7.88}$$

Equation (7.88) is the most used form of the equation, as activity coefficients have more practical applications than do activities or fugacities. For a binary mixture,

$$\left[x_A\,\partial(\ln \gamma_A) = -x_B\,\partial(\ln \gamma_B)\right]_{P,T} \tag{7.89}$$

Variable-pressure and variable-temperature forms of the Gibbs-Duhem equation can be derived. These more-complicated forms have been used extensively to test vapor-liquid equilibrium data for thermodynamic consistency. Abbott and Van Ness (1975) and Van Ness, Byer, and Gibbs (1973) discuss these methods in greater detail. Chapter 8 will show some of these applications.

The general form of the Gibbs-Duhem equation which is of value for derivation of any form is therefore

$$S\,dT - V\,dP + \sum_i n_i\,d\mu_i = 0 \tag{7.90}$$

Exercise 7.G Derive Eq. (7.88).

Example 7.8 For a binary mixture of A and B, activity coefficient data for A are available over the entire composition range, while only one point of data is available for B. Show how you would determine activity coefficients for B over the entire range of compositions.

SOLUTION Using Eq. (7.89),

$$x_A\,\partial(\ln \gamma_A) = -x_B\,\partial(\ln \gamma_B)$$

Choosing two states 1 and 2 with 1 the known state,

$$\int_1^2 \partial(\ln \gamma_B) = \int_1^2 -\frac{x_A}{x_B}\,\partial(\ln \gamma_A)$$

$$\ln\frac{\gamma_{B_2}}{\gamma_{B_1}} = -\int_1^2 \frac{x_A}{x_B}\,\partial(\ln \gamma_A)$$

Evaluate the integral graphically between the known state and any state by plotting x_A/x_B versus $\ln \gamma_A$. Successive application of this method will yield the entire range of activity coefficients for B.

7.5 REFERENCES

Abbott, M. M., and H. C. Van Ness: *AIChE J.*, **21**:62 (1975).
Denbigh, K.: "The Principles of Chemical Equilibrium," 4th ed., Cambridge, London, 1981.
Hildebrand, J. H., J. M. Prausnitz, and R. L. Scott: "Regular and Related Solutions," Van Nostrand Reinhold, New York, 1970.
_____ and R. L. Scott: "Solubility of Nonelectrolytes," 3d ed., Reinhold, New York, 1950.
Hougen, O. A., K. M. Watson, and R. A. Ragatz: "Chemical Process Principles, Part II, Thermodynamics," 2d ed., Wiley, New York, 1959.
Lee, B. I., and M. G. Kesler: *AIChE J.*, **21**:510 (1975).
Lewis, G. N., and M. Randall: "Thermodynamics and the Free Energy of Chemical Substances," McGraw-Hill, New York, 1923; 2d ed. with K. S. Pitzer and L. Brewer, 1961.
Lin, C. T., and T. E. Daubert: *Ind. Eng. Chem. Process Des. Dev.*, **17**:544 (1978).
Prausnitz, J. M., et al.: "Computer Calculations for Multicomponent Vapor-Liquid and Liquid-Liquid Equilibria," Prentice-Hall, Englewood Cliffs, N.J., 1980.
Van Ness, H. C., S. M. Byer, and R. E. Gibbs: *AIChE J.*, **19**:238 (1973).

7.6 PROBLEMS

1 Using the data below for the standard enthalpy of formation of HCl from the elements at 25°C, determine the partial molar enthalpies of HCl and water in a solution containing 10 kmol HCl/m^3 solution.

	Heat of formation, kcal/g mol		Heat of formation, kcal/g mol
Gas	−22.063	In $6H_2O$	−37.811
In $1H_2O$	−28.331	In $8H_2O$	−38.371
In $2H_2O$	−33.731	In $10H_2O$	−38.671
In $3H_2O$	−35.651	In $50H_2O$	−39.577
In $4H_2O$	−36.691	In $100H_2O$	−39.713
In $5H_2O$	−37.371	In ∞H_2O	−49.023

2 A liquid solution of two components contains n_1 mol of component 1 and n_2 mol of component 2. If the specific volume of the solution is plotted as the ordinate y and the number of moles of component 2 is plotted on the abscissa x, show that the two equations below are correct. MW = molecular weight.

$$\tilde{V}_1 = \text{MW}_1\left(y - \frac{n_1 n_2 + n_2^2\,\text{MW}_2}{\text{MW}_1}\frac{dy}{dx}\right)$$

$$\tilde{V}_2 = \text{MW}_2\left(y + \frac{n_1 n_2 + n_1^2\,\text{MW}_1}{\text{MW}_2}\frac{dy}{dx}\right)$$

3 If for a certain gas the van der Waals constants are $a = 5.46$ (atm · liter ^{2})/g mol^2 and $b = 0.0647$ liter/g mol, evaluate the fugacity coefficient and the Gibbs free energy and enthalpy differences between the ideal- and real-gas states at 25°C and 10 atm pressure.

4 Calculate the fugacity coefficient for a gas that follows the equation $P\overline{V} = RT(1 - 0.0052P)$, where P is in atmospheres, at pressures of 1, 5, and 10 atm.

5 Calculate the fugacity by using the Lydersen method for (*a*) 100°C and 1 atm liquid water and (*b*) water at the critical point.

6 For toluene at its normal boiling point and 10 MPa pressure, calculate its fugacity by both the Pitzer and the Redlich-Kwong-Soave methods. Repeat at conditions of $T_r = 0.9$ and 10 MPa pressure.

7 Calculate the fugacity of ice at 0°C and atmospheric pressure. Repeat for liquid water at 0°C and 10 MPa pressure.

8 Calculate the change in entropy and Gibbs free energy of mixing and the fugacity of A and B if 1 mol of component A and 2 mol of component B are mixed to form an ideal-liquid solution. At the temperature of the system, the vapor pressures of pure A and pure B are 432 and 124 mmHg, respectively.

9 The Henry's law constant for hydrogen in water at low pressures is 6.8×10^4 atm at 20°C. How much hydrogen can be dissolved in 1 kg of water at 20°C?

10 Determine the activity of water in a solution at 25°C and atmospheric pressure where its partial pressure is 20 mmHg. Do for three different standard states.

11 The following data apply to aqueous solutions of 1-propanol at 50°C.

x propanol	0	0.090	0.160	0.305	0.411	0.555	0.739	0.820	1.000
$\bar{P}$ H_2O, mmHg	92.0	87.2	87.0	84.5	83.0	78.2	60.9	49.2	0.0
$\bar{P}$ propanol, mmHg	0.0	49.2	51.7	54.6	57.4	60.2	68.4	72.1	90.0

Using a pure-liquid standard state for each component, calculate the activity and activity coefficient for $x = 0, 0.2, 0.5, 0.8$, and 1.0.

12 At 25°C and 1 atm, solid NaCl is in equilibrium with a 6.14 molal solution of NaCl. The volume of the NaCl in solution is 24.30 cm^3/g mol. The density of solid NaCl is 2.163 g/cm^3. Calculate the change in partial molar free energy of the solution at constant molality as the pressure increases by 1 atm.

13 Test the following isothermal data at 20°C for thermodynamic consistency by using the Gibbs-Duhem equation.

Wt% HNO_3	100	90	80	70	60	50
$\bar{P}$ H_2O, mmHg	0.0	0.46	1.35	2.83	4.93	7.53
$\bar{P}$ HNO_3, mmHg	48.0	26.03	10.49	3.08	0.89	0.49

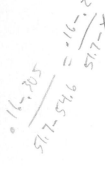

CHAPTER

EIGHT

PHYSICAL EQUILIBRIA AMONG PHASES

Based on the concepts developed in Chap. 7, equilibria among phases can be calculated. Starting with the definitions of equilibrium, the phase rule will be developed. Vapor-liquid equilibria as the most applied and calculable form of phase equilibria will be discussed in some detail for various types of systems. A brief treatment of liquid-liquid, gas-solid, and liquid-solid equilibria will follow.

8.1 EQUILIBRIUM—THE PHASE RULE

Equilibrium was defined in Chap. 1 as a state that after any short, small mechanical disturbance of external conditions will return to its initial state. Equilibrium was related to entropy and free energy in Chap. 4, Sec. 4.4.3. Phases were also defined in Chap. 1.

Within any closed system where phases exist, each phase is actually an open system, as it can change in composition or in mass, although the total closed system must have a constant overall composition and mass. For equilibrium to exist for the total system, it is necessary that equilibrium exist for each phase within the system. If the normal conditions of temperature and pressure constancy are maintained, then the change in free energy must be zero if no work is performed from outside forces. Thus the change in free energy for any j-phase n-component system is

$$dG = \sum_j \left(-S_j\, dT + V_j\, dP + \mu_j^{(1)}\, dn_j^{(1)} + \cdots + \mu_j^{(n)}\, dn_j^{(n)} \right) \tag{8.1}$$

At constant temperature and pressure,

$$dG = \sum_j \mu_j^{(1)}\, dn_j^{(1)} + \cdots + \mu_j^{(n)}\, dn_j^{(n)} = 0 \tag{8.2}$$

For each component, the total number of moles is constant:

$$\begin{aligned} dn_1^{(1)} &= -dn_2^{(1)} - dn_3^{(1)} - \cdots - dn_j^{(1)} \\ &\cdots\cdots\cdots\cdots\cdots\cdots \\ dn_i^{(n)} &= -dn_2^{(n)} - dn_3^{(n)} - \cdots - dn_j^{(n)} \end{aligned} \tag{8.3}$$

If Eq. (8.3) is substituted into Eq. (8.2), it can be shown that

$$\begin{aligned} \mu_1^{(1)} &= \mu_2^{(1)} = \mu_3^{(1)} = \cdots = \mu_j^{(1)} \\ &\cdots\cdots\cdots\cdots\cdots\cdots \\ \mu_1^{(n)} &= \mu_2^{(n)} = \mu_2^{(n)} = \cdots = \mu_j^{(n)} \end{aligned} \tag{8.4}$$

Thus, the chemical potential of any given component in every phase is equal at equilibrium. This is both an essential and mathematically sufficient condition for equilibrium.

As stated earlier, this can also be specified in terms of fugacities since $d\mu_i = RT\,d(\ln f_i)$, which if substituted into Eq. (8.2) and then solved will show that

$$\begin{aligned} f_1^{(1)} &= f_2^{(1)} = \cdots = f_j^{(1)} \\ &\cdots\cdots\cdots\cdots \\ f_1^{(n)} &= f_2^{(n)} = \cdots = f_j^{(n)} \end{aligned} \tag{8.5}$$

Thus the fugacity of each component must be identical in each phase. This criterion is certainly more useful than that of chemical potential.

In addition, the equilibrium criterion can be written in terms of activity, as fugacity is merely the product of the activity and the standard-state fugacity.

Gibbs in 1875 proposed his famous phase rule that governs the number of intensive variables which can be varied in any equilibrium system without changing the number of phases or number of components that will be present in a given system. This number of variables that can be varied without changing the number of phases or components is known as the *number of degrees of freedom* or variance V of the system.

For any phase the free-energy change can be written as

$$dG_i = f_i\left(T, P, \mu^{(1)}, \ldots, \mu^{(n)}\right) = 0 \tag{8.6}$$

Thus, the number of intensive properties which can be varied in any phase ϕ of C components is the number of components plus 2.

If Eq. (8.6) is written for each phase ϕ in $C + 2$ variables, solution of the equations will yield values for only ϕ variables. Thus $C + 2 - \phi$ variables are not fixed nor are the degrees of freedom V.

The phase rule: $$V = C + 2 - \phi \tag{8.7}$$

The phase rule then allows calculation of the number of intensive variables that can be independently fixed in any equilibrium system so long as only temperature, pressure, and chemical potential are significant variables. In systems where the *PVT* behavior is also dependent on another variable such as surface tension, another variable must be added and $V = C + 3 - \phi$. Cases of up to two additional variables are known but are usually not important for phase equilibria determinations.

The number of components to be used in the phase rule is defined as the least number of independent chemical compounds from which the system can be created. This becomes important where chemical reactions are involved. For example, if a system consisting of CH_3OH, H_2, and CO is at equilibrium, the number of components is only one, as the chemical reaction $CO + 2H_2 \rightarrow CH_3OH$ is occurring.

In this chapter only equilibrium situations without chemical reaction will be considered. Chapter 9 will consider systems with chemical reactions.

The phase rule can be applied to any equilibrium, and the number of degrees of freedom may vary from zero (e.g., the triple point of water where $C = 1$ and $\phi = 3$) to the number of components plus 1 (e.g., liquid water where $C = 1$ and $\phi = 1$).

Example 8.1 Determine the number of degrees of freedom for each of the following cases, where each phase is in equilibrium with all other phases and no chemical reactions occur.

(*a*) Benzene and toluene are undergoing a simple distillation at 1 atm pressure. Using Eq. (8.7),

$$V = C + 2 - \phi$$

$$C = 2 \qquad \phi = 2 \text{ (vapor and liquid)}$$

Therefore $V = 2 + 2 - 2 = 2$, but pressure is fixed, thus the available degrees of freedom are 1, either the temperature or the composition of either phase.

(*b*) Extraction of a binary mixture with a pure solvent soluble only in one of the components at 1 atm pressure and room temperature is carried out.

$$C = 3 \qquad \phi = 2 \text{ (both liquid)}$$

Therefore $V = 3 + 2 - 2 = 3$, but pressure and temperature are fixed; thus only 1 degree of freedom is available, the composition of either binary phase.

(*c*) Adsorption of gaseous methane from a mixture with air on a solid adsorbent at ambient pressure and a fixed temperature.

$$C = 3 \qquad \text{(two gases and solid)}$$

$$\phi = 2 \qquad \text{(gas and solid)}$$

Therefore $V = 3 + 2 - 2 = 3$, but pressure and temperature are fixed, leaving 1 degree of freedom available, the concentration of adsorbed methane or of methane in the air.

Exercise 8.A Determine the number of available degrees of freedom for the following nonreacting equilibrium systems:

1. Two completely miscible materials in vapor-liquid equilibrium, with vapor composition specified at a given pressure and temperature
2. Two partially miscible liquid phases, each containing the same three components
3. Two liquid phases and vapor phase, each containing the same two components at constant pressure
4. A vapor phase containing ammonia in air and a liquid phase containing ammonia in water at a given temperature

8.2. FUNDAMENTALS OF VAPOR-LIQUID EQUILIBRIA

For any system in a state of equilibrium between liquid and vapor, the fugacity of any component i is the same in each phase:

$$\bar{f}_i^V = \bar{f}_i^L \tag{8.8}$$

For any system, the number of degrees of freedom will equal $C + 2 - 2 = C$. Usually either the temperature or pressure and either the liquid- or vapor-phase composition are fixed. This will be sufficient to define the system.

The equilibrium relationship can be written in terms of activities or fugacity coefficients as

Pure	Mixture	
$a_i \equiv \dfrac{f_i}{f_i^\circ}$	$\bar{a}_i = \dfrac{\bar{f}_i}{f_i^\circ}$	(8.9)
$\phi_i = \dfrac{f_i}{P_i}$	$\bar{\phi}_i = \dfrac{\bar{f}_i}{x_i P_T}$	(8.10)

where f_i = fugacity of pure component at T, P
f_i° = standard-state fugacity of component i
$\bar{f}_i$ = fugacity of component i in mixture at T, P

An overbar always signifies "in solution"; no overbar signifies "pure state." Thus, for the vapor phase,

$$\bar{f}_i^V = \bar{\phi}_i^V y_i P_T \quad \text{or} \quad \bar{f}_i^V = \bar{a}_i^V f_i^{\circ V}$$

and for the liquid phase,

$$\bar{f}_i^L = \bar{\phi}_i^L x_i P_T \quad \text{or} \quad \bar{f}_i^L = \bar{a}_i^L f_i^{\circ L}$$

Equating

$$y_i \bar{\phi}_i^V = x_i \bar{\phi}_i^L \tag{8.11}$$

or

$$\bar{a}_i^V f_i^{\circ V} = \bar{a}_i^L f_i^{\circ L} \tag{8.12}$$

these equations define the equilibrium between a vapor and a liquid phase.

Rearranging Eq. (8.12), an expression for the ratio of activities called the *equilibrium vaporization constant* or *K value* is obtained for any component:

$$K_i = \frac{\bar{a}_i^V}{\bar{a}_i^L} = \frac{f_i^{\circ L}}{f_i^{\circ V}} \tag{8.13a}$$

The K value is a function of temperature and pressure and can be defined in several ways. While Eq. (8.13a) is one definition of an equilibrium constant, experimental K values are often defined as the ratio of y/x for each component. From Eq. (8.11),

$$K_i = \frac{y_i}{x_i} = \frac{\bar{\phi}_i^L}{\bar{\phi}_i^V} \tag{8.13b}$$

This latter definition is more often used for high-pressure equilibria. Equilibrium K values are most often used for multicomponent systems and will be discussed in detail in Sec. 8.3.1.

The vapor-liquid equilibria can also be described in terms of activity coefficients by Eq. (7.73):

$$\gamma_i = \frac{a_i}{x_i} = \frac{f_i}{f_i^\circ x_i}$$

Thus, for the vapor phase,

$$a_i^V = \gamma_i^V y_i = \frac{\bar{f}_i^V}{f_i^{\circ V}}$$

and for the liquid phase,

$$a_i^L = \gamma_i^L x_i = \frac{\bar{f}_i^L}{f_i^{\circ L}}$$

Since $\bar{f}_i^V = \bar{f}_i^L$,

$$\gamma_i^V y_i f_i^{\circ V} = \gamma_i^L x_i f_i^{\circ L} \tag{8.14}$$

Based on Eq. (8.13a),

$$K_i = \frac{f_i^{\circ L}}{f_i^{\circ V}} = \frac{\gamma_i^V y_i}{\gamma_i^L x_i} \tag{8.15a}$$

Based on Eq. (8.13b),

$$K_i = \frac{y_i}{x_i} = \frac{\gamma_i^L f_i^{\circ L}}{\gamma_i^V f_i^{\circ L}} \tag{8.15b}$$

Application of Eqs. (8.11) to (8.15) to various cases requires certain assumptions.

8.2.1 Idealized Low-Pressure Systems

Considering Eq. (8.11), $y_i \bar{\phi}_i^V = x_i \bar{\phi}_i^L$. If the vapor phase is ideal, Eq. (7.29) indicates that $\bar{\phi}_i^V = 1$. If the liquid phase is an ideal solution,

$$\bar{\phi}_i^L = \frac{\bar{f}_i^L}{x_i P_T} = \frac{f_i^L}{P_T} \tag{8.16}$$

If the fugacity of pure i can be assumed to be independent of pressure, then f_i^L will equal f_i at the saturation pressure and temperature of the system. This is a good approximation at low pressures, as the Poynting correction approaches zero. Since the vapor is also ideal, ϕ_i at saturation equals unity and thus the liquid saturation fugacity is equal to the vapor pressure P_i°. Thus

$$f_i^L = P_i^\circ \tag{8.17}$$

Substituting Eqs. (8.16) and (8.17) into Eq. (8.11),

$$y_i = \frac{x_i P_i^\circ}{P_T} \tag{8.18}$$

which is a form of Raoult's law as discussed in Chap. 7. This most simple equation for estimating vapor-liquid equilibria is usually only applicable to materials of like size and chemical family, e.g., toluene-ethylbenzene, methanol-ethanol, heptane-octane.

Exercise 8.B Isobutane and isobutylene are to be distilled at a constant pressure of 100 kPa. The following data apply to the system:

	T, °C	$P°$, kPa	T, °C	$P°$, kPa
Isobutylene	−7	100	−12	80
Isobutane	−12	100	−7	130

Calculate the vapor and liquid compositions over the entire range of compositions, assuming the system is ideal, and plot x vs. y for isobutane.

8.2.2 General Low-Pressure Systems

Calculations for low-pressure systems are often based on a combination of the fugacity coefficient concept used for the vapor phase and the activity coefficient concept used for the liquid phase.

Since by Eq. (8.8) $\bar{f}_i^V = \bar{f}_i^L$,

$$\bar{\phi}_i^V y_i P_T = \gamma_i^L x_i f_i^{\circ L} \tag{8.19}$$

This equation then makes the best use of both worlds for estimation of vapor-liquid equilibria. The vapor fugacity coefficient only depends on the vapor-phase composition, while the liquid-phase activity coefficient depends only on the liquid-phase composition, and the standard-state fugacity only depends on the pure liquid. At low pressures it can be assumed that both the liquid-phase activity coefficient and liquid standard-state fugacity coefficient are independent of pressure.

To apply Eq. (8.19) three parameters must be estimated.

1. The vapor-phase fugacity coefficient is estimated by the methods of Chap. 7.
2. The liquid standard-state fugacity can be calculated by taking the standard state as pure liquid at the temperature of the system. Thus

$$f_i^\circ = f_i = \phi_i^\circ P_i^\circ \tag{8.20}$$

where superscript ° refers to the standard state. Since both saturated vapor and saturated liquid will have the same fugacity coefficient, it can be evaluated by the same methods of Chap. 7 as for the vapor-phase fugacity coefficient.
3. The activity coefficient of the liquid must be evaluated from some correlation. Alternatively, the equation could be written in terms of the fugacity coefficient of the liquid and the methods of Chap. 7:

$$\bar{f}_i^V = \bar{\phi}_i^V y_i P_T = \bar{f}_i^L = \bar{\phi}_i^L x_i P_T \tag{8.21}$$

This latter method is advanced by Graboski and Daubert (1978*a*, *b*; 1979) for hydrocarbon and hydrocarbon-nonhydrocarbon gas systems by using the Redlich-Kwong-Soave equation of state which was determined to be the most accurate method for such systems and is discussed in the previous chapter. However, for polar associating systems, such a method is inaccurate and the method of activity coefficients must be used. Methods for prediction of activity coefficients are given in the next subsection.

Example 8.2 A mixture of isobutane and isobutylene is in equilibrium at 95°C and 10 atm total pressure. Write an equation which could be used to determine the liquid-phase activity coefficient of isobutane if composition data were also available.

SOLUTION Using Eq. (8.19) for a low-pressure system,

$$\bar{\phi}_i^V y_i P_T = \gamma_i^L x_i f_i^{\circ L}$$

Since the pressure is low, let

$$\bar{\phi}_i^V = \phi_i \text{ at } T_r \qquad \text{and} \qquad P_r = \frac{P}{P_c}$$

and

$$f_i^{\circ L} = P_i^\circ \phi_i^\circ$$

where

$$\phi_i^\circ = \phi_i \text{ at } T_r \qquad \text{and} \qquad P_r' = \frac{P_i^\circ}{P_c}$$

From App. A for isobutane, $T_c = 135°C$ and $P_c = 3648$ kPa. In Example 7.3, Z_c was calculated to be 0.283.

Using the simple Lydersen method,

$$T_r = \frac{368.1}{408.1} = 0.902 \qquad P_r = \frac{1013}{3648} = 0.278$$

Thus, using Fig. 7.1,

$$\phi_{Z_c=0.27} = 0.88$$

Using Eq. (7.39) and Fig. 7.3,

$$\phi_{Z_c=0.283} = 0.88 \times 10^{0.3(0.013)} = 0.888$$

Similarly, $P_r' = P_i^\circ / P_c$. From App. A, $P^\circ = 1800$ kPa. Therefore

$$P_r' = \frac{1800}{3648} = 0.493$$

Thus, using Fig. 7.1,

$$\phi_{Z_c=0.27} = 0.78$$

Using Eq. (7.39) and Fig. 7.3,

$$\phi_{Z_c=0.283} = 0.78 \times 10^{0.42(0.013)} = 0.790$$

Thus,

$$(0.888)\,y(1013\text{ kPa}) = \gamma^L x(0.790)(1800)$$

$$\gamma^L = 0.633\left(\frac{y}{x}\right)_{\text{isobutane}}$$

8.2.3 Prediction of Activity Coefficients

Activity coefficients cannot normally be accurately predicted with no experimental vapor-liquid equilibrium data, although some such attempts have been made and will be noted later in this section. Generally, the methods advanced are based on the ability to extend limited existing experimental data. Essentially all methods begin with the Gibbs-Duhem equation discussed in Chap. 7. In binary system form,

$$\left[x_A \frac{\partial(\ln \gamma_A)}{\partial x_A} = x_B \frac{\partial(\ln \gamma_B)}{\partial x_B}\right]_{T,P} \tag{7.89}$$

This equation as stated earlier allows one to test vapor-liquid equilibrium data for thermodynamic consistency. The most accurate methods currently available for such an application are those of Abbott and Van Ness (1975), also discussed by Van Ness and Abbott (1982).

Normally only spotty data for the activity coefficients as a function of composition exist. Thus some approximation which is consistent with the Gibbs-Duhem equation must be made. Such approximations usually start with the concept of excess Gibbs free energy, the free energy above the free energy calculated for an ideal solution at the pressure, temperature, and composition of the solution. This excess energy for a binary system is, as can be inferred from Eq. (7.63),

$$G^E_{T,P} = RT(n_A \ln \gamma_A + n_B \ln \gamma_B) \tag{8.22}$$

Differentiating with respect to the number of moles of A and B gives Eqs. (8.23):

$$\left(\frac{\partial G^E}{\partial n_A}\right)_{T,P,n_B} = RT \ln \gamma_A \qquad \left(\frac{\partial G^E}{\partial n_B}\right)_{T,P,n_A} = RT \ln \gamma_B \tag{8.23}$$

It remains to determine an equation which will represent $G^E_{T,P}$ as a function of composition. From experimental data then, the constants in such an equation can be fixed, after which values of γ_A and γ_B can be estimated for any particular composition. Many equations of varying degrees of exactness, none totally accurate, have been advanced. A few of the more noteworthy are given.

Wilson (1964) proposed the following general equation for the excess Gibbs free energy:

$$\frac{G^E}{RT} = -\sum_i \left[x_i \ln \left(\sum_j x_j G_{ij}\right)\right] \tag{8.24}$$

where i and j are general components in a mixture; e.g., for a binary, three G_{ij}'s exist; for a ternary, six. For a one-component system, $G_{ij} = 1$. Combining Eqs. (8.23) and (8.24) for any general component,

$$\ln \gamma_i = 1 - \ln\left(\sum_j x_j G_{ij}\right) - \sum_k \frac{x_k G_{ki}}{\sum_j x_j G_{kj}} \tag{8.25}$$

where i is the component being considered and j and k refer to all components except i.

The G_{ij} terms can be estimated from

$$G_{ij} = \frac{V_j}{V_i} e^{-c_{ij}/RT} \tag{8.26}$$

The c_{ij}'s are relatively independent of temperature and pressure and are treated as constants. Thus one set of data points at a given temperature for a binary mixture can be used to determine c_{AB} and c_{BA} for the mixtures and can be used at other temperatures and in multicomponent-system calculations. Methods for predicting Wilson parameters are given by Gothard (1976). The Wilson equation is only applicable to completely miscible systems but appears to be accurate for highly nonideal binaries.

Several equations which are somewhat simpler than the Wilson equations are available for the calculation of binary mixtures which are not radically nonideal. The van Laar and Margules equations are two-parameter equations based on Wohl's (1946) expression for excess free energy which was developed statistically in terms of the composition, effective molal volume, and effective volume fraction of the components in a mixture. The Margules equations are

$$\begin{aligned} \log \gamma_A &= x_B^2[A + 2(B - A)x_A] = (2B - A)x_B^2 + 2(A - B)x_B^3 \\ \log \gamma_B &= x_A^2[B + 2(A - B)x_B] = (2A - B)x_A^2 + 2(B - A)x_A^2 \end{aligned} \tag{8.27}$$

The van Laar equations are

$$\begin{aligned} \log \gamma_A &= \frac{Ax_B^2}{[(A/B)x_A + x_B]^2} \\ \log \gamma_B &= \frac{Bx_A^2}{[x_A + (B/A)x_B]^2} \end{aligned} \tag{8.28}$$

In both cases

$$\log \gamma_{A, x_A=0} = A$$

$$\log \gamma_{B, x_B=0} = B$$

Thus, at constant temperature and pressure, a small amount of data can be used to generate a complete data set. A single set of vapor and liquid compositions together with vapor pressures of the pure components are sufficient.

The van Laar equations are especially helpful where an azeotrope exists, since the vapor and liquid compositions are identical. Equations (8.28) can be rearranged to solve for A and B:

$$\begin{aligned} A &= \log \gamma_A \left(1 + \frac{x_B \log \gamma_B}{x_A \log \gamma_A}\right)^2 \\ B &= \log \gamma_B \left(1 + \frac{x_A \log \gamma_A}{x_B \log \gamma_B}\right)^2 \end{aligned} \tag{8.29}$$

Activity coefficients at infinite dilution are relatively easy to determine by modern analytical methods such as gas-liquid chromatography. Such data can be used with any excess-free-energy model such as that of van Laar. Considering Eqs. (8.28), as $x_A \to 0$, $B = \log \gamma_B^\infty$, and as $x_B \to 0$, $A = \log \gamma_A^\infty$. Similar procedures can be carried out by using any of the models, such as that of Wilson.

More-complex methods have been more recently developed. These include the nonrandom two-liquid (NRTL) equations with three parameters for a binary and the universal quasi-chemical (UNIQUAC) equations with two parameters. These methods are also widely used for more than one liquid-phase system. Prausnitz et al. (1980) describe the UNIQUAC method in detail. Null (1970) gives an excellent summary and discussion of the methods for predicting liquid-phase activity coefficients.

Example 8.3 At 45°C and 40.25 kPa total pressure, a vapor phase containing 43.4 mol % ethanol and 56.6 mol % benzene is in equilibrium with a liquid phase containing 61.6 mol % benzene. The system forms an azeotrope at 45°C. Assuming that few molecular interactions exist, determine the composition of the azeotrope and the total pressure of the azeotropic system. The vapor pressures of pure ethanol and benzene at 45°C are 22.9 and 29.6 kPa, respectively.

SOLUTION Assuming that the vapor phase is ideal at low pressure, $\phi_i^V = 1$, and that the standard states of liquid and vapor are pure components at their vapor pressures for the liquid and at the system pressure for the vapor, Eq. (8.19) reduces to

$$\gamma_i^L x_i P_i^\circ = y_i P_T$$

Using this equation, γ_i^L for each component can be calculated for the given composition.

If E = ethanol and B = benzene, then

$$\gamma_\mathrm{E} = \frac{y_\mathrm{E} P_T}{x_\mathrm{E} P_\mathrm{E}^\circ} = \frac{(0.434)(40.25)}{(0.616)(22.9)} = 1.238$$

$$\gamma_\mathrm{B} = \frac{y_\mathrm{B} P_T}{x_\mathrm{B} P_\mathrm{B}^\circ} = \frac{(0.566)(40.25)}{(0.384)(29.6)} = 2.004$$

The van Laar constant can then be determined by using Eqs. (8.29), after which the van Laar Eqs. (8.28) can be solved for various compositions for γ_E and γ_B. At the azeotrope $y_\mathrm{E} = x_\mathrm{E}$ and $y_\mathrm{B} = x_\mathrm{B}$. Thus

$$\frac{y_\mathrm{E}}{x_\mathrm{E}} = \frac{\gamma_\mathrm{E} P_\mathrm{E}^\circ}{P_T} = 1 \qquad \text{and} \qquad \frac{y_\mathrm{B}}{x_\mathrm{B}} = \frac{\gamma_\mathrm{B} P_\mathrm{B}^\circ}{P_T} = 1$$

Therefore, at the azeotrope $P_T = \gamma_\mathrm{E} P_\mathrm{E}^\circ = \gamma_\mathrm{B} P_\mathrm{B}^\circ$. Activity coefficients must be determined for various assumed compositions by trial and error until this equation is satisfied, yielding the azeotrope. Replacing A by E in Eqs. (8.29)

$$\mathrm{E} = \log \gamma_\mathrm{E}\left(1 + \frac{x_\mathrm{B} \log \gamma_\mathrm{B}}{x_\mathrm{E} \log \gamma_\mathrm{E}}\right)^2$$

$$= \log 1.238\left(1 + \frac{0.384 \log 2.004}{0.616 \log 1.238}\right)^2$$

$$= 0.851$$

$$B = \log \gamma_\mathrm{B}\left(1 + \frac{x_\mathrm{E} \log \gamma_\mathrm{E}}{x_\mathrm{B} \log \gamma_\mathrm{B}}\right)^2$$

$$= \log 2.004\left(1 + \frac{0.616 \log 1.238}{0.384 \log 2.004}\right)^2$$

$$= 0.673$$

$$\log \gamma_\mathrm{E} = \frac{\mathrm{E}}{[1 + (\mathrm{E}x_\mathrm{E}/\mathrm{B}x_\mathrm{B})]^2} \qquad \log \gamma_\mathrm{B} = \frac{\mathrm{B}}{[1 + (\mathrm{B}x_\mathrm{B}/\mathrm{E}x_\mathrm{E})]^2}$$

Assume that $x_E = 0.4$ and $x_B = 0.6$. Therefore

$$\gamma_E = 1.78 \qquad \gamma_B = 1.38$$

$$P_T = (1.78)(22.9) = 40.76 \text{ kPa}$$

$$P_T = (1.38)(29.6) = 40.85 \text{ kPa}$$

These are close to a balance; thus the azeotrope consists of 40% ethanol and 60% benzene at a total pressure of 41 kPa.

8.2.4 Multicomponent-System Activity Coefficients

Methods for estimation of multicomponent vapor-liquid equilibria for low-pressure systems are essentially identical to those for binaries and are based on the excess Gibbs free energy.

$$G^E_{T,P} = RT\sum_i n_i \ln \gamma_i \tag{8.30}$$

Differentiating,

$$\left(\frac{\partial G^E}{\partial n_i}\right)_{T,P,\text{all } n_j\text{'s}} = RT \ln \gamma_i \tag{8.31}$$

Where all n_j's indicate the constancy of moles of all components except component i, the Wilson (1964) equation can be used in its generalized form [Eq. (8.25)] and only involves binary interaction coefficients which can be established from experimental data as noted in Sec. 8.2.3.

A two-suffix Margules set of equations can also be developed:

$$RT \ln \gamma_l = \sum_i \sum_j \left(A_{il} - \tfrac{1}{2}A_{ij}\right) x_i x_j \tag{8.32}$$

The constants A can be developed from binary data.

The NRTL and UNIQUAC methods are also applicable to multicomponent systems, as will be indicated later. Hougen, Watson, and Ragatz (1959) indicate a general method for ternary systems which allows calculation of ratios of activity coefficients.

8.3 HIGH-PRESSURE VAPOR-LIQUID EQUILIBRIA

High pressure introduces various complexities into vapor-liquid equilibria calculations. The Poynting correction [Eq. (7.44)] is required, especially near the critical region. The partial molar volume and molar volume of the pure liquid are not equal. Supercritical components often are present, and their standard-state fugacities must be calculated in order to use standard methods; such calculations are not possible by using the normal standard state of pure liquid at system temperature and pressure as liquid cannot exist in the supercritical region.

Retrograde condensation, i.e., the existence of a liquid at a temperature higher than its critical temperature when in a mixture, may occur. Finally, vapor-phase fugacity coefficients for any component may differ substantially from unity, or $\phi_i^V \neq \bar{\phi}_i^V$.

Three general methods are used for dealing with high-pressure vapor-liquid equilibria: the K-value and convergence-pressure methods, use of equations of state, and use of activity coefficients.

8.3.1 *K*-Value Methods

The equilibrium vaporization constant or K value was defined in Eqs. (8.13*a*) and (8.15*b*) in terms of standard-state fugacities. By Eqs. (8.13*b*) and (8.15*b*), the definition was simply the ratio of y/x and related to actual fugacity coefficients in solution. The use of these constants at high pressures requires that fugacities or fugacity coefficients be estimated for conditions where the liquid state is hypothetical, $T > T_s$ or $T > T_c$, or where the vapor state is hypothetical, $P_T > P°$. Thus some sort of extrapolation procedures are necessary.

Various methods have been used to estimate values of K_i. Tabulations for hydrocarbon systems were made from experimental data. K charts were prepared by using the Lydersen type of corresponding-states approach (Hougen et al., 1959). Specific charts for various hydrocarbons were made as a function of temperature and pressure.

The most useful methods have been proposed for hydrocarbons, based on the concept of convergence pressure. Such methods are based on the fact that at the critical point, where the composition of the liquid and vapor phases are equal, all K values must equal unity. At low pressures the K-value ratios are equal to the ideal fugacity ratios or equilibrium constants, while near the critical they diverge sharply. Equilibrium ratios of components which are not at their critical converge toward unity at a single point. This convergence occurs at the true critical temperature of the mixture. However, at temperatures other than the critical, the observed K values deviate from the equilibrium constants y_i/x_i along curves which when extrapolated converge at a point other than the true critical temperature of the mixture. This point is called the *convergence pressure* and only equals the critical pressure when the temperature equals the critical temperature. Figure 8.1 illustrates this phenomena.

This method has been used extensively by many investigators for developing plots of K values for specific compounds as a function of temperature, pressure, and convergence pressure. Daubert, Graboski, and Danner (1978) showed that the modified method of Hadden and Grayson (1961) was the most accurate of the methods for hydrocarbon systems, both light and heavy. This procedure is given in its entirety in the API's (1982) "Technical Data Book—Petroleum Refining" (TDB-PR). Figures 8.2 to 8.4 illustrate the basic charts. These charts are directly applicable to systems where the convergence pressure is greater than about 30 MPa, using the operating temperature and pressure. Otherwise the procedures in the TDB-PR corrected by convergence pressure must be utilized.

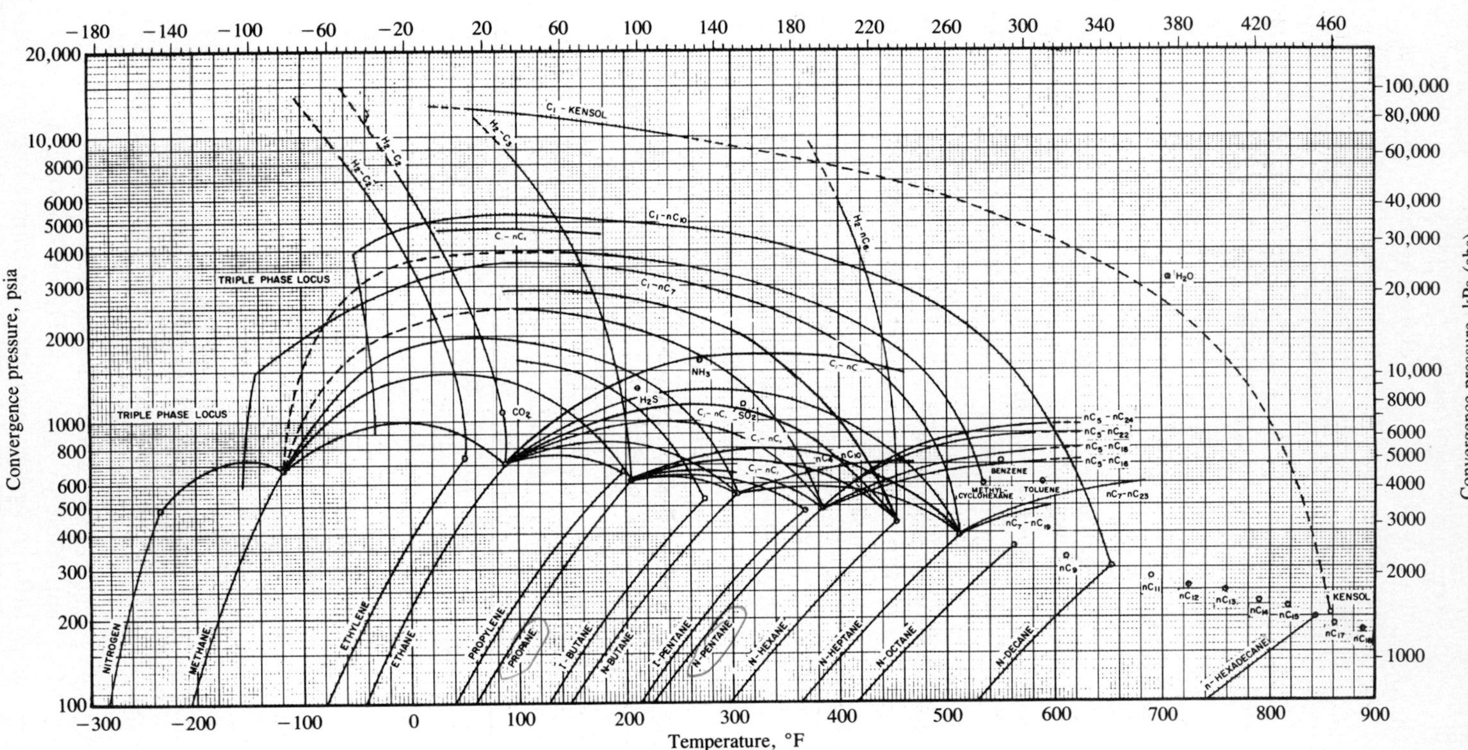

Figure 8.1 Convergence pressure for hydrocarbons. (*Courtesy of Gas Processors Association.*)

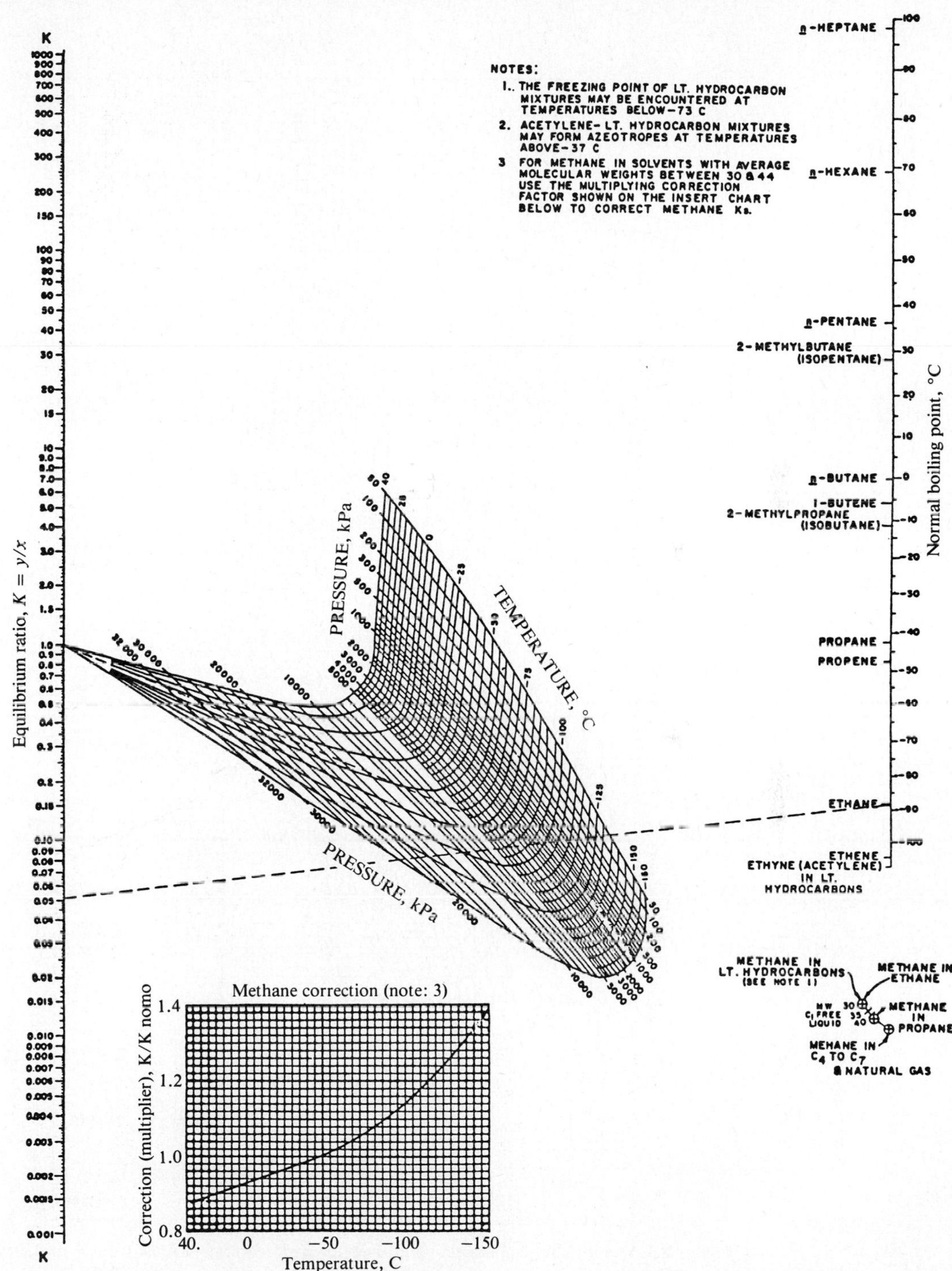

Figure 8.2 Vapor-liquid equilibrium K values for hydrocarbon systems from −160 to 40°C. *(Adapted from "Technical Data Book—Petroleum Refining," 4th ed., copyright 1982 by American Petroleum Institute. Used with permission.)*

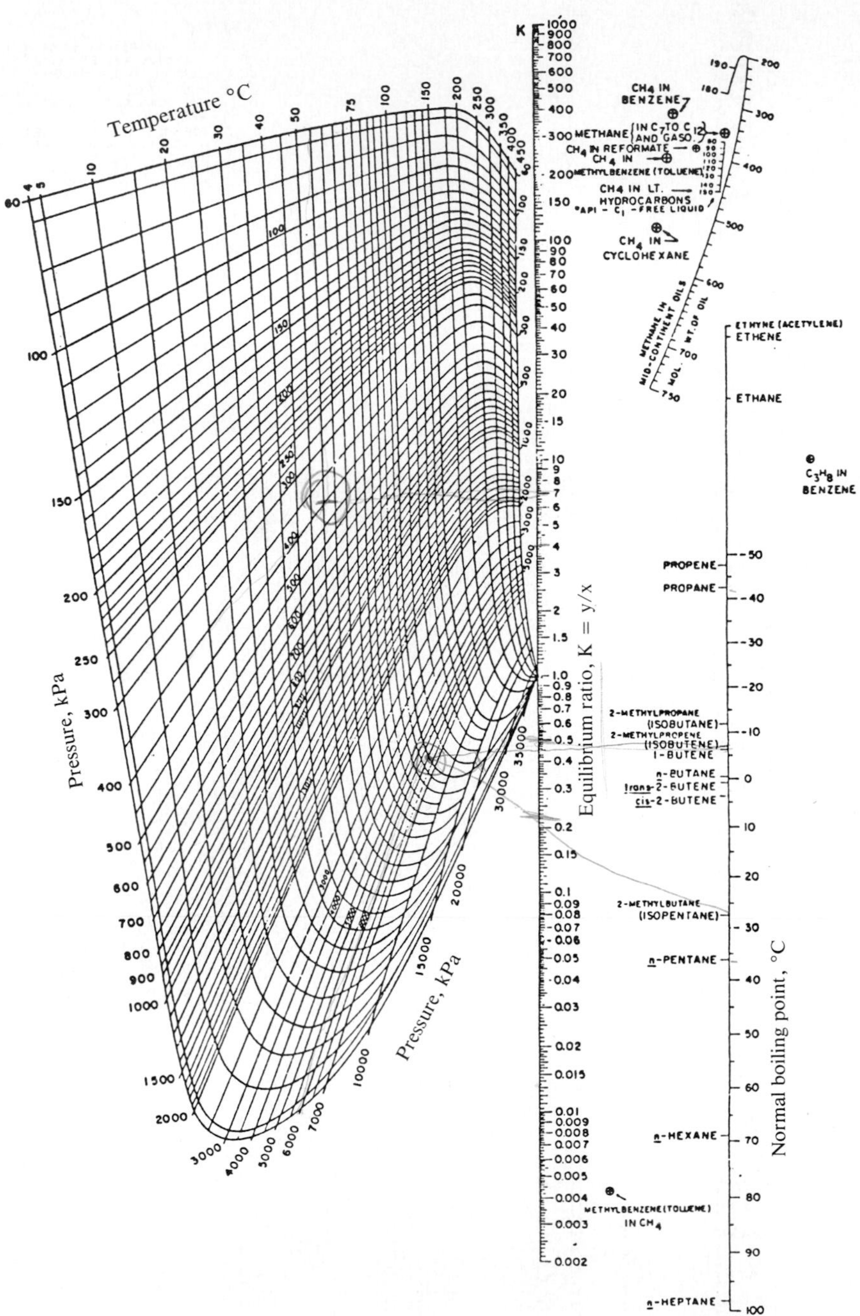

Figure 8.3 Vapor-liquid equilibrium K values for hydrocarbon systems from 4 to 450°C. *(Adapted from "Technical Data Book—Petroleum Refining," 4th ed., copyright 1982 by American Petroleum Institute. Used with permission.)*

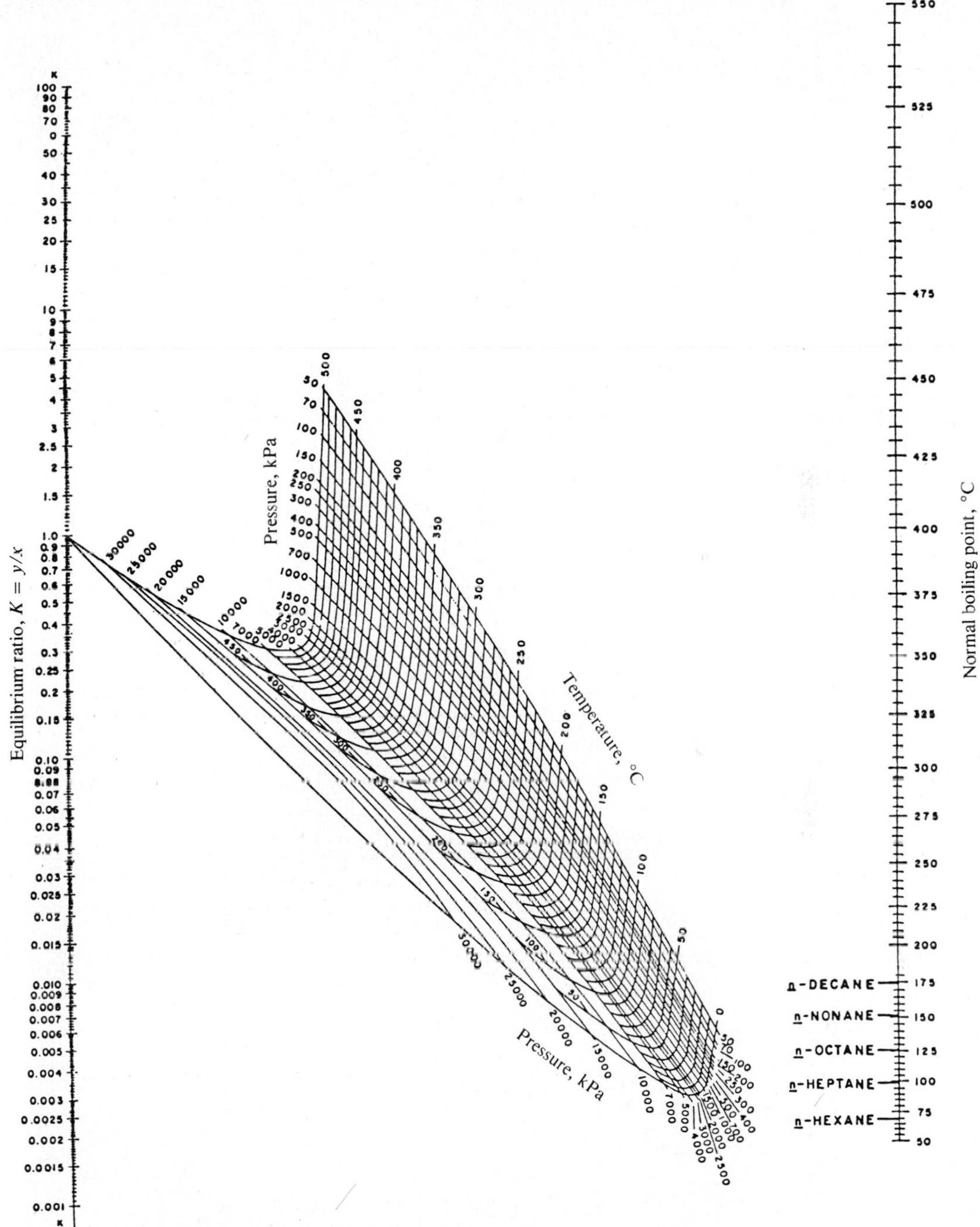

Figure 8.4 High-temperature vapor-liquid equilibrium K values for heavy hydrocarbons. *(Adapted from "Technical Data Book—Petroleum Refining," 4th ed., copyright 1982 by American Petroleum Institute. Used with permission.)*

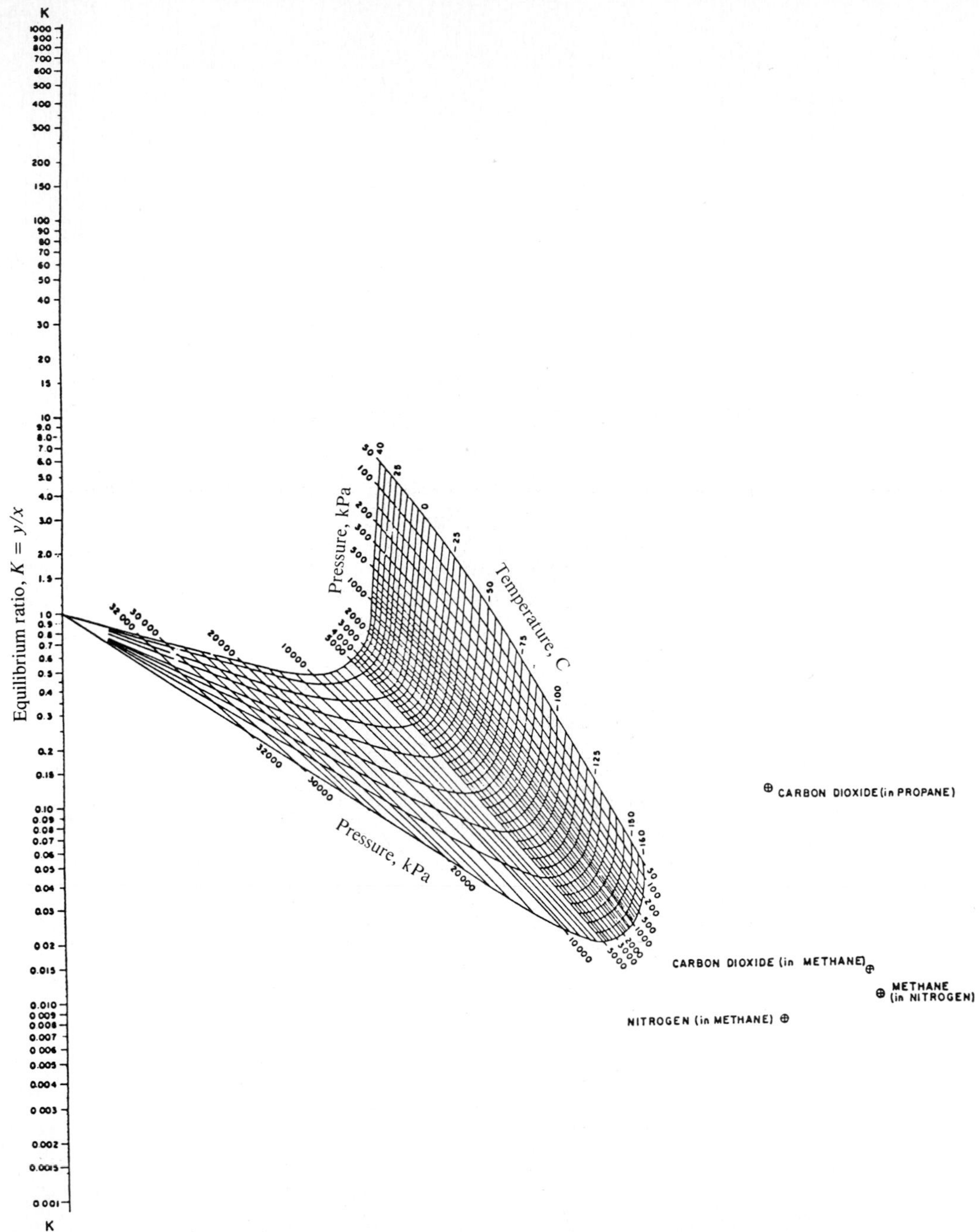

Figure 8.5 Vapor-liquid equilibrium K values for hydrocarbon-nonhydrocarbon systems from -160 to 40°C. *(Adapted from "Technical Data Book—Petroleum Refining," 4th ed., copyright 1982 by American Petroleum Institute. Used with permission.)*

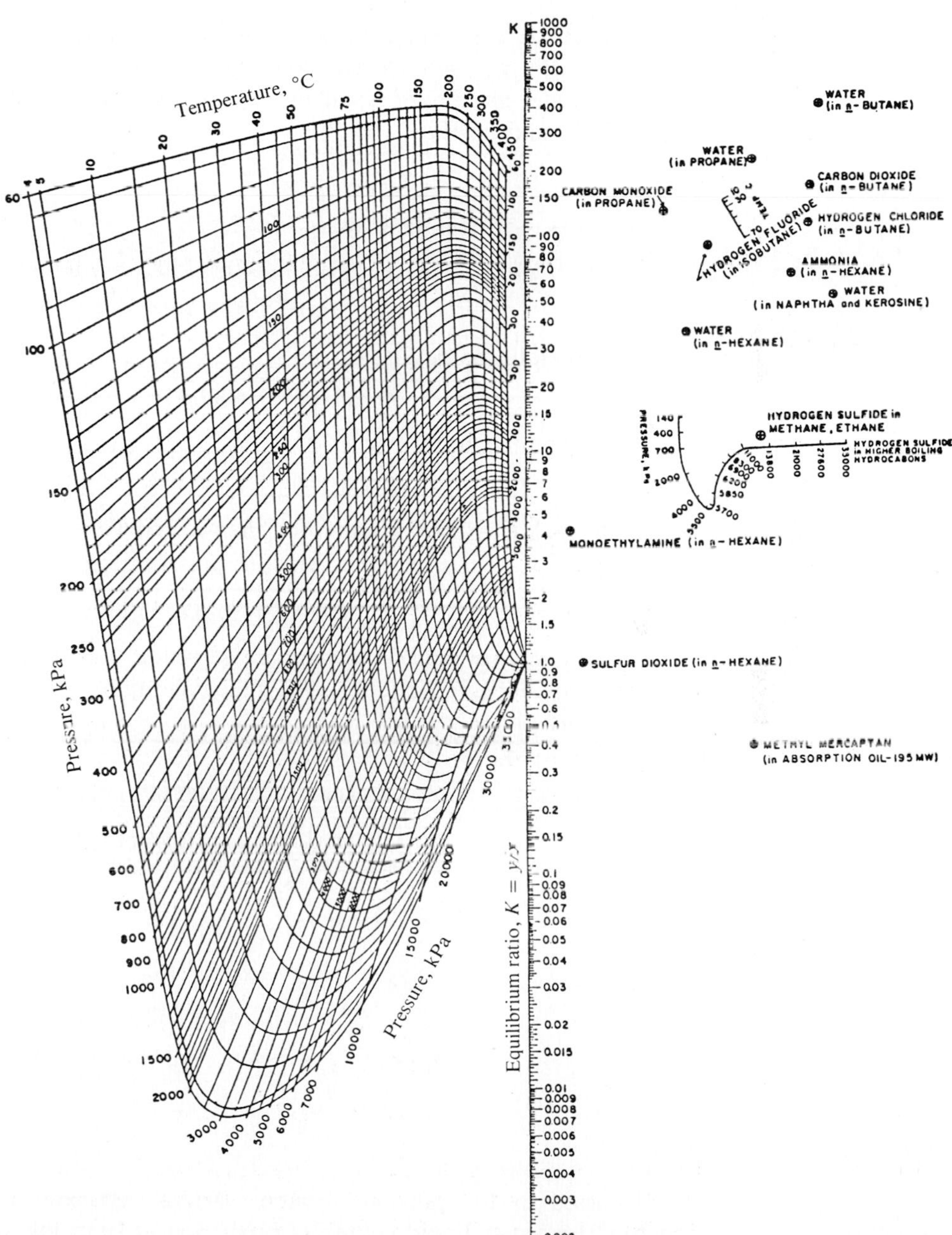

Figure 8.6 Vapor-liquid equilibrium K values for hydrocarbon-nonhydrocarbon systems from 4 to 450°C. *(Adapted from "Technical Data Book—Petroleum Refining," 4th ed., copyright 1982 by American Petroleum Institute. Used with permission.)*

For mixtures containing nonhydrocarbon gases other than hydrogen, graphical procedures have also been evaluated by Daubert, Graboski, and Danner (1978) and the method of Hadden and Grayson is given in detail in the TDB-PR; Figs. 8.5 and 8.6 illustrate this procedure and can be used directly within their range of conditions.

Example 8.4 Determine the K value for isobutane at -100, 0, and 100°C and 6 MPa pressure. What would be the equilibrium compositions of a vapor-liquid mixture of isobutane and isopentane at 100°C and 6 MPa *and* at 100°C and 1 MPa?

SOLUTION For isobutane, using Fig. 8.2,

$$K \text{ (at } -100°\text{C, 6000 kPa)} = 0.00118$$

$$K \text{ (at 0°C, 6000 kPa)} = 0.082$$

For isobutane, using Fig. 8.3,

$$K \text{ (at 100°C, 6000 kPa)} = 0.49$$

For isopentane, using Fig. 8.3,

$$K \text{ (at 100°C, 6000 kPa)} = 0.23$$

Since $\Sigma y = 1.0 = \Sigma Kx = 1.0$, then at 100°C,

$$K_4x_4 + K_5x_5 = 1.0$$

$$0.49x_4 + 0.23x_5 = 1.0$$

Also $x_4 + x_5 = 1.0$. Solving, x_4 is greater than 1, indicating that vapor-liquid equilibrium cannot exist at these conditions and all components are in the liquid phase.

From Fig. 8.3,

$$K_4 \text{ (at 100°C, 1000 kPa)} \approx 1.8$$

$$K_5 \text{ (at 100°C, 1000 kPa)} \approx 0.8$$

Therefore

$$K_4x_4 + K_5(1 - x_4) = 1.0$$

$$1.8x_4 + 0.8(1 - x_4) = 1.0$$

$$x_4 = 0.2 \qquad x_5 = 0.8$$

$$y_4 = K_4x_4 = 0.36 \qquad y_5 = K_5x_5 = 0.64$$

8.3.2 Equations of State

As discussed in the previous section, various equations of state have been proposed for vapor-liquid equilibria prediction at both low and high pressures. Daubert, Graboski, and Danner (1978) evaluated nine of these recent equations and determined that the modified Redlich-Kwong-Soave equation was the most applicable to hydrocarbon and hydrocarbon-nonhydrocarbon gas systems. The method was discussed in Chap. 7 as applicable to hydrocarbon systems. For hydrocarbon-nonhydrocarbon gas systems, an interaction coefficient as discussed in Chap. 7 is necessary. Estimation of this coefficient was discussed by Graboski and Daubert (1978*b*) in terms of solubility parameters, which will be discussed later in Sec. 8.5.

The general method of using this approach to calculate K values, which is best-adapted to calculations by digital computer, includes the steps given below:

1. Input data of temperature, pressure, and feed composition.
2. Make a flash calculation by using a set of assumed K values and the known feed composition to get an estimate of the vapor and liquid compositions.
3. Use the composition to calculate a_m and b_m for the Soave equation. Calculate the saturated-liquid and saturated-vapor volumes based on the known temperature and pressure from the Soave equation.
4. Calculate the fugacities of all components in each phase by using Eq. (7.36).
5. If the fugacity of each component is the same in each phase, stop the calculation and calculate the K_i's from Eq. (8.15). Otherwise, reestimate the K values and return to step 1.

This procedure is given in detail in the TDB-PR. In addition, it should be noted that binary or higher-interaction coefficients as discussed in Sec. 2.4.4 are not necessary for hydrocarbon systems but are useful for systems containing light nonhydrocarbon gases such as H_2S, N_2, and CO. Graboski and Daubert (1978*b*) discuss these methods in detail.

Subroutines for implementing the Soave equation of state method for calculating K-values, bubble points, and dew points are given in Appendix E.

Exercise 8.C Set up exactly how you would complete the calculation of the equilibrium composition of a vapor-liquid mixture of isobutane and isopentane at 100°C and 1 MPa by using an equation of state.

8.3.3 Activity Coefficients

The use of activity coefficients for calculating high-pressure vapor-liquid equilibria is based on Eq. (8.19):

$$\bar{\phi}_i^V y_i P_T = \gamma_i^L x_i f_i^{\circ L} \tag{8.19}$$

The estimation of $\bar{\phi}_i^V$ was discussed earlier. The activity coefficient γ_i^L can be estimated from solubility parameters (see Sec. 8.5) by the Scatchard-Hildebrand equation

$$\ln \gamma_i = \frac{V_i^L(\delta_i - \bar{\delta})^2}{RT} \tag{8.33}$$

where δ_i = solubility parameter of pure liquid i
$\bar{\delta}$ = average solubility parameter defined in Sec. 8.5.1

Note that this equation is independent of pressure. The pure-liquid fugacity $f_i^{\circ L}$ for supercritical components must be determined from binary data. Chao and Seader (1961) presented such a correlation which has been modified by Grayson and Streed (1963) and others. This type of correlation is simple to use, as only pure-component data are required, but has been determined to be not so accurate as the Redlich-Kwong-Soave equation-of-state method, especially in the critical region, by Daubert, Graboski, and Danner (1978). Lin (1979) has made a detailed analysis of these types of methods and has advanced a van Laar type of model.

Example 8.5 A 50 : 50 by volume liquid mixture of chloroform and carbon tetrachloride is in equilibrium with a vapor at 100°C. Estimate the activity coefficients for each. The following data are available:

Chloroform:	$\overline{V}_L = 81\ \text{cm}^3/\text{g mol}$	$\delta = 9.2\ (\text{cal/cm}^3)^{1/2}$
Carbon tetrachloride:	$\overline{V}_L = 97\ \text{cm}^3/\text{g mol}$	$\delta = 8.6\ (\text{cal/cm}^3)^{1/2}$

SOLUTION Using Eq. (8.33),

$$\ln \gamma_i = \frac{\overline{V}_i^L(\delta_i - \bar{\delta})^2}{RT}$$

$$\bar{\delta} = \sum \phi_i \delta_i = (0.5)(9.2 + 8.6) = 8.9\ (\text{cal/cm}^3)^{1/2}$$

For chloroform (1),

$$\ln \gamma_1 = \frac{(81)(9.2 - 8.9)^2}{(1.987)(373.1)} = 0.009833 \qquad \gamma_1 = 1.010$$

For carbon tetrachloride (2),

$$\ln \gamma_2 = \frac{(97)(8.6 - 8.9)^2}{(1.987)(373.1)} = 0.01178 \qquad \gamma_2 = 1.012$$

Exercise 8.D Tell exactly how you would complete the calculation of the vapor composition in Example 8.5 at a given pressure by using the most desirable method.

8.4 EQUILIBRIUM PHASE DIAGRAMS

To more clearly understand the concepts of physical equilibria, graphical presentation of the equilibrium relationships existing between phases are important for both understanding and for many design applications. This section will illustrate the various types of diagrams and how they are constructed. Vapor-liquid equilibria will be treated in detail for miscible, partially miscible, and immiscible systems. Liquid-liquid equilibrium diagrams, for which the theory will be discussed in later sections, will also be illustrated.

Figure 8.7 gives various equilibrium diagrams for water as a review of the various diagrams presented in earlier chapters. Many of the diagrams to be

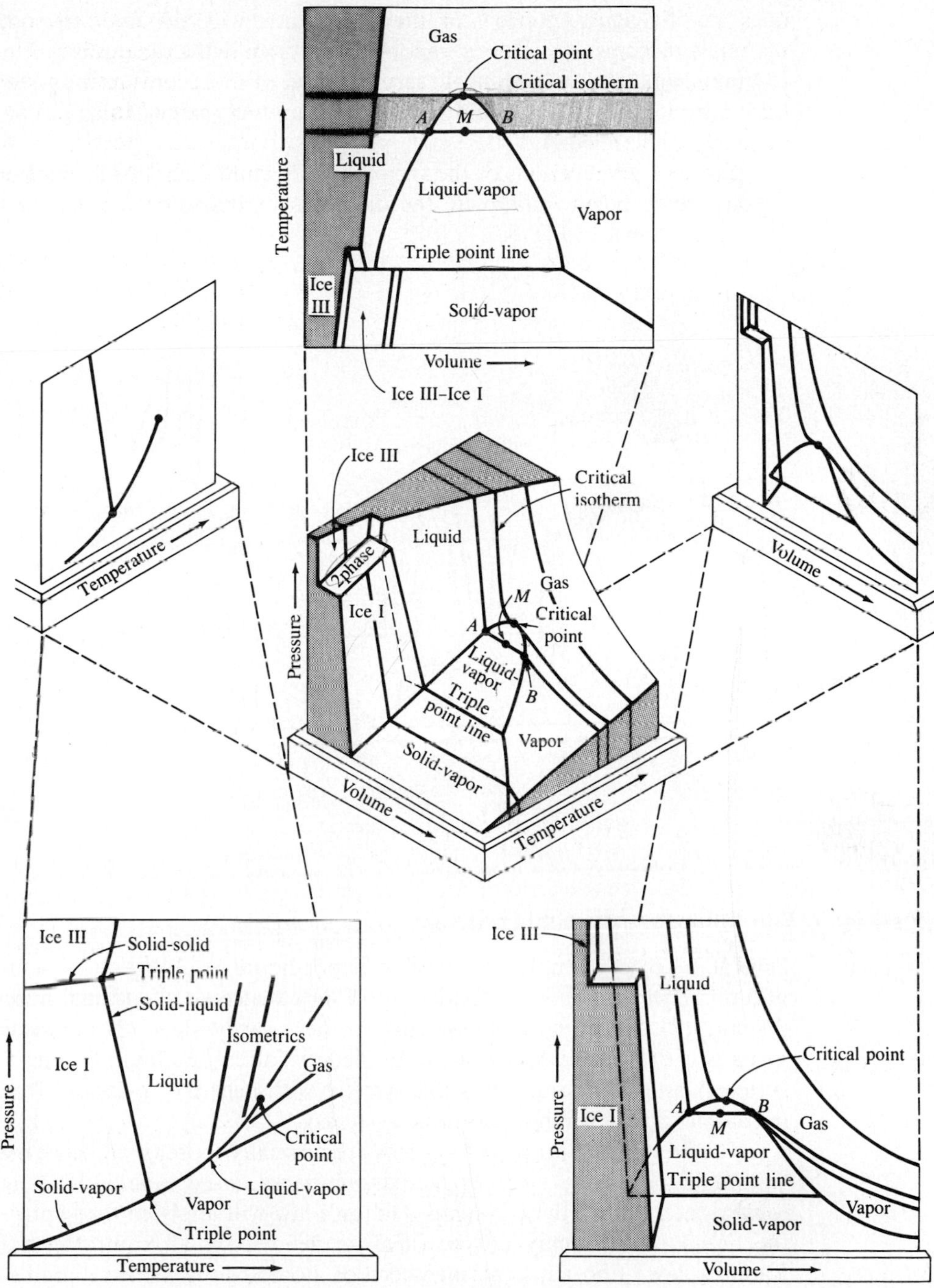

Figure 8.7 Phase diagrams for water.

presented later in this section are developed from parts of diagrams such as the ones in this figure. Review of these diagrams will aid in understanding the materials to be presented. For vapor-liquid systems, the diagrams will be limited to binary mixtures, as graphical representation of multicomponent systems serves little purpose. For liquid-liquid systems, ternary mixtures will also be considered.

Exercise 8.E Portions of the various equilibrium diagrams for carbon dioxide are given below. Complete the diagrams in the same manner as those for water shown in Fig. 8.7.

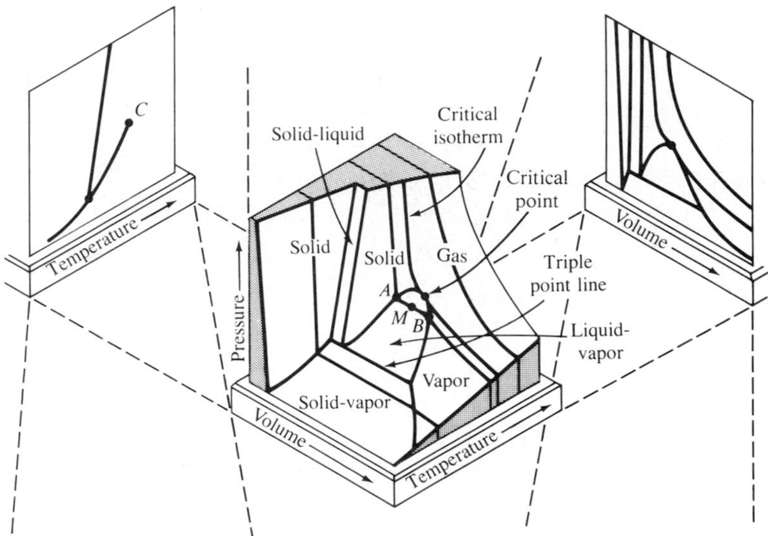

8.4.1 Vapor-Liquid Equilibria for Miscible Systems

Several types of completely miscible vapor-liquid equilibrium systems are important. These systems are divided into four categories: (1) normal miscible ideal systems, (2) normal miscible systems not necessarily ideal, (3) nonideal miscible systems with positive deviations from ideality (often minimum boiling azeotropic systems), and (4) nonideal miscible systems with negative deviations from ideality (often maximum boiling azeotropic systems).

The first two categories of systems are normally those of adjacent members of a homologous series such as methanol-ethanol, *n*-heptane–*n*-octane, benzene-toluene, etc. For all these systems, Dalton's law will apply to the vapor; however, the liquid may or may not be ideal as defined by all components following Raoult's law. For adjacent members of homologous series Raoult's law can usually be assumed to apply to the liquid over the entire composition range, while for slightly less similar systems Henry's law is followed by each liquid at low concentrations of that liquid, while Raoult's law is followed by the liquid at high concentrations.

Figure 8.8 shows a three-dimensional diagram for a normal miscible system of the first or second type. Sections of this diagram in the temperature-composition or pressure-composition plane yield diagrams as shown in Fig. 8.9*a* and *c*. Figure 8.9*b* is the common *xy* or equilibrium diagram which can be constructed by plotting the terminal points of the horizontal tie lines in Fig. 8.9*a*. Figure 8.9*b* is commonly useful for distillation processes, as distillation is normally carried out at constant pressure.

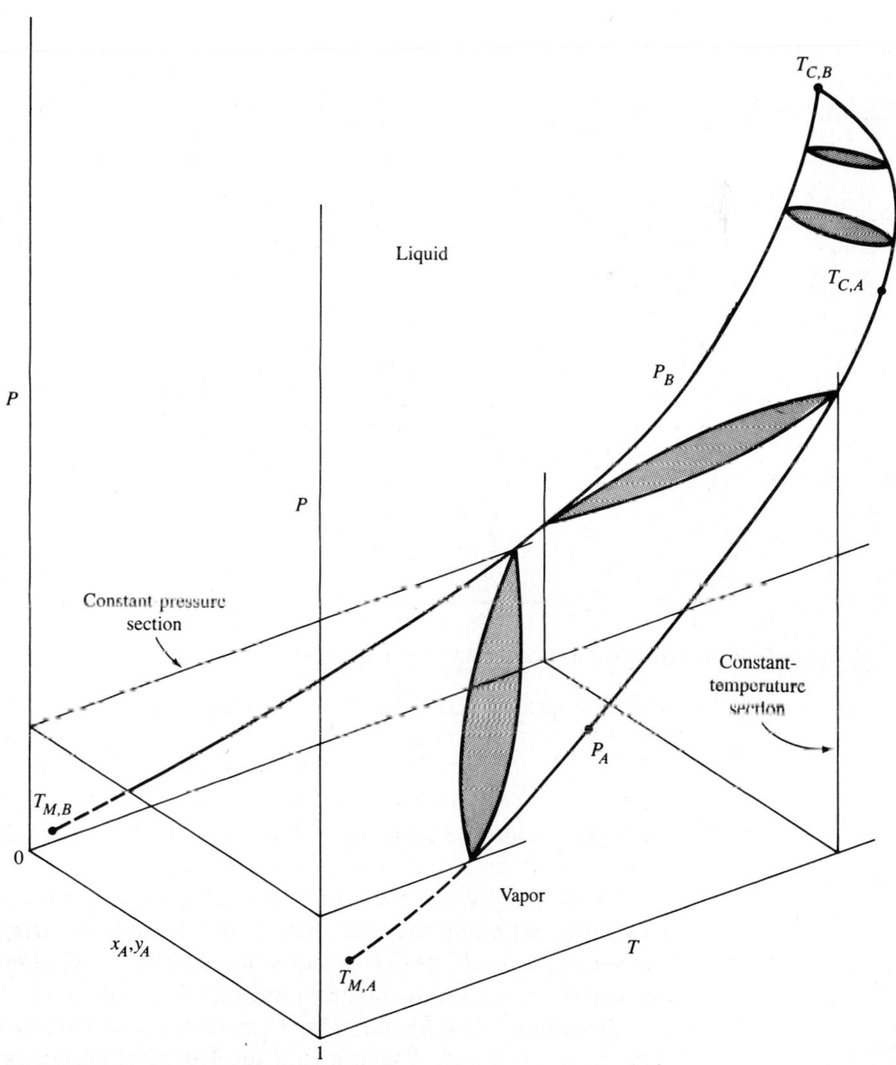

Figure 8.8 Binary vapor-liquid equilibria diagram.

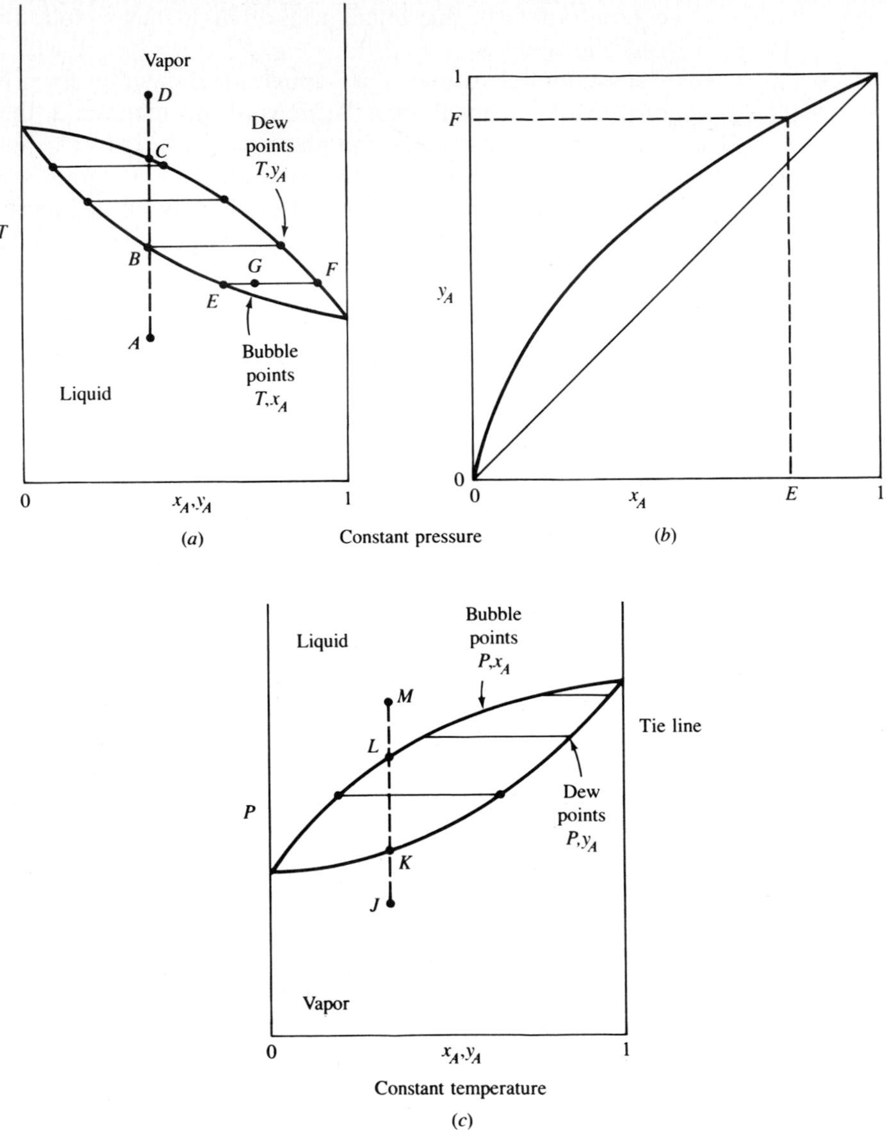

Figure 8.9 Constant-temperature and constant-pressure vapor-liquid equilibria.

For ideal systems the partial pressures can be calculated from Raoult's law and are directly additive. The activity coefficients are unity. Thus such diagrams can easily be constructed by equating Raoult's and Dalton's laws [Eq. (8.18)]. Typical diagrams for a common ideal system are shown in Fig. 8.10.

Deviations from ideality may be positive or negative; i.e., the total pressure of a given composition system at a given temperature is larger or smaller than the pressure if calculated from Raoult's law. In such systems, Raoult's law is almost followed at very high concentrations, while Henrys law is almost followed at low

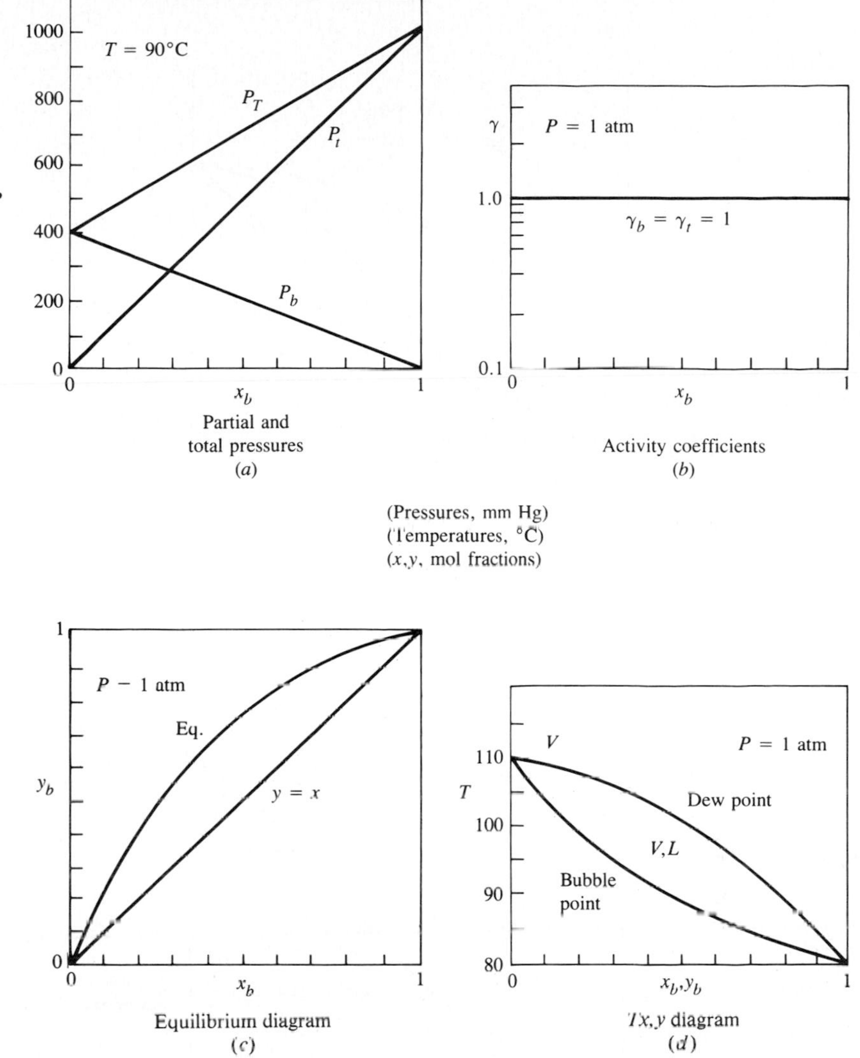

Figure 8.10 Ideal-solution vapor-liquid equilibria for benzene (*b*)-toluene (*t*) system.

concentrations. Pressure-composition diagrams for such systems are compared with ideal systems in Fig. 8.11. Such azeotropes or constant-boiling mixtures are homogeneous.

For the more-common positively deviating and minimum-boiling azeotropes, of which many examples such as ethanol-water, CCl_4–ethanol and acetone–CS_2 exist, the total pressure will rise through a maximum, while the activity coefficients become equal at the azeotropic point. The behavior of minimum-boiling azeotropes is illustrated for isopropyl ether–isopropyl alcohol in Fig. 8.12 at constant pressure. Calculation of such diagrams usually assumes Dalton's law applicable to the vapor, with the liquid phase being nonideal and requiring

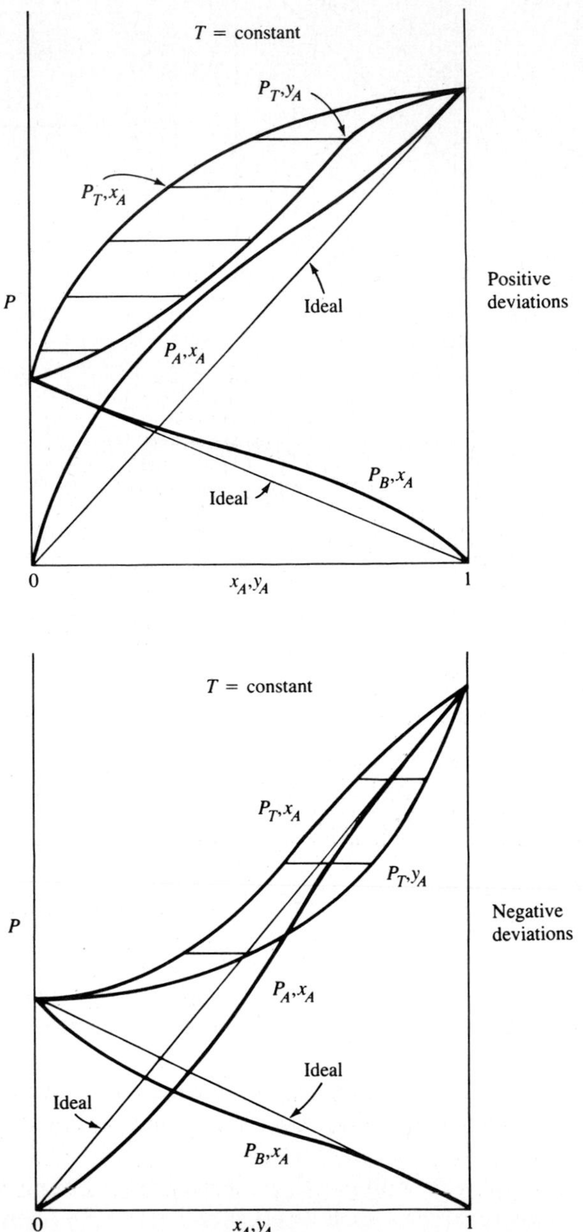

Figure 8.11 Deviations from ideality in vapor-liquid equilibria.

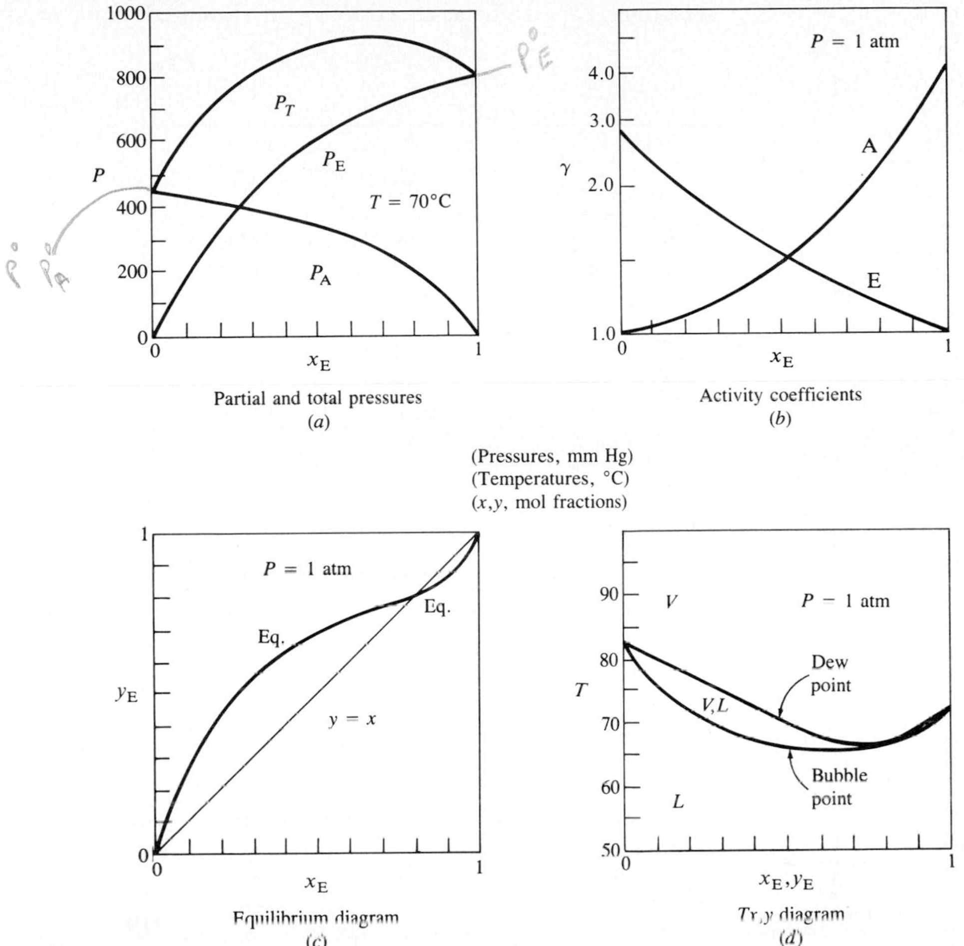

Figure 8.12 Minimum-boiling azeotropic vapor-liquid equilibria for isopropyl ether (E)–isopropyl alcohol (A) system.

prediction of the activity coefficient from an appropriate correlation before application of Eq. (8.19).

Example 8.6 Verify the equilibrium composition of the isopropyl ether (E)–isopropyl alcohol (A) mixture at 70°C shown in Fig. 8.12*c* by using only pure-component vapor pressures and activity coefficients from Fig. 8.12*a* and *b* at liquid mol fractions of isopropyl ether of 0.3 and 0.9.

SOLUTION Assuming an ideal-vapor phase, the vapor-liquid equilibria are represented by Eq. (8.19) with a vapor-phase fugacity coefficient of 1.0.

$$\overline{P}_i = y_i P_T = \gamma_i^L x_i f_{i,L}^\circ$$

Assuming $f^{\circ}_{i,L}$ can be assumed to be the saturated vapor pressure at 70°C,

$$\overline{P}_i = y_i P_T = \gamma_i^L x_i P_i^{\circ}$$

	Source	x_E 0.3	x_E 0.9
$\overline{P}_E$	Fig. 8.12*a*	450 mm	770 mm
$\overline{P}_A$	Fig. 8.12*a*	370 mm	130 mm
γ_E	Fig. 8.12*b*	~ 1.85	~ 1.05
γ_A	Fig. 8.12*b*	~ 1.15	~ 3.0
P_E°	Fig. 8.12*a*	810 mm	810 mm
P_A°	Fig. 8.12*b*	470 mm	470 mm
y_E	Fig. 8.12*c*	0.56–0.57	0.86–0.87
$P_T = \overline{P}_E + \overline{P}_A =$		820 mm	900 mm

At $x_E = 0.3$,

$$y_E = \frac{\gamma_E x_E P_E^{\circ}}{P_T} = \frac{(1.85)(0.3)(810)}{820} = 0.548$$

$$y_A = \frac{\gamma_A x_A P_A^{\circ}}{P_T} = \frac{(1.15)(0.7)(470)}{820} = 0.419$$

$$\Sigma y = 0.967$$

The sum of vapor-phase concentrations must equal unity; the discrepancy is caused by reading numbers from the plot. Thus normalize to $\Sigma y = 1.0$, and therefore

$$y_E = \frac{0.548}{0.967} = 0.567 \qquad y_A = \frac{0.419}{0.967} = 0.433$$

This corresponds to the tabulated answers.

At $x_E = 0.9$,

$$y_E = \frac{(1.05)(0.9)(810)}{900} = 0.851$$

$$y_A = \frac{(3.0)(0.1)(470)}{900} = 0.157$$

$$\Sigma y = 1.008$$

Normalizing, $y_E = 0.844$ and $y_A = 0.156$, corresponding reasonably to the tabulated answers.

Note that Fig. 8.12*c* and these calculations are for constant temperature. Since vapor-liquid equilibrium processes are more commonly carried out at constant pressure, the normal procedure for developing phase diagrams at constant pressure is to use the previous procedure, a geometric average of the

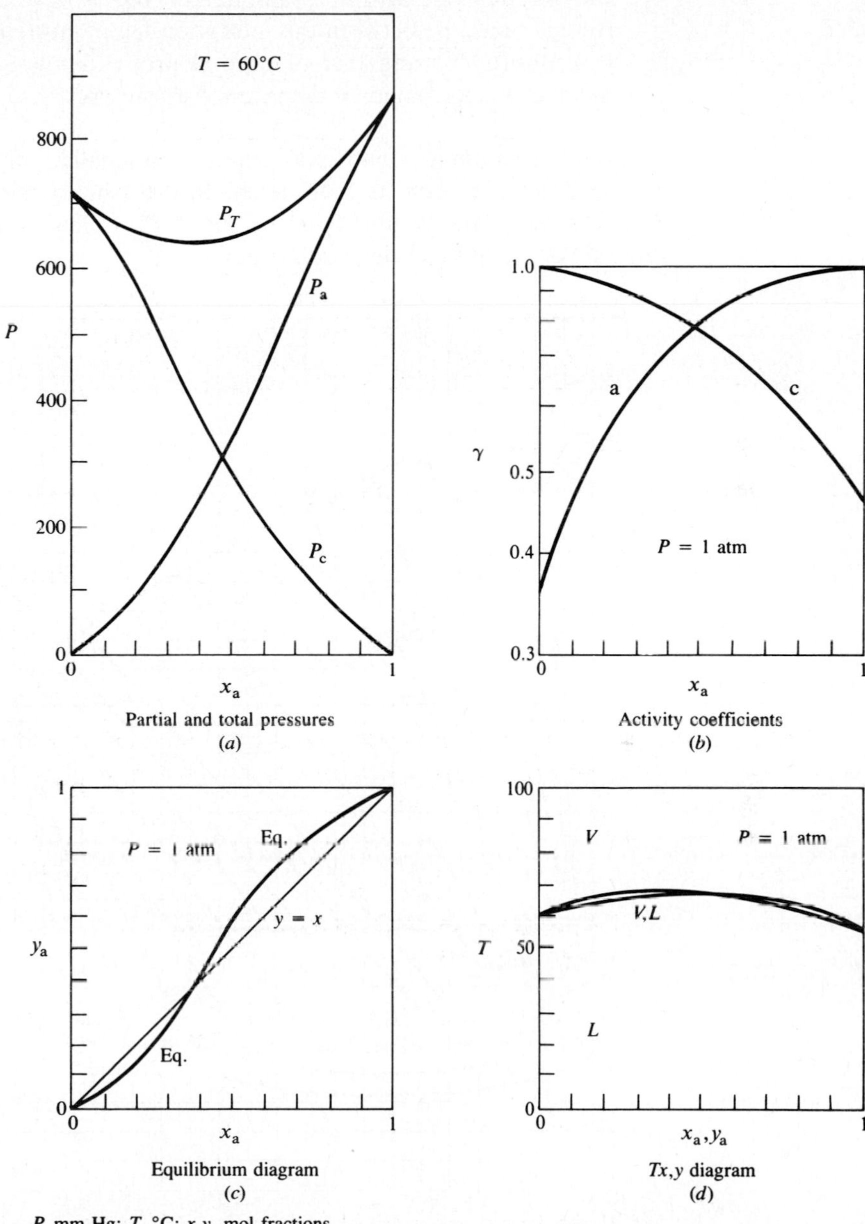

Figure 8.13 Maximum-boiling azeotropic vapor-liquid equilibria for acetone (a)-chloroform (c) system.

vapor pressures of the pure components at the saturation temperatures for the two pure components, and an activity coefficient at an average temperature for the system. For example, for Fig. 8.12*d* at 1 atm pressure, the average would be between the saturation temperature of pure isopropyl ether at 1 atm (69°C) and that of pure isopropyl alcohol at 1 atm (82°C). The vapor pressures would be determined at each of these temperatures.

The less-common negative deviation from ideality which yields a maximum-boiling azeotrope such as acetone-chloroform shows the total pressure falling though a minimum, as illustrated by Fig. 8.13. Calculational procedures are the same as for the positive-deviation systems.

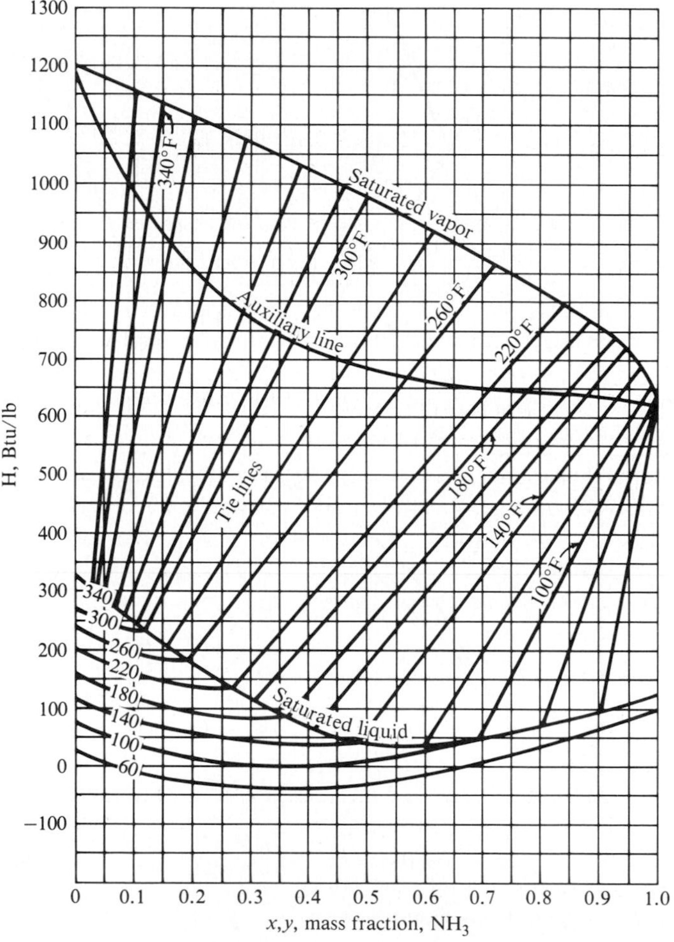

Ammonia-water (10 atm)

Figure 8.14 Enthalpy-concentration diagram for ammonia-water system at 10 atm total pressure. *(From McCabe and Smith, "Unit Operations in Chemical Engineering," 2d ed., copyright 1967 by McGraw-Hill, Inc. Used with permission.)*

Exercise 8.F Repeat Example 8.6 for the acetone-chloroform system shown in Fig. 8.13 at 60°C and the same mole fractions. Also, retrieve the information required to develop Fig. 8.13*d*.

For systems which are highly nonideal with respect to heat effects, enthalpy-concentration diagrams are often utilized, as illustrated for the ammonia-water system in Fig. 8.14. Such diagrams can be constructed by the methods discussed earlier. The tie lines connect the equilibrium points and can be drawn by using right triangles and the auxiliary line.

Thus far the diagrams noted are most useful for distillation calculations. Other vapor-liquid operations such as absorption normally begin with an equilibrium diagram plotting the concentration of a gas in a liquid against the partial pressure of the gas above the solution, as can be developed from thermodynamic relations and illustrated in Fig. 8.15. An equilibrium curve is then developed by assuming Henry's law at low-liquid composition, as also illustrated. At higher concentrations Henry's law is not applicable, and the methods utilized in the previous discussion for vapor-liquid equilibria are utilized.

Exercise 8.G Verify Fig. 8.15*b* by using the data of Fig. 8.15*a*.

As most vapor-liquid equilibrium processes are carried out at constant pressure, the selection of the total pressure to be utilized is important. Examination of Fig. 8.8 shows that as the critical point is approached, a pressure- or temperature-composition diagram will not extend across the entire composition range and the possible range of operation becomes smaller and smaller as the increase in pressure comes near the critical point.

Figure 8.16 illustrates this point by using a temperature-composition and equilibrium diagram for increasing pressure from P_{T1} to P_{T5}. Examination of this diagram shows the effect of pressure on the equilibrium compositions, with lower pressures favoring better separation as long as the pressure is not so low that any component or its mixture cannot exist in both vapor and liquid phases. The total pressure will also affect the composition and temperature of the azeotropic point for such mixtures. This effect can be obtained by application of the Clapeyron equation. Such analysis becomes important, since for some systems use of the correct pressure will cause operation to fall in a region where the azeotrope is of no importance for the desired process.

Katz and Kurata (1940) and Kay (1938) discuss pressure effects in some detail. Hougen, Watson, and Ragatz (1959) give a very lucid discussion of these effects as well as discuss in detail the important concept of retrograde condensation, a result of such effects.

8.4.2 Vapor-Liquid Equilibria for Immiscible Systems

True immiscible systems probably do not exist, although some systems are so close to immiscibility that its assumption is warranted, e.g., water and heavy hydrocarbons, water and bromobenzene. In this case each component will exert its own vapor pressure, and the composition of the vapor phase can be calculated

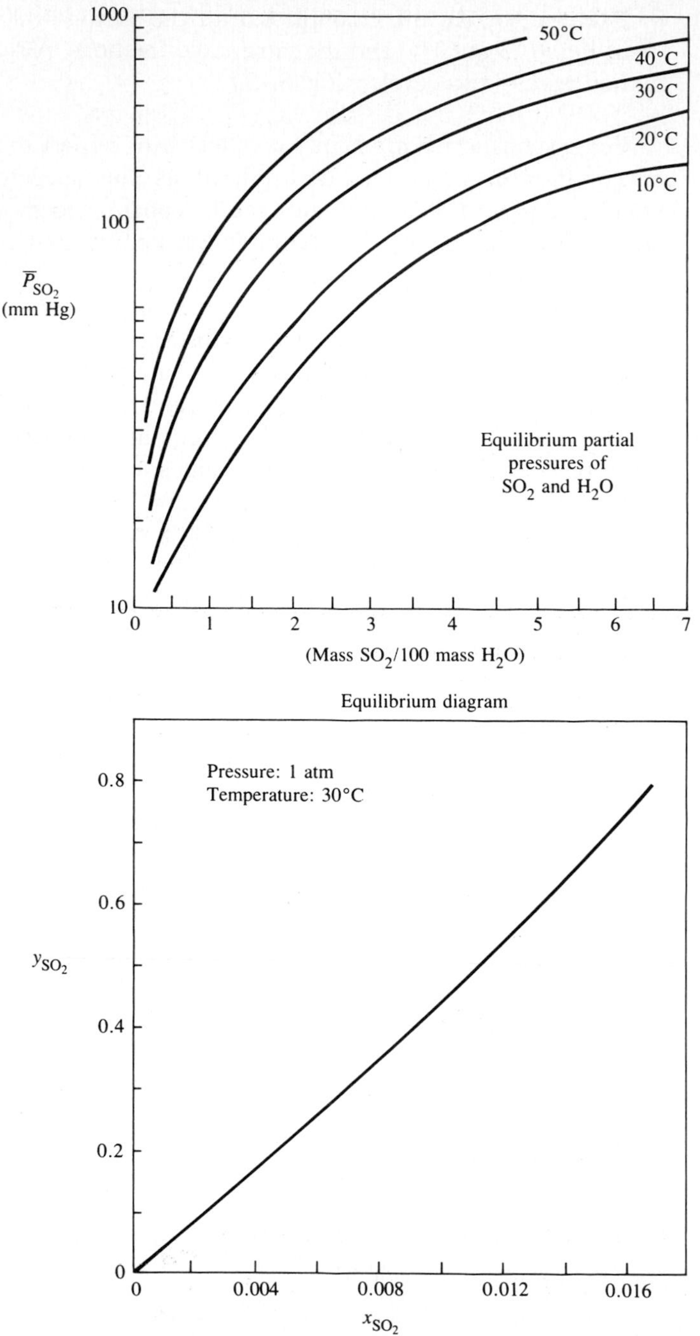

Figure 8.15 Low-concentration vapor-liquid equilibria for sulfur dioxide–water system.

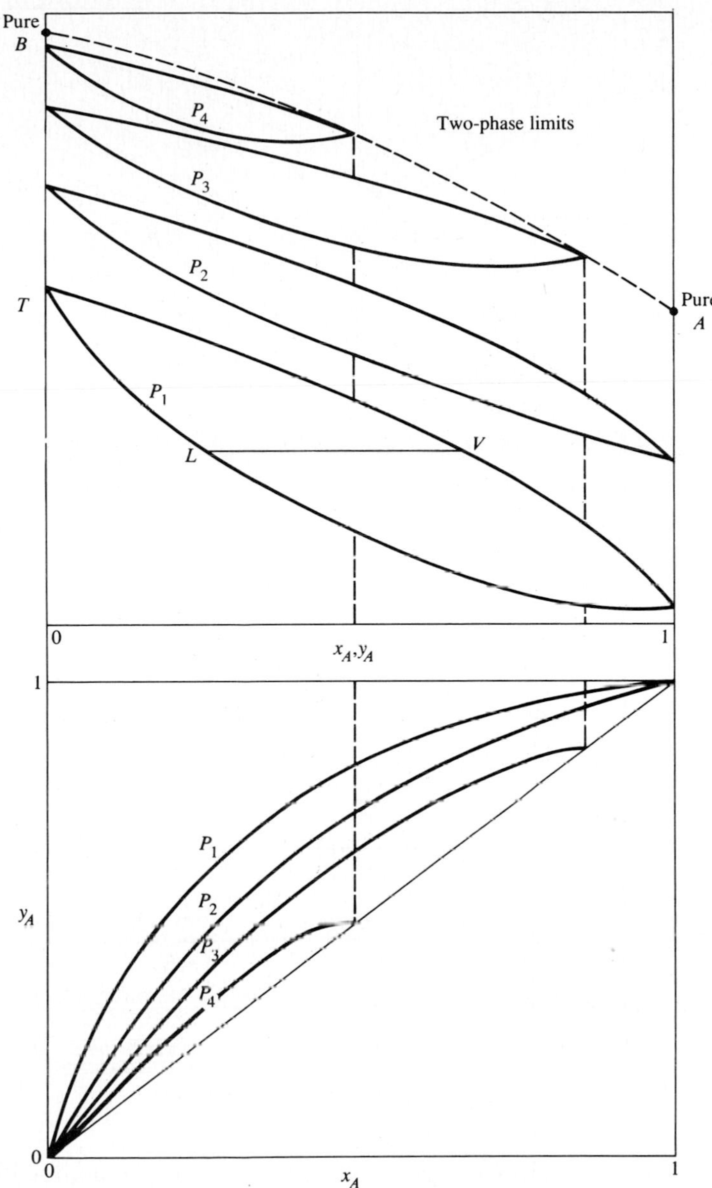

Figure 8.16 Effect of pressure on vapor-liquid equilibria.

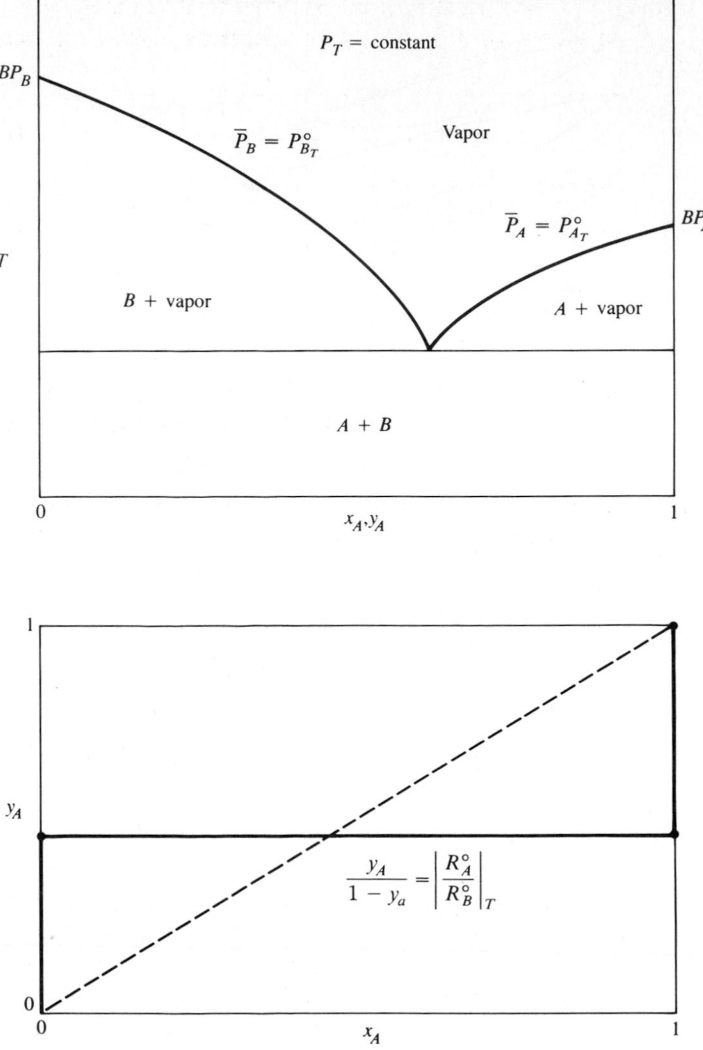

Figure 8.17 Vapor-liquid equilibria for immiscible systems.

from the ratio of the vapor pressures of the immiscible components at the temperature of the system. For a binary system of components, a temperature-composition diagram and an equilibrium diagram are given in Fig. 8.17. The sequence of events occurring depends on the starting composition and temperature. However, the mixture will vaporize at constant temperature until a single component is entirely vaporized.

Example 8.7 Dimethylaniline (DMA) can be considered to be completely immiscible with water (W). If dimethylaniline is being distilled with steam at

0.090 MPa total pressure, determine the composition of the organic material in the vapor phase. Vapor pressure data are given below.

T, °C	$P°$, MPa DMA	W
70	0.001333	0.03119
101.6	0.005332	0.10691
125.8	0.013329	0.28586

SOLUTION For purposes of the problem, plots of ln $P°$ vs. $1/T$ may be used for interpolation. Trial and error is necessary, as $P_T = P^\circ_{\text{DMA}} + P^\circ_{\text{W}}$.

$$P_T = \begin{cases} 0.00413 + 0.08455 = 0.08868 & \text{at } 95°\text{C} \\ 0.00427 + 0.08771 = 0.09198 & \text{at } 96°\text{C} \end{cases}$$

Interpolating to $T = 95.4°$,

$$P^\circ_{\text{DMA}} = 0.0042$$

$$P^\circ_{\text{W}} = \underline{0.0858}$$

$$P_T = 0.09000$$

$$\text{Mass fraction DMA} = \frac{(0.0042)(121)}{(0.0042)(121) + (0.0858)(18)} = 0.248$$

$$\text{Mass fraction W} = 0.752$$

This composition will be constant until all DMA has been distilled.

Exercise 8.H Assuming chlorobenzene and water are immiscible, determine the temperature of distillation and the composition of the distillate if the equilibrium vaporization is carried out with two liquid phases present at total pressures of 0.004 and 0.40 MPa. Draw an xy diagram for the latter pressure if a feed of 1 mol of chlorobenzene and 2 mol of water is used. For chlorobenzene:

T, °C	22.2	35.3	70.7	89.4	110.0	132.2	160.2
P, kPa	1.33	2.67	13.33	26.86	53.3	101.3	202.6

8.4.3 Vapor-Liquid Equilibria for Partially Miscible Systems

Partially miscible systems exhibit very large positive deviations from ideality and generally have the same characteristics as minimum-boiling azeotropic systems. Water and alcohols of 4-carbon atoms and higher are the most important examples of such systems. For binary systems, the two components do not dissolve completely in the liquid state, and in the two-liquid phase region, the partial and total pressures remain constant since the relative amounts of the two liquid phases change, as was the case for immiscible systems. Obviously the activity coefficients are very high. For a system of two liquid phases and one vapor phase at constant pressure, there are zero degrees of freedom, and the temperature and composition of all three phases are fixed as shown on a general temperature-composition diagram in Fig. 8.18. Points A and C are the liquid-phase compositions, while point B is the vapor-phase composition. Below temperature T and to the right of point C and to the left of point A, only one liquid phase exists, as solubility limits of the solute component in the solvent component in each region have not been exceeded (lines $\overline{AD}$ and $\overline{CE}$). As the pressure increases, the three-phase equilibrium temperature T will increase and lines $\overline{AD}$ and $\overline{CE}$ will meet at the critical point of the mixture, the upper critical-solution temperature. At this point the line $\overline{AC}$ is actually a point, and above this pressure only one liquid phase will exist. The system becomes a miscible, minimum-boiling azeo-

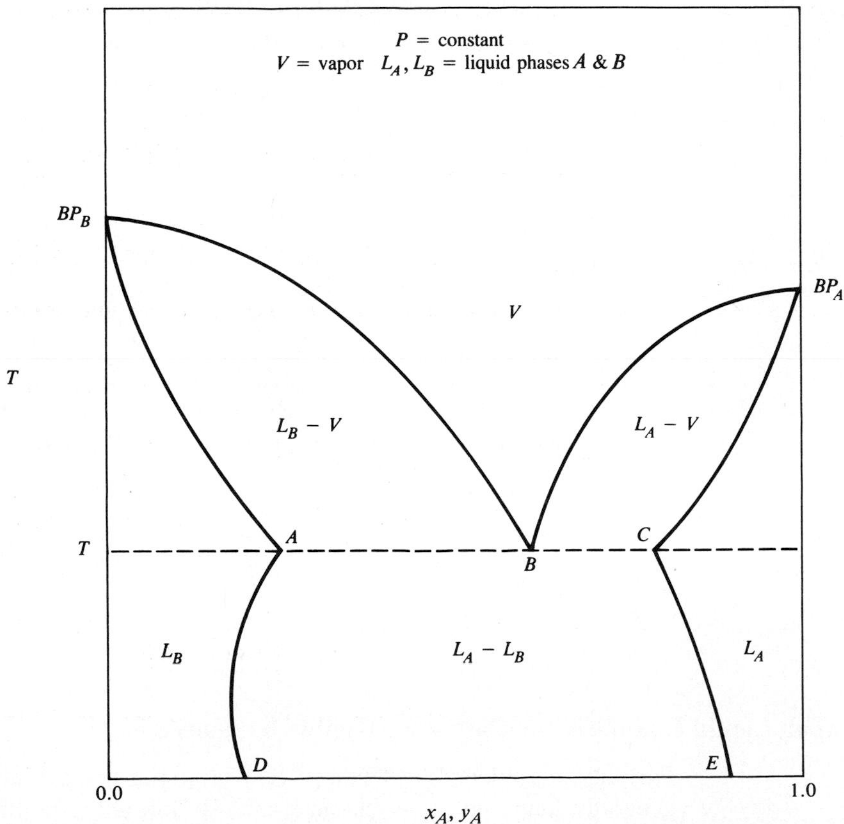

Figure 8.18 Vapor-liquid equilibria for partially miscible systems.

tropic system. Although this is the normal sequence of events, other phenomena can occur to change the behavior; e.g. the temperature is raised above a vapor-liquid critical point, a solid phase is formed, or a lower critical-solution temperature is reached.

Figure 8.19 illustrates the various diagrams for a common system: *n*-butanol and water.

Certain special systems such as the common phenol-water system exhibit partial miscible behavior only on one side of the three-phase point and behave like a usual minimum-boiling azeotropic system on the other side.

Calculations for partially miscible systems follow exactly the same procedures as for minimum-boiling azeotropic systems in the two-phase region. In the two-liquid-phase region, an approximate method assumes that Henry's law holds for the lower-concentration component in each liquid phase, Raoult's law applies

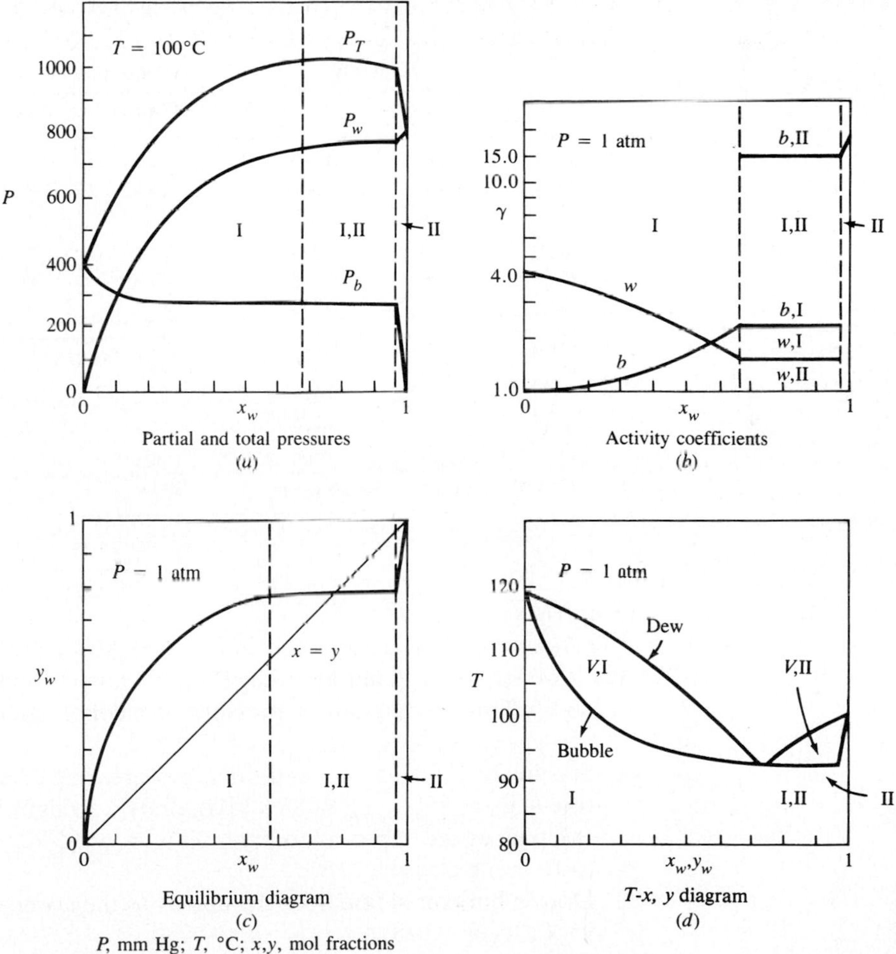

Figure 8.19 Partially miscible vapor-liquid equilibria for water (w)-butanol (b) system.

to the higher concentration, and Dalton's law applies to the vapor. Setting up these two equations for each of the two regions will yield the two-phase envelope. Each of these calculations for a constant total pressure will be trial and error for temperature since the vapor pressures required are functions of temperature. This procedure requires knowledge of the two limits of solubility and determination of the Henry's law constants.

Example 8.8 Secondary butyl alcohol (A) and water (W) are partially miscible liquids. Data on solubilities and vapor pressures are given below. Assuming that the procedure for estimating vapor-liquid equilibria discussed in this section hold, construct a diagram for the temperatures 95, 97, and 100°C such that the total pressure is 1 atm or more for (*a*) the partial pressure of alcohol vs. mole fraction in the total liquid mixture, (*b*) the partial pressure of water on the same diagram, and (*c*) the total pressure of the system.

	Solubility x_A		Vapor pressure, kPa	
T, °C	In water	In alcohol	Water	Alcohol
30	0.0495	0.289	4.24	6.29
50	0.0386	0.308	12.32	16.55
70	0.0386	0.298	31.15	37.51
90	0.0433	0.254	70.1	75.3
95	0.0480	0.235	83.2	88.6
96	0.0490	0.230	87.7	91.3
97	0.0500	0.227	90.9	94.6
98	0.0520	0.222	94.3	97.3
99	0.0530	0.217	97.7	100.6
100	0.0541	0.212	101.3	104.1
102	0.0625	0.200	108.4	
107	0.1194	0.1194	134.4	

SOLUTION

(*a*) Assuming that Henry's law is followed at alcohol concentrations below the lower-solubility limit and that Raoult's law is followed above the upper-solubility limit, a diagram of pressure of alcohol can be plotted as follows:

1. Starting at 100% alcohol and the vapor pressure of alcohol at the temperature (e.g., at 95°C, $P_A^\circ = 88.6$ kPa), draw a straight line toward the origin, stopping at the upper-solubility limit (e.g., at 95°C, $x_A = 0.235$) according to Raoult's law.
2. Draw a horizontal line from the upper- to the lower-solubility limit (e.g., at 95°C, $x_A = 0.048$).
3. Draw a straight line from the lower-solubility limit to the origin according to Henry's law.

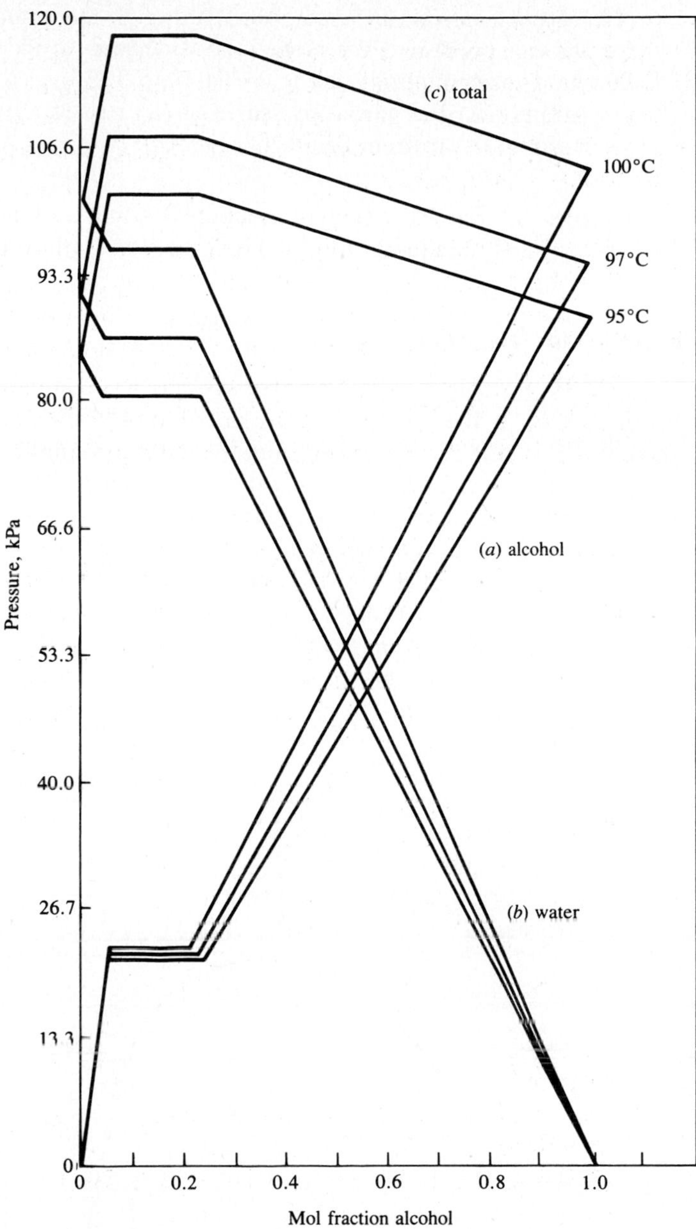

Figure 8.20 Results of Example 8.8.

(*b*) For water do the reverse of (*a*), although the assumption of Henry's law for water is poor since the range is so wide (e.g., at 95°C, $x_A = 0$, $P_W^\circ = 83.2$ kPa, and solubility limits are $x_A = 0.048$ to 0.235).

(*c*) Addition of the partial pressures of (*a*) and (*b*) yields the total pressure. Results are shown in Fig. 8.20

Exercise 8.I For the system of Example 8.8, make a temperature-composition (T-x, y) and an equilibrium (xy) diagram for 1 atm total pressure.

8.4.4 Liquid-Liquid Equilibrium Diagrams

Completely miscible two-liquid-phase systems are common and no phase equilibrium exists. If the components are only partially miscible, phase equilibria as shown in Fig. 8.21 are formed. Within the envelope equilibrium compositions are shown by horizontal lines in amounts which can be determined by the inverse lever rule. These curves, called *mutually solubility curves*, are determined from solubility data and the methods of Sec. 8.6. Point *A* on Fig. 8.21 is the critical-solution or upper-consolute temperature where both phases are identical. Alcohol–higher hydrocarbon systems exhibit such behavior.

Liquid extraction systems are at least ternaries, with, e.g., a solvent *S* added to separate a binary mixture of *A* and *B*. If only one component, say *A*, is soluble

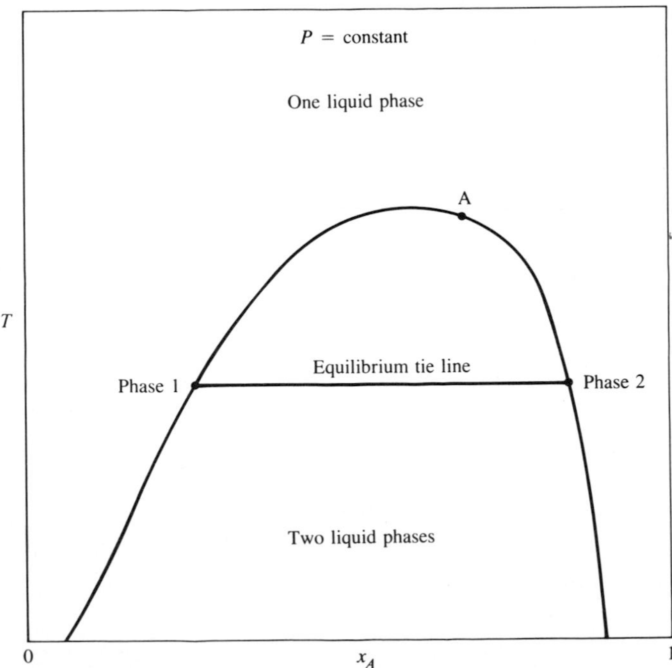

Figure 8.21 Liquid-liquid binary equilibria.

in the solvent, calculations are made by using a simple equilibrium diagram plotting solubility of A in B (raffinate) vs. the solubility of A in S (extract). If the solvent and components are all partially miscible, either a triangular diagram or a Janecke diagram are used sometimes in conjunction with a distribution diagram, each of which is illustrated in Fig. 8.22 for a type I system. Part a shows a triangular diagram with the two-liquid-phase envelope. Equilibrium tie lines such

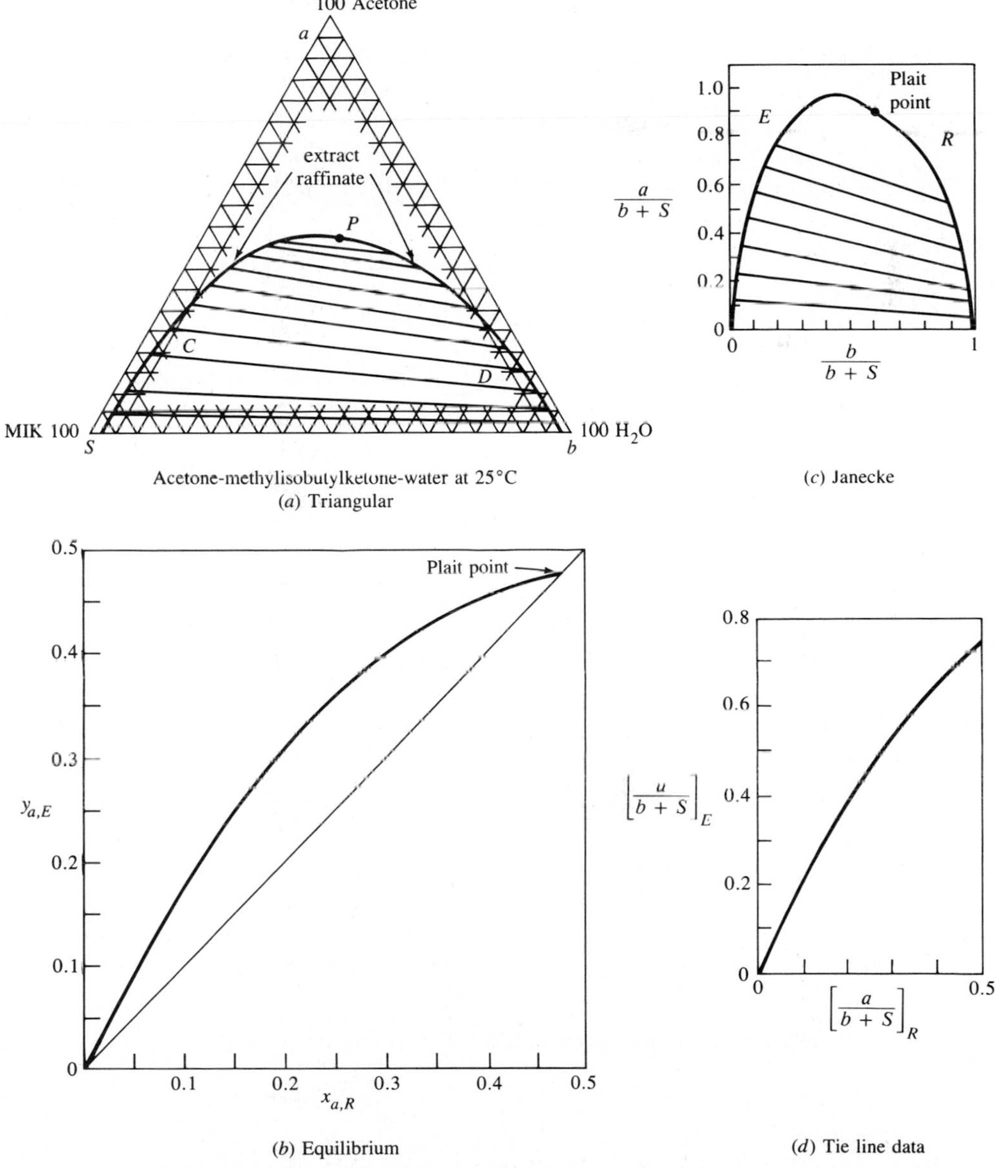

Figure 8.22 Ternary liquid-liquid equilibria for a type I system.

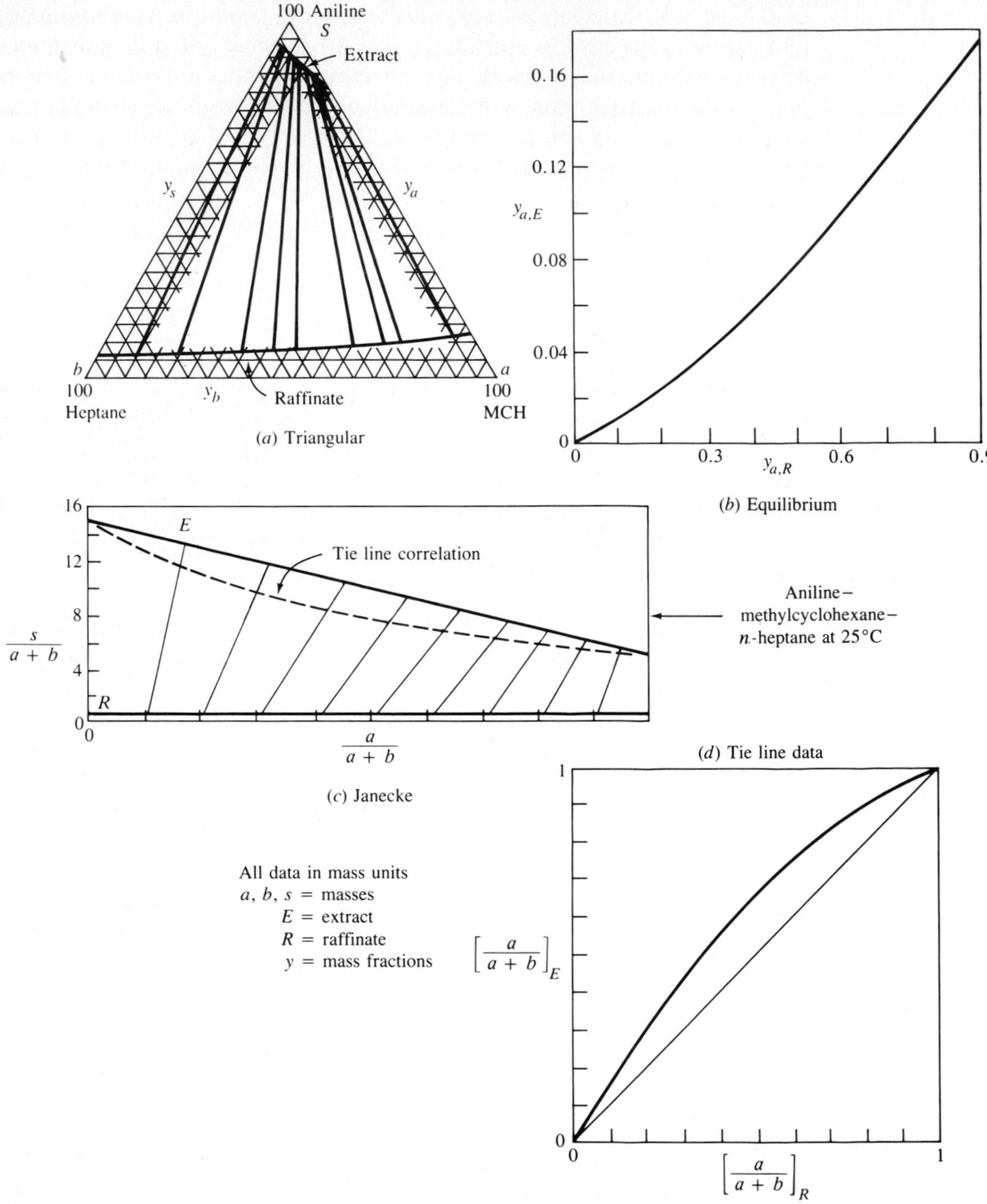

Figure 8.23 Ternary liquid-liquid equilibria for a type II system.

as $\overline{CD}$ connect equilibrium points which become shorter and reach a point at P, the plait point where both phases are identical. These are also illustrated in part b. As construction is difficult on a triangular diagram, the Janecke type of diagram (part c) is developed in rectangular coordinates where the amount of b in the sum of b and s is plotted against the amount of a in the sum of b and s. Usually for diagrams of this type a distribution diagram (Fig. 8.22d) is drawn, plotting the molar ratio of a in combined $b + s$ in the extract against the same ratio in the raffinate. From this diagram a tie line can be drawn at any point on the Janecke diagram.

In certain systems named type II, the solubilities extend over the entire concentration range and the tie lines do not reach a plait point. Such diagrams are illustrated in Fig. 8.23. The Janecke diagram shown (Fig. 8.23c) has a tie line correlation line constructed by making right triangles out of each tie line. From this line or by using the distribution diagram (Fig. 8.23d), additional tie lines can easily be constructed.

Calculations of liquid-liquid systems follow the procedures of Sec. 8.6, although often complete experimental data must be obtained. Treybal (1963), Francis (1963), Null (1970), and Sorensen et al. (1978) should be consulted for a complete discussion of such systems.

Example 8.9 For the system acetone–methyl isobutyl ketone–water at 25°C given in Fig. 8.22, determine (a) the equilibrium amount and composition of the two liquid phases formed by a feed of 15% acetone, 35% water, and 50% methyl isobutyl ketone (MIK) by mass, (b) the maximum possible concentration of acetone attainable in the solvent-free extract, (c) the equilibrium composition of the phases formed by contacting a raffinate containing 40 kg of water, 10 kg of MIK, and 35 kg of acetone with an extract containing 5 kg of water, 95 kg of MIK, and 30 kg of acetone.

SOLUTION

(a) Using Fig. 8.22a, locate the feed point F. As it falls on a tie line, follow to the ends of the tie line and read the compositions.

	Extract	Raffinate
MIK(s)	77.5	2.0
Acetone(a)	18.5	10.0
Water(b)	4.0	88.0
	100.0	100.0

E F R

By inverse lever law

$$\frac{\text{Extract}}{\text{Raffinate}} = \frac{\overline{FR}}{\overline{FE}} = \frac{25}{19} = \frac{1.32}{1}$$

Note that the mass fraction of acetone in the two phases at equilibrium check on Fig. 8.22b.

(*b*) Draw a tangent from 100% MIK along the extract portion of the binodal curve and extend it to the acetone-water binary line on the right. This gives the maximum possible concentration as 85% acetone, 15% water.
(*c*) Using the Janecke diagram, Fig. 8.22*c*, locate the feed points.

A feed	*B* feed
$b = 40$ kg	$b = 5$ kg
$s = 10$ kg	$s = 95$ kg
$a = 35$ kg	$a = 30$ kg
85 kg	130 kg
$\frac{b}{b+s} = 0.8$	$\frac{b}{b+s} = 0.05$
$\frac{a}{b+s} = 0.7$	$\frac{a}{b+s} = 0.30$

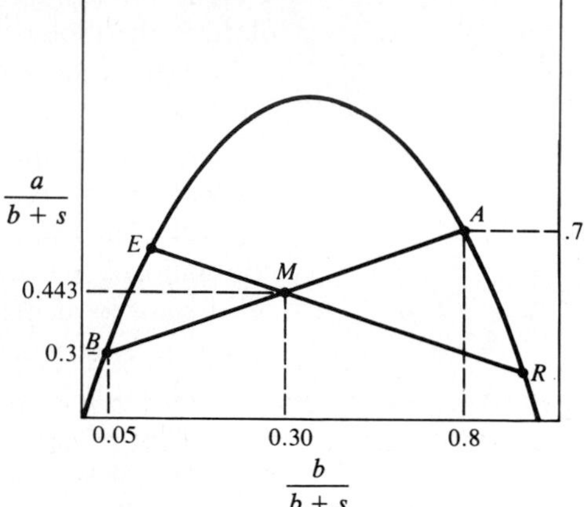

Connect the two points and locate the mixture point by using material balances or the inverse lever law.

$$\frac{\overline{AM}}{\overline{BM}} = \frac{130}{85} = 1.53$$

or

$$\left(\frac{b}{b+s}\right)_M = \frac{45}{150} = 0.30$$

$$\left(\frac{a}{b+s}\right)_M = \frac{65}{150} = 0.433$$

Locate the tie line through *M*. The ends of this line define the compositions of the phases formed at equilibrium.

	Extract	Raffinate
$\frac{a}{b+s}$	0.50	0.28
$\frac{b}{b+s}$	0.10	0.96

The total amount of each phase is obtained by the inverse lever rule:

$$\frac{E}{R} = \frac{\overline{MR}}{\overline{ME}} = \frac{18.5}{6.0} = \frac{3.08}{1}$$

From these readings, compositions can be determined by material balance.

$$(y_a)_E = \frac{0.50}{1.50} = 0.333 \qquad (y_a)_R = \frac{0.28}{1.28} = 0.219$$

$$(y_b)_E = \frac{0.10}{1.50} = 0.067 \qquad (y_b)_R = \frac{0.96}{1.28} = 0.750$$

$$(y_s)_E = \frac{0.90}{1.50} = \underline{0.600} \qquad (y_s)_R = \frac{0.04}{1.28} = \underline{0.031}$$

$$1.000 \qquad\qquad 1.000$$

Exercise 8.J For the aniline-heptane-methylcyclohexane system at 25°C given in Fig. 8.23, determine the composition and amount of the two liquid phases formed at equilibrium by a mixture of 50% aniline, 20% methylcyclohexane (MCH), and 30% heptane and the composition of the solvent-free extract. Also, determine by using the Janecke diagram the composition of the equilibrium phases formed when a mixture of 50 kg MCH and 50 kg heptane is contacted with 800 kg aniline.

8.4.5 Miscellaneous Equilibrium Diagrams

Equilibrium diagrams of various kinds can be constructed from thermodynamic principles for liquid-solid and solid-solid systems and are important in metallurgical operations and in hydrate systems. Figure 8.24 illustrates such a diagram for the $MgSO_4$–water system. Diagrams for mixed-metal systems are similar. The figure illustrates the various mixtures which can exist over the concentration range as listed below.

$\overline{abcdq}$ = solubility curve
a = solid which is mixture of ice and $MgSO_4 \cdot 12H_2O$ = eutectic
$\overline{fg}$ = $MgSO_4 \cdot 12H_2O$ (wt fraction ≈ 0.36)
$\overline{ih}$ = $MgSO_4 \cdot 7H_2O$ (wt fraction ≈ 0.49)
$MgSO_4 \cdot 6H_2O$ (wt fraction ≈ 0.53)
$MgSO_4 \cdot H_2O$ (wt fraction ≈ 0.87)
$\overline{cihb}$ = saturated solution and $MgSO_4 \cdot 7H_2O$
$\overline{dljc}$ = saturated solution and $MgSO_4 \cdot 6H_2O$
$\overline{qdk}$ = saturated solution and $MgSO_4 \cdot H_2O$
$\overline{pae}$ = saturated solution and ice

Gas-solid and liquid-solid systems in physical adsorption are diagrammed in the same manner as gas-liquid systems in absorption. Liquid-solid systems in extraction (leaching) are depicted by using distribution diagrams as discussed under liquid-liquid diagrams (Sec. 8.4.4).

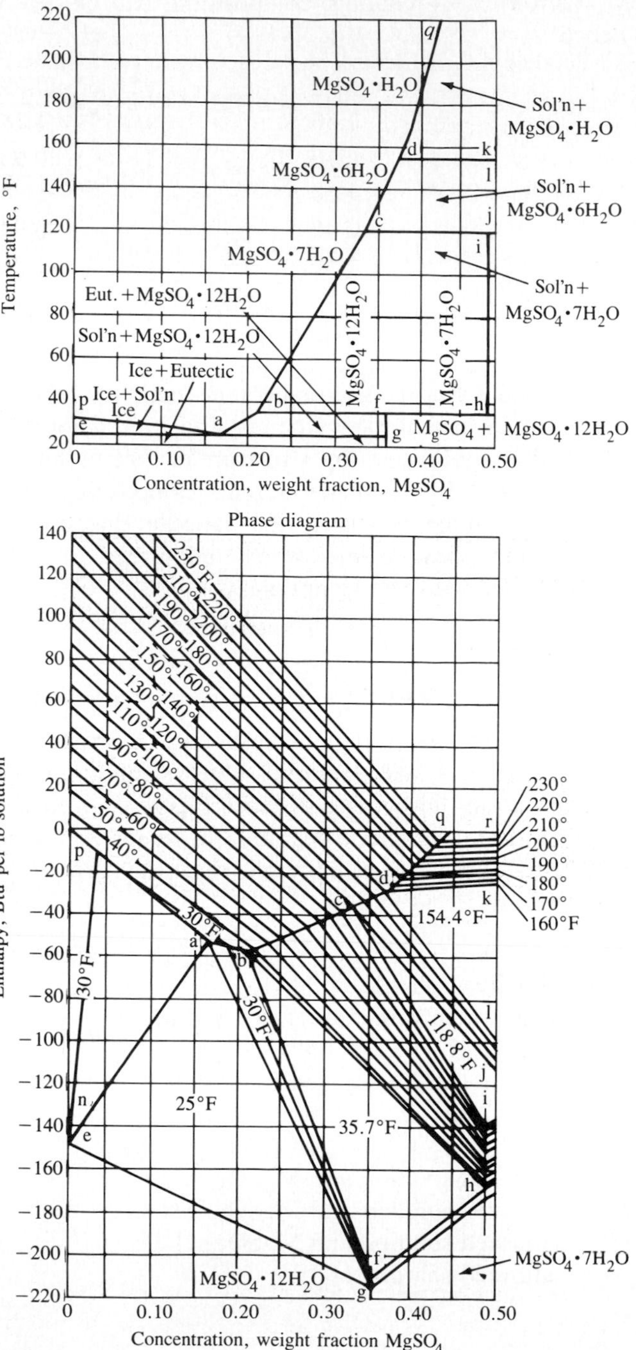

Figure 8.24 Liquid-solid equilibrium phase and enthalpy-concentration diagrams. *(From Perry et al., "Chemical Engineers Handbook," 4th ed., copyright 1963 by McGraw-Hill. Used with permission.)*

8.5 SOLUBILITY AND LIQUID-PHASE PARAMETERS

The treatment of the liquid phase is in general more complex than that of the vapor phase. A general discussion of the liquid phase, the estimation of solubility of liquid and liquid-gas systems, and methods for estimation of liquid-phase parameters will be discussed in this section. These parameters are useful for both the vapor-liquid systems discussed in Secs. 8.2 and 8.3 and the liquid-liquid systems discussed in Sec. 8.6. The state of the art for liquids is progressing but not yet fully developed; thus, most correlations and predictive methods are empirical or semiempirical.

8.5.1 Regular Solutions

Regular solutions are defined as solutions which have an excess entropy of mixing equal to zero and for which the nonideality can be attributed only to van der Waals forces of attraction. Thus

$$\Delta \tilde{H}_i^M = RT \ln \gamma_i \tag{8.34}$$

Hildebrand et al. (1962, 1970) express the excess molal free energy in terms of solubility parameters and volume fractions of the liquid for binary regular solutions of nonpolar materials:

$$\Delta \overline{G}^E = \Delta \overline{H}_L^M = V_M \phi_1 \phi_2 (\delta_1 - \delta_2)^2 \tag{8.35}$$

where

$$\phi_i = \frac{\overline{V}_i^L x_i}{V_M} = \text{molar liquid volume fraction of component } i$$

$$V_M = \sum_i \overline{V}_i^L x_i = \text{mean molal volume of the solution}$$

The solubility parameter suggested by Hildebrand, δ_i, is a parameter which takes into account the van der Waals attraction forces among the different species and is defined as

$$\delta_i = \left(\frac{\Delta E_i}{\overline{V}_{i,L}} \right)_S^{1/2} \tag{8.36}$$

where ΔE_i = internal energy of vaporization of saturated liquid to ideal gaseous state at temperature T

$\overline{V}_{i,L}$ = molal volume of saturated liquid at temperature T

At high temperatures and near the critical, ΔE_i can be estimated from correlations presented earlier in this book. At low temperatures, the following approximation can be made:

$$\Delta E_i \approx \overline{\lambda}_{v,i} - RT \tag{8.37}$$

where $\overline{\lambda}_{v,i}$ is the molal heat of vaporization of pure i at temperature T. Table 8.1

Table 8.1 Solubility parameters and effective molar volumes for hydrocarbons and nonhydrocarbons

Compound name	Solubility parameter, $(MJ/m^3)^{1/2}$	Molar volume, m^3/kmol	Footnote
Hydrogen	6.65	0.0310	3
Methane	11.62	0.0520	3
Ethane	12.38	0.0680	4
Propane	13.09	0.0840	4
i-Butane	12.56	0.1055	1
n-Butane	13.54	0.1014	1
i-Pentane	13.81	0.1174	1
n-Pentane	14.36	0.1161	1
Neopentane	12.54	0.1233	1
n-Hexane	14.87	0.1316	1
n-Heptane	15.20	0.1475	1
n-Octane	15.44	0.1636	1
n-Nonane	15.65	0.1797	1
n-Decane	15.79	0.1959	1
n-Undecane	15.93	0.2122	2
n-Dodecane	16.04	0.2286	2
n-Tridecane	16.14	0.2449	2
n-Tetradecane	16.22	0.2613	2
n-Pentadecane	16.28	0.2777	2
n-Hexadecane	16.34	0.2941	2
n-Heptadecane	16.40	0.3104	2
Ethylene	12.44	0.0610	4
Propylene	13.15	0.0790	4
1-Butene	13.60	0.0953	1
cis-2-Butene	14.69	0.0912	1
trans-2-Butene	14.18	0.0938	1
i-Butene	13.77	0.0954	1
1, 3-Butadiene	14.52	0.0880	1
1-Pentene	14.44	0.1104	1
cis-2-Pentene	15.03	0.1078	1
trans-2-Pentene	14.81	0.1106	1
2-Methyl-1-Butene	14.67	0.1087	1
3-Methyl-1-Butene	13.77	0.1128	1
2-Methyl-2-Butene	15.18	0.1068	1
1-Hexene	14.95	0.1259	1
Cyclopentane	16.59	0.0947	1
Methylcyclopentane	16.06	0.1131	1

Table 8.1 Continued

Compound name	Solubility parameter, $(MJ/m^3)^{1/2}$	Molar volume, $m^3/kmol$	Footnote
Cyclohexane	16.77	0.1088	1
Methylcyclohexane	16.02	0.1283	1
Benzene	18.74	0.0894	1
Toluene	18.23	0.1069	1
o-Xylene	18.39	0.1212	1
m-Xylene	18.04	0.1235	1
p-Xylene	17.94	0.1239	1
Ethylbenzene	17.98	0.1231	1
Nitrogen	9.08	0.0530	4
Oxygen	8.18	0.0284	4
Carbon monoxide	6.40	0.0352	4
Carbon dioxide	14.56	0.0440	4
Hydrogen sulfide	18.00	0.0431	4
Sulfur dioxide	12.27	0.0452	4
2-Methylpentane	14.46	0.1309	1
3-Methylpentane	14.58	0.1306	1
2, 2-Dimethylbutane	13.72	0.1337	1
2, 3-Dimetylbutane	14.26	0.1312	1
1-Heptene	15.30	0.1418	1
Propadiene	14.01	0.0616	4
1, 2-Butadiene	16.00	0.0837	1
Ethylcyclopentane	16.24	0.1288	1
Ethylcyclohexane	16.28	0.1432	1
Water	45.61	0.0180	4

Calculational procedure: The solubility parameters may be calculated from the relation

$$\delta = \sqrt{\left(\Delta H_v - \frac{RT}{MW}\right)\rho}$$

where ΔH_v = heat of vaporization at 25°C, MJ/kg
T = 298.15 K
MW = molecular weight
ρ = density at 25°C, kg/m^3
$R = 8.314 \times 10^{-3}$ MJ/(kmol · K)

[1]Calculated as noted above.
[2]Calculated as noted above, using the tables of API Research Report 44, 1982.
[3]Prausnitz and Shair, *AIChE J.*, **7**:682 (1961).
[4]Parameters derived for use with Chao-Seader correlation.

lists values of solubility parameters as calculated by Graboski and Daubert (1978*b*) together with molar volumes for common materials.

Regular solutions are endothermic, have zero excess entropies, and have entropies of mixing equal to $-\Sigma x_i R \ln x_i$. Activity coefficients are all positive.

For binary mixtures, the partial molal heat of mixing can be determined by differentiating Eq. (8.35) with respect to a component and combining with Eq. (8.34). For component 1,

$$\Delta \tilde{H}_1 = RT \ln \gamma_1 = \overline{V}_1^L \phi_2^2 (\delta_1 - \delta_2)^2$$
$$\ln \gamma_1 = \frac{\overline{V}_1^L}{RT} \phi_2^2 (\delta_1 - \delta_2)^2 \tag{8.38}$$

This equation can be extended to multicomponent regular solutions.

$$\ln \gamma_i = \frac{\overline{V}_i^L}{RT} (\delta_i - \bar{\delta})^2 \tag{8.39}$$

where $\qquad \bar{\delta} = \sum_i \phi_i \delta_i \qquad (\delta_i = \text{molar liquid volume fraction})$

These relations assume no binary interactions, and liquid-phase activity coefficients can be predicted from pure-component data at any composition and temperature. However, they are limited to nonpolar mixtures. The state of the art currently does not allow calculation of binary interaction parameters with much success as reported by Reid et al. (1977). Lin (1979) has also shown the problems involved in estimating such interactions.

Example 8.10 The following data were taken for the excess enthalpy of mixing of the system *n*-hexane–*n*-hexadecane at 40°C. Compare the results with those calculated from regular solution theory,

$\Delta \overline{H}^M$, kJ/kmol	39	66	47
x_2, mole fraction hexadecane	0.20	0.50	0.80

SOLUTION Use Eq. (8.35):

$$\Delta \overline{H}_L = V_M \phi_1 \phi_2 (\delta_1 - \delta_2)^2$$

where $\qquad \phi_i = \dfrac{\overline{V}_i^L x_i}{\Sigma \overline{V}_i^L x_i = V_M}$

From Table 8.1,

	δ, $(\text{MJ/m}^3)^{1/2}$	$\overline{V}_L$, m^3/kmol
$n-C_6$	14.87	0.1316
$n-C_{16}$	16.34	0.2941

At $x_2 = 0.50$:

$$V_M = (0.5)(0.1316) + (0.5)(0.2941) = 0.2129$$

$$\phi_1 = \frac{0.0658}{0.2129} = 0.309 \qquad \phi_2 = \frac{0.1471}{0.2129} = 0.691$$

$$\Delta \overline{H}_L = (0.2129)(0.309)(0.691)(14.87 - 16.34)^2(10^3)$$
$$= 98.2 \text{ kJ/kmol}$$

At $x_2 = 0.20$:

$$V_M = (0.8)(0.1316) + (0.2)(0.2941) = 0.1641$$

$$\phi_1 = \frac{0.1053}{0.1641} = 0.642 \qquad \phi_2 = \frac{0.0588}{0.1641} = 0.358$$

$$\Delta \overline{H}_L = (0.1641)(0.642)(0.358)(14.87 - 16.34)^2(10^3)$$
$$= 81.5 \text{ kJ/kmol}$$

At $x_2 = 0.80$:

$$V_M = 0.2616 \qquad \phi_1 = \frac{0.0263}{0.2616} = 0.100 \qquad \phi_2 = \frac{0.2353}{0.2616} = 0.900$$

$$\Delta \overline{H}_L = (0.2616)(0.100)(0.900)(14.87 - 16.34)^2(10^3)$$
$$= 50.9 \text{ kJ/kmol}$$

Although the shape of the curve is correct, all results are high. Postulate why.

Exercise 8.K Estimate the activity coefficients for both components over the composition range for Example 8.10.

8.5.2 Solubilities of Gases in Liquids

Most gases are slightly soluble in liquids, their solubilities decreasing as temperatures are increased except near the critical where increases in solubilities occur. Many tabulations of such solubilities exist as a function of concentration, although data are more usually reduced to Henry's law constants for low-solubility solutes. For a binary mixture, if A is the solvent and B is the solute,

$$f_B = H_B x_B \tag{8.40}$$

where
$$H_B = \lim_{x_B \to 0} \frac{f_B}{x_B}$$

with the pressure of the system equal to the vapor pressure of the solvent at the system temperature. At high pressures, the Poynting correction is also necessary:

$$\ln \frac{f_B}{x_B} = \ln \frac{\phi_B y_B P_T}{x_B} = \ln H_B + \frac{\tilde{V}_B^{\infty}(P_T - P_1^{\circ})}{RT} \tag{8.41}$$

where $\tilde{V}_2^{\infty}$ is the partial molal liquid volume of the solute at infinite dilution. These equations are only useful for low concentrations, which may be extremely small for molecules differing substantially in chemical structure.

Equation (7.64) gave Henry's law as $\overline{P}_i = H_i x_i$, where $\overline{P}_i$ is the partial pressure of the solute. For the solute gas, $f_B^\circ = 1.0$. If the gas is ideal, $\gamma_B^V = 1.0$. Thus $a_B^V = f_B = \phi_B y_B P_T$. For the gas dissolved in the liquid with standard state taken as pure component in an imaginary liquid state at T and P,

$$a_B^L = \gamma_B^L x_B = \frac{f_B}{f_B^{\circ L}}$$

The equilibrium constant is then

$$K_B = \frac{a_B^V}{a_B^L} \tag{8.42}$$

Thus

$$K_B = \frac{\phi_B y_B P_T}{\gamma_B^L x_B} = f_B^{\circ L} \tag{8.43}$$

Since $\overline{P}_B = H_B x_B = y_B P_T$, then

$$H_B = \frac{K_B \gamma_B^L}{\phi_B} \tag{8.44}$$

This expression can then be used to predict solubilities for an ideal solution if γ_B^L is assumed equal to 1 and for a nonideal solution by predicting γ_B^L by equations such as (8.39).

An ideal solubility would be calculable by equating Henry's and Raoult's laws at the system temperature which would give a first estimate of the solubility of the solute.

For a mixed solvent, ideal mixing would give a combining rule such as

$$\ln H_{B,M} = \sum_j x_j \ln H_{B,j} \tag{8.45}$$

where j = each solvent component. This equation is useful for reasonable estimates even for nonideal mixtures.

A survey of methods for estimation of solubilities of gases in liquids has been carried out by Kabadi et al. (1982) and suggests various correlation and prediction techniques. For hydrocarbon systems the TDB-PR should be consulted.

8.5.3 Liquid-Liquid Solubilities

Chemically similar liquids are usually completely soluble in one another with mutual solubility decreasing as they become more dissimilar. These differences increase the separate activity coefficients, and two liquid phases form if they are highly different. Since the two liquid phases are in equilibruim, their fugacities are equal:

$$(f_{L1} = f_{L2} = a_{L1} f_{L1}^\circ = a_{L2} f_{L2}^\circ)_i \tag{8.46}$$

Since the standard states are the same in both cases, pure liquid at system temperature and pressure, then

$$f_{L1}^\circ = f_{L2}^\circ$$

and $$(a_{L1} = a_{L2})_i \quad \text{or} \quad (x_{L1}\gamma_{L1} = x_{L2}\gamma_{L2})_i \tag{8.47}$$

Any two-constant expression could be used to estimate the activity coefficients $\gamma_{L1,i}$ and $\gamma_{L2,i}$ and hence the distribution of component i in the phases $x_{L1,i}/x_{L2,i}$. For any component, take an expression such as the van Laar or Margules equations with A and B parameters. Thus, two equations in four unknowns—$\gamma_{L1,i}\gamma_{L2,i}$, A, and B—exist. Using the material balances on each of the two phases as the other two equations, the four parameters can be estimated, and such solubility data have yielded estimates of the activity coefficients.

Methods for actual prediction of activity coefficients for such systems will be given in the next section.

It is often necessary to determine the temperature at which complete miscibility occurs—the critical-solution temperature. This temperature is determined by noting that where both phases merge, the first and second derivatives of activity are zero:

$$\frac{\partial(\ln \gamma_i x_i)}{\partial x_i} = 0 \qquad \frac{\partial^2(\ln \gamma_i x_i)}{\partial x_i^2} = 0 \tag{8.48}$$

This equation can be solved for any model.

For immiscible phases a given component has the same activity regardless of concentration.

8.5.4 Prediction of Liquid Activity Coefficients

Many investigators have recently attempted to predict activity coefficients by group contribution methods which have previously been successful for predicting properties such as volumes, critical temperatures, and critical pressures. Such methods have the inherent ability to predict properties easily but do not take into account the interactions between molecular groups in their simplest forms. These methods are of optimum value where correlations are limited to one type of compound. Second-order interactions such as the difference between primary, secondary, and tertiary hydrogens must be taken into account to improve the accuracy of the correlations.

Various methods have been reviewed by Reid et al. (1977) and by Lin (1979). The UNIFAC method derived from the UNIQUAC model is the most general and assumes that the activity coefficient can be divided into two parts: molecular-size contributions (C) and interaction contributions (R).

$$\ln \gamma_i = \ln \gamma_i^C + \ln \gamma_i^R \tag{8.49}$$

The first part of the equation only requires pure-component properties and the size and shape of the molecules, while the second part depends on group areas and group interactions. A succinct treatment of the method has been given by Reid et al. (1977), while a complete treatment is available by Fredenslund et al. (1977). New UNIFAC parameters have been published by Skold-Jorgensen et al. (1979), Gmehling et al. (1982), and Macedo et al. (1983) and continue to be published at irregular intervals.

8.6 LIQUID-LIQUID, SOLID-LIQUID, AND SOLID-VAPOR EQUILIBRIA

This section will give a brief treatment of this area and attempt to point the way rather than be inclusive in selecting methods for the estimation of phase equilibria. The state of the art is not nearly so well-developed in these areas as is the case for vapor-liquid equilibria, and hence the methods for treatment of such equilibria are not nearly so acccurate or numerous.

8.6.1 Liquid-Liquid Equilibria

Two or more liquid phases will form at temperatures below the critical-solution temperature or consolute temperature. In a few cases two such temperatures exist —a lower and an upper bound—only between which two liquid phases will occur. For a simple one-parameter model for a binary mixture, where $\Delta G^E = Ax_A x_B$, application of Eqs. (8.48) shows that two phases will occur if $A > 2RT$. If two phases do occur the procedures of Sec. 8.5.3 can be used to determine equilibrium compositions. A similar method can be used for ternary and higher mixtures. However, such equations normally have multiple roots, and it is often difficult to converge on a solution that is physically significant.

The methods discussed can be used if formulations for the excess Gibbs free energy such as those of van Laar are applicable. However, for liquid-liquid systems estimation of composition almost entirely depends on the activity coefficients; thus a more accurate formulation for the activity coefficient is required than is the case for vapor-liquid equilibria.

Both the NRTL and UNIQUAC type of formulations have been used with varying degrees of success. Some preliminary observations have been noted by Reid et al. (1977). Sorensen et al. (1978, 1979) have reviewed liquid-liquid equilibrium data and correlations in detail. Prausnitz et al. (1980) describe the UNIQUAC method in detail. Null (1970) presents a concise summary.

A class of liquid-liquid systems where great differences in molecular size exist, which are of great importance today, are polymer-solvent systems. Flory (1953) advanced a theory which has gained wide acceptance for such systems. If the solvent is A and the polymer solute is B, the activity of the solvent can be calculated as

$$\ln a_A = \ln \phi_A + \left(1 - \frac{\bar{V}_A^L}{\bar{V}_B^L}\right)\phi_B + \chi\phi_2^2 \tag{8.50}$$

where $\phi =$ volume fraction

$$\phi_A = \frac{xw_A/\rho_A}{(xw_A/\rho_A) + (xw_B/\rho_B)}$$

$$\phi_B = \frac{xw_B/\rho_B}{(xw_B/\rho_B) + (xw_A/\rho_A)}$$

$x =$ Flory interaction parameter dependent on temperature and in theory independent of concentration $\bar{V}_A^L/\bar{V}_B^L$ equals zero for typical solutions

This equation can be utilized for estimating the volatility of a solvent in a polymer solution given a value for χ. The method can be extended to multi-component systems.

The Flory theory has also been extended to calculation of solubilities in liquid-liquid equilibria.

8.6.2 Solid-Liquid Equilibria

Less is known of solid-liquid equilibria, although the general equilibrium relations presented earlier are applicable to any phase equilibrium situation. If the activity in the liquid phase of the solid solute is referred to pure solute in an imaginary liquid state at the system temperature and pressure *and* the activity of the solid in the solid phase is taken as unity for pure solid at the same conditions, the equilibrium solubility constant can be written as

$$K_B = \frac{a_{B,L}}{a_{B,S}} = \frac{f^\circ_{B,S}}{f^\circ_{B,L}} = \frac{\gamma_{B,L} x_{B,L}}{\gamma_{B,S} x_{B,S}} \tag{8.51}$$

This solubility constant will vary with temperature in the same manner as any equilibrium constant.

$$\left[\frac{\partial(\ln K_B)}{\partial T}\right]_P = \frac{\bar{H}^\circ_{B,L} - H^\circ_{B,S}}{RT^2} = \frac{\lambda_{\text{fus}}}{RT^2} \tag{8.52}$$

If the heat of fusion is independent of temperature, the equation can be integrated between the melting point where $K_B = 1$ and $f^\circ_{B,S} = f^\circ_{B,L}$ to give

$$\ln K_B = -\frac{\lambda_{\text{fus}}}{R}\left(\frac{1}{T} - \frac{1}{T_M}\right) \tag{8.53}$$

where T_M = melting point. Similarly, an equation can be developed for the effect of pressure on the equilibrium solubility constant.

$$\left[\frac{\partial(\ln K_B)}{\partial P}\right]_T = \frac{\bar{V}^\circ_{B,L} - \bar{V}^\circ_{B,S}}{RT} \tag{8.54}$$

which when integrated is

$$\ln K_B = -\frac{\bar{V}^\circ_{B,L} - \bar{V}^\circ_{B,S}}{RT}(P_T - P_M)$$

where P_M = vapor pressure at the melting point.

When only one variable is changed, the solute with the higher molecular weight will be the least soluble, the solute with the higher melting point will be the least soluble, and the solute with the higher heat of fusion will be the least soluble. Hildebrand and Scott (1962) show a complete analysis.

For an ideal solution, Eq. (8.51) shows that $K_B = x_{B,L}$. For a nonideal solution, $K_B = \gamma_{B,L} x_{B,L}$ if the activity of the solid phase is taken as unity. If this is substituted into Eq. (8.53),

$$\ln(\gamma_{B,L}x_{B,L}) = -\frac{\lambda_{\text{fus}}}{R}\left(\frac{1}{T} - \frac{1}{T_M}\right) \tag{8.55}$$

In the ideal case the solubility will depend only on properties of the solute, and intermolecular forces are neglected. In the nonideal case, the solution of Eq. (8.55) requires an expression for Gibbs free energy. The two equations can then be solved by trial and error for γ_B vs. x_B. For example, a one-constant correlation might be $\ln\gamma_B = (A/RT)(1 - x_B)^2$ which can be solved simultaneously with Eq. (8.55) if A has been determined from experimental data. A van Laar or Margules two-constant equation can also be used.

Equations (8.38) have also been used to determine activity coefficients for nonpolar mixtures with some success.

Hougen et al. (1959) suggest a method for calculation of solubilities for electrolytes provided some experimental data are known. Null (1970) summarizes both methods of presentation of experimental data and methods for solid-liquid equilibria process calculations.

8.6.3 Solid-Vapor Equilibria

Solid-vapor equilibria have become of increasing importance with the use of adsorption as a technique to remove gases from vapor streams, especially for air pollution control. Such equilibria are usually expressed as mass solubilities, mass of adsorbate per unit mass of adsorbent (m). The integral heat of adsorption per unit mass of adsorbent, ΔH_A, is the change in enthalpy per unit of adsorbed gas when adsorbed on gas-free adsorbent to form a given concentration of adsorbate and increases with increased concentration. It is composed for physical adsorption of both the heat of condensation of the adsorbate plus the heat of wetting caused by van der Waals forces of attraction and capillarity. For activated adsorption which often occurs in catalytic reactions, a heat of formation of a surface compound is also present.

The most-simplified form of such equilibria for the case of adsorption at constant temperature, composition, and pressure where the partial molal volume of the adsorbed gas can be neglected and the gas can be assumed to be ideal is

$$\Delta H = \frac{R}{\text{MW}}\int_0^m \frac{\partial(\ln P)}{\partial(1/T)_m}\,dm \tag{8.56}$$

where P = equilibrium pressure above adsorbent at given concentration and temperature

MW = molecular weight of gas

This expression can be solved by using experimental equilibrium isotherms as input data.

A more-advanced review and treatment of both pure-gas and mixed-gas adsorption is given by Suwanayuen (1980). A compilation of adsorption data and methods for prediction of pure- and mixed-gas adsorption are given in the "Technical Data Book—Petroleum Refining" (API, 1982).

8.7 REFERENCES

Abbott, M. M., and H. C. Van Ness: *AIChE J.*, **21**:62 (1975).

American Petroleum Institute (API): "Technical Data Book—Petroleum Refining," 4th ed., Washington, extant 1982.

Chao, K. C., and J. C. Seader: *AIChE J.*, **7**:598 (1961).

Daubert, T. E., M. S. Graboski, and R. P. Danner: "Technical Data Book—Petroleum Refining, Documentation Report No. 8-78," Order no. 2050355, Univ. Microfilms, Ann Arbor, Mich., 1978.

Flory, P. J.: "Principles of Polymer Chemistry," Cornell, Ithaca, N.Y., 1953.

Francis, A. W.: "Liquid-Liquid Equilibriums," Interscience-Wiley, New York, 1963.

Fredenslund, A., A. J. Gmehling, and P. Rasmussen: "Vapor-Liquid Equilibria Using UNIFAC," Elsevier, Amsterdam, 1977.

Gmehling, J., P. Rasmussen, and A. Fredenslund: *Ind. Eng. Chem. Process Des. Dev.*, **21**:118 (1982).

Gothard, F. A., M. F. C. Ciobanu, et al.: *Ind. Eng. Chem. Process Des. Dev.*, **15**:333 (1976).

Graboski, M. S., and T. E. Daubert: *Ind. Eng. Chem. Proc. Des. Dev.*, **17**:443 (1978*a*); **17**:448 (1978*b*); **18**:300 (1979).

Grayson, H. G., and C. W. Streed: Proceedings 6th World Petroleum Congress, III, Paper 20-PD7, pp. 233–245 (1963).

Hadden, S. T., and H. G. Grayson: *Hydrocarbon Process.*, **40**(9):207 (1961).

Hildebrand, J. H., and R. L. Scott: "Regular Solutions," Prentice-Hall, Englewood Cliffs, N.J., 1962.

______, J. M. Prausnitz, and R. L. Scott: "Regular and Related Solutions," Van Nostrand Reinhold, New York, 1970.

Hougen, O. A., K. M. Watson, and R. A. Ragatz: "Chemical Process Principles, Part Two: Thermodynamics," 2d ed., Wiley, New York, 1959.

Kabadi, V. N., R. P. Danner, and T. E. Daubert: "Technical Data Book—Petroleum Refining, Documentation Report No. 9-82," Order no. 2051962, Univ. Microfilms, Ann Arbor, Mich., 1982.

Katz, D. L., and R. Kurata: *Ind. Eng. Chem.*, **32**:817 (1940).

Kay, W. B.: *Ind. Eng. Chem.*, **30**:459 (1938).

Lin, C. T.: Ph.D. dissertation, The Pennsylvania State University, University Park, 1979.

Macedo, E., U. Weidlich, J. Gmehling, and P. Rasmussen: *Ind. Eng. Chem. Process Des. Dev.*, **22**:678 (1983).

Null, H. R.: "Phase Equilibrium in Process Design," Wiley-Interscience, New York, 1970.

Othmer, D. F.; R. E. White, and E. Trueger: *Ind. Eng. Chem.*, **33**:1240 (1941).

Prausnitz, J. M., et al.: "Computer Calculations for Multicomponent Vapor-Liquid and Liquid-Liquid Equilibria," Prentice-Hall, Englewood Cliffs, N.J., 1980.

Reid, R. C., J. M. Prausnitz, and T. K. Sherwood: "Properties of Gases and Liquids," 3d ed., McGraw-Hill, New York, 1977.

Skold-Jorgensen, S., B. Kolbe, J. Gmehling, and P. Rasmussen: *Ind. Eng. Chem Process Des. Dev.*, **18**:714 (1979).

Sorensen, J. M., T. Magnussen, P. Rasmussen, and A. Fredenslund: *Fluid Phase Equilibria*, **2**:297 (1978); **3**:47 (1979).

Suwanayuen, S., and R. P. Danner: *AIChE J.*, **26**:68, 76 (1980).

Treybal, R. E.: "Liquid Extraction," 2d ed., McGraw-Hill, New York, 1963.

Van Ness, H. C., and M. M. Abbott: "Classical Thermodynamics of Nonelectrolyte Solutions," McGraw Hill, New York, 1982.

Varteressian, K. A., and M. R. Fenske: *Ind. Eng. Chem.*, **29**:270 (1937).
Wilson, G. M.: *J. Am. Chem. Soc.*, **86**:127 (1964).
Wohl, K.: *Trans. AIChE*, **42**:215 (1946).

8.8 PROBLEMS

1 Consider the two-dimensional plots in Fig. 8.7; apply the phase rule to determine the number of degrees of freedom and specify from what variables they can be chosen for:

(*a*) *TV* plot, solid-vapor region
(*b*) *TV* plot, intersection of vapor, liquid-vapor, and solid-vapor regions
(*c*) *TV* plot, critical point
(*d*) *TP* plot, solid-liquid line
(*e*) *TP* plot, triple point—solid-liquid-vapor
(*f*) *TP* plot, liquid isometric line
(*g*) *TP* plot, ice III region
(*h*) *PV* plot, liquid isotherm
(*i*) *PV* plot, triple point line
(*j*) *PV* plot, ice I region

2 Consider a binary mixture of *A* and *B* in binary vapor-liquid equilibrium. At constant temperature and pressure, make a rough plot of a_A, a_B, γ_A^L, and γ_B^L as functions of x_A assuming an ideal solution in the vapor phase and the standard state for the liquid phase being pure components at the system temperature and the vapor pressure at the system temperature if the liquid phase is ideal, if the liquid phase deviates positively from ideality, and if the liquid phase deviates negatively from ideality.

3 A liquid mixture of 10 mol % *n*-hexane, 30 mol % benzene, 40 mol % *n*-heptane, and 20 mol % *n*-octane is in equilibrium with its vapor at 1 atm total pressure. Determine the equilibrium temperature and vapor composition if the vapor phase is ideal gas and the liquid phase is ideal solution.

4 A mixture of ethylene and propylene is brought to equilibrium at 0°C and 500 lb/in^2. The vapor phase can be taken as ideal solution (not ideal gas), as may the liquid phase. Determine the composition of each phase.

5 At 1 atm total pressure, ethyl acetate and ethyl alcohol form an azeotrope containing 53.9% ethyl acetate which boils at 71.8°C. Assuming that the van Laar model for activity coefficients is applicable, estimate the azeotropic composition and the total pressure at a boiling point of 56.3°C. Use the vapor pressure data given below:

	71.8°C	56.3°C
Ethyl alcohol	587 mmHg	298 mmHg
Ethyl acetate	636 mmHg	360 mmHg

6 Data for the vapor-liquid equilibrium for 1,3-butadiene–furfural at 100°F are given below:

x butadiene	0.046	0.097	0.17	0.29	0.51
P_T, mmHg	517	1034	1551	2069	2586

Assuming an ideal vapor phase and a liquid phase with Wilson parameters $G_{12} = 0.53685$ and $G_{21} = 0.21779$, determine the composition of the vapor phase at each point.

7 Vapor-liquid equilibrium data for nitrogen-oxygen at 1 atm total pressure are given below. Test the data for thermodynamic consistency.

T, K	x N_2	y N_2
89.5	0.0590	0.1545
88.5	0.1018	0.2563
87.5	0.1469	0.3515
86.5	0.1956	0.437
85.5	0.2490	0.5193
84.5	0.3069	0.5955
83.5	0.3707	0.6665
82.5	0.4406	0.7327
81.5	0.5208	0.7878
80.0	0.6147	0.8522
79.0	0.7233	0.9067
78.0	0.9190	0.9782

How would you determine the constants in the Margules equations (8.27) for this system? Assume ideal gas in the vapor phase.

8 Using Fig. 8.2 for determining K values, determine the composition of the vapor in equilibrium with the liquid and the pressure of the system for a liquid mixture of 5 mol CH_4, 10 mol C_2H_6, 30 mol C_3H_8, 25 mol iso-C_4H_{10}, and 30 mol n-C_4H_{10} at 40°C. If the mixture was a vapor, repeat the problem for the pressure and composition of the liquid phase.

9 Using K-value figures, determine:

(*a*) The K value for propylene at 40°C and 1500 kPa total pressure
(*b*) The K value for propylene at 40°C and 100 kPa total pressure
(*c*) The K value of propylene at its normal boiling point and saturation pressure
(*d*) The K value of propylene at its critical point
(*e*) The K value of n-decane at 250°C and 700 kPa total pressure
(*f*) A curve of the K value of n-hexane from −100 to 500°C at 100 kPa total pressure

10 Using Fig. 8.6, determine:

(*a*) The K value of carbon monoxide in propane at 200°C and 34 atm
(*b*) The K value of hydorgen sulfide–n-nonane at 200°C and 12 atm

Also determine by using Eq. (8.33) the activity coefficient for the carbon monoxide and hydrogen sulfide if the mixtures are equimolar.

11 Vapor-liquid equilibrium data for benzene (1)-cyclohexane (2) at 1 atm total pressure are given below. Compare the data with data calculated, assuming an ideal liquid phase by plotting both sets of data on an xy diagram. Calculate the Margules equation constants for the data.

T, °C	$x(1)$	$y(1)$	$\gamma(1)$	$\gamma(2)$
79.7	0.088	0.113	1.300	1.003
79.1	0.156	0.190	1.256	1.008
78.5	0.231	0.268	1.219	1.019
78.0	0.308	0.343	1.189	1.032
77.7	0.400	0.422	1.136	1.056
77.6	0.470	0.482	1.108	1.075
77.6	0.545	0.544	1.079	1.102
77.6	0.625	0.612	1.058	1.138
77.8	0.701	0.678	1.039	1.178
78.0	0.757	0.727	1.025	1.221
78.3	0.822	0.791	1.018	1.263
78.9	0.891	0.863	1.005	1.328
79.5	0.953	0.938	1.003	1.369

12 An example of a partially miscible system is diethyl ether–water. At 60°C ether dissolves to the extent of 0.9 mol % in water and water dissolves to the extent of 6.7 mol % in ether. The vapor pressures of ether and water at 60°C are 1730 and 149.4 mmHg, respectively.

Construct an *xy* diagram for the system at a constant temperature of 60°C. Assume Henry's law is followed by each component at concentrations below the lower-solubility limit and that Raoult's law is followed at concentrations above the upper-solubility limit.

13 Toluene and water are essentially immiscible in the liquid state. Calculate:

(*a*) The normal boiling point of an equimolar mixture

(*b*) The normal boiling point of a mixture of 20 mol % toluene and 80 mol % water

(*c*) The vapor composition at the bubble point for (*a*) and (*b*)

(*d*) If an equimolar mixture at 1 atm total pressure and 115°C is cooled, determine the temperature and composition of the phases at the dew point.

(*e*) Determine the highest possible temperature of complete condensation of an equimolar mixture at 1 atm total pressure.

14 Construct an enthalpy-concentration diagram including tie lines from the data given below for the ethanol (1)-water (2) system at 1 atm total pressure.

	Enthalpy data		Equilibrium data	
Wt % 1	Sat. liquid H, Btu/lb	Sat. vapor H, Btu/lb	x of (1), wt %	y of (1), wt %
0	180	1150	0	0
10	160	1080	5	37.5
20	145	1010	10	52.5
30	135	945	20	65.5
40	128	875	30	71.5
50	123	805	40	74.8
60	117	735	50	77.0
70	111	665	60	79.3
80	104	595	70	82.3
90	96	530	80	86.0
100	89	460	90	91.3

Determine the saturated-liquid and saturated-vapor enthalpies and the vapor composition in equilibrium with a 7.5 wt % ethanol solution. Determine the feed composition if equal weights of liquid and vapor are formed.

15 Plot an X (mol CO_2/mol solution)-Y (mol CO_2/mol inert gas) diagram at 77°F and an *xy* diagram at 122°F for a total pressure of 1.2 atm for aqueous solutions of monoethanolamine and carbon dioxide by using the data below. Assume monoethanolamine is essentially nonvolatile at these temperatures.

	$\bar{P}$ CO_2, mmHg	
(mol CO_2/mol sol.)	77°F	122°F
0.052	...	7.5
0.054	...	13.6
0.056	...	25.0
0.058	5.6	47.1
0.060	12.8	96.0
0.062	29.0	259
0.064	56.0	
0.066	98.7	
0.068	155	
0.070	232	

Determine the composition of the equilibrium vapor at $x = 0.061$ for both temperatures.

16 Equilibrium data for aqueous solutions of HCl at 30°C are given below:

Wt % HCl	$\bar{P}$, mmHg	
	HCl	H_2O
20	0.48	19.0
22	1.02	17.1
24	2.17	15.4
26	4.56	13.5
28	9.90	11.8
30	21.0	10.2
32	44.5	8.7
34	92.0	7.3
36	188	6.1
38	360	5.0

If 100 lb of 34.46 wt % HCl solution contacts 5000 ft^3 of dry air at 30°C at 725 mmHg and equilibrium is established at constant temperature and pressure, determine the partial pressure of each component in the gas phase, the volume of the gas phase, the weight of liquid phase, and the composition of the liquid phase.

17 Equilibrium data for isopropyl ether–acetic acid–water at 24°C and 1 atm total pressure are given below.

(*a*) Plot the triangular solubility curve with tie lines (acid at top apex, ether at right apex).

(*b*) Plot an equilibrium (xy) diagram.

(*c*) If a 50 : 50 mixture by weight of acetic acid and water is extracted with an equal weight of isopropyl ether at 24°C and 1 atm, determine the amount and composition of each phase formed at equilibrium.

Equilibrium data, wt %		
Isopropyl ether	Acetic acid	Water
50.3	36.8	13.0
31.05	45.1	23.8
16.7	48.4	34.9
13.25	48.1	38.6
10.85	47.3	41.85
3.5	37.6	58.8
4.6	40.0	55.4
5.12	41.5	53.4
69.0	25.1	6.19
60.2	30.7	9.1
53.4	35.0	11.6
45.2	39.5	15.3
11.1	48.0	40.9
99.37	0.0	0.63
1.2	0.0	98.8

Tie line data, wt % acetic acid	
In isopropyl ether phase	In water phase
9.4	21.9
6.1	16.1
16.75	33.5
30.2	45.3
39.0	48.2

18 Consider the pyridine (C)-chlorobenzene (B)-water (A) data below for equilibrium tie lines. Compute the selectivity of chlorobenzene for pyridine for each tie line and plot it against the concentration of pyridine in water.

Extract			Raffinate		
wt % C	wt % B	wt % A	wt % C	wt % B	wt % A
0	99.95	0.05	0	0.08	99.92
11.05	88.28	0.67	5.02	0.16	94.82
18.95	79.90	1.15	11.05	0.24	88.71
24.10	74.28	1.62	18.90	0.38	80.72
28.69	69.15	2.25	25.50	0.58	73.92
31.55	65.58	2.87	36.10	1.85	62.05
35.05	61.00	3.95	44.95	4.18	50.87
40.60	53.00	6.40	53.20	8.90	37.90
49.0	37.8	13.2	49.0	37.8	13.2

Selectivity is defined as the ratio of pyridine to water in the solvent chlorobenzene extract phase to the ratio of pyridine to water in the water raffinate phase.

19 Consider the $MgSO_4$ system shown in Fig. 8.24.

(*a*) A mixture of 1 mol of $MgSO_4$ and 10 mol of water is made and is heated to 35.7°F at atmospheric pressure. Determine the species formed and the amount of each.

(*b*) If 10 mol of a 20 wt % solution of $MgSO_4$ at 100°F is mixed adiabatically with 10 mol of a 35 wt % solution at 200°F, determine the composition, temperature, and state of the resulting mixture.

(*c*) Explain the triangular areas *deg* and *bfh* on the enthalpy-concentration diagram.

20 For regular solutions, a good correlation for binary mixtures is often given by $\Delta \tilde{H}^M/RT = Ax_1x_2$, where A is a constant. Plot a_1, $\ln \gamma_1$, and $\Delta \bar{G}^E/RT$ vs. x_1 for values of A of 0, 1, 2, and -1. Discuss the meaning of the curves in terms of deviations from ideality. Does partial miscibility occur in any of the cases? (*Hint*: In this case the first and second derivatives of chemical potential with respect to composition are zero, while the third derivative is positive.)

21 (*a*) Consider an equimolar mixture of methane and sulfur dioxide. Calculate the activity coefficient of each component by assuming regular solution.

(*b*) Consider an equimolar mixture of hydrogen, methane, and hydrogen sulfide. Estimate the activity coefficient of each component by assuming regular solution theory.

(*c*) Estimate the solubility parameter for acetone.

22 Using Eq. (8.44) to calculate the Henry's law constant for benzene, estimate the solubility of benzene in water at 100°C and 5 MPa total pressure.

23 Consider a binary mixture of aniline and methylcyclohexane having infinite-dilution activity coefficients at 25°C and 1 atm of 16.1 and 11.4, respectively. Using a van Laar activity coefficient model, determine their mutual solubility.

24 The concentration of benzoic acid dissolved in layers of benzene and water at equilibrium are, in units of grams acid per liter of solution:

In water	2.89	1.95	1.50	0.98	0.79
In benzene	97.0	41.2	25.2	10.5	7.37

Analyze the data and decide whether the benzoic acid is almost completely dissociated in the water or whether the benzoic acid is almost completely associated to dimer in the benzene. Any other conclusions?

25 Data for the equilibrium adsorption of acetone vapor on activated carbon at 30°C are given below:

g acetone adsorbed/g carbon	0	0.1	0.2	0.3	0.35
P acetone, mmHg	0	2	12	42	92

A 35% saturated air acetone vapor at 1 atm and 30°C is sealed in a liter flask with 2 g of activated carbon. What is the equilibrium vapor composition and the final pressure at 30°C? Assume air is not adsorbed and the vapor pressure of acetone is 283 mmHg. How would you determine the heat of adsorption?

CHAPTER

NINE

CHEMICAL EQUILIBRIA

A study of the thermodynamic equilibria of chemical reactions is of great importance to the chemical engineer. To design reactors to produce the myriad of industrial chemicals, petrochemicals, and plastics and polymers, knowledge of the thermodynamics of the reactions is absolutely essential. The maximum extent of a certain reaction is governed by thermodynamic equilibrium. Maximum conversion of the reactants to products is calculable for any reaction at a given temperature, pressure, and feed composition according to the laws of thermodynamics. It is impossible at any given set of conditions to obtain a greater conversion than that predicted by chemical equilibrium calculations. Thus such calculations serve as a screening device to determine if a certain reaction has any potential industrial use. If calculations show that the maximum possible conversion is 5 percent it is hopeless to attempt to attain 20 percent conversion at that given set of conditions.

Certain chemical reactors are operated as equilibrium reactors. However, most reactors are operated such that the residence time is not sufficient to reach thermodynamic equilibrium and the reactions are rate-controlled. These reactors are designed based on the principles of kinetics and reaction engineering rather than equilibrium principles, although, as stated earlier, such calculations are still necessary to set a range of operating conditions over which reasonable conversions can be attained.

The skills of determining equilibrium conversions and yields for a variety of operating conditions and types of reactions is the basic thrust of this chapter. The criteria for equilibrium, calculation of equilibrium constants, and the effects of variables on equilibrium constants and, hence, yields and conversions is the first topic of the chapter. The place of the phase rule is reviewed. Single- and multiple-reaction systems for vapor-phase, liquid-phase, mixed-phase, and heterogeneous systems are discussed. Solution methods for multireaction equilibria are discussed. Finally some more-complex reactions including electrochemical reactions are illustrated.

9.1 THE EQUILIBRIUM STATE

The basic criterion for equilibrium in a chemically reacting system is identical to that discussed in Chap. 8 for equilibrium among phases. To attain equilibrium, the Gibbs free energy is minimized for a given temperature and pressure. These criteria are given for physical equilibrium as Eqs. (8.1) to (8.5), which show that the criteria can be written as equality of chemical potentials for each component in each phase, which translates to equality of the fugacity of each component in each phase.

For chemical reaction equilibria, the concept of the reaction coordinate is useful. The *reaction coordinate* ϵ is defined as the amount, extent, or degree to which a reaction has taken place and has the advantage that the change in reaction coordinate $d\epsilon$ is the same for each compound participating in the reaction. This allows dn which varies for each compound to be replaced by the product of the reaction coordinate and the stoichiometric coefficient ν.

$$dn_i = \nu_i \, d\epsilon \tag{9.1}$$

Consider a general reaction

$$pP + qQ = rR = sS$$

The increase in number of moles of compounds is $-dn_P$, $-dn_Q$, dn_R, and dn_S. If a sign convention is adopted for the stoichiometric coefficients by using positive values for products and negative values for reactants, $\nu_P = -p, \nu_Q = -q, \nu_R = r$, and $\nu_S = s$.

If the reaction is now written $pP + qQ + rR + sS = 0$, ratios of the change in number of moles between any pair of compounds can be written

$$\frac{dn_P}{dn_Q} = \frac{\nu_P}{\nu_Q} \qquad \frac{dn_s}{dn_R} = \frac{\nu_s}{\nu_R} \qquad \cdots$$

Thus it is apparent that

$$\frac{dn_P}{\nu_p} = \frac{dn_Q}{\nu_Q} = \frac{dn_R}{\nu_R} = \frac{dn_s}{\nu_s} = d\epsilon \tag{9.2}$$

Thus Eq. (9.1) holds for any general reaction.

If the concept of reaction coordinate is used, Eq. (8.2) can be written for a reacting system as

$$\sum_i \mu_i \nu_i \, d\epsilon = 0 \tag{9.3}$$

Since $d\epsilon$ is independent of i, the criterion for equilibrium reduces to

$$\sum_i \mu_i \nu_i = 0 \tag{9.4}$$

As shown by Eq. (7.67), at constant T and P,

$$\Delta G_i = RT \ln \frac{\bar{f}_i}{f_i^\circ} = RT \ln \bar{a}_i$$

where $\Delta G_i = G_i - G_i^\circ$

G_i° = standard-state free energy taken at temperature of the system and a constant pressure, usually taken as 1 atm

Since for chemical equilibrium to exist $\Sigma \nu_i G_i = \Delta G = 0$, then

$$\sum \nu_i G_i - \sum \nu_i G_i^\circ = \sum_i \nu_i RT \ln \bar{a}_i$$

If $\Sigma \nu_i G_i^\circ = \Delta G^\circ$, called the *standard free-energy change*, then

$$\Delta G^\circ = -RT \sum_i \ln \bar{a}_i^{\nu_i}$$

or

$$\Delta G^\circ = -RT \ln \Pi \bar{a}_i^{\nu_i} \tag{9.5}$$

This equation then relates the standard free-energy change to the product of the activities of the compounds taking part in the reactions. Alternatively, this expression could be written in terms of the fugacity ratios $\bar{f}_i/f_i^\circ$, which produces the same resulting equation in terms of fugacities:

$$\Delta G^\circ = -RT \ln \Pi \left(\frac{\bar{f}_i}{f_i^\circ} \right)^{\nu_i} \tag{9.6}$$

Which of these formulations will be used depends on the standard states chosen. The states chosen are not critical as long as the pure-component fugacity, activity, and standard Gibbs free energy for any particular compound are based on the same standard state.

9.1.1 Standard-State Free Energy and Its Determination

The standard free energy as defined is at a fixed pressure and thus depends only on temperature and is not a function of composition. For any given reaction,

$$\Delta G^\circ = \sum_i \nu_i G_i^\circ \tag{9.7}$$

which implies the same additivity rules as apply to calculation of heats of reaction in Chap. 3,

$$\Delta H_r^\circ = \sum_{\text{Products}} n_p \Delta H_f - \sum_{\text{Reactants}} n_R \Delta H_r = \sum_i \nu_i H_i^\circ$$

and entropies of reaction in Chap. 4,

$$\Delta S_r^\circ = \sum_{\text{Products}} n_P S_P - \sum_{\text{Reactants}} n_R S_R = \sum_i \nu_i S_i^\circ$$

Here ΔG_f° might be called the *standard free energy of formation of compound i*, and an alternative form of Eq. (9.7) would be

$$\Delta G^\circ = \underset{\text{Products}}{\sum n_P \Delta G_f{}^\circ} - \underset{\text{Reactants}}{\sum n_R \Delta G_f{}^\circ} \tag{9.8}$$

Tabulations are sometimes made for standard free energies of formation, but since the standard free energy is a derived property, it is usually derived by application of the defining equation (5.3) for free energy, evaluated isothermally at the standard-state pressure. Appendix A lists values for standard free energies as well as for heats of formation and entropies.

$$\Delta G^\circ = \Delta H^\circ - T\Delta S^\circ \tag{9.9}$$

This equation can often be simplified by application of the relationship from Chap. 5 that

$$S = -\left(\frac{\partial G}{\partial T}\right)_P \qquad \text{or} \qquad \Delta S^\circ = -\frac{d(\Delta G^\circ)}{dT}$$

Combination of this relation with Eq. (9.9) yields

$$\Delta H^\circ = -RT^2 \frac{d(\Delta G^\circ/RT)}{dT} \tag{9.10}$$

This equation could then be evaluated to determine ΔG°. Another common method of tabulating standard free energies is by the free-energy function described in Chap. 4.

Thus, standard free energies can be estimated in various ways: (1) by Eq. (9.9), (2) by the tabulated values of free energies of formation or free-energy functions, Eq. (4.85), or (3) by calculation, using Eq. (9.10) if at least one value of ΔG° is known. In addition, equilibrium measurements allow calculation of ΔG° from experimental data by Eq. (9.5) or (9.6).

Example 9.1 Consider the low-pressure, gas-phase dehydrogenation of 1-butene to 1, 3 butadiene. Standard free-energy changes are as follows:

	ΔG_f°, kJ/kmol	
	At 700°K	At 800 K
1-butene (B)	178,900	207,037
1, 3-butadiene (BD)	211,852	228,097

(*a*) Determine the standard free-energy change at 700 K by using the data given.

Using Eq. (9.8) and noting that $\Delta G_f^\circ(H_2) = 0$,

$$\Delta G^\circ = \Delta G_f^\circ(\text{BD}) - \Delta G_f^\circ(\text{B})$$

Since $C_4H_8 = C_4H_6 + H_2$,

$$\Delta G^\circ = 211{,}852 - 178{,}900 = 32{,}952 \text{ kJ/kmol}$$

(*b*) Determine the standard free-energy change at 700 K by using heat of formation, enthalpy, and entropy data.

Using Eq. (9.9), $\Delta G^\circ = \Delta H^\circ - T\Delta S^\circ$,

$$\Delta H^\circ_{700} = \Delta H^\circ_{298} + \Delta C^\circ_P(700 - 298)$$

$$\Delta H^\circ_{298} = \Delta H^\circ_f(\text{BD}) - \Delta H^\circ_f(\text{B})$$

Using App. A,

$$\Delta H^\circ_{298} = 110{,}200 - (-126) = 110{,}326 \text{ kJ/kmol}$$

$$\Delta C^\circ_P = C^\circ_P(\text{BD}) - C^\circ_P(\text{B}) + C^\circ_P(\text{H}_2)$$

Using Table 3.1 at 200°C (673 K),

$$\Delta C^\circ_P = (2.122)(54.1) - (2.213)(56.1) + (14.499)(2)$$

$$= 19.65 \text{ kJ/(kmol} \cdot \text{K)}$$

Therefore

$$\Delta H^\circ_{700} = 110{,}326 + (19.65)(700 - 298) = 118{,}225 \text{ kJ/kmol}$$

Using Eqs. (4.13) and (4.15),

$$\Delta S^\circ_{700} = \Delta S^\circ_{298} + \Delta C^\circ_P \ln \tfrac{700}{298}$$

From App. A,

$$\Delta S^\circ_{298} = S^\circ(\text{BD}) + S^\circ(\text{H}_2) - S^\circ(\text{B})$$

$$= 278.7 + 130.6 - 305.6 = 103.7 \text{ kJ/(kmol} \cdot \text{K)}$$

$$\Delta S^\circ_{700} = 103.7 + 19.65 \ln \tfrac{700}{298} = 120.5 \text{ kJ/(kmol} \cdot \text{K)}$$

Therefore

$$\Delta G^\circ_{700} = \Delta H^\circ_{700} - T\Delta S^\circ_{700} = 118{,}225 - (700)(120.5)$$

$$= 33{,}875 \text{ kJ/kmol}$$

[Note that the difference from part (*a*) is small.]

Exercise 9.A Determine the standard free-energy change for the oxidation of sulfur dioxide to sulfur trioxide at 25, 500, and 700°C. Comment on the results and compare the trend noted with that of Example 9.1.

$$C_P(\text{SO}_3) = 33{,}720 + 53{,}740 \exp \frac{-1582.}{T^{1.269}}$$

where $C_P = \text{J/(kmol} \cdot \text{K)}$
$T = \text{K}$

9.1.2 The Equilibrium Constant

At constant temperature and pressure, Eqs. (9.5) and (9.6) describe the condition for chemical equilibrium. The equilibrium constant is defined as

$$K = \prod_i \bar{a}_i^{\nu_i} = \prod_i \left(\frac{\bar{f}_i}{f_i^\circ} \right)^{\nu_i} \tag{9.11}$$

Thus Eq. (9.5) becomes

$$\Delta G^\circ_{T,P} = -RT \ln K \tag{9.12}$$

In general form, the equilibrium constant then is, for the general reaction $pP + qQ = rR + sS$,

$$K = \frac{\bar{a}_R^r \bar{a}_S^s}{\bar{a}_P^p \bar{a}_Q^q} \tag{9.13}$$

Providing the standard-state fugacities of the pure materials can be equated to 1, as is the case for gases,

$$K = \frac{\bar{f}_R^r \bar{f}_S^s}{\bar{f}_P^p \bar{f}_Q^q} \tag{9.14}$$

Using these definitions, the expression for equilibrium constant can be simplified according to the phase condition of the reaction, the standard states chosen, and whether the solution is ideal or nonideal. Such simplification will be done in Secs. 9.2 and 9.4 of this chapter.

Inspection of Eq. (9.13) or (9.14) shows that for *net* reaction to proceed toward the right, the equilibrium constant must be greater than 1, which requires ΔG° in Eq. (9.12) to be negative. If a mixture of pure P and Q is reacted, a small amount of R and S will form to satisfy the equilibrium constant, no matter how small the constant. Conversely, *net* reaction will proceed to the left if the equilibrium constant is less than 1, requiring a positive ΔG°. The criteria that ΔG° be negative for spontaneous reaction in the direction the reaction is written is useful in preliminary thermodynamic analysis of any chemical reaction.

Equilibrium constants can be calculated (1) from Eq. (9.12) if ΔG° is first calculated, (2) from experimental measurement of the parameters in Eq. (9.13) or (9.14), or (3) by rearrangement of Eq. (9.10).

$$\int d(\ln K) = \int \frac{\Delta H^\circ}{RT^2}\, dT \tag{9.15}$$

where ΔH° can be calculated at any temperature by the methods of the next section. The first of these methods is the most utilized since no experimental equilibrium data are necessary and tabulations and established calculational methods can be utilized.

Example 9.2 Calculate the equilibrium constant for the reaction of Example 9.1 at 25°C and 700 K.

SOLUTION Using Eq. (9.12),

$$\Delta G^\circ = -RT \ln K$$

From App. A,

$$\Delta G^\circ_{298} = 150{,}670 - 71{,}500 = 79{,}170 \text{ kJ/kmol}$$

At 25°C,

$$79{,}170 = -(8.314)(298)\ln K$$
$$K = 1.32(10^{-14})$$

From Example 9.1, at 700 K,

$$33{,}875 = -(8.314)(700)\ln K$$
$$K = 0.00297$$

Exercise 9.B Determine the equilibrium constant for the reaction of Exercise 9.A at 25, 500, and 700°C. Comment on the results and compare the trend noted with that of Example 9.2.

9.1.3 Effect of Temperature and Pressure on ΔG° and K

The effect of temperature on the standard free energy at constant pressure is given by Eq. (9.10) and was rewritten in terms of the equilibrium constant in Eq. (9.15).

For the case of a small change in temperature, ΔH° will not change significantly; thus Eq. (9.15) can be integrated:

$$\int_{K_1}^{K_2} d(\ln K) = \frac{\Delta H^\circ}{R}\int_{T_1}^{T_2}\frac{dT}{T^2}$$

$$\ln\frac{K_2}{K_1} = -\frac{\Delta H^\circ}{R}\left(\frac{1}{T_2}-\frac{1}{T_1}\right) \tag{9.16}$$

Thus this equation is the same form as that of the Clausius-Clapeyron equation for vapor pressure.

Both Eqs. (9.15) and (9.16) indicate that the equilibrium constant will decrease with increasing temperature if ΔH° is negative, i.e., an exothermic reaction. Conversely, the equilibrium constant will increase with increasing temperature if ΔH° is positive, i.e., an endothermic reaction. This observation is important in establishing the practical temperature level at which a reaction will be carried out.

Often the heat of reaction cannot be assumed to be constant over the temperature range being considered. Thus an expression for the heat of reaction can be developed if the heat of reaction at one temperature as well as polynomial functions for the heat capacities of all reactants and products are known:

$$\Delta H_T^\circ = \Delta H_{T_0}^\circ = \int_{T_0}^{T}\Delta C_P^*\, dT \tag{9.17}$$

where $\Delta H_{T_0}^\circ$ = base enthalpy change at T_0 which can be determined by solution of Eq. (9.17) for the known value of ΔH_T°.

If, for example, $C_P^* = A + BT + CT^2$,

$$\Delta A = \underset{\text{Products}}{\sum n_i A_i} - \underset{\text{Reactants}}{\sum n_i A_i}$$

and similarly for ΔB, ΔC. Then

$$\Delta H_T^\circ = \Delta H_0 + \Delta A T + \Delta B \frac{T^2}{2} + \Delta C \frac{T^3}{3} \tag{9.18}$$

where $$\Delta H_0 = \Delta H_{T_0}^\circ - \Delta A T_0 - \Delta B \frac{T_0^2}{2} - \Delta C \frac{T_0^3}{3}$$

which is calculable given one point of enthalpy data. Substituting into Eq. (9.15) and solving,

$$\ln K = -\frac{\Delta H_0}{RT} + \frac{\Delta A}{R} \ln T + \frac{\Delta B}{2R} T + \frac{\Delta C}{6R} T^3 + \text{constant} \tag{9.19}$$

The constant can be obtained by solving (9.19) for a known value of K. A similar expression can be written for ΔG° by multiplying Eq. (9.19) by $-RT$.

The effect of pressure on an equilibrium constant at constant temperature can be determined by remembering that the change in free energy due to pressure is $dG = V\,dP$. Since free energy is a state function, a process may be envisioned whereby the reactants at a pressure P_1 can be reduced to any arbitrary pressure P and reacted to form products, after which the products can be raised to some pressure P_2. The change in free energy for reaction is then calculable as

$$\Delta G = \underset{\text{Products}}{\sum \int_P^{P_2} V\,dP} + \underset{\text{Reactants}}{\sum \int_{P_1}^{P} V\,dP}$$

No matter which pressure P is taken, the same result will occur and will be equal to the result if P is taken as the standard-state pressure. Thus all equilibrium constants obtained are equal, certifying that equilibrium constants are independent of the total pressure of the reaction vessel. Thus Eq. (9.12) is truly

$$\Delta G_T^\circ = -RT \ln K$$

and is applicable at any total pressure. It should be noted that although K is independent of total pressure, the composition of the reacting mixture at equilibrium is not usually independent of total pressure.

Example 9.3 The equilibrium constant for the dissociation of water at 2000°C is 0.0018 atm$^{1/2}$. What is the equilibrium constant at 1000°C?

SOLUTION

$$H_2O(g) = H_2(g) + \tfrac{1}{2}O_2(g)$$

Using Eq. (9.15),

$$\int d(\ln K) = \int \frac{\Delta H^\circ}{RT^2}\, dT$$

Using Table 3.2, and assuming the equations cover the entire range of temperature,

$$H^* = A + BT + CT^2 + DT^3 + ET^4 + FT^5 \tag{3.13}$$

	A	B	$C(10^3)$	$D(10^6)$	$E(10^9)$	$F(10^{13})$
H_2	28.672	13.396	2.960	−3.981	2.662	−6.100
O_2	−2.284	0.952	−0.281	0.655	−0.452	1.088
H_2O	5.730	1.915	−0.396	0.876	−0.495	1.039

Since units are kJ/(kg · K), molecular weights must be used to convert to moles.

$$\begin{aligned}\Delta A &= 2(28.672) + \tfrac{1}{2}(32)(-2.284) - (18)(-5.730)\\ &= 123.94\end{aligned}$$

Similarly,

$$\begin{aligned}\Delta B &= 7.554\\ \Delta C &= 8.552(10^{-3})\\ \Delta D &= -13.25(10^{-6})\\ \Delta E &= 7.002(10^{-9})\\ \Delta F &= 13.494(10^{-13})\end{aligned}$$

Therefore

$$\begin{aligned}\Delta H^\circ = {} & 123.94 + 7.554T + 8.552 \times 10^{-3}T^2 - 13.25 \times 10^{-6}T^3\\ & +7.002 \times 10^{-9}T^4 - 13.494 \times 10^{-13}T^5\end{aligned}$$

$$\int_{K_1}^{K_2} d(\ln K) = \frac{1}{R}\int_{1273}^{2273}\left(\frac{123.94}{T^2} + \frac{7.554}{T} + 8.552 \times 10^{-3} - 13.25 \times 10^{-6}T + 7.002 \times 10^{-9}T^2 - 13.494 \times 10^{-13}T^3\right) dT$$

$$\begin{aligned}\ln\frac{K_2}{K_1} = {} & \frac{1}{R}\left[123.94\left(\frac{1}{1273} - \frac{1}{2273}\right) + 7.554\ln\frac{2273}{1273} + 8.552 \times 10^{-3}(1000)\right.\\ & - \frac{13.25 \times 10^{-6}}{2}(2273^2 - 1273^2)\\ & \left. + \frac{7.002 \times 10^{-9}}{3}(2273^3 - 1273^3) - \frac{13.494 \times 10^{-13}}{4}(2273^4 - 1273^4)\right]\end{aligned}$$

$$\begin{aligned}\ln\frac{0.0018}{K_1} &= \frac{1}{8.3143}(0.04263 + 4.3792 + 8.552 - 23.492 + 22.594 - 8.119)\\ &= \frac{3.597}{8.3143} = 0.4759\end{aligned}$$

$$\frac{0.0018}{K_1} = 1.6055$$

$$K_1 = 0.00112$$

Exercise 9.C Derive equations similar to Eqs. (9.18) and (9.19) for ΔH_T° and K if the polynomials of Eq. (3.12) hold. Derive an additional equation for ΔG°.

9.2 GAS-PHASE REACTION EQUILIBRIA

Consider the general gas-phase equilibrium reaction

$$pP + qQ = rR + sS$$

The general equilibrium constant for this reaction as given by Eq. (9.11) is

$$K = \prod_i \left(\frac{\bar{f}_i}{f_i^{\circ}} \right)^{\nu_i} \tag{9.11}$$

where ν_i = stoichiometric coefficients, + for products, − for reactants. As stated earlier, the standard-state gas-phase fugacities can be taken equal to unity, reducing the equation to

$$K = \prod_i \bar{f}_i^{\nu_i} \tag{9.20}$$

which can also be written as Eq. (9.14):

$$K = \frac{\bar{f}_R^r \bar{f}_S^s}{\bar{f}_P^p \bar{f}_Q^q} \tag{9.14}$$

The fugacities of the various components in the mixture are functions of temperature, pressure, and composition. The equilibrium constants are only functions of temperature.

In terms of fugacity coefficients,

$$\bar{f}_i = \bar{\phi}_i y_i P_T \tag{9.21}$$

where $\bar{\phi}_i$ = fugacity coefficient of i in mixture
y_i = mole fraction of i in mixture
P_T = total pressure, atm

Thus, Eq. (9.20) can be written

$$K = P_T^{\nu} \prod_i (y_i \bar{\phi}_i)^{\nu_i} \tag{9.22}$$

where $$\nu = \sum_i \nu_i$$

This general equation can be solved provided the fugacity coefficients can be estimated. The fugacity coefficients are estimable for the components in mixtures by the methods of Sec. 7.2.1. However, this would normally require trial-and-error solution since the mole fractions are generally unknown.

Most often for gaseous mixtures, *ideal solutions* can be assumed and $\bar{\phi}_i$ is then the fugacity coefficient of pure i at the temperature and pressure of the system and is easily evaluated from the methods of Secs. 7.2.1 and 7.2.2.

$$K = P_T^{\nu} \prod_i (y_i \phi_i)^{\nu_i} \tag{9.23}$$

If in addition the gaseous mixture can be assumed to be an ideal gas, as occurs at high temperature or low pressure, the fugacity coefficients all approach unity and

$$K = K_P = P_T^{\nu} \prod_i (y_i)^{\nu_i} \tag{9.24}$$

This equation can then be easily used together with material balance equations to solve for the composition of the mixture. In Eq. (9.24), K is equated to K_P, which indicates that the equilibrium is based on ideal gas.

An alternative formulation of Eq. (9.24) for the general-example reaction is

$$K_P = \frac{y_R^r y_S^s}{y_P^p y_Q^q} P_T^{r+s-p-q}$$

$$= K_y P_T^{r+s-p-q} \tag{9.25}$$

This formulation defines K_y, which is a shorthand method of writing the mole fraction product. Similarly, Eq. (9.23) can be written

$$K = K_\phi K_y P_T^\nu \tag{9.26}$$

It is important to determine the effects of varying reaction conditions on the equilibrium conversions of chemical reactions. This can be approached in two ways: use of reaction coordinates, to be explored in a later problem, or general application of Eq. (9.26) when combined with a material balance on the system.

Again, consider the general gas-phase reaction $pP + qQ = rR + sS$ occurring where nonreacting materials in the amount of n_n may also be present. The total moles of material in the reactor, n_T, are then

$$n_T = n_P + n_Q + n_R + n_S + n_n$$

Substituting into Eq. (9.26),

$$K = K_\phi = P_T^\nu \frac{(n_R/n_T)^r (n_S/n_T)^s}{(n_P/n_T)^p (n_Q/n_T)^q}$$

Rearranging,

$$\frac{n_R^r n_S^s}{n_P^p n_Q^q} = \frac{K n_T^\nu}{K_\phi P_T^\nu} \tag{9.27}$$

Consider this equation with respect to the equilibrium conversion, i.e., changes that will increase the right side of Eq. (9.27) and thus increase the ratio of R and S to the ratio of P and Q.

Temperature. A change in temperature will primarily affect the equilibrium constant K, as K_ϕ is a weak function of temperature and also often approaches unity. Thus, as discussed in Sec. 9.1.3, the equilibrium constant will increase if temperature is increased for an endothermic reaction and will decrease if temperature is increased for an exothermic reaction. Thus the equilibrium conversion follows a similar pattern.

Pressure. Pressure will primarily affect the term P_T^ν although K_ϕ varies with pressure and, as discussed earlier, such effects can be calculated. As equilibrium conversion is inversely proportional to P_T^ν, the effect on conversion will depend on the value of ν whether positive, zero, or negative. If there is no change in the number of moles of reactants and products, $\nu = 0$, total pressure will have no effect on the equilibrium conversion. If the number of moles of products is greater than the number of moles of reactants, the equilibrium conversion will decrease with rise in pressure. Conversely, a decrease in the number of moles upon reaction will cause an increase in the equilibrium conversion. This latter statement is the well-known *principle of Le Chatelier*.

Addition of nonreacting material. The addition of a nonreacting material will only increase n_T. Thus the effect on n_T^{ν} is the effect of inert addition on the conversion and is opposite to that of an increase in pressure. If $\nu = 0$, there is no effect. If ν is positive, the conversion will be increased. If ν is negative, the conversion will be decreased. This principle would often be important as air rather than oxygen is used as an oxidant for many combustion reactions.

Products in feed. If products are present in the feed, the numerator of the left side of Eq. (9.27) requires less conversion to satisfy the equilibrium constant and thus decreases the equilibrium conversion.

Reactant excess in feed. If one reactant is in stoichiometric excess, the denominator on the left side of Eq. (9.27) is affected and will force additional conversion of the other limiting reactant, increasing the equilibrium conversion.

For calculations of equilibrium conversion to be made for a reaction at various conditions of temperature and pressure such as often occurs in multistage adiabatic-equilibrium reactors (for example, SO_2 catalytic oxidation), charts are often developed by plotting equilibrium conversion vs. temperature at constant-feed composition with total pressure as a parameter in the possible range of applicability. By calculating the equilibrium constant at a given conversion and pressure by the defining equation, after which the relation between K and $\Delta G°$ is used to determine the temperature of the point, a chart such as Fig. 9.1 is then developed.

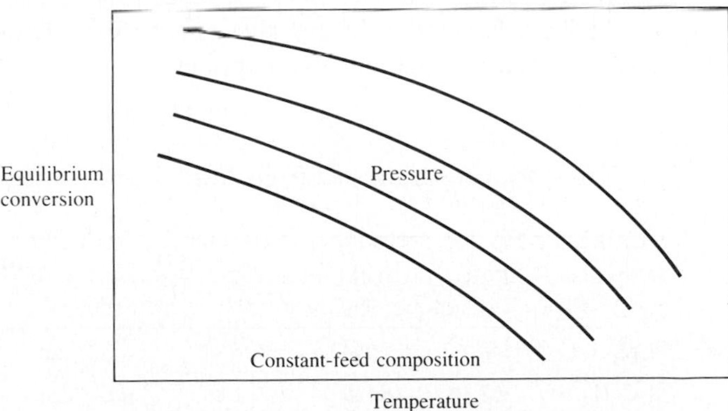

Figure 9.1 Schematic of equilibrium conversion vs. temperature plot.

Example 9.4 Gaseous nitrogen and acetylene will combine stoichiometrically to form 2 mol of gaseous HCN in a reactor operating at 300°C where the standard free-energy change is 30,050 kJ/kmol. Assume for HCN that $T_c = 457$ K, $P_c = 4.965$ MPa, and that Z_c is close to 0.27. Estimate the mole fraction of HCN in the product at equilibrium if the reactor total pressure is at (*a*) atmospheric pressure and (*b*) 20 MPa.

$$N_2(g) + C_2H_2(g) = 2HCN(g)$$

SOLUTION

(*a*) At atmospheric pressure the mixture can be assumed to be ideal with all fugacity coefficients approaching unity. Thus Eqs. (9.24) and (9.25) apply:

$$K = K_P = p_T{}^{\nu} \prod_i (y_i)^{\nu_i}$$

As $\nu = 0$

$$K_P = K_y = \frac{y_{HCN}^2}{y_{N_2} y_{C_2H_2}}$$

Since $\Delta G° = -RT \ln K$,

$$\ln K = \frac{-30{,}050}{(8.314)(573)} = -6.308$$

$$K = 1.822 \times 10^{-3} = K_p = K_y$$

Taking a basis of a feed of 2 mol containing 1 mol N_2 and 1 mol C_2H_2 and assuming a fractional conversion of each of x, the equilibrium mixture contains: $1 - x$ mol N_2, $1 - x$ mol C_2H_2, and $2x$ mol HCN for a total of 2 mol of product. Thus

$$1.822 \times 10^{-3} = \frac{(2x)^2}{(1-x)^2}$$

Solving directly

$$x = 0.0209$$

Thus, the product contains $2x/2$, or 0.0209 mol fraction HCN.

(*b*) At 20 MPa total pressure, the assumption of unit fugacity coefficients cannot be made. Equation (9.23) applies:

$$K = P_T^{\nu} \prod_i (y_i \phi_i)^{\nu_i}$$

Using critical parameters from App. A and Lydersen's method (Fig. 7.1):

	T_c, K	P_c, MPa	$T_r = \frac{573}{T_c}$	$p_r = \frac{20}{P_c}$	ϕ
N_2	126.2	3.399	4.54	5.88	0.980
C_2H_2	308.3	6.139	1.86	3.26	0.915
HCN	457	4.965	1.25	4.03	0.555

As $\nu = 0$

$$1.882 \times 10^{-3} = \frac{(2x)^2(0.555)^2}{(1-x)(0.980)(1-x)(0.915)}$$

$$1.326 \times 10^{-3} = \frac{x^2}{(1-x)^2}$$

Solving directly,

$$x = 0.0351$$

Thus, the product contains 0.0351 mol fraction HCN. Note that the nonideality changes the composition by 75 percent over the result if assumed ideal.

(If Pitzer's method is used for estimating ϕ, the result is slightly higher, with $x = 0.038$.)

Exercise 9.D Pure ethanol is to be decomposed to ethylene and water in the gas phase at 125°C, where the standard free-energy change is -4530 kJ/kmol and the heat of reaction is $+46{,}000$ kJ/kmol.

Quantitatively, show (1) whether the reaction should be carried out at 0.1 or 0.5 MPa pressure, (2) whether the ethanol should be diluted with nitrogen, (3) whether the reaction temperature should be increased to 200°C regardless of the pressure, and (4) how to estimate the equilibrium constant at 200°C given only the data in the problem statement.

For adiabatic reactions, coupling of the heat transfer equation relating heat of reaction to sensible heat transfer change in the gas to this plot will allow simple graphical solution of a complex-reaction calculation. Such coupling would lead to plots for a constant pressure and feed composition as shown in Fig. 9.2.

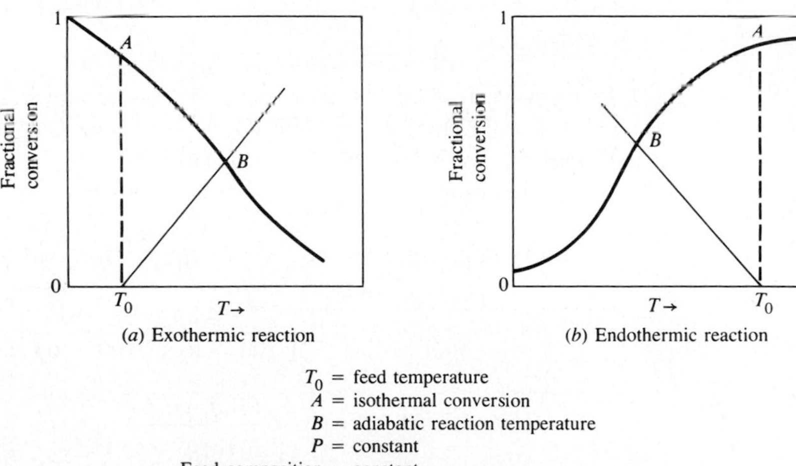

Figure 9.2 Conversion-temperature plots for adiabatic reactions.

Such a method is quite useful for simple calculation of adiabatic-reaction temperature for a constant-pressure system.

Many gas-phase reaction solutions require trial-and-error solutions since the fractional conversion will often appear in the equilibrium equation raised to several different fractional powers.

Extension of the simple solutions to multiple gas-phase reactions which require simultaneous solution of several equilibrium equations will be deferred to Sec. 9.5.

Example 9.5 Consider the oxidation of sulfur dioxide to sulfur trioxide. The equilibrium constant can be calculated from

$$K = \exp\left[\frac{-1}{R}\left(\frac{-22{,}630}{T} + 5.281 \ln T - 0.959 \times 10^{-2} T + 0.280 \times 10^{-5} T^2 - 7.68\right)\right]$$

where $R = 1.987$ cal/(mol · K)
$T =$ K
$K =$ atm$^{-1/2}$

A feed containing 12% SO_2, 9% O_2, and 79% N_2 is reacted at atmospheric pressure. Calculate the fractional conversion of SO_2 if reaction takes place (*a*) isothermally at 749 K and (*b*) adiabatically with a feed temperature of 400°C.

$$SO_2(g) + \tfrac{1}{2}O_2(g) = SO_3(g)$$

SOLUTION

(*a*) At atmospheric pressure, $K = K_P$. Using the equation above, $K = K_P = 74.0$ atm$^{-1/2}$. Using Eq. (9.24) with $\nu = 0$,

$$K_P = \frac{y_{SO_3}}{y_{SO_2} y_{O_2}^{1/2}}$$

Assuming a feed of 100 mol and a fractional conversion f, the equilibrium mixture will contain $12f$ mol SO_3, $9 - 6f$ mol O_2, $12 - 12f$ mol SO_2, and 79 mol N_2, for a total of $100 - 6f$ mol.

Substituting,

$$74.0 = \frac{\dfrac{12f}{100 - 6f}}{\left(\dfrac{12 - 12f}{100 - 6f}\right)\left(\dfrac{9 - 6f}{100 - 6f}\right)^{1/2}}$$

$$= \frac{f(100 - 6f)^{1/2}}{(1 - f)(9 - 6f)^{1/2}}$$

Solving by trial and error,

$$f = 0.934$$

(b) For an adiabatic reaction, Sec. 3.6.3 and Eq. (3.43) applies as an enthalpy balance:

$$\sum H_{\text{Re}} = \sum H_{\text{Pr}} + \Delta H_{r,T_0}$$

Since no latent heats are involved if the base temperature T_0 is taken as 298 K,

$$\sum H_{\text{Re}} = \sum n_{\text{Re}} \int_{T_0}^{T_i} (C_P)_{\text{Re}}\, dT \qquad \sum H_{\text{Pr}} \int_{T_0}^{T} (C_P)_{\text{Pr}}\, dT$$

$$T_i = 400°\text{C} = 673 \text{ K}$$

Using standard values of heats of formation and heat capacities as given in Chap. 3,

$$\sum H_{\text{Re}} = 12 \int_{298}^{673} (C_P)_{SO_2}\, dT + 9 \int_{298}^{673} (C_P)_{O_2}\, dT + 79 \int_{298}^{673} (C_P)_{N_2}\, dT$$

$$\sum H_{\text{Pr}} = 12f \int_{298}^{T} (C_P)_{SO_3}\, dT + 12(1-f) \int_{298}^{T} (C_P)_{SO_2}\, dT$$

$$+ (9 - 6f) \int_{298}^{T} (C_P)_{O_2}\, dT + 79 \int_{298}^{T} (C_P)_{N_2}\, dT$$

$$\Delta H_{r,T_0} = 12f \Delta H^{\circ}_{r,298}$$

Solving Eq. (3.43) and the expression for the equilibrium constant simultaneously for T and f by trial and error, $T \approx 870$ K and $f = 0.69$, a considerably lower conversion than the isothermal case in part (a).

An alternative graphical solution would be to plot the two equations and determine their intersection. This method is especially used when several adiabatic reactors are used in series.

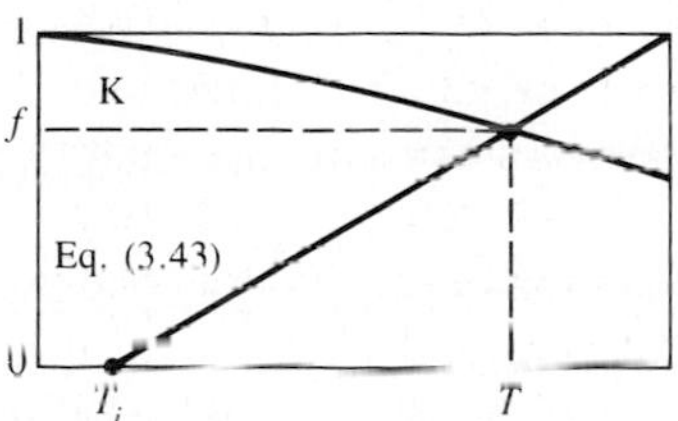

9.3 PHASE-RULE CRITERIA FOR CHEMICAL EQUILIBRIA

The phase rule was developed in Sec. 8.1 as Eq. (8.7) for systems in which only chemical potential, temperature, and pressure were significant variables:

$$V = C + 2 - \phi \tag{8.7}$$

where C = minimum number of independent chemical compounds required to define system

ϕ = number of phases

V = degrees of freedom equal to number of variables that can be varied independently: temperature, pressure, and compositions

The phase rule can be modified in a form very useful for situations in which chemical reactions occur. Assuming ϕ phases and C *total* chemical species, the variables are temperature, pressure, and $C - 1$ mol fractions in each phase for a total of $2 + \phi(C - 1)$ variables. There will be $C(\phi - 1)$ phase-equilibrium relationships, as is the case for nonreacting systems. However, Eq. (9.4) provides an additional equation for each independent chemical reaction that must be satisfied at equilibrium. For R independent equilibrium reactions, there then must be $C(\phi - 1) + R$ equations which independently relate phase-rule variables. The degrees of freedom then must be the difference between the number of variables and the number of equations which results in a phase-rule statement for such systems:

$$\begin{aligned} V &= [2 + \phi(C - 1)] - [C(\phi - 1) + R] \\ &= 2 + C - \phi - R \end{aligned} \tag{9.28}$$

To determine the R independent chemical reactions, Duhem's theory gives a systematic method which makes certain that the correct number of reactions is used. Such a method gives a set of reactions which describes the system at equilibrium although the set may not be unique, and according to the method used for reduction of equations, a completely equivalent set of equations may be developed. The method consists of three steps:

1. Write an equation for the formation of each chemical compound present in the system from the elements.
2. Algebraically combine the equations to eliminate any element not also present in the system by choosing one element to be eliminated and combine one equation containing this element with every other equation containing the same element.
3. Repeat step 2 with the new set of equations to eliminate another element. Repeat until all such elements are eliminated. Note that sometimes two elements may be eliminated in the same procedure.

The resulting set of equations is an independent set of equilibrium equations defining the system.

If a special constraint such as a feed composition is given, the number of degrees of freedom must be reduced for each special constraint SC. Thus the phase rule becomes

$$V = 2 + C - \phi - R - SC \tag{9.29}$$

Example 9.6 Determine the number of degrees of freedom for each of the following systems: (*a*) A system formed by decomposing pure solid $CaCO_3$ into a vacuum.

$$CaCO_3(s) = CaO(s) + CO_2(g)$$

Using Eq. (9.28),

$$V = 2 + C - \phi - R$$

where

$$C = 3$$
$$\phi = 3 \text{ (2 solid, 1 gas)}$$
$$R = 1 \text{ equation}$$

Therefore

$$V = 2 + 3 - 3 - 1 = 1$$

Thus, if temperature is fixed, pressure is fixed, or vice versa.

(*b*) A system formed by decomposing pure methanol vapor to only CO and H_2 into a vacuum.

$$CH_3OH(g) = CO(g) + 2H_2(g)$$

Using Eq. (9.29),

$$V = 2 + C - \phi - R - SC$$

where

$$C = 3$$
$$\phi = 1 \text{ (gas)}$$
$$R = 1 \text{ equation}$$
$$SC = 1 \text{ (that is, the ratio of } CO/H_2 \text{ in the product)}$$

Therefore

$$V = 2 + 3 - 1 - 1 - 1 = 2$$

Thus temperature and pressure can be fixed.

(*c*) A system occurring in the gas-phase catalytic oxidation of ammonia containing NH_3, NO_2, NO, H_2O, O_2, and N_2.

Here the number of equations must be determined by writing formation reactions.

(i) $$\tfrac{1}{2}N_2(g) + \tfrac{3}{2}H_2(g) = NH_3(g)$$

(ii) $$\tfrac{1}{2}N_2(g) + O_2(g) = NO_2(g)$$

(iii) $$\tfrac{1}{2}N_2(g) + \tfrac{1}{2}O_2(g) = NO(g)$$

(iv) $$\tfrac{1}{2}O_2(g) + H_2(g) = H_2O(g)$$

From these four formation reactions, hydrogen must be eliminated as it is not present in the system. Algebraically combining (i) and (iv) to eliminate H_2,

(v) $$\tfrac{3}{2}H_2O(g) + \tfrac{1}{2}N_2(g) = NH_3(g) + \tfrac{3}{4}O_2(g)$$

Thus, Eqs. (ii), (iii), and (v) form a set of three independent reactions. No special conditions exist; thus, using Eq. (9.28),

$$V = 2 + C - \phi - R$$

where $C = 6$

$$\phi = 1$$
$$R = 3$$

Therefore

$$V = 2 + 6 - 1 - 3 = 4$$

Thus temperature, pressure, and the amount of any two components can be set.

Exercise 9.E Determine the number of degrees of freedom for each of the following systems:

1. The gaseous system of H_2O, Cl_2, HCl, and O_2 formed by reacting HCl and O_2 fed in a ratio of 6HCl:1O_2 at 5 atm total pressure
2. The gaseous system formed by reacting methanol and oxygen to give formaldehyde, water, hydrogen, and unreacted feed components
3. The equilibrium system formed by decomposing solid ammonium chloride to gaseous ammonia and hydrogen chloride into a vacuum
4. An equilibrium gas-phase mixture of CO, CH_4, H_2O, and O_2 at a given temperature and pressure

For each case specify the variables which would most likely be set by the processor.

9.4 REACTIONS INVOLVING LIQUIDS AND SOLIDS

Section 9.2 discussed the gas-phase reaction for which rigorous analysis is possible in most situations. For multiphase, liquid, and solid reactions, less data are usually available and prediction methods are not so precise. However, the basic principles are not changed although the calculations often must be carried out after making certain simplifying assumptions.

It is important to note that unless the total pressure of a system is above the critical for all components of a system and their possible mixtures, an isothermal increase in total pressure will result in the formation of liquid or solid phases and will convert homogeneous reactions to heterogeneous reactions. Often there is an option as to whether a homogeneous or a heterogeneous reaction situation will occur according to the choice of conditions. However, often a reaction must be heterogeneous at any reasonable set of conditions.

9.4.1 Liquid-Phase Reactions

Equations (9.11) and (9.13) define the equilibrium constant for any reaction type. Normally for liquid-phase reactions the standard-state fugacity f_i° is taken as pure liquid at the temperature of the system and at 1 atm total pressure. Generally, the concept of activity coefficient is used.

$$\gamma_i = \frac{\bar{f}_i}{f_i x_i} \tag{9.30}$$

where $\bar{f}_i$ = fugacity of liquid in mixture
f_i = fugacity of pure liquid at temperature and pressure of equilibrium

Since $\bar{a}_i = \bar{f}_i/f_i^\circ$, then

$$\bar{a}_i = \frac{\gamma_i f_i x_i}{f_i^\circ} \tag{9.31}$$

As liquid fugacities normally are only weakly dependent on pressure, f_i/f_i° is normally taken as unity although it can be evaluated from Eq. (7.44). In the normal case where f_i/f_i° is unity,

$$K = \prod_i (\gamma_i x_i)^{\nu_i} \tag{9.32a}$$

This equation can then be solved by estimating values for activity coefficients by the methods of Chap. 8. This normally requires an iterative procedure, as activity coefficients depend upon compositions which have not yet been determined and must be estimated.

Analogous to the discussion for gas-phase reactions, an alternative expression for Eq. (9.32a) is

$$K = K_\gamma K_x \tag{9.32b}$$

where $$K_\gamma = \sum_i \gamma_i^{\nu_i} \qquad K_x = \sum_i x_i^{\nu_i}$$

If a liquid solution is ideal, i.e., Raoult's law applies to a component, which is the usual case for high-concentration components, then by Eq. (7.45), $\bar{f}_i = x_i f_i^\circ$, and by Eq. (7.72), $\gamma_i = 1$. Thus

$$a_i = x_i \tag{9.33}$$

This simplifies Eq. (9.31) for such components.

If all components of a liquid solution are ideal, then it follows that $K_\gamma = 1$ and

$$K = K_x \tag{9.34}$$

For components present in low concentration, the standard state of the solute is usually taken as the hypothetical state which would exist if the solute obeyed Henry's law to a molality of unity. Thus

$$\bar{f}_i = H_i m_i \tag{9.35}$$

where H_i = Henry's law constant
m = molality, kmol/kg solvent

This is shown in Fig. 9.3, where the slope of the curve is equal to the Henry's law constant. Point A is the hypothetical fugacity.

Using this standard state,

$$f_i^\circ = H_i(1) = H_i$$

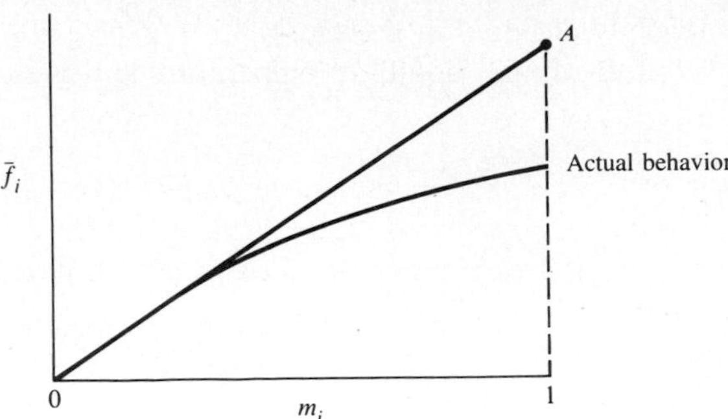

Figure 9.3 Henry's law plot.

Thus, from Eq. (9.35),

$$\bar{f}_i = H_i m_i = f_i^\circ m_i$$

From the definition of activity,

$$a_i = \frac{\bar{f}_i}{f_i^\circ} = \frac{f_i^\circ m}{f_i^\circ} = m_i \tag{9.36}$$

This provides a simple relation which can be used at any time when standard free-energy data are available based on this hypothetical standard state.

9.4.2 Gas-Liquid Reactions

Consider an equilibrium reaction between a gas G and a liquid L to form a solution S.

$$gG + lL = sS$$

Treatment of such a system can be carried out in three different ways:

1. Assume reactions all occur in the gas phase by using the standard state for gases for each component to evaluate the free-energy changes. Couple these equilibrium expressions with the equations for mass transfer between the phases to maintain phase equilibrium.
2. Assume reactions all occur in the liquid phase by using the standard state for liquids and the phase equilibrium relations as in (1).
3. Assume G is a pure gas at unit fugacity, L is a pure liquid at 1 atm pressure, and S is in its standard state as a hypothetical 1 m solution. This leads to an expression for K:

$$K = \frac{a_S^{(s)}}{a_G^{(g)} a_L^{(l)}} = \frac{m_S^{(s)}}{f_G^{(g)} (\gamma_L x_L)^{(l)}} \tag{9.37}$$

The various parameters in Eq. (9.37) can be evaluated by the methods shown previously, making this formulation quite useful for reactions which occur in aqueous solutions.

Each of the three methods should yield the same results. The method chosen should be based on ease of calculation.

Example 9.7 Consider the reaction of ethylene gas and liquid water to form ethanol in aqueous solution. Experimental data at 527 K and 8.37 MPa pressure at equilibrium show the liquid phase to contain 0.014 mol fraction ethanol and 0.956 mol fraction water and the vapor phase to contain 0.475 mol fraction ethylene. Estimate the equilibrium constant by using the experimental data and its definition.

$$C_2H_4(g) + H_2O(l) + C_2H_5OH(aq)$$

SOLUTION Assuming standard states as discussed, Eq. (9.37) will apply:

$$K = \frac{a_{C_2H_5OH}}{a_{C_2H_4}a_{H_2O}} = \frac{m_{C_2H_4OH}}{\bar{f}_{C_2H_4}\gamma_{H_2O}x_{H_2O}}$$

As the standard state for ethanol is 1 m solution, $a_{C_2H_5OH} = m_{C_2H_5OH}$.

Since the standard state for water is pure liquid at atmospheric pressure where $a = 1$ and water is in very high concentration, the assumption of $\gamma_{H_2O} = 1$ is justified and $a_{H_2O} = x_{H_2O}$.

The standard state for ethylene is the pure gas at unit fugacity; the fugacity can be assumed to be $\bar{f}_{C_2H_4} = y_{C_2H_4}f_{C_2H_4,T_r,P_r}$.

$$m_{C_2H_5OH}\frac{\text{mol}}{1000\text{ g H}_2\text{O}} = \frac{0.014\text{ mol}}{(0.956)(18)\text{g H}_2\text{O}} \times 10^3 = 0.8136$$

$$a_{H_2O} = 0.956$$

$$\bar{f}_{C_2H_4} = 0.475\phi_{C_2H_4} \qquad (8.37\text{ MPa})$$

Use the Pitzer procedure for ethylene. From App. A for ethylene,

$$T_c = 9.2°\text{C} \qquad P_c = 5.032\text{ MPa}$$

$$T_r = \frac{527}{282.3} = 1.867 \qquad P_c = \frac{8.37}{5.032} = 1.663$$

From Table 6.3,

$$\omega = 0.0856$$

Using Eq. (7.41),

$$\log\phi = (\log\phi)^{(0)} + \omega(\log\phi)^{(1)}$$

From Fig. 7.4,

$$(\log\phi)^{(0)} \approx -0.04$$

From Fig. 7.6,

$$(\log\phi)^{(1)} \approx 0.06$$

Thus

$$\log(\phi) = -0.04 + (0.0856)(0.06) = -0.035$$

$$\phi = 0.92$$

Therefore

$$\bar{f}_{C_2H_4} = (0.475)(0.92)(8.37) = 3.66 \text{ MPa}$$

and

$$K = \frac{0.8136}{(3.66)(0.956)} = 0.233$$

Exercise 9.F Consider the reaction of Example 9.7. From generalized data estimate the equilibrium constant for the reaction and compare your result with that obtained from the experimental data.

9.4.3 Heterogeneous Reactions with Solids

There are no known reactions between any two solids; thus all reactions involving solids also involve gases or liquids. Solid-gas reactions are common and include combustion of solid fuels, reduction of metal oxides, decomposition of solids, and decomposition of certain gases. Liquid-solid reactions take place in ore-leaching systems and in precipitation of pigments.

If the standard state for solids and liquids is taken at atmospheric pressure or at a low-equilibrium vapor pressure, the activities of pure liquids or solids may normally be taken as unity when in equilibrium with gases. Thus the gas phase at equilibrium is not affected by the liquid or solid presence. At high pressures, however, the activities of pure liquids or solids are not unity and the gaseous equilibrium is affected. For solid or liquid solutions, the activities of the components depart from unity even at relatively low pressures, and the equilibrium composition of the gas is substantially affected.

To be in a standard state with an activity of unity, a solid or liquid must be pure and at 1 atm total pressure at the system temperature. The fugacity of a solid in its standard state can normally be taken as its vapor pressure. Thus it will exist as a vapor if the partial pressure of the component is less than the vapor pressure, and its activity will equal the ratio of the partial pressure to the vapor pressure.

Usually analysis of gas-solid reaction systems is accomplished by assuming activities of solids are equal to unity and that the gases form ideal solutions such that $K = K_P$. Phase-rule analysis of such systems is important in determining what phases will exist for both single and multiple solid-gas reactions. Such analyses are presented below for common types of reactions.

Solid = solid + gas Such reactions include common decomposition reactions such as occur for inorganics, for example, $Al_2(SO_4)_3$, $Na_2SO_4 \cdot 10H_2O$, $NaHCO_3$, $CaCO_3$. Consider the decomposition of $CaCO_3$ as an example:

$$CaCO_3(s) = CaO(s) + CO_2(g)$$

If the activities of both solids are unity,

$$K = K_P = \overline{P}_{CO_2} = y_{CO_2} P_T$$

For phase-rule analysis,

$$\phi = 3 \qquad C = 3 \qquad R = 1 \qquad SC = 0$$

Thus

$$V = 1$$

The equilibrium constant is plotted vs. the reciprocal of absolute temperature on semilog coordinates in Fig. 9.4*a*. Analysis of the system shows that if the temperature is increased to the point where the decomposition pressure is greater than the pressure of the system, then only calcium oxide and carbon dioxide exist in the system and 2 degrees of freedom exist. Conversely, if the temperature is decreased to a point where the system pressure is greater than the decomposition pressure, only calcium carbonate and carbon dioxide exist in the system. Equilibrium is established by fixing either temperature or pressure.

Thus, to cause the reaction to proceed quantitatively to the right, operate below the equilibrium line, i.e., at such a temperature that the decomposition pressure of calcium carbonate is greater than the actual pressure in the system.

The diagram used for this analysis, plotting an equilibrium relation vs. reciprocal of absolute temperature, is useful as a phase diagram to show the regions of stability of the various phases at equilibrium and is illustrated further below.

Solid + solid = solid + gas Common reactions of this type are reduction of metal oxides with solid carbon to form a metal and carbon oxides. Consider, as an example,

$$\mathrm{Si}(c) + 4\mathrm{MgO}(c) = 2\mathrm{Mg}(g) + \mathrm{Mg_2SiO_4}(c)$$

where (c) = crystalline state. If the solid activities are unity,

$$K = K_P = \overline{P}_{Mg}$$

Phase-rule variables are

$$C = 4 \qquad \phi = 4 \qquad R = 1 \qquad SC = 0$$

leading to

$$V = 1$$

An identical analysis as in Fig. 9.4*a* is shown in Fig. 9.4*b*.

Solid = gases Technically a sublimation would fall in this category, but no reaction occurs, eliminating it as a possibility. Take, as an example, the reaction

$$\mathrm{NH_4Cl}(c) = \mathrm{NH_3}(g) + \mathrm{HCl}(g)$$

If the ammonium chloride activity can be taken as unity,

$$K = K_P = \overline{P}_{NH_3} \overline{P}_{HCl}$$

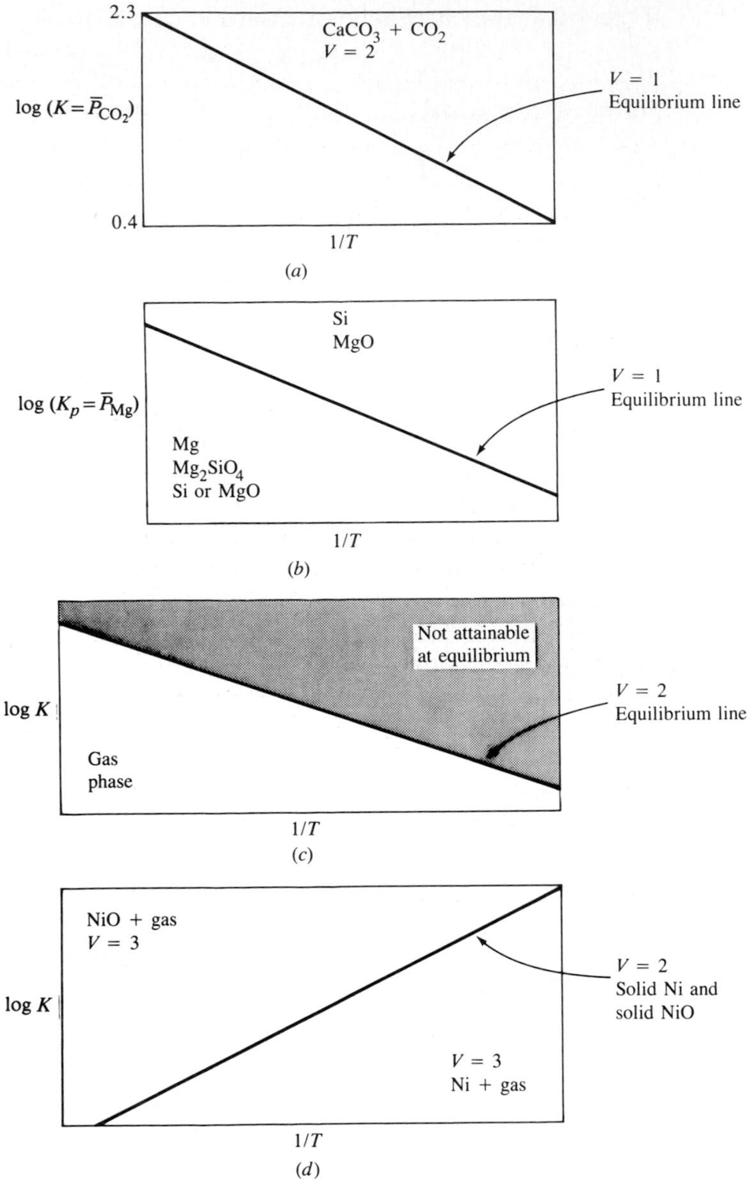

Figure 9.4 Solid-gas equilibrium. (*a*) Solid = solid + gas; (*b*) solid + solid = solid + gas; (*c*) solid + gas = gas; (*d*) solid + gas = solid + gas.

If solid ammonium chloride is placed in a low-pressure container, dissociation will occur and both solid and gas phases will be present if the container is not large enough so that complete gasification occurs. The phase variables are $\phi = 2$, $C = 3$, $R = 1$, and $SC = 1$ as $\bar{P}_{NH_3} = \bar{P}_{HCl}$ if only NH_4Cl is used as feed. Thus $V = 1$. The same situation would occur if equimolar amounts of ammonia and hydrogen chloride are reacted. If nonequal amounts of gases are reacted, the system becomes divariant and two of the three variables of temperature and the two partial pressures must be fixed.

Solid + gas = gas Such reactions occur in reactions of carbon and sulfur. For example,

$$C(c) + CO_2(g) = 2CO(g)$$

If the activity of carbon can be taken as unity,

$$K = K_P = \frac{\bar{P}_{CO}^2}{\bar{P}_{CO_2}}$$

The phase-rule variables are $C = 3$, $\phi = 2$, $R = 1$, $SC = 0$. Thus $V = 2$.

As shown in Fig. 9.4*c*, the region above the line is not attainable at equilibrium, as any attempt to increase the partial pressure of carbon monoxide will result in its decomposition by the given reaction. Below the line only a gas phase will exist, as solid carbon can only exist in contact with the gas on the equilibrium line.

Solid + gas = solid + gas Such reactions occur in metallurgical reductions such as

$$NiO(c) + H_2(g) = Ni(c) + H_2O(g)$$

If solid activities are unity,

$$K = K_P = \frac{\bar{P}_{H_2O}}{\bar{P}_{H_2}}$$

The phase-rule variables are $C = 4$, $\phi = 3$, $R = 1$, and $SC = 0$. Thus $V = 2$—temperature and one partial pressure. As shown in Fig. 9.4*d*, if the ratio of partial pressures K is adjusted to a value above the equilibrium line, the reaction is forced to completion to the left. Conversely, if the ratio is adjusted to a value below the line, the reaction is forced to the right. Both above and below the line, temperature and both partial pressures can be varied arbitrarily and independently with no phase appearance or disappearance.

Gas-solid multiple reactions Reactions of any of the five types shown can exist as multiple reactions occurring in the same system. Consider, as an example,

(1) $$Fe_3O_4 + H_2 = 3FeO + H_2O$$

(2) $$FeO + H_2 = Fe + H_2O$$

(3) $$\tfrac{1}{4}Fe_3O_4 + H_2 = \tfrac{3}{4}Fe + H_2O$$

Each has an equilibrium constant

$$K = K_P = \frac{\bar{P}_{H_2O}}{\bar{P}_{H_2}}$$

In this case, each reaction is also of the type solid + gas = solid + gas and hence will proceed to the left if the partial-pressure ratio is above the equilibrium line and will go to the right if the ratio is below the equilibrium line as shown in Fig. 9.5*a*.

In area *A* reactions (2) and (3) go to the left and reaction (1) goes to the right. Thus the surviving solid phase is FeO regardless of whether the starting solid phase was Fe_3O_4, FeO, or metallic iron.

In area *B* reactions (1) and (3) go to the right, while reaction (2) goes to the left. Thus the surviving solid is FeO.

Below these lines all reactions go to the right and only metallic iron is stable, while above these lines all reactions go to the left and only Fe_3O_4 is stable. Similarly area *C* is the Fe region and area *D* is the Fe_3O_4 region. A region summary is shown in Fig. 9.5*b*. A phase-rule analysis shows that $C = 5$, $R = 2$, and $SC = 0$. On the lines, $\phi = 3$ and the system is divariant. In the areas, $\phi = 2$

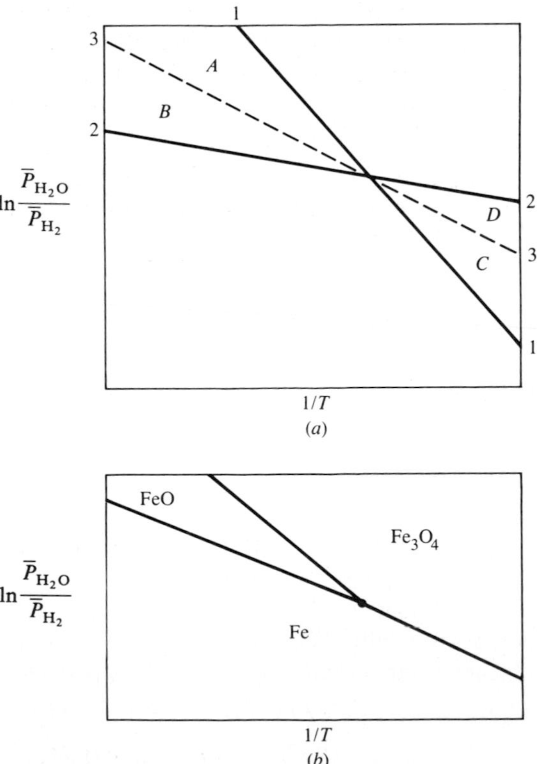

Figure 9.5 Gas-solid multiple reactions; equilibrium constant; temperature plots.

and the system is trivariant. Use of such a diagram gives much insight into what range of operating conditions are reasonable to attain a certain goal with a given set of reactions.

If the equilibrium constants for the various reactions are different ratios, the diagram must be plotted by using pseudoequilibrium constants for all but one reaction. For example, consider the following two reactions:

(1) $$A(s) + B(g) = C(g) \qquad K_1 = \frac{\bar{P}_c}{\bar{P}_B}$$

(2) $$D(s) + B(g) = 2C(g) \qquad K_2' = \frac{\bar{P}_c^2}{\bar{P}_B}$$

Since $P_T = \bar{P}_B + \bar{P}_c$, choose the ratio to be plotted as $\bar{P}_c/\bar{P}_B$. Thus, for reaction (1), plot K_1, and for reaction (2), plot $K_2' = K_2/\bar{P}_c$.

For reactions also involving liquids the same phase-rule-type analyses can be carried out to shed light on complex-reaction situations.

Example 9.8 In the production of powdered iron, powdered FeO is mixed with high-grade carbon powder and is pelleted and dried. The pellets are charged to an externally fired retort, the retort is evacuated, and the retort is set so gases can escape when the internal pressure exceeds 1 atm. The pellets contain 25 wt % carbon and 75 wt % FeO. Determine the lowest temperature for complete reaction at 1 atm. Some relevant reactions are

(1) $$C + O_2 = CO_2 \qquad \Delta G° = -94{,}200 - 0.2T(\text{K})$$

(2) $$2C + O_2 = 2CO \qquad \Delta G° = -53{,}400 - 41.9T(\text{K})$$

(3) $$2Fe + O_2 = 2FeO \qquad \Delta G° = -126{,}400 + 30.9T(\text{K})$$

where $\Delta G°$ is in kcal/kmol and the T range is 298–1600 K.

SOLUTION It is desired for FeO to react with C to give Fe. Do a phase-rule analysis to determine how many independent reactions can exist. Using Eq. (9.29), $V = 2 + C - \phi - R - SC$, if only pressure is to be set,

$$V = 1$$

$$C = 5 \text{ (Fe, FeO, C, } CO_2\text{, CO) assuming no oxygen present}$$

$$\phi = 4 \text{ (gas, Fe, FeO, C)}$$

$$SC = 0$$

Therefore

$$1 = 2 + 5 - 4 - R - 0$$

$$R = 2$$

Reconstruct reactions of FeO + C and eliminate O_2.

(2) – (3): $$2C + O_2 = 2CO$$

$$-2Fe - O_2 = -2FeO$$

Thus

(A) $$C + FeO = CO + Fe$$

$$\Delta G^\circ_{(A)} = \tfrac{1}{2}(\Delta G^\circ_{(2)} - \Delta G^\circ_{(3)}) = 36{,}500 - 36.4T$$

(1) – (3): $$C + O_2 = CO_2$$

$$-2Fe - O_2 = -2FeO$$

Thus

(B) $$C + 2FeO = CO_2 + 2Fe$$

$$\Delta G^\circ_{(B)} = \Delta G^\circ_{(1)} - \Delta G^\circ_{(3)} = 32{,}200 - 31.1T$$

The lowest temperature at which equilibrium will exist is where the total pressure of the system will be 1 atm. Thus

$$\bar{P}_{CO} + \bar{P}_{CO_2} = 1$$

Assuming $K = K_P$,

$$K_{P_A} = \bar{P}_{CO} K_{P_B} \qquad = \bar{P}_{CO_2}$$

$$= e - \frac{\Delta G^\circ_{(A)}}{RT} \qquad = e - \frac{\Delta G^\circ_{(B)}}{RT}$$

Therefore

$$e\left(\frac{-18{,}369}{T} + 18.32\right) + e\left(\frac{-16{,}205}{T} + 15.65\right) = 1$$

By trial and error,

$$T = 976.5 \text{ K}$$

Thus, the lowest temperature for complete reaction to equilibrium is 976.5 K, where $\bar{P}_{CO} = 0.61$ and $\bar{P}_{CO_2} = 0.39$.

The assumption that no O_2 exists appears valid, as any reaction to produce oxygen is highly unfavorable at these conditions. Note that if the system had not been evacuated, oxygen would react with carbon until all oxygen was consumed and then the reaction would proceed to the same equilibrium condition. Also, remember that reactions (A) and (B) need not be the actual reactions to determine the equilibrium composition but just an independent set of reactions.

Exercise 9.G A mixture of 10% H_2, 70% CH_4, and 20% N_2 is to be preheated at 2 atm total pressure such that no carbon is deposited by the reaction

$$C(s) + 2H_2(g) = CH_4(g)$$

$$K_P(482°C) = 3.046\ \text{atm}^{-1} \qquad K_P(760°C) = 0.0695\ \text{atm}^{-1}$$

Determine the temperature below which no solid carbon will form.

Do a phase-rule analysis of the system if solid carbon is absent and if solid carbon is present.

9.5 SOLUTION OF MULTIREACTION EQUILIBRIA

Solution of multireaction equilibrium situations requires the simultaneous solutions of a system of equilibrium and material balance relationships. Prerequisite to solution is often the identification of the *independent* reactions which are occurring in the system by the methods of Sec. 9.3. If at the temperature of the system an independent reaction has an equilibrium constant very much lower than the other reactions occurring, it can probably be eliminated. For example, if three independent reactions have equilibrium constants of 1.0, 0.1, and 0.0001, the third reaction is most likely insignificant. Three methods of solution are widely used: trial and error, series reactor, and lagrangian multiplier. The first is traditional, the second simpler for hand calculator or computer use, and the third completely systematized. Each of these is discussed in this section.

9.5.1 Trial and Error

For any complex equilibria, the equilibrium relationships can be written and the number of variables reduced to the number of equations by use of material balances. Thus n equations in n unknowns will exist. However, these equations are usually highly nonlinear and will require multiple trial-and-error solution. Thus this method is normally not used for solution of more than two equations in two unknowns. Calculation of a highly nonlinear set of two equations by trial and error can be aided by a simple computer program which can be programmed with only a dozen statements.

Example 9.9 Consider the steam-methane reaction at 593 K. If only CO, CO_2, and H_2 are formed, two independent reactions exist. A consistent set of reactions is

(1) $\quad CH_4 + H_2O = CO + 3H_2 \qquad K_P(593\ \text{K}) = 0.41$

(2) $\quad CH_4 + 2H_2O = CO_2 + 4H_2 \qquad K_P(593\ \text{K}) = 1.09$

Determine the equilibrium composition if the feed contains 5 mol steam/mol methane and reaction is carried out at 2 atm total pressure. Equation (9.24) applies to each reaction:

$$K_P = P_T^{\nu} \prod_i (y_i)^{\nu_i}$$

BASIS 6 mol feed. Assume Y = mol CO formed, Z = mol CO_2 formed.

	Feed	Product
CH_4	1	$1 - Y - Z$
H_2O	5	$5 - Y - 2Z$
CO	0	Y
H_2	0	$3Y + 4Z$
CO_2	0	Z
Total	6	$6 + 2Y + 2Z$

$$(1)\quad K_{P1} = \frac{y\text{CO}\, y\text{H}_2^3 P_T^2}{y\text{CH}_4\, y\text{H}_2\text{O}} = \frac{(Y)(3Y + 4Z)^3 P_T^2}{(1 - Y - Z)(5 - Y - 2Z)(6 + 2Y + 2Z)^2}$$

$$(2)\quad K_{P2} = \frac{y\text{CO}_2\, y\text{H}_2^4 P_T^2}{y\text{CH}_4\, y^2\text{H}_2\text{O}} = \frac{(Z)(3Y + 4Z)^4 P_T^2}{(1 - Y - Z)(5 - Y - 2Z)^2(6 + 2Y + 2Z)^2}$$

Since all variables are known except Y and Z, the two equations are solved by double trial and error for Y and Z [that is, assume Y, calculate Z from (1), calculate Y from (2) by using this Z, go back to (1) until assumed Y and calculated Y are equal]. Note that the sum of Y and Z cannot be greater than unity and $Z > Y$ as $K_{P2} > K_{P1}$. Solving, $Y \approx 0.18$, $Z \approx 0.60$. Thus, the product composition is

	Moles	Mole fraction
CH_4	0.22	0.029
H_2O	3.62	0.479
CO	0.18	0.024
H_2	2.94	0.389
CO_2	0.60	0.079
	7.56	1.000

Exercise 9.H Consider the following two reactions to be carried out at 10 atm pressure and constant temperature with a molar feed ratio of 2 : 1 of A/B.

(1) $\quad A + B = C + D \qquad K_{P1} = 2.67$

(2) $\quad A + C = 2E \qquad K_{P2} = 3.20$

Using the standard trial-and-error method, calculate the composition of the equilibrium product.

9.5.2 Series-Reactor Solution Technique

This technique reduces a multiple-reaction system to a series of single-equation solutions at the same temperature and pressure by assuming a set of reactors in series. Consider reactions A, B, C, and the following reaction train:

Reactor:	$1 \rightarrow 2 \rightarrow 3 \rightarrow 4 \rightarrow 5 \rightarrow \cdots$
Reaction:	$A \quad B \quad C \quad A \quad B$

Consecutive application of this principle until no additional reaction occurs yields the equilibrium composition.

Regardless of the independent series of reactions chosen, the composition of the mixture will always converge on the equilibrium conversion, as the second law requires that the change in free energy of an isothermal system undergoing spontaneous reaction at constant pressure is always negative. At equilibrium the free-energy change is zero. Thus the system always proceeds toward equilibrium.

This method has a major advantage over trial and error when there are more than two reactions and is even useful for two reactions, as trial and error never exceeds a single variable.

To use this method effectively and converge easily, the reactions must be arranged in the correct order. If this is not done, a reaction may be starved for reactants and converge so slowly that the method becomes useless. The arrangement is made by using a set of rules that depends on the three various types of reactions which occur in the system. These are:

Parallel reactions. Each reaction uses the same reactants. For example,

(1) $$CH_4 + H_2O = CO + 3H_2$$

(2) $$CH_4 + 2H_2O = CO_2 + 4H_2$$

Sequential reactions. The product of one reaction is consumed in another reaction. For example,

(3) $$CH_4 + CO_2 = 2CO + 2H_2$$

(4) $$2CO + 4H_2 = C_2H_4 + 2H_2O$$

Mixed reactions. The reactant of an earlier reaction combines with the product of the same or an earlier reaction. For example,

(2) $$CH_4 + 2H_2O = CO_2 + 4H_2$$

(3) $$CH_4 + CO_2 = 2CO + 2H_2$$

To illustrate why certain rules are necessary to allow a reaction sequence to converge, consider the following sequential reactions:

$$2CH_4 = C_2H_4 + 2H_2 \qquad K_P = 1.23 \times 10^{-6}$$

$$C_2H_4 + 2H_2O = 2CO + 4H_2 \qquad K_P = 1.4 \times 10^5$$

It is apparent that the second reaction is starved since the equilibrium constant for the first reaction is very small.

Consider the following parallel reactions.

(1) $$CS_2 + 3O_2 = CO_2 + 2SO_2 \qquad K_P = 3.8(10^{74})$$

(2) $$CS_2 + 4O_2 = CO_2 + 2SO_3 \qquad K_P = 2.55(10^{79})$$

(3) $$SO_2 + \tfrac{1}{2}O_2 = SO_3 \qquad K_P = 2.6(10^{2})$$

Here the first reaction starves the second reaction of CS_2; thus, a better order would be (1), (3), (2).

To avoid reagent starvation and slow convergence, proceed as follows:

1. Decide on the number of independent reactions to be recognized as occurring in the equilibrium system under study.
2. Formulate various possible groups of independent reactions and select the group which will converge rapidly.
   a. As far as possible, make the compounds present by sequential reactions having equilibrium constants $K > 0.1$ and preferably $K > 1$.
   b. When one of the feed components is in stoichiometric excess with reference to the first reaction occurring, write a later reaction of the mixed type in which this excess reactant reacts with a product of an earlier reaction which if possible should have a $K > 1$.
   c. As it may not be possible to produce all or any of the compounds present by sequential reactions with K's greater than 1, make these compounds by parallel reactions having $K < 1$ and insert these reactions in the series at those points where the reactants for these parallel reactions are not yet reduced to starvation levels.

Thus, as an example, consider the following reactions:

(1) $$CH_4 + H_2O = CO + 3H_2 \qquad K_P = 0.41$$

(2) $$CH_4 + 2H_2O = CO_2 + 4H_2 \qquad K_P = 1.09$$

(3) $$CH_4 + CO_2 = 2CO + 2H_2 \qquad K_P = 0.154$$

(4) $$CO + H_2O = CO_2 + H_2 \qquad K_P = 2.65$$

Step 1 shows that there are two independent reactions.

The *first priority* is to choose sequential reactions having an equilibrium constant > 0.1 (rule 2*a*). Thus reaction (2) followed by reaction (3) would be the two preferable reactions. An alternate formulation of *second priority* from rule 2*c* would be to use reaction (1) followed by reaction (2). Use of rule 2*b* would yield reaction (1) followed by reaction (4) and would also be of second priority.

Example 9.10 Consider the steam-methane reaction at 593 K, where only CO, CO_2, and H_2 are formed and where two independent reactions exist. The possible reactions and K_P's are listed in the text. For the same conditions and feed as Example 9.9, determine the equilibrium composition by using the series-reactor technique.

SOLUTION As discussed in the text the first priority on reactions is sequential reactions with K_P's > 0.1. Thus the following reaction should be used:

(a) $$CH_4 + 2H_2O = CO_2 + 4H_2 \qquad K_P = 1.09$$

(b) $$CH_4 + CO_2 = 2CO + 2H_2 \qquad K_P = 0.154$$

Therefore assume only reaction (a) occurs in the first reactor.

Let x = conversion of methane.

	Feed	Product
CH_x	1	$1 - x$
H_2O	5	$5 - 2x$
CO_2	0	x
H_2	0	$4x$
CO	0	0
	6	$6 + 2x$

Using Eq. (9.24) therefore,

$$1.09 = \frac{(x)(4x)^4(2)^2}{(1-x)(5-2x)^2(6+2x)^2}$$

(A) $$1.0645 \times 10^{-3} = \frac{x^5}{(1-x)(5-2x)^2(6+2x)^2}$$

By trial and error,

$$x = 0.727$$

Thus the product from the first reactor becomes the feed to the second reactor where only reaction (b) occurs.

Let y = conversion of methane.

	Feed	Product
CH_4	0.273	$0.273 - y$
H_2O	3.546	3.546
CO_2	0.727	$0.727 - y$
H_2	2.908	$2.908 + 2y$
CO	0	$2y$
	7.454	$7.454 + 2y$

Using Eq. (9.24) therefore,

$$\text{(B)} \qquad 0.154 = \frac{(2y)^2(2.908 + 2y)^2(2)^2}{(0.273 - y)(0.727 - y)(7.454 + 2y)^2}$$

By trial and error,

$$y = 0.084$$

This product enters the third reactor, where reaction (a) occurs:

	Feed	Product
CH_4	0.189	$0.189 - x$
H_2O	3.546	$3.546 - 2x$
CO_2	0.643	$0.643 + x$
H_2	3.076	$3.076 + 4x$
CO	0.168	0.168
	7.622	$7.622 + 2x$

$$1.09 = \frac{(0.643 + x)(3.076 + 4x)^4(2)^2}{(0.189 - x)(3.546 - 2x)^2(7.622 + 2x)^2}$$

By trial and error,

$$y = -0.04$$

This product enters the fourth reactor, where reaction (b) occurs:

	Feed	Product
CH_4	0.229	$0.229 - y = 0.219$
H_2O	3.626	$3.626 = 3.626$
CO_2	0.603	$0.603 - y = 0.593$
H_2	2.916	$2.916 + 2y = 2.936$
CO	0.168	$0.168 + 2y = 0.188$
	7.542	$7.542 + 2y$

$$0.154 = \frac{(2.916 + 2y)^2(0.168 + 2y)^2(2)^2}{(0.229 - y)(0.603 - y)(7.542 + 2y)}$$

By trial and error,

$$y = +0.01$$

Note that convergence is almost achieved and the moles of CO_2 and CO are essentially those obtained in Example 9.9. The solution can be continued according to the accuracy desired.

Note that if just any two independent reactions are taken, convergence will be much slower; e.g., if the parallel reactions

(c) $$CH_4 + 2H_2O = CO_2 + 4H_2 \qquad K_P = 1.09$$

(d) $$CH_4 + H_2O = CO + 3H_2 \qquad K_P = 0.41$$

are taken in this order with x and y the conversion of methane by reactions (c) and (d), respectively, the results through four reactors are:

Reactor (reaction)	1(c)	2(d)	3(c)	4(d)
	$x = 0.73$	$y = 0.10$	$x = -0.07$	$y = +0.04$

as compared to this solution:

Reactor (reaction)	1(a)	2(b)	3(c)	4(d)
	$x = 0.727$	$y = 0.084$	$x = -0.04$	$y = +0.01$

Thus more work is required for solution.

The definite advantage of this method comes with three or more reactions with three or more unknowns.

Exercise 9.I Repeat Exercise 9.H by using the series-reactor technique.

9.5.3 Lagrangian-Multiplier Technique

This method is based on the fact that at equilibrium the total free energy of a system is at a minimum:

$$G_{T,P} = f(n_1, n_2, \ldots, n_n)$$

where n_i = number of moles of component i. Thus the problem is to determine the set of n's which minimize the total free energy for any specific temperature and pressure subject to the constraints of the material balances.

For gas-phase systems:

1. Formulate the constraining equations. Note that the total number of atoms of each element is constant. Let

   A_k = total number of atomic weights of kth element present in the system

   a_{ik} = total number of atoms of kth element present in each molecule of chemical species i

   Thus the material balance on each element k is

$$\sum_i n_i a_{ik} = A_k$$

or

$$\sum_i n_i a_{ik} - A_k = 0 \qquad (k = 1, 2, \ldots, m) \tag{9.38}$$

2. Introduce lagrangian multipliers λ_k, one for each element, by multiplying each element balance by its λ_k:

$$\lambda_k\left(\sum_i n_i a_{ik} - A_k\right) = 0 \qquad (k = 1, 2, \ldots, \omega)$$

Summing over k elements

$$\sum_k \lambda_k\left(\sum_i n_i a_{ik} - A_k\right) = 0$$

3. Form a new function F by adding the previous sum to the total free energy:

$$F = G + \sum_k \lambda_k\left(\sum_i n_i a_{ik} - A_k\right)$$

As the Σ term is zero, $F \equiv G$. The partial derivatives of F and G with respect to n_i will differ as F incorporates material balance constraints.

4. Minimize F by taking the partial derivatives and setting them to zero.

$$\left(\frac{\partial F}{\partial n_i}\right)_{T,P,n_j} = \left(\frac{\partial G^t}{\partial n_i}\right)_{T,P,n_j} + \sum_k \lambda_k a_{ik} = 0$$

$$= \mu_i + \sum_k \lambda_k a_{ik} = 0 \qquad (i = 1, 2, \ldots, n) \qquad (9.39)$$

Since $\mu_i = G_i^\circ + RT \ln \bar{a}_i$.

For gas-phase reactions (standard state = ideal gas at 1 atm)

$$\mu_i = G_i^\circ + RT \ln \bar{f}_i$$

If G_i° is set to zero for all *elements* in their standard states, then $G_i^\circ = \Delta G_{fi}^\circ$. Introducing the fugacity coefficient $\bar{f}_i = y_i \bar{\phi}_i P$, then

$$\mu_i = \Delta G_{f_i}^\circ + RT \ln(y_i \bar{\phi}_i P)$$

Combining with Eq. (9.39),

$$\Delta G_{f_i}^\circ + RT \ln(y_i \bar{\phi}_i P_T) + \sum_k \lambda_k a_{ik} = 0 \qquad (i = 1, 2, \ldots, n) \qquad (9.40)$$

where $\Delta G_{f_i}^\circ = 0$ for elements.

There are now n equilibrium equations (9.40), one for each species. There are m material balance equations (9.38), one for each element. There are a total of $n + m$ equations. The unknowns are the

$$n_i\text{'s} - n \text{ of them} \qquad \left(y_i = \frac{n_i}{\Sigma n_i}\right)$$

$$\lambda_k\text{'s} = \qquad m \text{ of them}$$

$$\text{Total} = n + m$$

Remember that:
If ideal gas,

$$\bar{\phi}_i\text{'s} = 1$$

If ideal solution,

$$\bar{\phi}_i\text{'s} = \phi_i$$

and can be estimated.

If real gases,

$$\bar{\phi}_i\text{'s} = f(y_i\text{'s})$$

and iteration is required. Start by setting $\bar{\phi}_i$'s to 1 to get an initial set of y_i's which is good enough for low pressures or high temperatures. Otherwise, use these initial y_i's with an equation of state to get a better set of $\bar{\phi}_i$'s for use in Eq. (9.40). Repeat until convergence is achieved.

This method has the advantage that the actual chemical reactions involved are immaterial. Choosing a set of chemical compounds is equivalent to choosing a set of reactions. Obviously, it is important that each species present in reasonable amount (say 0.5 percent or more) must be considered or the resulting compositions will be different.

Example 9.11 Repeat Example 9.9 by using the lagrangian-multiplier method.

COMPONENTS $CH_4, H_2O, CO, CO_2, H_2$

ELEMENTS C, H_2, O_2

SET UP TABLE

$$\text{Feed} = 5 \text{ mol } H_2O + 1 \text{ mol } CH_4$$

$$\text{Pressure} = 2 \text{ atm, temperature} = 593 \text{ K}$$

	Carbon	Oxygen	Hydrogen	k
i	$A_C = 1$	$A_O = 5$	$A_H = 14$	A_k
CH_4	1	0	4	a_{ik}'s
H_2O	0	1	2	
CO	1	1	0	
CO_2	1	2	0	
H_2	0	0	2	

The three material balance equations (9.38) are

C: $\quad n_{CH_4} + n_{CO} + n_{CO_2} = 1 \qquad n = \text{mol of component}$

O: $\quad n_{H_2O} + n_{CO} + 2n_{CO_2} = 5$

H: $\quad 4n_{CH_4} + 2n_{H_2O} + 2n_{H_2} = 14$

The five equilibrium equations (9.40), assuming all ϕ's equal unity at low pressure,

$$\Delta G^\circ_{f_i} + RT\ln(y_i P_T) + \sum_k \lambda_k a_{ik} = 0$$

Using the procedures of Chap. 3 and the definition $\Delta G^\circ_f = \Delta H^\circ_f - T(\Delta S^\circ_f)$ at 593 K,

$$\Delta G^\circ_f(H_2) = 0$$

$$\Delta G^\circ_f(CO_2) = -1989 \text{ kJ/kmol}$$

$$\Delta G^\circ_f(CO) = -50{,}277 \text{ kJ/kmol}$$

$$\Delta G^\circ_f(CH_4) = -74{,}412 \text{ kJ/kmol}$$

$$\Delta G^\circ_f(H_2O) = -35{,}416 \text{ kJ/kmol}$$

$$R = 8.314 \text{ kJ/(kmol}\cdot\text{K)} \qquad T = 593 \text{ K} \qquad P_T = 2 \text{ atm} \qquad y_i = \frac{n_i}{n_T}$$

H_2: $$0 + (8.314)(593)\ln\frac{n_{H_2}}{n_T}(2) + 2\lambda_H = 0$$

CO: $$-50{,}277 + (8.314)(593)\ln\frac{n_{CO}}{n_T}(2) + \lambda_C + \lambda_O = 0$$

CO_2: $$-1989 + (8.314)(593)\ln\frac{n_{CO_2}}{n_T}(2) + \lambda_C + 2\lambda_O = 0$$

CH_4: $$-74{,}412 + (8.314)(593)\ln\frac{n_{CH_4}}{n_T}(2) + \lambda_C + 4\lambda_H = 0$$

H_2O: $$-35{,}416 + (8.314)(593)\ln\frac{n_{H_2O}}{n_T}(2) + 2\lambda_H + \lambda_O = 0$$

where

$$n_T = n_{H_2} + n_{CO} + n_{CO_2} + n_{CH_4} + n_{H_2O}$$

The five equilibrium relations and three material balances can then be solved for the five n's and the three λ's, after which the mole fraction can be easily calculated.

Exercise 9.J Consider a system at equilibrium at 600°C and 10 atm pressure containing naphthalene, CO, CO_2, CH_4, H_2, H_2S, and SO_2. Set up *all* the equations necessary and tell how to solve for the equilibrium composition by using the lagrangian-multiplier technique.

9.6 UNFAVORABLE AND COMPLEX EQUILIBRIUM REACTIONS

There are many important industrial reactions for which there is no temperature at which the rates and equilibrium are simultaneously favorable or where equilibrium is unfavorable at all attainable temperatures and pressures. Such reactions have positive free energies, and, to cause reaction to occur, work must be supplied to the reacting system. Several ways have been used to circumvent these difficulties. Use of higher or lower pressures is most common. Reaction coupling and the Solvay technique have been used. In very unfavorable circumstances electrolytic processes have been used. The manipulation of pressure in the proper direction to aid gas-phase reactions or to cause a liquid or solid phase to form or disappear where advantageous has been discussed throughout this chapter. The other techniques available will be discussed in this section. In addition, some comments will be made on some important common complex industrial reaction situations.

9.6.1 Reaction Coupling

Reaction coupling is merely consumption of a by-product from a primary reaction with an unfavorable free-energy change by introducing a new compound into the system which will react with the by-product in a reaction with a favorable free-energy change. For example, consider the reaction of P and Q to produce R, the desired product, and S, a by-product, by a reaction with a positive free-energy change coupled with the reaction of T with S by a reaction with a negative free-energy change.

$$pP + qQ = R + sS$$

$$tT + sS = V$$

$$\text{Overall} \quad pP + qQ + tT = R + V$$

Such coupling will increase the yield of R over that which would occur by the first reaction.

For example, reduction of metal oxides is a common use of reaction coupling. For the case of iron oxide,

Primary reaction:	$FeO = Fe + \frac{1}{2}O_2$	$\Delta G^\circ_{1000\text{ K}} = 47.55$ kcal
Secondary reaction:	$C + \frac{1}{2}O_2 = CO$	$\Delta G^\circ_{1000\text{ K}} = -47.94$ kcal
Overall reaction:	$FeO + C = Fe + CO$	$\Delta G^\circ_{1000\text{ K}} = -0.39$ kcal

The use of phase diagrams as discussed in Sec. 9.4 can be used to show the equilibrium situation at any point. Obviously, the reaction $\frac{1}{2}C + \frac{1}{2}O_2 = \frac{1}{2}CO_2$ must also be taken into account. In this case carbon was introduced, which entailed additional cost. Often it is attempted to use air, or a very low cost material must be used. Obviously, the production of excess products V, which may be of low value, must be considered with respect to their separation from the primary product and their disposal.

Another important industrial coupled reaction is the Deacon reaction for converting HCl, a by-product of chlorination reactions, back to chlorine.

Primary reaction:	$HCl = \frac{1}{2}H_2 + \frac{1}{2}Cl_2$	$\Delta G^\circ_{900\text{ K}} = 23.9$ kcal
Secondary reaction:	$\frac{1}{2}H_2 + \frac{1}{4}O_2 = \frac{1}{2}H_2O$	$\Delta G^\circ_{900\text{ K}} = -23.4$ kcal
Overall reaction:	$HCl + \frac{1}{4}O_2 = \frac{1}{2}Cl_2 + \frac{1}{2}H_2O$	$\Delta G^\circ_{900\text{ K}} = 0.5$ kcal

Either air or oxygen can be used, and an analysis as to which should be used depends on the relative cost of the oxygen vs. the increased separation costs of the product if diluted with nitrogen. At lower temperatures the overall standard free-energy change will be more favorable; however, rates become slow rapidly.

The Solvay technique is merely a set of coupled reactions carried out sequentially rather than simultaneously. Use of a set of sequential reactions, all of which exhibit favorable rates and equilibria at reasonable temperatures and pressures and which add up to the desired overall reaction for which rates and equilibria are both unfavorable, is the purpose of this technique. These sequential reactions are normally carried out in different pieces of equipment.

Solvay's original process is the substitution of the six sequential reactions below for the overall reaction

$$2NaCl + CaCO_3 = CaCl_2 + Na_2CO_3 \qquad \Delta G^\circ_{298\,K} = 9.6 \text{ kcal}$$

$$\text{(1)} \qquad CaCO_3 = CaO + CO_2 \qquad \text{at } 1000°C$$

$$\text{(2)} \qquad CaO + H_2O = Ca(OH)_2 \qquad \text{at } 100°C$$

$$\text{(3)} \qquad Ca(OH)_2 + 2NH_4Cl = CaCl_2 + 2NH_3 + 2H_2O \qquad \text{at } 120°C$$

$$\text{(4)} \qquad 2NH_3 + 2H_2O + 2CO_2 = 2NH_4HCO_3 \qquad \text{at } 60°C$$

$$\text{(5)} \qquad 2NH_4HCO_3 + 2NaCl = 2NaHCO_3 + 2NH_4Cl \qquad \text{at } 60°C$$

$$\text{(6)} \qquad 2NaHCO_3 = Na_2CO_3 + H_2O + CO_2 \qquad \text{at } 200°C$$

Obviously, intermediates must be recycled, and each process may be carried out at a different set of reaction conditions in such a process and energy consumption could be high if not properly designed. Such processes have mainly been developed for inorganic processes, although a few organic processes exist.

Reaction coupling, in general, is a technique which should be considered and used for reactions with unfavorable free-energy relationships when possible and economical.

9.6.2 Electrolytic Processes

Electrolytic processes add energy inefficiently as electricity and often show very poor economics, causing them to be used only if no other process can be utilized for the task at hand. In addition, they can only be used when reactants and products can be made to participate in two complementary, near-reversible, half-cell reactions whose sum equals the desired overall reaction. Few reactions thus qualify.

The most common industrial reaction is that for sodium hydroxide production:

$$2NaCl + 2H_2O = 2NaOH + H_2 + Cl_2 \qquad \Delta G^\circ_{25°C} = 102 \text{ kcal}$$

At least part of any reaction's free energy is directly convertible with electric energy in an isothermal cell. Reactions with positive free energies accept free energy as electric energy, while reactions with a negative free energy release the energy. Reactions can be made to go to completion regardless of their equilibrium constants. Because of inefficiencies in cell designs, the amount of energy required for reaction will be greater than the theoretical standard free energy. Conversely, a fuel cell or a battery will produce less than the theoretical free energy because of losses in the cell. The anode reaction releases electrons, while the cathode reaction adds electrons.

An electrochemical reaction is isothermal and reversible and thus

$$\Delta G = -W = -nFE \tag{9.41}$$

where n = number of kilogram-equivalents or electrons transferred
F = faraday = 96.493×10^6 abs C/kg-equivalents
= number of coulombs that will deposit 1 kg-equivalent of material
E = electrode potential, V
ΔG = free-energy change, J

If the reactants and the products are in their standard states, pure compounds with activities equal to unity and at 298 K, the change in free energy per mole will be the standard free-energy change of the reaction and a standard cell voltage $E°$ can be obtained.

$$\Delta G° = -nFE° \tag{9.42}$$

Values of standard cell voltages are given in Table 9.1 and are arranged in descending order of activity as a cathode.

If products and/or reactants are not in their standard states, an analogous expression for free-energy change results:

$$\Delta G = \Delta G° + RT \ln \Pi \bar{a}_i^{\nu} \tag{9.43}$$

$$E = E° + \frac{RT}{nF} \ln \Pi \bar{a}_i^{\nu} \tag{9.44}$$

Reactants are evaluated at their initial activities and products at their final activities. Equation (9.44) is the *Nernst equation*.

For any cell in a closed circuit, current will flow. Heat will be generated because of electric resistance in the cell, from membranes separating the compartments, from gases accumulating on the electrodes, from the electrodes themselves, and from the external circuit. This causes irreversibilities in the cell which can be decreased by applying an external opposing voltage until current flow goes to zero, hence establishing reversibility. Industrial cells use large currents to establish reasonable yields and hence are always irreversible. The efficiency of a cell is defined as the equivalents of recoverable product divided by the faradays of current consumed and is not always equal for anode and cathode.

Actual cells require four components: (1) the theoretical voltage, (2) the *IR* drop through the electrolyte, (3) concentration polarization, and (4) overvoltages at the electrodes. The first has been discussed. The second, electrolytic resistance, varies with the electrolyte—the more conductive the less resistive. In aqueous electrolytes, conductivities generally decrease with the salt concentration. Also, conductivities generally decrease with increasing temperature.

Table 9.1 Standard electrode potentials

$E°$ = electrode potential at 25°C relative to the standard hydrogen electrode; ions at unit activity.

Electrode	Reaction when electrode is cathode in cell; ions at unit activity	$E°$, V
K; K^+	$K^+ + e = K$	−2.922
Ca; Ca^{2+}	$\frac{1}{2}Ca^{2+} + e = \frac{1}{2}Ca$	−2.87
Na; Na^+	$Na^+ + e = Na$	−2.712
Mg; Mg^{2+}	$\frac{1}{2}Mg^{2+} + e = \frac{1}{2}Mg$	−2.34
Be; Be^{2+}	$\frac{1}{2}Be^{2+} + e = \frac{1}{2}Be$	−1.70
Al; Al^{3+}	$\frac{1}{3}Al^{3+} + e = \frac{1}{3}Al$	−1.67
Mn; Mn^{2+}	$\frac{1}{2}Mn^{2+} + e = \frac{1}{2}Mn$	−1.05
Zn; Zn^{2+}	$\frac{1}{2}Zn^{2+} + e = \frac{1}{2}Zn$	−0.762
Cr; Cr^{3+}	$\frac{1}{3}Cr^{3+} + e = \frac{1}{3}Cr$	−0.71
Ga; Ga^{3+}	$\frac{1}{3}Ga^{3+} + e = \frac{1}{3}Ga$	−0.52
Fe; Fe^{2+}	$\frac{1}{2}Fe^{2+} + e = \frac{1}{2}Fe$	−0.440
Cd; Cd^{2+}	$\frac{1}{2}Cd^{2+} + e = \frac{1}{2}Cd$	−0.402
In; In^{3+}	$\frac{1}{3}In^{3+} + e = \frac{1}{3}In$	−0.340
Tl; Tl^+	$Tl^+ + e = Tl$	−0.336
Co; Co^{2+}	$\frac{1}{2}Co^{2+} + e = \frac{1}{2}Co$	−0.277
Ni; Ni^{2+}	$\frac{1}{2}Ni^{2+} + e = \frac{1}{2}Ni$	−0.250
Sn; Sn^{2+}	$\frac{1}{2}Sn^{2+} + e = \frac{1}{2}Sn$	−0.136
Pb; Pb^{2+}	$\frac{1}{2}Pb^{2+} + e = \frac{1}{2}Pb$	−0.126
Pt; $H_2(g)$; H^+	$H^+ + e = \frac{1}{2}H_2$	−0.000
Normal calomel electrode	$\frac{1}{2}Hg_2Cl_2 + e = Hg + Cl^-$	−0.2802
Cu; Cu^{2+}	$\frac{1}{2}Cu^{2+} + e = \frac{1}{2}Cu$	−0.345
Cu; Cu^+	$Cu^+ + e = Cu$	−0.522
Pt; Fe^{2+}, Fe^{3+}	$Fe^{3+} + e = Fe^{2+}$	−0.771
Hg; Hg_2^{3+}	$\frac{1}{2}Hg_2^{2+} + e = Hg$	−0.799
Ag; Ag^+	$Ag^+ + e = Ag$	−0.800
Pd; Pd^{2+}	$\frac{1}{2}Pd^{2+} + e = \frac{1}{2}Pd$	−0.83
Hg; Hg^{2+}	$\frac{1}{2}Hg^{2+} + e = \frac{1}{2}Hg$	−0.854
Pt; $Cl_2(g)$; Cl^-	$\frac{1}{2}Cl_2(g) + e = Cl^-$	−1.358
Pt; Pt^{2+}	$\frac{1}{2}Pt^{2+} + e = \frac{1}{2}Pt$	~ −1.2
Au; Au^{3+}	$\frac{1}{3}Au^{3+} + e = \frac{1}{3}Au$	−1.42
Au; Au^+	$Au^+ + e = Au$	−1.68

Basis: Hydrogen electrode with Pt saturated with H_2 at 1 atm pressure and H^+ ions at unit activity.

The third, concentration polarization, is caused by cations traveling toward the cathode and anions traveling toward the anode and setting up concentration gradients. Movement is slow unless mixing occurs to minimize the gradients and remove the changes occurring in the amount or charge of ions at each electrode. Electrical neutrality must be maintained. The back voltage, caused by concentration gradients, opposes the net operating voltage and is reduced by vigorous stirring. The fourth, overvoltage, is caused by ions and electrons having difficulty in entering into half-cell reactions.

Various types of cells are available and are discussed by Meissner (1971). The most common are diaphragm cells and mercury cells. A diaphragm cell uses a membrane between anode and cathode to keep products separate to avoid undesired chemical reactions, while keeping the spacing between anode and cathode small to minimize energy losses from resistive heating of the electrolyte.

The mercury cell electrolyzes a salt brine between a graphite anode and a mercury cathode.

As an example of the operation of these two cells, consider the electrolysis of sodium chloride.

Diaphragm cell:

Anode: $$Cl^- = \tfrac{1}{2}Cl_2 + e$$

Cathode: $$e + H_2O = OH^- + \tfrac{1}{2}H_2$$

Overall: $$Cl^- + H_2O = \tfrac{1}{2}H_2 + OH^- + \tfrac{1}{2}Cl_2$$

Side reactions:

Electrolyte: $$Cl_2 + 2OH^- = Cl^- + OCl^- + H_2O$$

$$CO_2 + 2OH^- = CO_3^{2-} + H_2O$$

$$3OCl^- = ClO_3^- + 2Cl^-$$

Anode: $$OH^- = \tfrac{1}{2}H_2O + \tfrac{1}{4}O_2 + e$$

$$OCl^- = \tfrac{1}{2}O_2 + \tfrac{1}{2}Cl_2 + e$$

$$ClO_2^- = \tfrac{1}{2}Cl_2 + \tfrac{3}{2}O_2 + e$$

$$C + O_2 = CO_2$$

Mercury cell:

Anode: $$Cl^- = \tfrac{1}{2}Cl_2 + e$$

Cathode: $$e + Na^+ = Na(\text{Hg amalgam})$$

Overall: $$Na^+ + Cl^- = Na(\text{Hg amalgam}) + \tfrac{1}{2}Cl_2$$

The reference cited gives many more details on the NaCl cell and others.

9.6.3 Complex Equilibria Situations

For many reaction systems the resultant equilibrium product is made up of the equilibria of all reactions occurring. However, such analysis is often impossible as for many petroleum conversion processes involving hundreds of components. Catalysts are used to speed some specific reaction toward equilibrium while being able to neglect other reactions even though their free-energy changes are negative. At infinite time all reactions will reach equilibrium at the conditions employed, but such a situation is not of real interest. Petroleum isomerizations and reforming reactions are a good example of important reactions which must be analyzed. Usually only the most important reactions are considered.

Steam reforming of methane is another example where multiple reactions exist; however, reactions with very small equilibrium constants at the reaction conditions can be eliminated from such analyses. For example, at 600°C,

$$CH_4 + H_2O = CO + 3H_2 \qquad K = 0.574$$

$$CO + H_2O = CO_2 + H_2 \qquad K = 2.21$$

$$2CO = C + CO_2 \qquad K = 8.14$$

$$2CH_4 = C_2H_6 + H_2 \qquad K = 5.5 \times 10^{-5}$$

$$H_2O = H_2 + \tfrac{1}{2}O_2 \qquad K = 1.1 \times 10^{-12}$$

$$CO_2 = CO + \tfrac{1}{2}O_2 \qquad K = 5 \times 10^{-13}$$

Only the first three reactions would normally be considered.

The complex reactions in internal-combustion engines is another prime example of a complex system which must be analyzed. In this case, at such high temperatures free-radical reactions also must be taken into account. Atmospheric chemistry and industrial gas-phase partial oxidations are similar.

Metallurgical reactions involving solids as discussed in Sec. 9.4 may be quite complex and have fewer degrees of freedom than for gas-phase reactions.

Complex liquid-phase reactions are usually assumed to be independent of pressure and if possible in ideal solution.

Hougen, Watson, and Ragatz (1959) give examples of each of these situations with numerical examples. With the advent of the high-speed, large-capacity computer, calculations for equilibrium systems with a large number of simultaneous reactions occurring has become possible, and many examples of such calculations have been published. The lagrangian-multiplier method of Sec. 9.5.3 is easily adaptable to computer calculations.

9.7 REFERENCES

Hougen, O. A., K. M. Watson, and R. A. Ragatz: "Chemical Process Principles, Part Two: Thermodynamics," 2d ed., Wiley, New York, 1959.

Meissner, H. P.: "Processes and Systems in Industrial Chemistry," Prentice-Hall, Englewood Cliffs, N.J., 1971.

9.8 PROBLEMS

1 Consider the reaction $CO_2(g) + H_2(g) = H_2O(g) + CO(g)$.

(*a*) Using the data below, calculate ΔG°_{298}.

$$CO_2(g) + 4H_2(g) = CH_4(g) + 2H_2O(g) \qquad \Delta G^\circ_{298} = -26{,}912 \text{ cal}$$

$$2H_2(g) + O_2(g) = 2H_2O(g) \qquad \Delta G^\circ_{298} = -109{,}014 \text{ cal}$$

$$2C(s) + O_2(g) = 2CO(g) \qquad \Delta G^\circ_{298} = -65{,}020 \text{ cal}$$

$$C(s) + 2H_2(g) = CH_4(g) \qquad \Delta G^\circ_{298} = -12{,}206 \text{ cal}$$

(*b*) Determine ΔG°_{298} by using free-energy functions.

(*c*) Calculate ΔS°_{298}.

2 Consider the reaction $C_2H_4(g) + H_2O(g) = C_2H_5OH(g)$. Calculate the standard free-energy change at 500°C if the standard free-energy change at 25°C is -2030 cal/mol C_2H_4 and if:

(*a*) The heat of reaction is constant at $-11{,}000$ cal.

(*b*) The heat of reaction varies with temperature.

3 At a pressure of 750 atm and a temperature of 3400°F, carbon dioxide is 40 percent dissociated to carbon monoxide and oxygen. Assuming ideal-vapor phase but not ideal gases, calculate K for the following reactions.

(*a*) $2CO_2 = 2CO + O_2$

(*b*) $CO_2 = CO + \frac{1}{2}O_2$

(*c*) $CO + \frac{1}{2}O_2 = CO_2$

4 Consider the reaction $C_2H_4(g) + H_2O(g) = C_2H_5OH(g)$.

(*a*) Calculate the equilibrium constant at 25°C for the reaction by using the following data:

	S°_{298}, cal/(mol · K)	H°_{298}, cal/mol
$C_2H_4(g)$	52.75	11,700
$H_2O(g)$	45.17	−57,800
$C_2H_5OH(g)$	66.40	−57,070

(*b*) Quantitatively discuss the feasibility of the reaction over a temperature range of 25 to 500°C if the expression for standard free energy shown below applies.

$$\Delta G^\circ_T = -9670 + 6.43T \ln T - 9.01T - 0.00665T$$

where ΔG° is in cal/mol of ethanol and T is in K.

(*c*) At 200°C calculate the equilibrium conversion of ethylene at 1 and 5 atm total pressure, assuming ideal gas behavior using (i) an equimolar feed of ethylene and water and (ii) 100 percent excess water.

(*d*) For reaction at 200°C and 1 atm total pressure calculate the equilibrium conversion for a feed mixture of 1 mol ethylene, 2 mol water, and 1 mol inert gas. Compare the result with the corresponding result from (*c*) with no inert gas.

5 Consider the reaction $C_4H_8(g) = C_4H_6(g) + H_2(g)$. The standard free-energy change is given by:

$$\Delta G^\circ_T = 24{,}760 - 5.01T \ln T + 3.09T$$

where ΔG°_T is in cal/mol butene and T is in K.

(*a*) Over what range of temperature is the reaction promising from a thermodynamic viewpoint?

(*b*) For reaction of pure butene at 800 K, calculate the equilibrium conversion for operation at 1 and 5 atm total pressure, assuming ideal gases.

(*c*) Repeat part (*b*) at 1 atm for a feed of 50 mol % butene and 50 mol % inert gas and compare the results.

6 The commercial ammonia synthesis reaction $\frac{1}{2}N_2(g) + \frac{3}{2}H_2(g) = NH_3(g)$ operates at 500°C and 300 atm total pressure.

$$\Delta G^\circ_T = 12.31T \log T - 12.03T + 0.0002526T^2 - 8.45 \times 10^{-7}T^3 - 9170$$

where ΔG°_T is in cal/mol NH_3 and T is in K.

Determine the maximum conversion to ammonia possible, assuming ideal-vapor phase but nonideal gases if:

(*a*) Stoichiometric quantities of gas are used.

(*b*) 100 percent excess nitrogen is used.

(*c*) The feed is 20 mol % N_2, 60 mol % H_2, and 20 mol % argon.

7 Prepare an equilibrium conversion chart such as Fig. 9.1 for the ammonia synthesis reaction for a feed of 20 mol % N_2, 60 mol % H_2, and 20 mol % argon at total pressures of 0.5, 1, 2, and 3 atm. Determine the adiabatic-reaction temperature for the 0.5- and 2-atm cases if the reactants enter the reactor at 25°C and no pressure drop takes place in the reactor.

8 A gas-phase system reacts H_2O and COS to form an equilibrium mixture of COS, H_2O, CO, O_2, and H_2S at 1 atm total pressure. Quantitatively determine the number of independent reactions and the number of additional variables that can be set by the processor.

9 Determine the number of independent reactions and the degrees of freedom for the following systems and equilibrium. Discuss the degrees of freedom.

(*a*) A liquid-vapor system of HNO_3, HNO_2, N_2O_3, NO, N_2O_4, NO_2, and H_2O.

(*b*) MNO_2, MNO_3, M_2O, N_2, O_2, NO, and NO_2 where M is a metal. Consider the cases where the metallic compounds are all present as immiscible phases and where the metallic compounds are miscible. Gas solubility in condensed phases is negligible.

10 Consider the liquid-phase reaction of ethanol and acetic acid to form ethyl acetate and water at 25°C and 1 atm total pressure. For reaction of equimolar amounts of ethanol and acetic acid, what yield would be obtained at equilibrium? What would be the yield at 200°C and a pressure high enough to maintain liquid-phase reaction? In both cases, the assumption of ideal-liquid solution is made. How would the calculations be made and what additional information would be necessary if nonideal-liquid solution was assumed?

11 Carbon monoxide and liquid benzene are to be reacted over a catalyst to form liquid benzaldehyde at 50°C and 500 atm total pressure. Discuss quantitatively the calculational procedure you would use to determine the maximum yield of benzaldehyde and any assumptions necessary given standard free energies, enthalpies of formation, and heat capacities for each compound, liquid densities, either *PVT* data or an appropriate equation of state for carbon monoxide, and critical constants for each compound.

12 Consider an equilibrium system containing solid iron, ferrous oxide (FeO), magnetic iron oxide (Fe_3O_4), and gaseous CO and CO_2. Determine the number of degrees of freedom in the system and discuss them.

Using the data given below, estimate the temperature at which all the compounds are at equilibrium at 25°C and 1 atm total pressure, noting any assumptions made. Comment on whether this temperature is dependent on total pressure.

	ΔH_f (298 K), cal/mol	ΔG_f° (298 K), cal/mol	C_P, cal/(mol · K)
Fe(*s*)	0	0	6.0
FeO(*s*)	−63,700	−58,400	10.4
$Fe_3O_4(s)$	−267,000	−242,400	36.4
CO	−26,400	−32,800	7.0
CO_2	−94,100	−94,300	8.9

13 A certain porous ore contains iron and nickel orthosilicates (Fe_2SiO_4 and Ni_2SiO_4), which are reducible with CO, as follows:

$$\tfrac{1}{2}Fe_2SiO_4 + CO = Fe + \tfrac{1}{2}SiO_2 + CO_2 \tag{1}$$

$$\tfrac{1}{2}Ni_2SiO_4 + CO = Ni + \tfrac{1}{2}SiO_2 + CO_2 \tag{2}$$

The following reactions are also relevant:

$$FeO + CO = Fe + CO_2 \tag{3}$$

$$NiO + CO = Ni + CO_2 \tag{4}$$

$$C + CO_2 = 2CO \tag{5}$$

The silicates in this ore are so intimately admixed that separation by physical means is not feasible. It is proposed to reduce the nickel selectively to metal, leaving the iron as the silicate. The nickel metal can then be separated from the partly reduced ore mass by leaching. To avoid losing porosity by fusion, temperature during reduction should be kept below 1000°C. Again, to maintain porosity, no solid carbon is to form in the reduced product.

Equilibrium constants are as follows at around 900–1000°C.

$$\log K_P = \frac{58}{T} - 0.841 \qquad \text{for (1)}$$

$$\log K_P = \frac{1661}{T} - 0.0277 \qquad \text{for (2)}$$

$$\log K_P = \frac{950}{T} - 1.145 \qquad \text{for (3)}$$

$$\log K_P = \frac{2201}{T} + 0.0014 \qquad \text{for (4)}$$

$$\log K_P = -\frac{8950}{T} + 9.37 \qquad \text{for (5)}$$

In the foregoing expressions, K_P is in atm and T in K.

(*a*) Plot $\log(P_{CO_2}/P_{CO})$ vs. $1/T$ at a total pressure of 1 atm for each of the above reactions, including reaction (5), and show the range of conditions at 1 atm total pressure for a gas system containing only CO and CO_2 over which a nickel silicate–iron silicate mixture can be selectively reduced to metallic nickel without risking formation of metallic iron or the deposition of carbon.

(*b*) How could operations be conducted to keep the temperature at around 900°C and the gas composition within the safe range? It will be assumed that there is a source of pure CO within the plant.

14 If carbon disulfide and carbon dioxide are reacted over magnesia catalyst above 400°C, carbonyl sulfide is formed. If CS_2 and CO_2 in equimolar proportions are reacted to equilibrium at 900 K and 1 atm pressure, determine the composition of the product stream and the heat requirement per mole of COS formed, assuming isothermal reaction. Reactions which may occur are:

	K at 900 K	ΔH°, kcal, at 900 K
$C + S_2(g) = CS_2$	13.5	−3.1
$CO_2 + CS_2 = 2COS$	9.4	−1.7
$2COS = 2CO + S_2$	0.0015	45.5
$5CO_2 + CS_2 = 2SO_2 + 6CO$	10^{-15}	141.7
$3S_2(g) = S_6(g)$	0.89	−64.4
$4S_2(g) = S_8(g)$	0.193	−95.5

15 A 50 wt % solution of nitric acid in water is reacted at 400 K and 1 atm total pressure in the vapor phase. A set of independent reactions for the system is given. Determine the equilibrium gas composition by using the series-reactor technique.

$$NO + \tfrac{1}{2}O_2 = NO_2 \qquad K = 3950$$

$$2NO_2(g) = N_2O_4(g) \qquad K = 0.196$$

$$2HNO_3(g) + NO(g) = 3NO_2(g) + H_2O(g) \qquad K = 5348$$

(Rearrange the equations as necessary.)

16 Pure *p*-xylene is heated at 1 atm to 1000 K and allowed to form an equilibrium mixture of xylenes, trimethylbenzenes, benzene, and toluene only. Using minimization of free-energy techniques with lagrangian multipliers, determine the composition at equilibrium. (Assume all xylenes have the thermodynamic properties of *p*-xylene, and all trimethylbenzenes have standard free energies 15 percent higher than the xylenes when obtaining data from the appendixes.)

17 A mixture of $BaCO_3$ and solid C are reacted at 1482 K, where the following reactions occur:

$$BaCO_3 = BaO + CO_2 \qquad K_a = 0.1 \text{ atm}$$

$$C + CO_2 = 2CO$$

(*a*) Determine the product-gas composition of CO and CO_2 and its total pressure at equilibrium if the reaction occurs in an evacuated batch container.

(*b*) At what temperature should the reactions be run to make the total pressure in the container 1 atm at equilibrium K_a at 1558 K = 0.25 atm, at 1622 K = 0.5 atm, at 1694 K = 1.0 atm, and at 1994 K = 10 atm.

(*c*) Why would the second reaction be included if the purpose of the reaction is to form BaO?

18 A galvanic cell has one platinum electrode with hydrogen at a partial pressure of 0.9 atm, another electrode of mercury with solid mercurous chloride, and an aqueous electrolyte over which the pressure of gaseous hydrogen chloride is 0.01 atm. The electrode potential of this cell is 0.0110 V at 25°C. Assuming ideal gases and the reaction below, calculate ΔG, $\Delta G°$, and $E°$.

$$\tfrac{1}{2}H_2(g) + HgCl(c) = Hg(l) + HCl(g)$$

19 A diaphragm cell is operating such that the product caustic soda liquor contains chlorate and carbonate and the anode gas is impure. The cell consumes 3700 A at 4.6 V when the cell is processing 70 liters/h of saturated brine (sp gr = 1.18), containing 24 wt % NaCl and 0.1 wt % HCl. The cathode product liquor (sp gr = 1.14) contains 16.3% NaCl, 5.6% NaOH, 0.12% Na_2CO_3, and 0.14% $NaClO_3$. Dry anode gas contains 72% Cl_2, 7.1% O_2, 5.3% H_2, 0.8% CO_2, and 14.8% N_2 by volume. The cathode hydrogen is pure.

(*a*) Determine the gram-equivalents of Cl_2/h in the off-gas.

(*b*) Explain each impurity in the anode gas and the cell liquor. What causes the high cell voltage? What improvements should be made?

(*c*) Do a current balance by tabulating the equivalents per hour at the anode for Na_2CO_3, $NaClO_3$ in the product liquor, and the Cl_2, O_2, and CO_2 in the anode gas. Reconcile with the total current flow.

(*d*) Do a current balance at the cathode to all products and compare with the total current flow.

20 Consider the reactions for the steam reforming of methane at 600°C in Sec. 9.6.3. For a system beginning with a methane and water feed in which no carbon is to be formed:

(*a*) Determine the range of pressures which can be utilized.

(*b*) Determine at any chosen pressure how you would determine the optimum ratio of CH_4/H_2O in the feed for obtaining a H_2/CO ratio close to 3.

(*c*) Determine the composition of the equilibrium mixture formed at a pressure of 1 atm with a feed ratio of H_2O/CH_4 of unity.

(*d*) Discuss how you would completely optimize such a reaction system toward a maximum equilibrium yield of H_2 and CO.

APPENDIX

A

PHYSICAL PROPERTY TABLES (INCLUDING VAPOR PRESSURE) FOR COMMON SUBSTANCES

Common physical properties for 96 pure compounds are tabulated. The values given are those values recommended by the American Petroleum Institute's "Technical Data Book—Petroleum Refining" and the American Institute of Chemical Engineers' Design Institute for Physical Property Data in the "Data Compilation." (See Chap. 1 for references.)

Definitions and units for each of the properties are given below:

MW	molecular weight, IUPAC
NBP	normal boiling point, K
FP	freezing point at 1 atm, K
TC	critical temperature, K
PC	critical pressure, kPA
VC	critical volume, $m^3/kmol$
ZC	critical compressibility factor as defined by $(PC)(VC)/(R)(TC)$
ω	acentric factor as defined by Eq. (2.9)
$\bar{V}_L$	liquid molar volume at 25°C and 1 atm, $(m^3/kmol)(10^2)$
C_P^*	ideal-gas heat capacity at 25°C, kJ/(kmol · K)
ΔH_f	heat of formation from the elements in their standard states to an ideal gas at 25°C, $(kJ/kmol)(10^{-4})$
ΔG_f	Gibbs free energy of formation from the elements in their standard states to an ideal gas at 25°C, $(kJ/kmol)(10^{-4})$
S	absolute entropy of the ideal gas at 25°C, $[kJ/(kmol \cdot K)](10^{-2})$
$(C_P)_l$	liquid heat capacity at 25°C and 1 atm if NBP greater than 25°C, at 25°C and saturation pressure in all other cases, kJ/(kmol · K)
ΔH_C	standard net (water gaseous) heat of combustion at 25°C, $(kJ/kmol)(10^{-5})$
ΔH_{fus}	heat of fusion at the freezing point, $(kJ/kmol)(10^{-3})$
ΔH_{vap}	heat of vaporization at the normal boiling point, $(kJ/kmol)(10^{-3})$

Note: ΔH_f, ΔG_f, S, and ΔH_C are totally thermodynamically consistent for all compounds in the table.

Physical property data

No.	Compound	MW	NBP, K	*FP*, K	*TC*, K	*PC*, kPa	*VC*, m^3/kmol	*ZC*	ω	$\overline{V}_L$, $(m^3/kmol)(10)^2$
	Paraffins									
1	Methane	16.04	111.7	90.7	190.6	4604.	0.0992	0.288	0.01083	3.780
2	Ethane	30.54	184.6	89.9	305.4	4880.	0.1479	0.284	0.0989	5.474
3	Propane	44.10	231.1	85.5	369.8	4249.	0.2029	0.280	0.1517	7.561
4	*n*-Butane	58.12	272.7	134.8	425.2	3797.	0.2550	0.274	0.1931	9.665
5	Isobutane	58.12	261.4	113.5	408.1	3648.	0.2627	0.282	0.1770	9.770
6	*n*-Pentane	72.15	309.2	143.4	469.7	3369.	0.3040	0.262	0.2486	11.610
7	*n*-Hexane	86.18	341.9	177.8	507.4	3012.	0.3701	0.264	0.3047	13.129
8	*n*-Heptane	100.2	371.6	182.6	540.3	2736.	0.4323	0.263	0.3494	14.711
9	*n*-Octane	114.2	398.8	216.4	568.8	2486.	0.4920	0.259	0.3962	16.366
10	*n*-Nonane	128.3	424.0	219.7	595.6	2306.	0.5477	0.255	0.4368	17.932
11	*n*-Decane	142.3	447.3	243.5	618.5	2123.	0.6031	0.249	0.4842	19.534
12	*n*-Hexadecane	226.4	560.0	291.3	720.6	1419.	0.9300	0.220	0.7471	29.406
13	2,2,4-Trimethyl pentane	114.2	372.4	165.8	544.0	2568.	0.4680	0.266	0.3031	16.536
	Cycloparaffins									
14	Cyclopentane	70.13	322.4	179.3	511.8	4502.	0.2583	0.273	0.1943	9.351
15	Methylcyclopentane	84.16	345.0	130.7	532.8	3784.	0.319	0.272	0.2302	11.304
16	Cyclohexane	84.16	353.9	279.7	553.5	4075.	0.3079	0.273	0.2149	10.886
17	Methylcyclohexane	98.19	374.1	146.6	572.2	3471.	0.3680	0.268	0.2350	12.819
18	*cis*-Decalin	138.3	469.0	230.2	677.2	2498.	0.416	0.267	0.2942	15.462
19	*trans*-Decalin	138.3	460.5	242.8	664.2	2616.	0.393	0.238	0.2536	15.933
	Unsaturates									
20	Ethylene	28.05	169.5	104.0	282.4	5032.	0.1291	0.277	0.0852	4.924
21	Propylene	42.08	225.4	87.9	364.8	4613.	0.1810	0.275	0.1424	6.880
22	Isobutylene	56.11	266.3	132.8	417.9	3999.	0.2389	0.275	0.1893	8.943
23	1-Butene	56.11	266.9	87.8	419.6	4020.	0.2399	0.276	0.1867	8.962
24	*c*-2-Butene	56.11	276.9	134.2	435.6	4206.	0.2340	0.272	0.2030	8.745
25	*t*-2-Butene	56.11	274.0	167.6	428.6	4102.	0.2382	0.274	0.2182	8.941
26	1-Hexene	84.16	336.6	133.3	504.0	3140.	0.3540	0.265	0.2800	12.610
27	1-Decene	140.3	443.8	206.9	615.2	2200.	0.650	...	...	19.035
28	1,3-Butadiene	54.09	268.7	164.3	425.4	4330.	0.2208	0.270	0.1932	8.311
29	Acetylene	26.04	188.7	192.0	308.3	6139.	0.1130	0.271	0.1873	4.214
30	Cyclohexene	82.1	356.2	169.7	...	...	...	...	...	10.19
	Aromatics									
31	Benzene	78.11	353.2	278.7	562.2	4898.	0.2589	0.271	0.2108	8.950
32	Toluene	92.14	383.8	178.2	591.8	4109.	0.3158	0.264	0.2641	10.656
33	Ethylbenzene	106.2	409.4	178.2	617.2	3609.	0.3738	0.264	0.3036	12.269
34	*o*-Xylene	106.2	417.6	248.0	630.4	3733.	0.3692	0.263	0.3127	12.125
35	*m*-Xylene	106.2	412.3	225.3	617.1	3541.	0.3758	0.259	0.3260	12.341
36	*p*-Xylene	106.2	411.5	286.5	616.3	3511.	0.3791	0.260	0.3259	12.385
37	Cumene	120.2	425.6	177.1	631.2	3209.	0.4277	0.262	0.3482	13.977

Physical property data (Continued)

No.	Compound	MW	NBP, K	FP, K	TC, K	PC, kPa	VC, m^3/kmol	ZC	ω	$\overline{V}_L$, (m^3/kmol)(10)^2
	Aromatics									
38	Naphthalene	128.2	491.1	353.4	748.4	4051.	0.4130	0.269	0.3019	13.083
39	Tetralin	132.2	480.8	237.4	720.2	3300.	0.4410	0.243	0.2981	13.670
40	Styrene	104.2	418.3	242.5	647.6	3999.	0.3518	0.261	0.2339	11.584
41	Phenanthrene	178.2	613.5	372.3	869.3	2900.	0.6180	0.248	0.4856	16.575
42	Indene	116.2	455.8	271.7	687.0	3820.	0.3920	0.262	0.3352	11.715
43	Biphenyl	154.2	528.2	342.4	789.3	3847.	0.5016	0.294	0.3659	15.588
	Gases									
44	Carbon monoxide	28.01	81.7	68.1	132.9	3499.	0.0931	0.296	0.0663	3.545
45	Carbon dioxide	44.01	194.6	216.6	304.2	7382.	0.0940	0.274	0.2276	3.742
46	Hydrogen sulfide	34.08	212.8	187.7	373.5	8937.	0.0985	0.283	0.0814	3.614
47	Carbon disulfide	76.13	319.4	161.5	552.0	7903.	0.1600	0.276	0.1079	6.062
48	Sulfur dioxide	64.06	263.1	197.7	430.8	7884.	0.1220	0.268	0.2451	4.386
49	Sulfur trioxide	80.06	317.9	290.0	490.9	8207.	0.1271	0.256	0.4215	4.235
50	Water	18.015	373.15	273.15	647.3	22090.	0.0568	0.233	0.3442	1.807
51	Ammonia	17.03	239.7	195.4	405.7	1128.	0.0725	0.242	0.2515	2.499
52	Chlorine	70.91	239.1	172.2	417.2	7711.	0.1238	0.275	0.0690	4.539
53	Hydrogen chloride	36.47	188.1	159.0	324.7	8309.	0.0810	0.250	0.1322	3.053
54	Nitric oxide	30.01	121.4	112.2	180.2	6485.	0.0577	0.250	0.5846	2.319
55	Nitrous oxide	44.01	184.7	182.3	309.6	7245.	0.0974	0.274	0.1418	3.476
56	Hydrogen	2.016	20.38	13.95	33.25	1297.	0.0650	0.305	−0.2252	2.856
57	Oxygen	32.00	90.2	54.4	154.6	5043.	0.0734	0.288	0.0218	2.804
58	Nitrogen	28.	77.4	63.1	126.1	3394.	0.0901	0.292	0.0403	3.471
59	Argon	39.95	87.3	83.8	150.9	4898.	0.0746	0.291	−0.0038	2.859
60	Helium-4	4.00	4.2	1.8	5.2	228.	0.0573	0.301	−0.3948	3.201
	Alcohols									
61	Methanol	32.04	337.9	175.5	512.6	8096.	0.1178	0.224	0.5688	4.069
62	Ethanol	46.07	351.4	159.1	516.3	6383.	0.1667	0.248	0.6371	5.852
63	Isopropanol	60.10	355.4	184.7	508.3	4764.	0.2201	0.248	0.6689	7.692
64	n-Propanol	60.10	370.4	147.0	536.7	5170.	0.2185	0.253	0.6279	7.494
65	1-Hexanol	102.2	430.2	228.6	611.4	3510.	0.3813	0.263	0.5803	12.521
66	Ethylene glycol	62.07	470.5	260.2	645.1	7530.	0.1910	0.268	1.1360	5.591
67	1,2-Propylene glycol	76.10	460.8	213.2	626.0	6100.	0.2390	0.280	1.1058	7.369
	Other oxygenated compounds									
68	Formaldehyde	30.03	254.1	181.2	408.0	6586.	0.1050	0.204	0.2816	3.674
69	Acetaldehyde	44.05	293.6	150.2	461.0	5550.	0.1570	0.227	0.3167	5.652
70	Acetone	58.08	329.4	178.5	508.2	4701.	0.2090	0.233	0.3064	7.393
71	Methyl ethyl ketone	72.11	352.8	186.5	535.5	4154.	0.2670	0.249	0.3241	9.020
72	Acetic acid	60.05	391.1	289.8	592.7	5786.	0.1710	0.201	0.4624	5.758
73	Terephthalic acid	166.1	sub	700.	1392.	3952.	0.4240	0.145	...	8.517
74	Methyl acetate	74.08	330.1	175.2	506.8	4691.	0.2280	0.254	0.3254	7.983

Physical property data (Continued)

No.	Compound	MW	NBP, K	FP, K	TC, K	PC, kPa	VC, m^3/kmol	ZC	ω	$\overline{V}_L$, $(m^3/kmol)(10)^2$
	Other oxygenated compounds									
75	Ethyl acetate	88.11	350.2	189.6	523.3	3830.	0.2860	0.252	0.3611	9.861
76	Phenol	94.11	455.0	314.1	694.3	6130.	0.2290	0.243	0.4259	8.909
77	Dimethyl terephthalate	194.2	561.2	413.8	766.	2785.	0.5290	0.231	0.6887	17.887
78	Diethyl ether	74.12	307.6	156.9	466.7	3638.	0.2800	0.262	0.2846	10.467
79	Isopropyl ether	102.2	341.5	187.7	500.0	2878.	0.3860	0.267	0.3238	14.178
80	Benzoic acid	122.1	522.4	395.5	751.	4470.	0.3440	0.246	0.6310	11.291
	Nitrogen and sulfur compounds									
81	Methylamine	31.06	266.8	179.7	430.1	7458.	0.1540	0.322	0.2813	4.483
82	Ethylamine	45.08	289.7	192.2	456.2	5624.	0.1820	0.270	0.2851	6.563
83	Methylmercaptan	48.10	279.1	150.2	470.0	7235.	0.1450	0.268	0.1486	5.414
84	Ethylmercaptan	62.13	308.2	125.3	499.2	5490.	0.2070	0.274	0.1921	7.461
85	Dimethylsulfide	62.13	310.5	174.9	503.0	5530.	0.2009	0.266	0.1893	7.309
86	Urea	60.06	478.	405.9	725.0	6750.	0.2180	0.244	...	...
	Halogenated organics									
87	Methyl chloride	50.49	248.9	175.4	416.3	6679.	0.1390	0.268	0.1603	5.018
88	Ethyl chloride	64.51	285.4	136.8	460.4	5269.	0.2000	0.275	0.1905	7.118
89	Methylene chloride	84.93	312.9	178.0	510.0	6080.	0.1850	0.265	0.1916	6.443
90	Chloroform	119.4	334.3	209.6	536.4	5472.	0.2390	0.293	0.2209	8.066
91	Carbon tetrachloride	153.8	349.8	250.3	556.4	4560.	0.2760	0.272	0.1926	9.711
92	Dichlorodi-fluoromethane	120.9	243.4	115.2	385.0	4125.	0.2170	0.280	0.1796	8.138
93	Dichloroethylene	99.0	356.6	237.5	561.0	5370.	0.2200	0.253	0.2876	7.944
	Inorganics									
94	Sodium hydroxide	40.00	1830.	596.0	2815.	25,331.	0.2	0.22	1.015	2.243
95	Sulfuric acid	98.07	610.	283.5	925.	5066.	0.3	0.2	0.467	5.346
96	Phosphoric acid	98.00	680.	315.5	1030.	5066.	0.34	0.2	...	5.278

Physical property data (Continued)

	Ideal-gas thermal properties							
No.	C_P^*, kJ/(kmol · K)	ΔH_f, (kJ/kmol)(10^{-4})	ΔG_f, (kJ/kmol)(10^{-4})	S, [kJ/(kmol · K)] (10^{-2})	$(C_P)_l$, kJ/(kmol · K)	Net ΔH_C, (kJ/kmol) (10^{-5})	ΔH_{fus}, (kJ/kmol) (10^{-3})	ΔH_{vap}, (kJ/kmol) (10^{-3})
1	35.73	−7.485	−5.082	1.8627	83.42*	−8.023	0.9414	8.165
2	52.48	−8.385	−3.195	2.2912	142.4	−14.278	2.859	14.70
3	73.59	−10.468	−2.440	2.7020	117.7	−20.440	3.524	18.80
4	98.85	−12.577	−1.607	3.1012	140.6	−26.571	4.661	22.50
5	97.23	−13.418	−2.076	2.9539	141.6	−26.487	4.540	21.40
6	120.1	−14.671	−0.877	3.495	167.5	−32.454	8.393	25.99
7	143.1	−16.694	−0.008	3.887	195.2	−38.552	13.079	29.12
8	165.9	−18.765	+0.815	4.280	225.5	−44.649	14.054	31.73
9	188.7	−20.882	+1.592	4.672	253.8	−50.746	20.740	34.77
10	211.6	−22.887	+2.473	5.064	283.9	−56.851	15.468	37.69
11	234.5	−24.953	+3.297	5.457	314.5	−62.942	28.715	40.02
12	370.3	−37.334	+8.368	7.783	501.5	−99.528	53.359	52.52
13	187.6	−22.401	+1.393	4.230	238.3	−50.653	9.196	31.02
14	82.92	−7.703	+3.887	2.929	127.2	−30.709	0.6088	27.26
15	...	−10.67	+3.577	3.399	...	−36.74	...	...
16	105.9	−12.314	+3.176	2.982	154.8	−36.558	2.677	29.89
17	135.1	−15.477	+2.728	3.4334	184.8	−42.571	6.751	31.82
18	...	−16.924	+8.552	3.7773	...	−58.92	...	39.80
19	...	−18.217	+7.355	3.7455	...	−58.81	...	38.62
20	43.73	5.223	6.812	2.1945	179.6	−13.226	3.351	13.46
21	63.92	1.971	6.214	2.6660	112.6	−19.257	3.003	18.49
22	89.16	−1.590	5.807	2.9359	130.7	−25.240	5.931	22.21
23	85.68	−0.054	7.024	3.0783	127.8	−25.412	3.848	22.45
24	78.94	−0.699	6.586	3.0083	128.0	−25.344	7.309	23.43
25	87.86	−1.117	6.297	2.9648	127.9	−25.287	9.758	22.91
26	128.7	−4.167	8.761	3.8464	183.3	−37.394	9.347	28.64
27	214.1	−12.33	12.20	...	300.3	−61.79	...	61.79
28	79.48	11.017	15.067	2.7874	123.6	−24.097	7.985	22.49
29	44.23	22.675	20.920	2.0082	...	−12.556	3.766	17.02
30	105.	...	...	...	149.	−35.32	...	35.32
31	81.66	8.293	12.966	2.6920	136.3	−31.356	9.866	30.75
32	103.8	5.000	12.229	3.2067	157.3	−37.339	6.636	33.60
33	128.4	2.979	13.057	3.6045	183.2	−43.448	9.184	35.91

Physical property data (Continued)

No.	Ideal-gas thermal properties: C_p^*, kJ/(kmol · K)	ΔH_f, (kJ/kmol)(10^{-4})	ΔG_f, (kJ/kmol)(10^{-4})	S, [kJ/(kmol · K)] (10^{-2})	$(C_p)_l$, kJ/(kmol · K)	Net ΔH_C, (kJ/kmol) (10^{-5})	ΔH_{fus}, (kJ/kmol) (10^{-3})	ΔH_{vap}, (kJ/kmol) (10^{-3})
34	133.3	1.900	12.208	3.5275	186.1	−43.328	13.598	37.00
35	127.6	1.724	11.885	3.5769	183.1	−43.318	11.569	36.33
36	126.9	1.803	12.127	3.5210	181.5	−43.328	17.113	35.82
37	151.7	0.393	13.698	3.8857	210.7	−49.513	7.786	38.07
38	133.3	15.058	22.410	3.3315	176.4	−49.809	18.979	43.42
39	154.9	2.661	16.710	3.6964	217.5	−53.575	12.450	42.35
40	122.1	14.736	21.380	3.4510	182.0	−42.193	10.950	36.62
41	189.8	20.710	30.810	3.9450	260.5	−68.344	16.463	54.38
42	125.1	16.328	23.397	3.3687	187.1	−46.195	10.200	40.94
43	162.3	18.209	28.008	3.9267	226.2	−60.317	18.577	49.05
44	29.14	11.053	−13.716	1.9754	76.13	−2.830	0.841	5.910
45	37.05	−39.352	−39.441	2.1369	209.3	0.0	8.652	15.30†
46	34.16	−2.063	−3.344	2.0559	67.7†	−6.170	2.377	18.89
47	45.48	11.707	6.691	2.3779	76.43	−11.042	4.393	27.03
48	39.87	−29.684	−30.016	2.4811	89.00	0.0	7.402	24.69
49	50.81	−39.572	−37.095	2.5651	258.1	0.989	1.968	40.23
50	33.60	−24.182	−22.859	1.8872	75.41	0.0	6.002	40.67
51	35.66	−4.590	−1.640	1.9267	80.88	−3.168	5.657	23.33
52	33.94	0.0	0.0	2.2297	67.0†	0.0	6.406	20.41
53	29.13	−9.231	−9.530	1.8680	62.†	−0.286	1.998	16.21
54	29.83	9.025	8.657	2.1060	67.6†	−0.572	2.301	13.55
55	38.52	8.205	10.417	2.1985	77.5†	−0.159	6.540	17.21
56	29.16*	0.0	0.0	1.3057	13.93†	−2.418	1.172	0.935†
57	29.34	0.0	0.0	2.0504	53.32†	0.0	4.435	6.825
58	29.10	0.0	0.0	1.9150	55.52†	0.0	7.196	5.543
59	20.78	0.0	0.0	1.5473	41.66†	0.0	1.182	6.440
60	20.79	0.0	0.0	1.2604	~ 12†	0.0	0.500	0.083
61	43.92	−20.108	−16.242	2.3970	80.92	−6.381	3.205	35.95
62	65.51	−23.443	−16.790	2.8259	112.1	−12.355	5.012	39.40
63	88.88	−27.242	−17.339	3.0991	156.5	−18.300	5.410	39.87
64	87.26	−25.640	−16.180	3.2472	144.6	−18.194	5.197	41.38
65	155.8	−31.757	−13.593	4.4150	244.1	−36.744	15.397	45.93

66	97.10	−38.932	−30.447	3.2355	148.6	−10.576	11.623	52.49
67	109.0	−42.150	−30.448	3.5187	190.8	−16.476	...	54.48
68	35.38	−11.590	−10.990	2.1866	...	−5.194	...	23.16
69	54.61	−16.619	−12.891	2.5020	109.4	−11.046	3.222	27.60
70	74.49	−21.715	−15.272	2.9535	126.3	−16.592	5.691	29.79
71	104.4	−23.836	−14.606	3.3811	158.6	−22.616	8.439	31.22
72	64.50	−43.225	−37.405	2.8250	123.3	−7.864	11.715	23.33
73	124.8	−71.789	−59.718	4.4237	...	−30.576	...	
74	85.41	−40.914	−32.154	3.1983	142.7	−14.609	...	30.61
75	113.4	−44.292	−32.740	3.6275	170.7	−20.548	10.480	32.32
76	103.8	−9.640	−3.263	3.1481	196.4	−29.215	11.514	47.31
77	168.0	−64.358	−47.365	5.4978	318.5	−44.115	31.631	53.35
78	112.5	−25.213	−12.175	3.4100	175.6	−25.035	7.301	27.09
79	157.9	−31.899	−11.924	3.8158	216.8	−37.023	11.046	29.04
80	103.4	−29.020	−21.041	3.6899	221.2	−30.951	18.075	50.63
81	50.07	−2.297	3.209	2.4330	96.95	−9.751	6.134	26.11
82	72.63	−4.602	3.728	2.8485	123.7	−15.874	...	27.35
83	50.29	−2.226	−0.916	2.5501	91.07	−11.517	5.904	25.48
84	72.70	−4.577	−0.427	2.9606	117.9	−17.357	4.975	26.87
85	74.09	−3.720	0.736	2.8585	118.1	−17.440	7.985	26.91
86	55.68	−24.560	−15.272	2.4937	120.5	−5.442	14.853	87.86†
87	40.77	−3.644	−6.295	2.3425	81.17	−6.754	6.548	21.55
88	62.73	−11.226	−6.050	2.7578	104.3	−12.849	4.452	24.75
89	51.23	−9.552	−6.897	2.7018	101.2	−5.139	4.602	28.38
90	65.68	−10.318	−7.041	2.9550	114.2	−3.799	9.540	29.79
91	83.80	−9.598	−5.367	3.0970	130.7	−2.581	2.535	29.84
92	73.10	−49.162	−45.272	3.0038	117.9	...	...	20.61
93	78.85	12.979	−7.393	3.0828	128.4	−11.050	8.828	32.16
94	48.66	−19.776	−20.046	2.2833	87.18*	...	6.611	
95	80.81	−73.513	−65.347	2.9370	139.1	...	10.711	50.1
96	...	125.42	−111.17	1.5062	145.0*	...	12.979	

*Extrapolated.

†At freezing point.

Vapor pressure

Vapor pressures are given for the range between the triple point and the critical point wherever possible. For each compound the temperature range in kelvins is given below the compound, e.g.,

Methane
90.67 190.58

The vapor pressures are given within this range in 10 equal intervals in units of pascals. For methane, e.g.,

90.67 . . . 190.58
0.117E + 05 . . . 0.459E + 07

Methane 90.67 190.58	90.67 0.117E + 05	100.66 0.370E + 05	110.65 0.935E + 05	120.64 0.201E + 06	130.63 0.382E + 06	140.62 0.663E + 06	150.62 0.107E + 07	160.61 0.163E + 07	170.60 0.238E + 07	180.59 0.336E + 07	190.58 0.459E + 07
Ethane 89.88 305.42	89.88 0.998E + 00	111.43 0.952E + 02	132.99 0.182E + 04	154.54 0.141E + 05	176.10 0.627E + 05	197.65 0.195E + 06	219.20 0.478E + 06	240.76 0.989E + 06	262.31 0.182E + 07	283.87 0.307E + 07	305.42 0.487E + 07
Propane 85.44 369.82	85.44 0.207E − 03	113.88 0.905E + 00	142.32 0.110E + 03	170.75 0.234E + 04	199.19 0.191E + 05	227.63 0.867E + 05	256.07 0.272E + 06	284.51 0.664E + 06	312.94 0.136E + 07	341.38 0.248E + 07	369.82 0.414E + 07
n-Butane 134.86 425.18	134.86 0.568E + 00	163.89 0.509E + 02	192.92 0.102E + 04	221.96 0.853E + 04	250.99 0.407E + 05	280.02 0.134E + 06	309.05 0.343E + 06	338.08 0.732E + 06	367.12 0.137E + 07	396.15 0.232E + 07	425.18 0.367E + 07
Isobutane 113.54 408.14	113.54 0.142E − 01	143.00 0.721E + 01	172.46 0.337E + 03	201.92 0.437E + 04	231.38 0.266E + 05	260.84 0.101E + 06	290.30 0.280E + 06	319.76 0.633E + 06	349.22 0.124E + 07	378.68 0.219E + 07	408.14 0.363E + 07
n-Pentane 143.42 469.65	143.42 0.529E − 01	176.04 0.119E + 02	208.67 0.395E + 03	241.29 0.440E + 04	273.91 0.251E + 05	306.53 0.928E + 05	339.16 0.256E + 06	371.78 0.578E + 06	404.40 0.113E + 07	437.03 0.201E + 07	469.65 0.332E + 07
n-Hexane 177.84 507.43	177.84 0.804E + 00	210.80 0.562E + 02	243.76 0.977E + 03	276.72 0.736E + 04	309.68 0.327E + 05	342.63 0.103E + 06	375.59 0.257E + 06	408.55 0.546E + 06	441.51 0.104E + 07	474.47 0.182E + 07	507.43 0.303E + 07
n-Heptane 182.56 540.26	182.56 0.115E + 00	218.33 0.158E + 02	254.10 0.411E + 03	289.87 0.400E + 04	325.64 0.211E + 05	361.41 0.746E + 05	397.18 0.202E + 06	432.95 0.456E + 06	468.72 0.905E + 06	504.49 0.164E + 07	540.26 0.280E + 07
n-Octane 216.38 568.83	216.38 0.254E + 01	251.62 0.800E + 02	286.87 0.971E + 03	322.11 0.634E + 04	357.36 0.270E + 05	392.60 0.851E + 05	427.85 0.215E + 06	463.09 0.462E + 06	498.34 0.877E + 06	533.58 0.151E + 07	568.83 0.243E + 07
2,2,4-Tri- methylpentane 165.78 543.96	165.78 0.145E − 01	203.60 0.476E + 01	241.42 0.195E + 03	279.23 0.246E + 04	317.05 0.153E + 05	354.87 0.600E + 05	392.69 0.174E + 06	430.51 0.409E + 06	468.32 0.829E + 06	506.14 0.151E + 07	543.96 0.255E + 07
n-Nonane 219.63 595.65	219.63 0.639E + 00	257.23 0.303E + 02	294.83 0.482E + 03	332.44 0.380E + 04	370.04 0.185E + 05	407.64 0.645E + 05	445.24 0.175E + 06	482.84 0.397E + 06	520.44 0.780E + 06	558.05 0.138E + 07	595.65 0.223E + 07
n-Decane 243.49 618.45	243.49 0.188E + 01	280.99 0.554E + 02	318.48 0.675E + 03	355.98 0.453E + 04	393.47 0.201E + 05	430.97 0.659E + 05	468.47 0.173E + 06	505.96 0.382E + 06	543.46 0.739E + 06	580.95 0.129E + 07	618.45 0.208E + 07
n-Hexadecane 291.32 720.60	291.32 0.875E − 01	334.25 0.611E + 01	377.18 0.128E + 03	420.10 0.122E + 04	463.03 0.686E + 04	505.96 0.265E + 05	548.89 0.787E + 05	591.82 0.193E + 06	634.74 0.412E + 06	677.67 0.792E + 06	720.60 0.141E + 07

Cyclopentane 179.28 511.76	179.28 0.834E + 01	212.53 0.278E + 03	245.78 0.318E + 04	279.02 0.188E + 05	312.27 0.717E + 05	345.52 0.204E + 06	378.77 0.474E + 06	412.02 0.950E + 06	445.26 0.171E + 07	478.51 0.286E + 07	511.76 0.450E + 07
Methyl- cyclopentane 130.72 532.79	130.72 0.669E − 04	170.93 0.395E + 00	211.13 0.609E + 02	251.34 0.152E + 04	291.55 0.137E + 05	331.75 0.663E + 05	371.96 0.217E + 06	412.17 0.549E + 06	452.37 0.116E + 07	492.58 0.219E + 07	532.79 0.378E + 07
Cyclohexane 279.69 553.54	279.69 0.536E + 04	307.07 0.193E + 05	334.46 0.543E + 05	361.84 0.127E + 06	389.23 0.259E + 06	416.61 0.477E + 06	444.00 0.810E + 06	471.39 0.129E + 07	498.77 0.197E + 07	526.16 0.288E + 07	553.54 0.409E + 07
Methylcyclohexane 146.58 572.19	146.58 0.153E − 03	189.14 0.589E + 00	231.70 0.753E + 02	274.26 0.169E + 04	316.82 0.142E + 05	359.38 0.660E + 05	401.95 0.209E + 06	444.51 0.516E + 06	487.07 0.108E + 07	529.63 0.201E + 07	572.19 0.348E + 07
cis-Decalin 230.20 702.25	230.20 0.157E + 00	277.40 0.216E + 02	324.61 0.553E + 03	371.81 0.530E + 04	419.02 0.274E + 05	466.22 0.950E + 05	513.43 0.252E + 06	560.63 0.555E + 06	607.84 0.108E + 07	655.04 0.193E + 07	702.25 0.324E + 07
trans-Decalin 242.79 687.05	242.79 0.115E + 01	287.22 0.739E + 02	331.64 0.121E + 04	376.07 0.867E − 04	420.49 0.370E + 05	464.92 0.113E + 06	509.34 0.272E + 06	553.77 0.562E + 06	598.20 0.104E + 07	642.62 0.179E + 07	687.05 0.291E + 07
Ethylene 103.97 282.36	103.97 0.140E + 03	121.81 0.184E + 04	139.65 0.119E + 05	157.49 0.480E + 05	175.33 0.143E + 06	193.17 0.341E + 06	211.00 0.698E + 06	228.84 0.127E + 07	246.68 0.213E + 07	264.52 0.335E + 07	282.36 0.502E + 07
Propylene 87.90 364.76	87.90 0.918E − 03	115.59 0.214E + 01	143.27 0.193E + 03	170.96 0.346E + 04	198.64 0.252E + 05	226.33 0.107E + 06	254.02 0.318E + 06	281.70 0.751E + 06	309.39 0.151E + 07	337.07 0.274E + 07	364.76 0.459E + 07
Isobutylene 132.81 417.90	132.81 0.654E + 00	161.32 0.605E + 02	189.83 0.120E + 04	218.34 0.967E + 04	246.85 0.448E + 05	275.35 0.145E + 06	303.86 0.365E + 06	332.37 0.773E + 06	360.88 0.145E + 07	389.39 0.248E + 07	417.90 0.397E + 07
1-Butene 87.80 419.59	87.80 0.356E − 06	120.98 0.382E − 01	154.16 0.190E + 02	187.34 0.831E + 03	220.52 0.101E − 05	253.70 0.579E + 05	286.87 0.209E + 06	320.05 0.558E + 06	353.23 0.121E + 07	386.41 0.229E + 07	419.59 0.392E + 07
cis-2-Butene 134.26 435.58	134.26 0.247E + 00	164.39 0.347E + 02	194.52 0.865E + 03	224.66 0.801E + 04	254.79 0.403E + 05	284.92 0.137E + 06	315.05 0.355E + 06	345.18 0.770E + 06	375.32 0.147E + 07	405.45 0.257E + 07	435.58 0.421E + 07
1-Hexene 133.39 504.03	133.39 0.299E − 03	170.45 0.652E + 00	207.52 0.675E + 02	244.58 0.141E + 04	281.65 0.118E + 05	318.71 0.553E + 05	355.77 0.179E + 06	392.84 0.452E + 06	429.90 0.963E + 06	466.97 0.182E + 07	504.03 0.317E + 07
1-Decene 206.89 617.05	206.89 0.152E − 01	247.91 0.312E + 01	288.92 0.116E + 03	329.94 0.151E + 04	370.95 0.100E + 05	411.97 0.419E + 05	452.99 0.127E + 06	494.00 0.310E + 06	535.02 0.647E + 06	576.03 0.122E + 07	617.05 0.217E + 07
1,3-Butadiene 164.25 425.37	164.25 0.691E + 02	190.36 0.104E + 04	216.47 0.739E + 04	242.59 0.323E + 05	268.70 0.101E + 06	294.81 0.253E + 06	320.92 0.536E + 06	347.03 0.101E + 07	373.15 0.174E + 07	399.26 0.281E + 07	425.37 0.433E + 07
Acetylene 192.40 308.32	192.40 0.126E + 06	203.99 0.232E + 06	215.58 0.395E + 06	227.18 0.629E + 06	238.77 0.953E + 06	250.36 0.138E + 07	261.95 0.195E + 07	273.54 0.267E + 07	285.14 0.358E + 07	296.73 0.472E + 07	308.32 0.615E + 07
Cyclohexene 169.67 560.40	169.67 0.888E − 01	208.74 0.207E + 02	247.82 0.657E + 03	286.89 0.689E + 04	325.96 0.371E + 05	365.03 0.131E + 06	404.11 0.349E + 06	443.18 0.771E + 06	482.25 0.149E + 07	521.33 0.263E + 07	560.40 0.434E + 07

Vapor pressure

Vapor pressures are given for the range between the triple point and the critical point wherever possible. For each compound the temperature range in kelvins is given below the compound, e.g.,

Methane
90.67 190.58

The vapor pressures are given within this range in 10 equal intervals in units of pascals. For methane, e.g.,

90.67 . . . 190.58
0.117E + 05 . . . 0.459E + 07

Benzene 278.68 562.16	278.68 0.480E + 04	307.03 0.188E + 05	335.38 0.562E + 05	363.72 0.138E + 06	392.07 0.291E + 06	420.42 0.549E + 06	448.77 0.948E + 06	477.12 0.153E + 07	505.46 0.234E + 07	533.81 0.343E + 07	562.16 0.487E + 07
Toluene 178.18 591.79	178.18 0.410E − 01	219.54 0.109E + 02	260.90 0.397E + 03	302.26 0.468E + 04	343.62 0.276E + 05	384.98 0.105E + 06	426.35 0.295E + 06	467.71 0.679E + 06	509.07 0.135E + 07	550.43 0.244E + 07	591.79 0.410E + 07
Ethylbenzene 178.15 617.17	178.15 0.404E − 02	222.05 0.264E + 01	265.95 0.156E + 03	309.86 0.243E + 04	353.76 0.172E + 05	397.66 0.733E + 05	441.56 0.224E + 06	485.46 0.543E + 06	529.36 0.113E + 07	573.27 0.209E + 07	617.17 0.360E + 07
o-Xylene 247.98 630.37	247.98 0.218E + 02	286.22 0.421E + 03	324.46 0.361E + 04	362.70 0.182E + 05	400.94 0.637E + 05	439.17 0.173E + 06	477.41 0.392E + 06	515.65 0.777E + 06	553.89 0.140E + 07	592.13 0.234E + 07	630.37 0.372E + 07
m-Xylene 225.30 617.05	225.30 0.323E + 01	264.47 0.120E + 03	303.65 0.154E + 04	342.82 0.100E + 05	382.00 0.418E + 05	421.17 0.128E + 06	460.35 0.316E + 06	499.52 0.666E + 06	538.70 0.125E + 07	577.87 0.217E + 07	617.05 0.353E + 07
p-Xylene 286.41 616.26	286.41 0.581E + 03	319.39 0.367E + 04	352.38 0.153E + 05	385.36 0.477E + 05	418.35 0.120E + 06	451.33 0.260E + 06	484.32 0.499E + 06	517.30 0.880E + 06	550.29 0.145E + 07	583.27 0.227E + 07	616.26 0.341E + 07
Cumene 177.14 631.15	177.14 0.397E − 03	222.54 0.692E + 00	267.94 0.697E + 02	313.34 0.147E + 04	358.74 0.122E + 05	404.14 0.561E + 05	449.55 0.175E + 06	494.95 0.426E + 06	540.35 0.886E + 06	585.75 0.170E + 07	631.15 0.321E + 07
Naphthalene 353.43 748.35	353.43 0.999E + 03	392.92 0.559E + 04	432.41 0.217E + 05	471.91 0.645E + 05	511.40 0.158E + 06	550.89 0.332E + 06	590.38 0.626E + 06	629.87 0.108E + 07	669.37 0.175E + 07	708.86 0.269E + 07	748.35 0.398E + 07
1,2,3,4-Tetralin 237.38 720.15	237.38 0.191E + 00	285.66 0.206E + 02	333.93 0.494E + 03	382.21 0.480E + 04	430.49 0.259E + 05	478.76 0.946E + 05	527.04 0.262E + 06	575.32 0.593E + 06	623.59 0.116E + 07	671.87 0.204E + 07	720.15 0.329E + 07
Styrene 242.54 648.00	242.54 0.106E + 02	283.09 0.308E + 03	323.63 0.329E + 04	364.18 0.186E + 05	404.72 0.692E + 05	445.27 0.193E + 06	485.82 0.442E + 06	526.36 0.874E + 06	566.91 0.155E + 07	607.45 0.256E + 07	648.00 0.398E + 07
Indene 271.70 687.00	271.70 0.169E + 02	313.23 0.400E + 03	354.76 0.368E + 04	396.29 0.187E + 05	437.82 0.644E + 05	479.35 0.171E + 06	520.88 0.380E + 06	562.41 0.749E + 06	603.94 0.136E + 07	645.47 0.233E + 07	687.00 0.384E + 07
Biphenyl 342.37 789.26	342.37 0.843E + 02	387.06 0.103E + 04	431.75 0.665E + 04	476.44 0.278E + 05	521.13 0.859E + 05	565.81 0.213E + 06	610.50 0.452E + 06	655.19 0.855E + 06	699.88 0.149E + 07	744.57 0.243E + 07	789.26 0.382E + 07
Carbon monoxide 68.15 132.92	68.15 0.152E + 05	74.63 0.418E + 05	81.10 0.944E + 05	87.58 0.185E + 06	94.06 0.327E + 06	100.53 0.535E + 06	107.01 0.828E + 06	113.49 0.123E + 07	119.97 0.177E + 07	126.44 0.250E + 07	132.92 0.347E + 07

Carbon dioxide 216.58 304.19	216.58 0.514E + 06	225.34 0.739E + 06	234.10 0.103E + 07	242.86 0.140E + 07	251.62 0.187E + 07	260.38 0.244E + 07	269.15 0.313E + 07	277.91 0.395E + 07	286.67 0.492E + 07	295.43 0.605E + 07	304.19 0.736E + 07
Hydrogen sulfide 187.68 373.53	187.68 0.228E + 05	206.26 0.712E + 05	224.85 0.181E + 06	243.43 0.392E + 06	262.02 0.754E + 06	280.60 0.132E + 07	299.19 0.214E + 07	317.77 0.326E + 07	336.36 0.475E + 07	354.94 0.663E + 07	373.53 0.896E + 07
Carbon disulfide 161.11 552.00	161.11 0.153E + 01	200.20 0.144E + 03	239.29 0.273E + 04	278.38 0.212E + 05	317.47 0.955E + 05	356.55 0.302E + 06	395.64 0.752E + 06	434.73 0.158E + 07	473.82 0.295E + 07	512.91 0.503E + 07	552.00 0.801E + 07
Sulfur dioxide 197.67 430.75	197.67 0.162E + 04	220.98 0.963E + 04	244.29 0.394E + 05	267.59 0.122E + 06	290.90 0.310E + 06	314.21 0.671E + 06	337.52 0.129E + 07	360.83 0.224E + 07	384.13 0.361E + 07	407.44 0.545E + 07	430.75 0.783E + 07
Sulfur trioxide 289.95 490.85	289.95 0.209E + 05	310.04 0.688E + 05	330.13 0.180E + 06	350.22 0.396E + 06	370.31 0.760E + 06	390.40 0.131E + 07	410.49 0.207E + 07	430.58 0.309E + 07	450.67 0.438E + 07	470.76 0.598E + 07	490.85 0.797E + 07
Water 273.16 647.29	273.16 0.615E + 03	310.57 0.645E + 04	347.99 0.384E + 05	385.40 0.154E + 06	422.81 0.471E + 06	460.22 0.117E + 07	497.64 0.252E + 07	535.05 0.484E + 07	572.46 0.852E + 07	609.88 0.140E + 08	647.29 0.220E + 08
Ammonia 195.41 405.65	195.41 0.612E + 04	216.43 0.273E + 05	237.46 0.904E + 05	258.48 0.240E + 06	279.51 0.541E + 06	300.53 0.108E + 07	321.55 0.195E + 07	342.58 0.327E + 07	363.60 0.517E + 07	384.63 0.779E + 07	405.65 0.113E + 08
Chlorine 172.12 417.15	172.12 0.137E + 04	196.62 0.970E + 04	221.13 0.424E + 05	245.63 0.134E + 06	270.13 0.334E + 06	294.63 0.708E + 06	319.14 0.133E + 07	343.64 0.226E + 07	368.14 0.360E + 07	392.65 0.541E + 07	417.15 0.779E + 07
Hydrogen chloride 158.97 324.65	158.97 0.135E + 05	175.54 0.468E + 05	192.11 0.127E + 06	208.67 0.286E + 06	225.24 0.568E + 06	241.81 0.102E + 07	258.38 0.170E + 07	274.95 0.268E + 07	291.51 0.404E + 07	308.08 0.589E + 07	324.65 0.836E + 07
Nitric oxide 109.50 180.15	109.50 0.219E + 05	116.56 0.571E + 05	123.63 0.130E + 06	130.69 0.265E + 06	137.76 0.494E + 06	144.82 0.856E + 06	151.89 0.140E + 07	158.95 0.217E + 07	166.02 0.322E + 07	173.08 0.464E + 07	180.15 0.648E + 07
Nitrous oxide 182.30 309.57	182.30 0.867E + 05	195.03 0.185E + 06	207.75 0.352E + 06	220.48 0.611E + 06	233.21 0.985E + 06	245.94 0.150E + 07	258.66 0.218E + 07	271.39 0.307E + 07	284.12 0.418E + 07	296.84 0.556E + 07	309.57 0.726E + 07
Hydrogen (normal) 13.95 33.25	13.95 0.662E + 04	15.88 0.193E + 05	17.81 0.437E + 05	19.74 0.839E + 05	21.67 0.144E + 06	23.60 0.229E + 06	25.53 0.346E + 06	27.46 0.503E + 06	29.39 0.713E + 06	31.32 0.997E + 06	33.25 0.138E + 07
Oxygen 54.35 154.58	54.35 0.146E + 03	64.37 0.205E + 04	74.40 0.132E + 05	84.42 0.530E + 05	94.44 0.155E + 06	104.46 0.364E + 06	114.49 0.733E + 06	124.51 0.132E + 07	134.53 0.219E + 07	144.56 0.340E + 07	154.58 0.504E + 07
Nitrogen 63.15 126.10	63.15 0.125E + 05	69.44 0.355E + 05	75.74 0.834E + 05	82.03 0.170E + 06	88.33 0.311E + 06	94.62 0.524E + 06	100.92 0.823E + 06	107.21 0.124E + 07	113.51 0.179E + 07	119.80 0.249E + 07	126.10 0.338E + 07
Argon 83.78 150.86	83.78 0.688E + 05	90.49 0.140E + 06	97.20 0.258E + 06	103.90 0.437E + 06	110.61 0.694E + 06	117.32 0.104E + 07	124.03 0.151E + 07	130.74 0.210E + 07	137.44 0.285E + 07	144.15 0.377E + 07	150.86 0.489E + 07
Helium-4 2.20 5.20	2.20 0.536E + 04	2.50 0.104E + 05	2.80 0.179E + 05	3.10 0.285E + 05	3.40 0.424E + 05	3.70 0.603E + 05	4.00 0.827E + 05	4.30 0.110E + 06	4.60 0.143E + 06	4.90 0.182E + 06	5.20 0.229E + 06

Vapor pressure

Vapor pressures are given for the range between the triple point and the critical point wherever possible. For each compound the temperature range in kelvins is given below the compound, e.g.,

Methane
90.67 190.58

The vapor pressures are given within this range in 10 equal intervals in units of pascals. For methane, e.g.,

90.67 . . . 190.58
0.117E + 05 . . . 0.459E + 07

Methanol 175.59 512.58	175.59 0.158E + 00	209.29 0.171E + 02	242.99 0.451E + 03	276.69 0.497E + 04	310.39 0.308E + 05	344.08 0.129E + 06	377.78 0.405E + 06	411.48 0.104E + 07	445.18 0.228E + 07	478.88 0.446E + 07	512.58 0.798E + 07
Ethanol 159.05 516.25	159.05 0.718E − 03	194.77 0.522E + 00	230.49 0.428E + 02	266.21 0.973E + 03	301.93 0.979E + 04	337.65 0.569E + 05	373.37 0.226E + 06	409.09 0.677E + 06	444.81 0.165E + 07	480.53 0.344E + 07	516.25 0.633E + 07
n-Propanol 146.95 536.71	146.95 0.651E − 06	185.93 0.951E − 02	224.90 0.381E + 01	263.88 0.214E + 03	302.85 0.372E + 04	341.83 0.302E + 05	380.81 0.146E + 06	419.78 0.494E + 06	458.76 0.128E + 07	497.73 0.273E + 07	536.71 0.501E + 07
Isopropanol 185.28 508.31	185.28 0.320E − 01	217.58 0.471E + 01	249.89 0.161E + 03	282.19 0.215E + 04	314.49 0.154E + 05	346.79 0.708E + 05	379.10 0.237E + 06	411.40 0.630E + 06	443.70 0.140E + 07	476.01 0.272E + 07	508.31 0.477E + 07
1-Hexanol 228.55 611.35	228.55 0.506E − 01	266.83 0.682E + 01	305.11 0.213E + 03	343.39 0.260E + 04	381.67 0.170E + 05	419.95 0.713E + 05	458.23 0.220E + 06	496.51 0.539E + 06	534.79 0.112E + 07	573.07 0.204E + 07	611.35 0.340E + 07
Ethylene glycol 260.15 645.00	260.15 0.199E + 00	298.63 0.128E + 02	337.12 0.256E + 03	375.60 0.242E + 04	414.09 0.139E + 05	452.57 0.570E + 05	491.06 0.187E + 06	529.54 0.526E + 06	568.03 0.134E + 07	606.51 0.322E + 07	645.00 0.748E + 07
Formaldehyde 181.15 408.00	181.15 0.887E + 03	203.83 0.591E + 04	226.52 0.255E + 05	249.20 0.816E + 05	271.89 0.210E + 06	294.57 0.463E + 06	317.26 0.907E + 06	339.94 0.163E + 07	362.63 0.272E + 07	385.31 0.432E + 07	408.00 0.659E + 07
Acetaldehyde 150.15 461.00	150.15 0.212E + 00	181.23 0.380E + 02	212.32 0.103E + 04	243.40 0.953E + 04	274.49 0.469E + 05	305.57 0.155E + 06	336.66 0.399E + 06	367.74 0.873E + 06	398.83 0.172E + 07	429.91 0.316E + 07	461.00 0.557E + 07
Acetone 178.45 508.20	178.45 0.259E + 01	211.42 0.116E + 03	244.40 0.166E + 04	277.37 0.116E + 05	310.35 0.506E + 05	343.32 0.160E + 06	376.30 0.404E + 06	409.27 0.867E + 06	442.25 0.165E + 07	475.22 0.287E + 07	508.20 0.467E + 07
Methyl ethyl ketone 186.48 535.50	186.48 0.139E + 01	221.38 0.780E + 02	256.28 0.124E + 04	291.19 0.913E + 04	326.09 0.410E + 05	360.99 0.132E + 06	395.89 0.339E + 06	430.79 0.739E + 06	465.70 0.143E + 07	500.60 0.253E + 07	535.50 0.420E + 07
Acetic acid 289.81 592.71	289.81 0.129E + 04	320.10 0.657E + 04	350.39 0.245E + 05	380.68 0.723E + 05	410.97 0.179E + 06	441.26 0.388E + 06	471.55 0.758E + 06	501.84 0.136E + 07	532.13 0.230E + 07	562.42 0.367E + 07	592.71 0.564E + 07
Terephthalic acid 523.00 700.15	523.00 0.976E + 02	540.71 0.228E + 03	558.43 0.548E + 03	576.14 0.132E + 04	593.86 0.313E + 04	611.57 0.719E + 04	629.29 0.158E + 05	647.00 0.330E + 05	664.72 0.647E + 05	682.43 0.118E + 06	700.15 0.200E + 06
Methyl acetate 175.15 506.80	175.15 0.917E + 00	208.31 0.617E + 02	241.48 0.112E + 04	274.64 0.903E + 04	307.81 0.432E + 05	340.97 0.145E + 06	374.14 0.379E + 06	407.30 0.833E + 06	440.47 0.161E + 07	473.63 0.283E + 07	506.80 0.465E + 07

Ethyl acetate 189.60 523.25	189.60 0.132E + 01	222.96 0.682E + 02	256.33 0.108E + 04	289.69 0.815E + 04	323.06 0.375E + 05	356.42 0.123E + 06	389.79 0.319E + 06	423.15 0.695E + 06	456.52 0.134E + 07	489.88 0.234E + 07	523.25 0.383E + 07
Phenol 314.06 694.25	314.06 0.215E + 03	352.08 0.197E + 04	390.10 0.112E + 05	428.12 0.449E + 05	466.14 0.139E + 06	504.15 0.351E + 06	542.17 0.761E + 06	580.19 0.146E + 07	618.21 0.254E + 07	656.23 0.407E + 07	694.25 0.611E + 07
Methylamine 179.69 430.05	179.69 0.177E + 03	204.73 0.211E + 04	229.76 0.136E + 05	254.80 0.570E + 05	279.83 0.177E + 06	304.87 0.444E + 06	329.91 0.945E + 06	354.94 0.178E + 07	379.98 0.306E + 07	405.01 0.490E + 07	430.05 0.741E + 07
Ethylamine 192.15 456.15	192.15 0.152E + 03	218.55 0.175E + 04	244.95 0.109E + 05	271.35 0.450E + 05	297.75 0.138E + 06	324.15 0.340E + 06	350.55 0.717E + 06	376.95 0.134E + 07	403.35 0.230E + 07	429.75 0.368E + 07	456.15 0.559E + 07
Methyl mercaptan 150.18 469.95	150.18 0.130E + 01	182.16 0.129E + 03	214.13 0.262E + 04	246.11 0.210E + 05	278.09 0.952E + 05	310.06 0.297E + 06	342.04 0.721E + 06	374.02 0.148E + 07	406.00 0.269E + 07	437.97 0.453E + 07	469.95 0.720E + 07
Ethyl mercaptan 125.26 499.15	125.26 0.999E − 03	162.65 0.212E + 01	200.04 0.193E + 03	237.43 0.358E + 04	274.82 0.269E + 05	312.20 0.116E + 06	349.59 0.355E + 06	386.98 0.854E + 06	424.37 0.175E + 07	461.76 0.323E + 07	499.15 0.551E + 07
Dimethyl sulfide 174.88 503.04	174.88 0.101E + 02	207.70 0.343E + 03	240.51 0.392E + 04	273.33 0.230E + 05	306.14 0.878E + 05	338.96 0.250E + 06	371.78 0.580E + 06	404.59 0.116E + 07	437.41 0.209E + 07	470.22 0.346E + 07	503.04 0.541E + 07
Methyl chloride 175.43 416.25	175.43 0.874E + 03	199.51 0.647E + 04	223.59 0.294E + 05	247.68 0.958E + 05	271.76 0.247E + 06	295.84 0.537E + 06	319.92 0.103E + 07	344.00 0.180E + 07	368.08 0.293E + 07	392.17 0.452E + 07	416.25 0.669E + 07
Ethyl chloride 134.80 460.35	134.80 0.117E + 00	167.35 0.250E + 02	199.91 0.772E + 03	232.46 0.807E + 04	265.02 0.439E + 05	297.57 0.157E + 06	330.13 0.424E + 06	362.68 0.946E + 06	395.24 0.185E + 07	427.79 0.328E + 07	460.35 0.546E + 07
Dichloromethane 178.01 510.00	178.01 0.702E + 01	211.21 0.287E + 03	244.41 0.370E + 04	277.61 0.235E + 05	310.81 0.937E + 05	344.00 0.274E + 06	377.20 0.644E + 06	410.40 0.130E + 07	443.60 0.234E + 07	476.80 0.388E + 07	510.00 0.608E + 07
Chloroform 209.63 536.40	209.63 0.693E + 02	242.31 0.116E + 04	274.98 0.359E + 04	307.66 0.391E + 05	340.34 0.125E + 06	373.01 0.315E + 06	405.69 0.673E + 06	438.37 0.128E + 07	471.04 0.222E + 07	503.72 0.363E + 07	536.40 0.567E + 07
Carbon tetra- chloride 250.33 556.35	250.33 0.112E + 04	280.93 0.671E + 04	311.53 0.267E + 05	342.14 0.796E + 05	372.74 0.193E + 06	403.34 0.402E + 06	433.94 0.746E + 06	464.54 0.127E + 07	495.14 0.204E + 07	525.75 0.310E + 07	556.35 0.454E + 07
1,2-Dichloroethane 237.49 561.00	237.49 0.237E + 03	269.84 0.231E + 04	302.19 0.127E + 05	334.54 0.479E + 05	366.89 0.137E + 06	399.24 0.325E + 06	431.59 0.669E + 06	463.95 0.124E + 07	496.30 0.212E + 07	528.65 0.341E + 07	561.00 0.523E + 07
Dichlorodi- fluoromethane 115.15 384.95	115.15 0.930E − 01	142.13 0.231E + 02	169.11 0.734E + 03	196.09 0.750E + 04	223.07 0.392E + 05	250.05 0.134E + 06	277.03 0.349E + 06	304.01 0.755E + 06	330.99 0.144E + 07	357.97 0.251E + 07	384.95 0.412E + 07
Sodium hydroxide 596.00 830.00	596.00 0.296E − 02	619.40 0.713E − 02	642.80 0.164E − 01	666.20 0.361E − 01	689.60 0.764E − 01	713.00 0.156E + 00	736.40 0.307E + 00	759.80 0.587E + 00	783.20 0.109E + 01	806.60 0.197E + 01	830.00 0.346E + 01
Sulfuric acid 383.15 500.00	383.15 0.446E + 02	394.83 0.873E + 02	406.52 0.153E + 03	418.20 0.299E + 03	429.89 0.497E + 03	441.57 0.819E + 03	453.26 0.130E + 04	464.94 0.201E + 04	476.63 0.300E + 04	488.31 0.436E + 04	500.00 0.616E + 04

APPENDIX

B

THERMODYNAMIC PROPERTIES OF STEAM

Symbols used in the steam tables:

P pressure, kPa (abs)
T thermodynamic temperature, °C
$\bar{V}$ specific volume, m^3/kg
U specific internal energy, kJ/kg
H specific enthalpy, kJ/kg
S specific entropy, kJ/(kg · K)

Subscripts

f property of liquid in equilibrium with vapor
g property of vapor in equilibrium with liquid
fg change by evaporation

Saturated steam: temperature table

		Specific volume		Internal energy			Enthalpy			Entropy		
Temp. T, °C	Press. P, kPa	Sat. liquid $\overline{V}_f$	Sat. vapor $\overline{V}_g$	Sat. liquid U_f	Evap. U_{fg}	Sat. vapor U_g	Sat. liquid H_f	Evap. H_{fg}	Sat. Vapor H_g	Sat. liquid S_f	Evap. S_{fg}	Sat. vapor S_g
0.01	0.6113	0.001 000	206.14	0.00	2375.3	2375.3	0.01	2501.3	2501.4	0.0000	9.1562	9.1562
5	0.8721	0.001 000	147.12	20.97	2361.3	2382.3	20.98	2489.6	2510.6	0.0761	8.9496	9.0257
10	1.2276	0.001 000	106.38	42.00	2347.2	2389.2	42.01	2477.7	2519.8	0.1510	8.7498	8.9008
15	1.7051	0.001 001	77.93	62.99	2333.1	2396.1	62.99	2465.9	2528.9	0.2245	8.5569	8.7814
20	2.339	0.001 002	57.79	83.95	2319.0	2402.9	83.96	2454.1	2538.1	0.2966	8.3706	8.6672
25	3.169	0.001 003	43.36	104.88	2304.9	2409.8	104.89	2442.3	2547.2	0.3674	8.1905	8.5580
30	4.246	0.001 004	32.89	125.78	2290.8	2416.6	125.79	2430.5	2556.3	0.4369	8.0164	8.4533
35	5.628	0.001 006	25.22	146.67	2276.7	2423.4	146.68	2418.6	2565.3	0.5053	7.8478	8.3531
40	7.384	0.001 008	19.52	167.56	2262.6	2430.1	167.57	2406.7	2574.3	0.5725	7.6845	8.2570
45	9.593	0.001 010	15.26	188.44	2248.4	2436.8	188.45	2394.8	2583.2	0.6387	7.5261	8.1648
50	12.349	0.001 012	12.03	209.32	2234.2	2443.5	209.33	2382.7	2592.1	0.7038	7.3725	8.0763
55	15.758	0.001 015	9.568	230.21	2219.9	2450.1	230.23	2370.7	2600.9	0.7679	7.2234	7.9913
60	19.940	0.001 017	7.671	251.11	2205.5	2456.6	251.13	2358.5	2609.6	0.8312	7.0784	7.9096
65	25.03	0.001 020	6.197	272.02	2191.1	2463.1	272.06	2346.2	2618.3	0.8935	6.9375	7.8310
70	31.19	0.001 023	5.042	292.95	2176.6	2469.6	292.98	2333.8	2626.8	0.9549	6.8004	7.7553
75	38.58	0.001 026	4.131	313.90	2162.0	2475.9	313.93	2321.4	2635.3	1.0155	6.6669	7.6824
80	47.39	0.001 029	3.407	334.86	2147.4	2482.2	334.91	2308.8	2643.7	1.0753	6.5369	7.6122
85	57.83	0.001 033	2.828	355.84	2132.6	2488.4	355.90	2296.0	2651.9	1.1343	6.4102	7.5445
90	70.14	0.001 036	2.361	376.85	2117.7	2494.5	376.92	2283.2	2660.1	1.1925	6.2866	7.4791
95	84.55	0.001 040	1.982	397.88	2102.7	2500.6	397.96	2270.2	2668.1	1.2500	6.1659	7.4159
100	101.35	0.001 044	1.6729	418.94	2087.6	2506.5	419.04	2257.0	2676.1	1.3069	6.0480	7.3549
105	120.82	0.001 048	1.4194	440.02	2072.3	2512.4	440.15	2243.7	2683.8	1.3630	5.9328	7.2958
110	143.27	0.001 052	1.2102	461.14	2057.0	2518.1	461.30	2230.2	2691.5	1.4185	5.8202	7.2387
115	169.06	0.001 056	1.0366	482.30	2041.4	2523.7	482.48	2216.5	2699.0	1.4734	5.7100	7.1833
120	198.53	0.001 060	0.8919	503.50	2025.8	2529.3	503.71	2202.6	2706.3	1.5276	5.6020	7.1296
125	232.1	0.001 065	0.7706	524.74	2009.9	2534.6	524.99	2188.5	2713.5	1.5813	5.4962	7.0775
130	270.1	0.001 070	0.6685	546.02	1993.9	2539.9	546.31	2174.2	2720.5	1.6344	5.3925	7.0269
135	313.0	0.001 075	0.5822	567.35	1977.7	2545.0	567.69	2159.6	2727.3	1.6870	5.2907	6.9777
140	361.3	0.001 080	0.5089	588.74	1961.3	2550.0	589.13	2144.7	2733.9	1.7391	5.1908	6.9299
145	415.4	0.001 085	0.4463	610.18	1944.7	2554.9	610.63	2129.6	2740.3	1.7907	5.0926	6.8833
150	475.8	0.001 091	0.3928	631.68	1927.9	2559.5	632.20	2114.3	2746.5	1.8418	4.9960	6.8379
155	543.1	0.001 096	0.3468	653.24	1910.8	2564.1	653.84	2098.6	2752.4	1.8925	4.9010	6.7935
160	617.8	0.001 102	0.3071	674.87	1893.5	2568.4	675.55	2082.6	2758.1	1.9427	4.8075	6.7502
165	700.5	0.001 108	0.2727	696.56	1876.0	2572.5	697.34	2066.2	2763.5	1.9925	4.7153	6.7078
170	791.7	0.001 114	0.2428	718.33	1858.1	2576.5	719.21	2049.5	2768.7	2.0419	4.6244	6.6663
175	892.0	0.001 121	0.2168	740.17	1840.0	2580.2	741.17	2032.4	2773.6	2.0909	4.5347	6.6256
180	1002.1	0.001 127	0.194 05	762.09	1821.6	2583.7	763.22	2015.0	2778.2	2.1396	4.4461	6.5857
185	1122.7	0.001 134	0.174 09	784.10	1802.9	2587.0	785.37	1997.1	2782.4	2.1879	4.3586	6.5465
190	1254.4	0.001 141	0.156 54	806.19	1783.8	2590.0	807.62	1978.8	2786.4	2.2359	4.2720	6.5079
195	1397.8	0.001 149	0.141 05	828.37	1764.4	2592.8	829.98	1960.0	2790.0	2.2835	4.1863	6.4698
200	1553.8	0.001 157	0.127 36	850.65	1744.7	2595.3	852.45	1940.7	2793.2	2.3309	4.1014	6.4323
210	1906.2	0.001 173	0.104 41	895.53	1703.9	2599.5	897.76	1900.7	2798.5	2.4248	3.9337	6.3585
220	2318.	0.001 190	0.086 19	940.87	1661.5	2602.4	943.62	1858.5	2802.1	2.5178	3.7683	6.2861
230	2795.	0.001 209	0.071 58	986.74	1617.2	2603.9	990.12	1813.8	2804.0	2.6099	3.6047	6.2146
240	3344.	0.001 229	0.059 76	1033.21	1570.8	2604.0	1037.32	1766.5	2803.8	2.7015	3.4422	6.1437
250	3973.	0.001 251	0.050 13	1080.39	1522.0	2602.4	1085.36	1716.2	2801.5	2.7927	3.2802	6.0730

Saturated steam: temperature table (Continued)

		Specific volume		Internal energy			Enthalpy			Entropy		
Temp. T, °C	Press. P, kPa	Sat. liquid $\overline{V}_f$	Sat. vapor $\overline{V}_g$	Sat. liquid U_f	Evap. U_{fg}	Sat. vapor U_g	Sat. liquid H_f	Evap. H_{fg}	Sat. Vapor H_g	Sat. liquid S_f	Evap. S_{fg}	Sat. vapor S_g
260	4 688.	0.001 276	0.042 21	1128.39	1470.6	2599.0	1134.37	1662.5	2796.9	2.8838	3.1181	6.0019
270	5 499.	0.001 302	0.035 64	1177.36	1416.3	2593.7	1184.51	1605.2	2789.7	2.9751	2.9551	5.9301
280	6 412.	0.001 332	0.030 17	1227.46	1358.7	2586.1	1235.99	1543.6	2779.6	3.0668	2.7903	5.8571
290	7 436.	0.001 366	0.025 57	1278.92	1297.1	2576.0	1289.07	1477.1	2766.2	3.1594	2.6227	5.7821
300	8 581.	0.001 404	0.021 67	1332.0	1231.0	2563.0	1344.0	1404.9	2749.0	3.2534	2.4511	5.7045
310	9 856.	0.001 447	0.018 350	1387.1	1159.4	2546.4	1401.3	1326.0	2727.3	3.3493	2.2737	5.6230
320	11 274.	0.001 499	0.015 488	1444.6	1080.9	2525.5	1461.5	1238.6	2700.1	3.4480	2.0882	5.5362
330	12 845.	0.001 561	0.012 996	1505.3	993.7	2498.9	1525.3	1140.6	2665.9	3.5507	1.8909	5.4417
340	14 586.	0.001 638	0.010 797	1570.3	894.3	2464.6	1594.2	1027.9	2622.0	3.6594	1.6763	5.3357
350	16 513.	0.001 740	0.008 813	1641.9	776.6	2418.4	1670.6	893.4	2563.9	3.7777	1.4335	5.2112
360	18 651.	0.001 893	0.006 945	1725.2	626.3	2351.5	1760.5	720.5	2481.0	3.9147	1.1379	5.0526
370	21 030.	0.002 213	0.004 925	1844.0	384.5	2228.5	1890.5	441.6	2332.1	4.1106	0.6865	4.7971
374.14	22 090.	0.003 155	0.003 155	2029.6	0	2029.6	2099.3	0	2099.3	4.4298	0	4.4298

Source: Adapted from Joseph H. Keenan, Frederick G. Keyes, Philip G. Hill, and Joan G. Moore, "Steam Tables," Wiley, New York, 1978.

Saturated steam: pressure table

Press. P, kPa	Temp. T, °C	Specific volume: Sat. liquid $\overline{V}_f$	Specific volume: Sat. vapor $\overline{V}_g$	Internal energy: Sat liquid U_f	Internal energy: Evap. U_{fg}	Internal energy: Sat. vapor U_g	Enthalpy: Sat. liquid H_f	Enthalpy: Evap. H_{fg}	Enthalpy: Sat. vapor h_g	Entropy: Sat. liquid S_f	Entropy: Evap. S_{fg}	Entropy: Sat. vapor S_g
0.6113	0.01	0.001 000	206.14	0.00	2375.3	2375.3	0.01	2501.3	2501.4	0.0000	9.1562	9.1562
1.0	6.98	0.001 000	129.21	29.30	2355.7	2385.0	29.30	2484.9	2514.2	0.1059	8.8697	8.9756
1.5	13.03	0.001 001	87.98	54.71	2338.6	2393.3	54.71	2470.6	2525.3	0.1957	8.6322	8.8279
2.0	17.50	0.001 001	67.00	73.48	2326.0	2399.5	73.48	2460.0	2533.5	0.2607	8.4629	8.7237
2.5	21.08	0.001 002	54.25	88.48	2315.9	2404.4	88.49	2451.6	2540.0	0.3120	8.3311	8.6432
3.0	24.08	0.001 003	45.67	101.04	2307.5	2408.5	101.05	2444.5	2545.5	0.3545	8.2231	8.5776
4.0	28.96	0.001 004	34.80	121.45	2293.7	2415.2	121.46	2432.9	2554.4	0.4226	8.0520	8.4746
5.0	32.88	0.001 005	28.19	137.81	2282.7	2420.5	137.82	2423.7	2561.5	0.4764	7.9187	8.3951
7.5	40.29	0.001 008	19.24	168.78	2261.7	2430.5	168.79	2406.0	2574.8	0.5764	7.6750	8.2515
10	45.81	0.001 010	14.67	191.82	2246.1	2437.9	191.83	2392.8	2584.7	0.6493	7.5009	8.1502
15	53.97	0.001 014	10.02	225.92	2222.8	2448.7	225.94	2373.1	2599.1	0.7549	7.2536	8.0085
20	60.06	0.001 017	7.649	251.38	2205.4	2456.7	251.40	2358.3	2609.7	0.8320	7.0766	7.9085
25	64.97	0.001 020	6.204	271.90	2191.2	2463.1	271.93	2346.3	2618.2	0.8931	6.9383	7.8314
30	69.10	0.001 022	5.229	289.20	2179.2	2468.4	289.23	2336.1	2625.3	0.9439	6.8247	7.7686
40	75.87	0.001 027	3.993	317.53	2159.5	2477.0	317.58	2319.2	2636.8	1.0259	6.6441	7.6700
50	81.33	0.001 030	3.240	340.44	2143.4	2483.9	340.49	2305.4	2645.9	1.0910	6.5029	7.5939
75	91.78	0.001 037	2.217	384.31	2112.4	2496.7	384.39	2278.6	2663.0	1.2130	6.2434	7.4564
100.	99.63	0.001 043	1.6940	417.36	2088.7	2506.1	417.46	2258.0	2675.5	1.3026	6.0568	7.3594
125.	105.99	0.001 048	1.3749	444.19	2069.3	2513.5	444.32	2241.0	2685.4	1.3740	5.9104	7.2844
150.	111.37	0.001 053	1.1593	466.94	2052.7	2519.7	467.11	2226.5	2693.6	1.4336	5.7897	7.2233
175.	116.06	0.001 057	1.0036	486.80	2038.1	2524.9	486.99	2213.6	2700.6	1.4849	5.6868	7.1717
200.	120.23	0.001 061	0.8857	504.49	2025.0	2529.5	504.70	2201.9	2706.7	1.5301	5.5970	7.1271
225.	124.00	0.001 064	0.7933	520.47	2013.1	2533.6	520.72	2191.3	2712.1	1.5706	5.5173	7.0878
250.	127.44	0.001 067	0.7187	535.10	2002.1	2537.2	535.37	2181.5	2716.9	1.6072	5.4455	7.0527
275.	130.60	0.001 070	0.6573	548.59	1991.9	2540.5	548.89	2172.4	2721.3	1.6408	5.3801	7.0209
300.	133.55	0.001 073	0.6058	561.15	1982.4	2543.6	561.47	2163.8	2725.3	1.6718	5.3201	6.9919
325.	136.30	0.001 076	0.5620	572.90	1973.5	2546.4	573.25	2155.8	2729.0	1.7006	5.2646	6.9652
350.	138.88	0.001 079	0.5243	583.95	1965.0	2548.9	584.33	2148.1	2732.4	1.7275	5.2130	6.9405
375.	141.32	0.001 081	0.4914	594.40	1956.9	2551.3	594.81	2140.8	2735.6	1.7528	5.1647	6.9175
400.	143.63	0.001 084	0.4625	604.31	1949.3	2553.6	604.74	2133.8	2738.6	1.7766	5.1193	6.8959
450.	147.93	0.001 088	0.4140	622.77	1934.9	2557.6	623.25	2120.7	2743.9	1.8207	5.0359	6.8565
500.	151.86	0.001 093	0.3749	639.68	1921.6	2561.2	640.23	2108.5	2748.7	1.8607	4.9606	6.8213
550.	155.48	0.001 097	0.3427	655.32	1909.2	2564.5	655.93	2097.0	2753.0	1.8973	4.8920	6.7893
600.	158.85	0.001 101	0.3157	669.90	1897.5	2567.4	670.56	2086.3	2756.8	1.9312	4.8288	6.7600
650.	162.01	0.001 104	0.2927	683.56	1886.5	2570.1	684.28	2076.0	2760.3	1.9627	4.7703	6.7331
700.	164.97	0.001 108	0.2729	696.44	1876.1	2572.5	697.22	2066.3	2763.5	1.9922	4.7158	6.7080

Saturated steam: pressure table (Continued)

		Specific volume		Internal energy			Enthalpy			Entropy		
Press. P, kPa	Temp. T, °C	Sat. liquid $\overline{V}_f$	Sat. vapor $\overline{V}_g$	Sat. liquid U_f	Evap. U_{fg}	Sat. vapor U_g	Sat. liquid H_f	Evap. H_{fg}	Sat. vapor h_g	Sat. liquid S_f	Evap. S_{fg}	Sat. vapor S_g
750.	167.78	0.001 112	0.2556	708.64	1866.1	2574.7	709.47	2057.0	2766.4	2.0200	4.6647	6.6847
800.	170.43	0.001 115	0.2404	720.22	1856.6	2576.8	721.11	2048.0	2769.1	2.0462	4.6166	6.6628
850.	172.96	0.001 118	0.2270	731.27	1847.4	2578.7	732.22	2039.4	2771.6	2.0710	4.5711	6.6421
900.	175.38	0.001 121	0.2150	741.83	1838.6	2580.5	742.83	2031.1	2773.9	2.0946	4.5280	6.6226
950.	177.69	0.001 124	0.2042	751.95	1830.2	2582.1	753.02	2023.1	2776.1	2.1172	4.4869	6.6041
1000.	179.91	0.001 127	0.194 44	761.68	1822.0	2583.6	762.81	2015.3	2778.1	2.1387	4.4478	6.5865
1100.	184.09	0.001 133	0.177 53	780.09	1806.3	2586.4	781.34	2000.4	2781.7	2.1792	4.3744	6.5536
1 200.	187.99	0.001 139	0.163 33	797.29	1791.5	2588.8	798.65	1986.2	2784.8	2.2166	4.3067	6.5233
1 300.	191.64	0.001 144	0.151 25	813.44	1777.5	2591.0	814.93	1972.7	2787.6	2.2515	4.2438	6.4953
1 400.	195.07	0.001 149	0.140 84	828.70	1764.1	2592.8	830.30	1959.7	2790.0	2.2842	4.1850	6.4693
1 500.	198.32	0.001 154	0.131 77	843.16	1751.3	2594.5	844.89	1947.3	2792.2	2.3150	4.1298	6.4448
1 750.	205.76	0.001 166	0.113 49	876.46	1721.4	2597.8	878.50	1917.9	2796.4	2.3851	4.0044	6.3896
2 000.	212.42	0.001 177	0.099 63	906.44	1693.8	2600.3	908.79	1890.7	2799.5	2.4474	3.8935	6.3409
2 250.	218.45	0.001 187	0.088 75	933.83	1668.2	2602.0	936.49	1865.2	2801.7	2.5035	3.7937	6.2972
2 500.	223.99	0.001 197	0.079 98	959.11	1644.0	2603.1	962.11	1841.0	2803.1	2.5547	3.7028	6.2575
3 000.	233.90	0.001 217	0.066 68	1004.78	1599.3	2604.1	1008.42	1795.7	2804.2	2.6457	3.5412	6.1869
3 500.	242.60	0.001 235	0.057 07	1045.43	1558.3	2603.7	1049.75	1753.7	2803.4	2.7253	3.4000	6.1253
4 000.	250.40	0.001 252	0.049 78	1082.31	1520.0	2602.3	1087.31	1714.1	2801.4	2.7964	3.2737	6.0701
5 000.	263.99	0.001 286	0.039 44	1147.81	1449.3	2597.1	1154.23	1640.1	2794.3	2.9202	3.0532	5.9734
6 000.	275.64	0.001 319	0.032 44	1205.44	1384.3	2589.7	1213.35	1571.0	2784.3	3.0267	2.8625	5.8892
7 000.	285.88	0.001 351	0.027 37	1257.55	1323.0	2580.5	1267.00	1505.1	2772.1	3.1211	2.6922	5.8133
8 000.	295.06	0.001 384	0.023 52	1305.57	1264.2	2569.8	1316.64	1441.3	2758.0	3.2068	2.5364	5.7432
9 000.	303.40	0.001 418	0.020 48	1350.51	1207.3	2557.8	1363.26	1378.9	2742.1	3.2858	2.3915	5.6772
10 000.	311.06	0.001 452	0.018 026	1393.04	1151.4	2544.4	1407.56	1317.1	2724.7	3.3596	2.2544	5.6141
12 000	324.75	0.001 527	0.014 263	1473.0	1040.7	2513.7	1491.3	1193.6	2684.9	3.4962	1.9962	5.4924
14 000.	336.75	0.001 611	0.011 485	1548.6	928.2	2476.8	1571.1	1066.5	2637.6	3.6232	1.7485	5.3717
16 000.	347.44	0.001 711	0.009 306	1622.7	809.0	2431.7	1650.1	930.6	2580.6	3.7461	1.4994	5.2455
18 000.	357.06	0.001 840	0.007 489	1698.9	675.4	2374.3	1732.0	777.1	2509.1	3.8715	1.2329	5.1044
20 000.	365.81	0.002 036	0.005 834	1785.6	507.5	2293.0	1826.3	583.4	2409.7	4.0139	0.9130	4.9269
22 000.	373.80	0.002 742	0.003 568	1961.9	125.2	2087.1	2022.2	143.4	2165.6	4.3110	0.2216	4.5327
22 090.	374.14	0.003 155	0.003 155	2029.6	0	2029.6	2099.3	0	2099.3	4.4298	0	4.4298

Source: Adapted from Joseph H. Keenan, Frederick G. Keyes, Philip G. Hill, and Joan G. Moore, "Steam Tables," Wiley, New York, 1978.

Superheated vapor

T	$\bar{V}$	U	H	S	$\bar{V}$	U	H	S	$\bar{V}$	U	H	S
	P = 10 kPa (45.81°C)				P = 50 kPa (81.33°C)				P = 100 kPa (99.63°C)			
Sat.	14.674	2437.9	2584.7	8.1502	3.240	2483.9	2645.9	7.5939	1.6940	2506.1	2675.5	7.3594
50	14.869	2443.9	2592.6	8.1749								
100	17.196	2515.5	2687.5	8.4479	3.418	2511.6	2682.5	7.6947	1.6958	2506.7	2676.2	7.3614
150	19.512	2587.9	2783.0	8.6882	3.889	2585.6	2780.1	7.9401	1.9364	2582.8	2776.4	7.6134
200	21.825	2661.3	2879.5	8.9038	4.356	2659.9	2877.7	8.1580	2.172	2658.1	2875.3	7.8343
250	24.136	2736.0	2977.3	9.1002	4.820	2735.0	2976.0	8.3556	2.406	2733.7	2974.3	8.0333
300	26.445	2812.1	3076.5	8.2813	5.284	2811.3	3075.5	8.5373	2.639	2810.4	3074.3	8.2158
400	31.063	2968.9	3279.6	9.6077	6.209	2968.5	3278.9	8.8642	3.103	2967.9	3278.2	8.5435
500	35.679	3132.3	3489.1	9.8978	7.134	3132.0	3488.7	9.1546	3.565	3131.6	3488.1	8.8342
600	40.295	3302.5	3705.4	10.1608	8.057	3302.2	3705.1	9.4178	4.028	3301.9	3704.7	9.0976
700	44.911	3479.6	3928.7	10.4028	8.981	3479.4	3928.5	9.6599	4.490	3479.2	3928.8	9.3398
800	49.526	3663.8	4159.0	10.6281	9.904	3663.6	4158.9	9.8852	4.952	3663.5	4158.6	9.5652
900	54.141	3855.0	4396.4	10.8396	10.828	3854.9	4396.3	10.0967	5.414	3854.8	4396.1	9.7767
1000	58.757	4053.0	4640.6	11.0393	11.751	4052.9	4640.5	10.2964	5.875	4052.8	4640.3	9.9764
1100	63.372	4257.5	4891.2	11.2287	12.674	4257.4	4891.1	10.4859	6.337	4257.3	4891.0	10.1659
1200	67.987	4467.9	5147.8	11.4091	13.597	4467.8	5147.7	10.6662	6.799	4467.7	5147.6	10.3463
1300	72.602	4683.7	5409.7	11.5811	14.521	4683.6	5409.6	10.8382	7.260	4683.5	5409.5	10.5183

T	$\bar{V}$	U	H	S	$\bar{V}$	U	H	S
	P = 200 kPa (120.23°C)				P = 400 kPa (143.63°C)			
Sat.	0.8857	2529.5	2706.7	7.1272	0.4625	2553.6	2738.6	6.8959
150	0.9596	2576.9	2768.8	7.2795	0.4708	2564.5	2752.8	6.9299
200	1.0803	2654.4	2870.5	7.5066	0.5342	2646.8	2860.5	7.1706
250	1.1988	2731.2	2971.0	7.7086	0.5951	2726.1	2964.2	7.3789
300	1.3162	2808.6	3071.8	7.8926	0.6548	2804.8	3066.8	7.5662
400	1.5493	2966.7	3276.6	8.2218	0.7726	2964.4	3273.4	7.8985
500	1.7814	3130.8	3487.1	8.5133	0.8893	3129.2	3484.9	8.1913
600	2.013	3301.4	3704.0	8.7770	1.0055	3300.2	3702.4	8.4558
700	2.244	3478.8	3927.6	9.0194	1.1215	3477.9	3926.5	8.6987
800	2.475	3663.1	4158.2	9.2449	1.2372	3662.4	4157.3	8.9244
900	2.706	3854.5	4395.8	9.4566	1.3529	3853.9	4395.1	9.1362
1000	2.937	4052.5	4640.0	9.6563	1.4685	4052.0	4639.4	9.3360
1100	3.168	4257.0	4890.7	9.8458	1.5840	4256.5	4890.2	9.5236
1200	3.399	4467.5	5147.3	10.0262	1.6996	4467.0	5146.8	9.7060
1300	3.630	4683.2	5409.3	10.1982	1.8151	4682.8	5408.8	9.8780

T	$\bar{V}$	U	H	S	$\bar{V}$	U	H	S
	P = 600 kPa (158.85°C)				P = 800 kPa (170.43°C)			
Sat.	0.3157	2567.4	2756.8	6.7600	0.2404	2576.8	2769.1	6.6628
200	0.3520	2638.9	2850.1	6.9665	0.2608	2630.6	2839.3	6.8158
250	0.3938	2720.9	2957.2	7.1816	0.2931	2715.5	2950.0	7.0384
300	0.4344	2801.0	3061.6	7.3724	0.3241	2797.2	3056.5	7.2328
350	0.4742	2881.2	3165.7	7.5464	0.3544	2878.2	3161.7	7.4089
400	0.5137	2962.1	3270.3	7.7079	0.3843	2959.7	3267.1	7.5716
500	0.5920	3127.6	3482.8	8.0021	0.4433	3126.0	3480.6	7.8673
600	0.6697	3299.1	3700.9	8.2674	0.5018	3297.9	3699.4	8.1333
700	0.7472	3477.0	3925.3	8.5107	0.5601	3476.2	3924.2	8.3770
800	0.8245	3661.8	4156.5	8.7367	0.6181	3661.1	4155.6	8.6033
900	0.9017	3853.4	4394.4	8.9486	0.6761	3852.8	4393.7	8.8153
1000	0.9788	4051.5	4638.8	9.1485	0.7340	4051.0	4638.2	9.0153
1100	1.0559	4256.1	4889.6	9.3381	0.7919	4255.6	4889.1	9.2050
1200	1.1330	4466.5	5146.3	9.5185	0.8497	4466.1	5145.9	9.3855
1300	1.2101	4682.3	5408.3	9.6906	0.9076	4681.8	5407.9	9.5575

Superheated vapor (Continued)

T	$\overline{V}$	U	H	S	$\overline{V}$	U	H	S	$\overline{V}$	U	H	S
	P = 1000 kPa (179.91°C)				P = 1200 kPa (187.99°C)				P = 1400 kPa (195.07°C)			
Sat.	0.194 44	2583.6	2778.1	6.5865	0.163 33	2588.8	2784.8	6.5233	0.140 84	2592.8	2790.0	6.4693
200	0.2060	2621.9	2827.9	6.6940	0.169 30	2612.8	2815.9	6.5898	0.143 02	2603.1	2803.3	6.4975
250	0.2327	2709.9	2942.6	6.9247	0.192 34	2704.2	2935.0	6.8294	0.163 50	2698.3	2927.2	6.7467
300	0.2579	2793.2	3051.2	7.1229	0.2138	2789.2	3045.8	7.0317	0.182 28	2785.2	3040.4	6.9534
350	0.2825	2875.2	3157.7	7.3011	0.2345	2872.2	3153.6	7.2121	0.2003	2869.2	3149.5	7.1360
400	0.3066	2957.3	3263.9	7.4651	0.2548	2954.9	3260.7	7.3774	0.2178	2952.5	3257.5	7.3026
500	0.3541	3124.4	3478.5	7.7622	0.2946	3122.8	3476.3	7.6759	0.2521	3121.1	3474.1	7.6027
600	0.4011	3296.8	3697.9	8.0290	0.3339	3295.6	3696.3	7.9435	0.2860	3294.4	3694.8	7.8710
700	0.4478	3475.3	3923.1	8.2731	0.3729	3474.4	3922.0	8.1881	0.3195	3473.6	3920.8	8.1160
800	0.4943	3660.4	4154.7	8.4996	0.4118	3659.7	4153.8	8.4148	0.3528	3659.0	4153.0	8.3431
900	0.5407	3852.2	4392.9	8.7118	0.4505	3851.6	4392.2	8.6272	0.3861	3851.1	4391.5	8.5556
1000	0.5871	4050.5	4637.6	8.9119	0.4892	4050.0	4637.0	8.8274	0.4192	4049.5	4636.4	8.7559
1100	0.6335	4255.1	4888.6	9.1017	0.5278	4254.6	4888.0	9.0172	0.4524	4254.1	4887.5	8.9457
1200	0.6798	4465.6	5145.4	9.2822	0.5665	4465.1	5144.9	9.1977	0.4855	4464.7	5144.4	9.1262
1300	0.7261	4681.3	5407.4	9.4543	0.6051	4680.9	5407.0	9.3698	0.5186	4680.4	5406.5	9.2984
	P = 1600 kPa (201.41°C)				P = 1800 kPa (207.15°C)				P = 2000 kPa (212.42°C)			
Sat.	0.123 80	2596.0	2794.0	6.4218	0.110 42	2598.4	2797.1	6.3794	0.099 63	2600.3	2799.5	6.3409
225	0.132 87	2644.7	2857.3	6.5518	0.116 73	2636.6	2846.7	6.4808	0.103 77	2628.3	2835.8	6.4147
250	0.141 84	2692.3	2919.2	6.6732	0.124 97	2686.0	2911.0	6.6066	0.111 44	2679.6	2902.5	6.5453
300	0.158 62	2781.1	3034.8	6.8844	0.140 21	2776.9	3029.2	6.8226	0.125 47	2772.6	3023.5	6.7664
350	0.174 56	2866.1	3145.4	7.0694	0.154 57	2863.0	3141.2	7.0100	0.138 57	2859.8	3137.0	6.9563
400	0.190 05	2950.1	3254.2	7.2374	0.168 47	2947.7	3250.9	7.1794	0.151 20	2945.2	3247.6	7.1271
500	0.2203	3119.5	3472.0	7.5390	0.195 50	3117.9	3469.8	7.4825	0.175 68	3116.2	3467.6	7.4317
600	0.2500	3293.3	3693.2	7.8080	0.2220	3292.1	3691.7	7.7523	0.199 60	3290.9	3690.1	7.7024
700	0.2794	3472.7	3919.7	8.0535	0.2482	3471.8	3918.5	7.9983	0.2232	3470.9	3917.4	7.9487
800	0.3086	3658.3	4152.1	8.2808	0.2742	3657.6	4151.2	8.2258	0.2467	3657.0	4150.3	8.1765
900	0.3377	3850.5	4390.8	8.4935	0.3001	3849.9	4390.1	8.4386	0.2700	3849.3	4389.4	8.3895
1000	0.3668	4049.0	4635.8	8.6938	0.3260	4048.5	4635.2	8.6391	0.2933	4048.0	4634.6	8.5901
1100	0.3958	4253.7	4887.0	8.8837	0.3518	4253.2	4886.4	8.8290	0.3166	4252.7	4885.9	8.7800
1200	0.4248	4464.2	5143.9	9.0643	0.3776	4463.7	5143.4	9.0096	0.3398	4463.3	5142.9	8.9607
1300	0.4538	4679.9	5406.0	9.2364	0.4034	4679.5	5405.6	9.1818	0.3631	4679.0	5405.1	9.1329

T	$\overline{V}$	U	H	S
	P = 3000 kPa (233.90°C)			
Sat.	0.066 68	2604.1	2804.2	6.1869
225				
250	0.070 58	2644.0	2855.8	6.2872
300	0.081 14	2750.1	2993.5	6.5390
350	0.090 53	2843.7	3115.3	6.7428
400	0.099 36	2932.8	3230.9	6.9212
450	0.107 87	3020.4	3344.0	7.0834
500	0.116 19	3108.0	3456.5	7.2338
600	0.132 43	3285.0	3682.3	7.5085
700	0.148 38	3466.5	3911.7	7.7571
800	0.164 14	3653.5	4145.9	7.9862
900	0.179 80	3846.5	4385.9	8.1999
1000	0.195 41	4045.4	4631.6	8.4009
1100	0.210 98	4250.3	4883.3	8.5912
1200	0.226 52	4460.9	5140.5	8.7720
1300	0.242 06	4676.6	5402.8	8.9442

Superheated vapor (Continued)

T	$\bar{V}$	U	H	S	$\bar{V}$	U	H	S
	P = 4000 kPa (250.40°C)				P = 5000 kPa (263.99°C)			
Sat.	0.049 78	2602.3	2801.4	6.0701	0.039 44	2597.1	2794.3	5.9734
275	0.054 57	2667.9	2886.2	6.2285	0.041 41	2631.3	2838.3	6.0544
300	0.058 84	2725.3	2960.7	6.3615	0.045 32	2698.0	2924.5	6.2084
350	0.066 45	2826.7	3092.5	6.5821	0.051 94	2808.7	3068.4	6.4493
400	0.073 41	2919.9	3213.6	6.7690	0.057 81	2906.6	3195.7	6.6459
450	0.080 02	3010.2	3330.3	6.9363	0.063 30	2999.7	3316.2	6.8186
500	0.086 43	3099.5	3445.3	7.0901	0.068 57	3091.0	3433.8	6.9759
600	0.098 85	3279.1	3674.4	7.3688	0.078 69	3273.0	3666.5	7.2589
700	0.110 95	3462.1	3905.9	7.6198	0.088 49	3457.6	3900.1	7.5122
800	0.122 87	3650.0	4141.5	7.8502	0.098 11	3646.6	4137.1	7.7440
900	0.134 69	3843.6	4382.3	8.0647	0.107 62	3840.7	4378.8	7.9593
1000	0.146 45	4042.9	4628.7	8.2662	0.117 07	4040.4	4625.7	8.1612
1100	0.158 17	4248.0	4880.6	8.4567	0.126 48	4245.6	4878.0	8.3520
1200	0.169 87	4458.6	5138.1	8.6376	0.135 87	4456.3	5135.7	8.5331
1300	0.181 56	4674.3	5400.5	8.8100	0.145 26	4672.0	5398.2	8.7055

T	$\bar{V}$	U	H	S	$\bar{V}$	U	H	S
	P = 6000 kPa (275.64°C)				P = 8000 kPa (295.06°C)			
Sat.	0.032 44	2589.7	2784.3	5.8892	0.023 52	2569.8	2758.0	5.7432
300	0.036 16	2667.2	2884.2	6.0674	0.024 26	2590.9	2785.0	5.7906
350	0.042 23	2789.6	3043.0	6.3335	0.029 95	2747.7	2987.3	6.1301
400	0.047 39	2892.9	3177.2	6.5408	0.034 32	2863.8	3138.3	6.3634
450	0.052 11	2988.9	3301.8	6.7193	0.038 17	2966.7	3272.0	6.5551
500	0.056 65	3082.2	3422.2	6.8803	0.041 75	3064.3	3398.3	6.7240
550	0.061 01	3174.6	3540.6	7.0288	0.045 16	3159.8	3521.0	6.8778
600	0.065 25	3266.9	3658.4	7.1677	0.048 45	3254.4	3642.0	7.0206
700	0.073 52	3453.1	3894.2	7.4234	0.054 81	3443.9	3882.4	7.2812
800	0.081 60	3643.1	4132.7	7.6566	0.060 97	3636.0	4123.8	7.5173
900	0.089 58	3837.8	4375.3	7.8727	0.067 02	3832.1	4368.3	7.7351
1000	0.097 49	4037.8	4622.7	8.0751	0.073 01	4032.8	4616.9	7.9384
1100	0.105 36	4243.3	4875.4	8.2661	0.078 96	4238.6	4870.3	8.1300
1200	0.113 21	4454.0	5133.3	8.4474	0.084 89	4449.5	5128.5	8.3115
1300	0.121 06	4669.6	5396.0	8.6199	0.090 80	4665.0	5391.5	8.4812

T	$\bar{V}$	U	H	S
	P = 10,000 kPa (311.06°C)			
Sat.	0.018 026	2544.4	2724.7	5.6141
325	0.019 861	2610.4	2809.1	5.7568
350	0.022 42	2699.2	2923.4	5.9443
400	0.026 41	2832.4	3096.5	6.2120
450	0.029 75	2943.4	3240.9	6.4190
500	0.032 79	3045.8	3373.7	6.5966
550	0.035 64	3144.6	3500.9	6.7561
600	0.038 37	3241.7	3625.3	6.9029
650	0.041 01	3338.2	3748.2	7.0398
700	0.043 58	3434.7	3870.5	7.1687
800	0.048 59	3628.9	4114.8	7.4077
900	0.053 49	3826.3	4361.2	7.6272
1000	0.058 32	4027.8	4611.0	7.8315
1100	0.063 12	4234.0	4865.1	8.0237
1200	0.067 89	4444.9	5123.8	8.2055
1300	0.072 65	4460.5	5387.0	8.3783

Superheated vapor (Continued)

T	$\overline{V}$	U	H	S	$\overline{V}$	U	H	S
	P = 15,000 kPa (342.24°C)*				P = 20,000 kPa (365.81°C)			
Sat.	0.010 337	2455.5	2610.5	5.3098	0.005 834	2293.0	2409.7	4.9269
350	0.011 470	2520.4	2692.4	5.4421				
400	0.015 649	2740.7	2975.5	5.8811	0.009 942	2619.3	2818.1	5.5540
450	0.018 445	2879.5	3156.2	6.1404	0.012 695	2806.2	3060.1	5.9017
500	0.020 80	2996.6	3308.6	6.3443	0.014 768	2942.9	3238.2	6.1401
550	0.022 93	3104.7	3448.6	6.5199	0.016 555	3062.4	3393.5	6.3348
600	0.024 91	3208.6	3582.3	6.6776	0.018 178	3174.0	3537.6	6.5048
650	0.026 80	3310.3	3712.3	6.8224	0.019 693	3281.4	3675.3	6.6582
700	0.028 61	3410.9	3840.1	6.9572	0.021 13	3386.4	3809.0	6.7993
800	0.032 10	3610.9	4092.4	7.2040	0.023 85	3592.7	4069.7	7.0544
900	0.035 46	3811.9	4343.8	7.4279	0.026 45	3797.5	4326.4	7.2830
1000	0.038 75	4015.4	4596.6	7.6348	0.028 97	4003.1	4582.5	7.4925
1100	0.042 00	4222.6	4852.6	7.8283	0.031 45	4211.3	4840.2	7.6874
1200	0.045 23	4433.8	5112.3	8.0108	0.033 91	4422.8	5101.0	7.8707
1300	0.048 45	4649.1	5376.0	8.1840	0.036 36	4638.0	5365.1	8.0442
	P = 30,000 kPa							
375	0.001 789 2	1737.8	1791.5	3.9305				
400	0.002 790	2067.4	2151.1	4.4728				
425	0.005 303	2455.1	2614.2	5.1504				
450	0.006 735	2619.3	2821.4	5.4424				
500	0.008 678	2820.7	3081.1	5.7905				
550	0.010 168	2970.3	3275.4	6.0342				
600	0.011 446	3100.5	3443.9	6.2331				
650	0.012 596	3221.0	3598.9	6.4058				
700	0.013 661	3335.8	3745.6	6.5606				
800	0.015 623	3555.5	4024.2	6.8332				
900	0.017 448	3768.5	4291.9	7.0718				
1000	0.019 196	3978.8	4554.7	7.2867				
1100	0.020 903	4189.2	4816.3	7.4845				
1200	0.022 589	4401.3	5079.0	7.6692				
1300	0.024 266	4616.0	5344.0	7.8432				
	P = 40,000 kPa				P = 60,000 kPa			
375	0.001 640 7	1677.1	1742.8	3.8290	0.001 502 8	1609.4	1699.5	3.7141
400	0.001 907 7	1854.6	1930.9	4.1135	0.001 633 5	1745.4	1843.4	3.9318
425	0.002 532	2096.9	2198.1	4.5029	0.001 816 5	1892.7	2001.7	4.1626
450	0.003 693	2365.1	2512.8	4.9459	0.002 085	2053.9	2179.0	4.4121
500	0.005 622	2678.4	2903.3	5.4700	0.002 956	2390.6	2567.9	4.9321
550	0.006 984	2869.7	3149.1	5.7785	0.003 956	2658.8	2896.2	5.3441
600	0.008 094	3022.6	3346.4	6.0114	0.004 834	2861.1	3151.2	5.6452
650	0.009 063	3158.0	3520.6	6.2054	0.005 595	3028.8	3364.5	5.8829
700	0.009 941	3283.6	3681.2	6.3750	0.006 272	3177.2	3553.5	6.0824
800	0.011 523	3517.8	3978.7	6.6662	0.007 459	3441.5	3889.1	6.4109
900	0.012 962	3739.4	4257.9	6.9150	0.008 508	3681.0	4191.5	6.6805
1000	0.014 324	3954.6	4527.6	7.1356	0.009 480	3906.4	4475.2	6.9127
1100	0.015 642	4167.4	4793.1	7.3364	0.010 409	4124.1	4748.6	7.1195
1200	0.016 940	4380.1	5057.7	7.5224	0.011 317	4338.2	5017.2	7.3083
1300	0.018 229	4594.3	5323.5	7.6969	0.012 215	4551.4	5284.3	7.4837

*( ) = saturation temperature at given pressure.

APPENDIX

C

ENTHALPY AND ENTROPY DATA FOR REPRESENTATIVE COMMON SUBSTANCES

Enthalpy of methane

P = kPa, T = °C, H = kJ/kg
Base: H = 0 at 0 K

Ideal gas		Saturation			P = 1000		P = 3000		P = 5000		P = 10000		P = 30000	
T	H	P	T	H	T	H	T	H	T	H	T	H	T	H
−130.0	296.3	761.3	−130.0	−170.5	−124.3	268.4	−96.0	255.8	−80.0	138.8	−85.0	12.8	−90.0	−0.7
		917.7	−126.1	−155.3										
−77.5	404.9	1042.6	−123.3	−144.3	−101.4	326.7	−54.7	395.2	−61.5	317.4	−57.2	169.8	−53.0	126.1
		1179.0	−120.6	−133.0										
−25.0	515.5	1327.6	−117.8	−121.7	−78.5	380.8	−13.3	501.7	−43.0	385.1	−29.4	327.8	−6.7	291.9
		1489.1	−115.0	−110.2										
27.5	630.6	1664.0	−112.2	−98.3	−55.7	432.9	28.0	602.9	−24.4	441.6	−1.7	437.2	21.1	391.9
		1853.1	−109.4	−86.2										
80.0	752.0	2057.2	−106.7	−73.7	−32.8	484.3	69.3	704.5	−5.9	493.4	26.1	527.1	61.0	530.4
		2187.1	−105.0	−65.9										
132.5	881.4	2322.9	−103.3	−58.1	−10.0	535.7	110.7	808.9	12.6	542.9	57.3	619.1	100.9	662.6
		2464.7	−101.7	−50.0										
185.0	1019.9	2612.8	−100.0	−41.7	12.9	587.6	152.0	917.5	31.1	591.2	88.4	707.4	140.8	791.0
		2767.4	−98.3	−33.0										
237.5	1168.2	2928.2	−96.7	−24.1	35.7	640.1	193.3	1031.1	50.9	642.3	119.6	794.9	180.7	918.6
		3097.1	−95.0	−14.7										
290.0	1327.0	3272.6	−93.3	−4.9	58.6	693.7	234.7	1150.4	70.7	693.4	150.7	883.1	220.6	1047.3
		3455.7	−91.7	5.6										
342.5	1496.2	3646.6	−90.0	16.9	81.4	748.5	276.0	1275.7	90.6	744.7	181.9	972.9	260.6	1178.6
		3778.4	−88.9	25.1										
395.0	1676.0	3913.9	−87.8	34.0	104.3	804.7	317.3	1407.3	110.4	796.6	213.0	1065.0	300.5	1313.5
		4053.3	−86.7	43.9										
447.5	1865.9	4196.6	−85.6	55.3	127.1	862.4	358.7	1545.1	130.2	849.3	244.2	1159.8	340.4	1452.6
		4344.0	−84.4	69.5										
500.0	2065.6	4495.6	−83.3*	90.4	150.0	921.7	400.0	1689.2	150.0	902.9	275.4	1257.6	380.3	1596.3
		4419.3	−83.9	190.4										
		4344.0	−84.4	198.7							306.5	1358.4	420.2	1744.8
		4269.8	−85.0	205.3										
		4196.6	−85.6	210.9							337.7	1462.6	460.1	1898.2
		4053.3	−86.7	220.0										
		3913.9	−87.8	227.3							368.8	1570.0	500.0	2056.4
		3778.4	−88.9	233.2										
		3646.6	−90.0	238.3							400.0	1680.9		
		3455.7	−91.7	244.6										
		3272.6	−93.3	249.7										
		3097.1	−95.0	253.9										
		2820.5	−97.8	259.5										
		2612.8	−100.0	262.9										
		2416.8	−102.2	265.6										
		2231.7	−104.4	267.6										
		2057.2	−106.7	269.2										
		1892.7	−108.9	270.2										
		1737.9	−111.1	270.9										
		1592.4	−113.3	271.2										
		1455.7	−115.6	271.1										
		1327.6	−117.8	270.8										
		1179.0	−120.6	270.0										
		1042.0	−123.3	268.8										
		917.7	−126.1	267.3										
		761.3	−130.0	264.6										

* Critical (approx.)

Enthalpy of propane

P = kPa, T = °C, H = kJ/kg
Base: H = 0 at 0 K

Ideal gas		Saturation			P = 1000		P = 3000		P = 5000		P = 10000		P = 30000	
T	*H*	*P*	*T*	*H*	*T*	*H*	*T*	*H*	*T*	*H*	*T*	*H*	*T*	*H*
−130.0	123.8	0.1	−130.0	−384.9	26.4	307.1	77.5	336.1	90.0	167.0	90.0	151.40	70.0	100.4
		2.7	−101.1	−326.7										
−98.5	159.3	20.2	−73.3	−266.8	54.4	359.3	123.2	462.1	117.8	373.3	127.0	277.7	107.0	197.8
		67.9	−51.1	−216.9										
−67.1	198.4	142.9	−34.4	−178.3	82.3	416.4	168.8	577.6	145.6	473.8	164.1	425.5	144.1	300.8
		269.8	−17.8	−138.7										
−35.6	241.3	467.1	−1.1	−97.8	110.3	475.4	214.4	695.9	173.3	555.4	201.1	559.6	181.1	409.8
		647.5	10.0	−69.6										
−4.2	288.1	874.0	21.1	−40.6	138.2	536.8	260.0	819.8	201.1	633.7	236.4	676.2	218.1	524.1
		1153.2	32.2	−10.7										
27.3	339.1	1492.0	43.3	20.5	166.2	601.0	305.7	950.2	229.6	713.8	271.8	788.8	255.1	643.2
		1898.2	54.4	53.2										
58.8	394.2	2381.2	65.6	88.1	194.1	668.1	351.3	1087.5	250.0	771.7	307.1	901.0	292.1	765.9
		2654.9	71.1	106.7										
90.2	453.6	2952.1	76.7	126.4	222.1	738.2	396.9	1231.3			342.4	1014.7	329.1	891.6
		3175.3	80.6	141.2										
121.7	517.2	3377.0	83.9	154.7	250.0	811.3	442.5	1381.5			377.7	1130.7	366.1	1020.0
		3552.8	86.7	167.0										
153.1	585.1	3698.8	88.9	177.7			488.1	1537.6			413.1	1249.4	403.1	1151.0
		3849.6	91.1	189.7										
184.6	657.2	3926.9	92.2	196.5			533.8	1699.2			448.4	1370.9	440.0	1284.5
		4005.5	93.3	204.1										
216.0	733.5	4085.4	94.4	213.2			579.4	1865.7			483.7	1495.3	477.0	1420.5
		4125.9	95.0	218.7										
247.5	813.9	4166.7	95.6*	225.3			625.0	2036.7			519.0	1622.6	514.0	1559.0
		4125.9	95.0	297.7										
279.0	898.0	4085.4	94.4	302.6							554.4	1752.4	551.0	1699.8
		4005.5	93.3	309.7										
310.4	986.5	3926.9	92.2	314.8							589.7	1884.9	588.0	1843.0
		3811.5	90.6	320.5										
341.9	1078.5	3625.2	87.8	326.9							625.0	2019.6	625.0	1988.2
		3274.9	82.2	333.6										
373.3	1173.9	2952.1	76.7	336.3										
		2654.9	71.1	336.7										
404.8	1272.8	2381.2	65.6	335.5										
		2129.5	60.0	333.2										
436.2	1374.9	1898.2	54.4	330.2										
		1686.1	48.9	326.5										
467.7	1480.0	1492.0	43.3	322.2										
		1297.8	37.2	317.1										
499.2	1588.0	1122.7	31.1	311.5										
		938.6	23.9	304.5										
530.6	1698.7	647.5	10.0	289.9										
		428.5	−3.9	274.4										
562.1	1811.8	269.8	−17.8	258.0										
		142.9	−34.4	237.7										
593.5	1927.3	67.9	−51.1	217.0										
		20.2	−73.3	189.4										
625.0	2044.9	2.7	−101.1	156.1										
		0.1	−130.0	123.8										

* Critical (approx.)

Enthalpy of ethylene

P = kPa, T = °C, H = kJ/kg

Base: H = 0 at 0 K

Ideal Gas		Saturation			P = 1000		P = 3000		P = 5000		P = 10000		P = 30000	
T	H	P	T	H	T	H	T	H	T	H	T	H	T	H
		15.6	−130.0	−342.6										
−130.0	176.0	84.8	−106.7	−289.1	−51.9	235.6	−13.1	230.9	8.9	137.2	−15.0	−55.3	−20.0	−59.2
		213.2	−90.0	−249.9										
−101.7	206.4	453.3	−73.3	−209.6	−26.7	273.7	25.5	319.1	27.4	267.7	12.8	30.6	17.0	30.7
		850.6	−56.7	−167.6										
−73.3	239.4	1123.0	−48.3	−145.8	−1.4	314.7	64.1	396.5	46.0	319.7	40.6	154.0	54.1	126.6
		1577.1	−37.2	−115.3										
−45.0	275.1	1847.7	−31.7	−99.3	23.8	356.2	102.7	473.0	63.8	362.9	68.3	275.3	91.1	228.8
		2118.6	−26.7	−84.4										
−16.7	313.5	2417.4	−21.7	−68.9	49.1	398.1	141.3	551.5	83.7	407.8	96.1	365.1	129.4	338.0
		2708.1	−17.2	−54.4										
11.7	354.7	2942.7	−13.9	−43.0	74.3	443.1	179.9	633.3	103.6	451.3	125.6	445.6	167.6	447.8
		3149.6	−11.1	−33.1										
40.0	398.6	3367.3	−8.3	−22.6	99.5	489.1	218.5	718.7	123.5	494.5	155.1	520.3	205.8	556.5
		3596.6	−5.6	−11.6										
68.3	445.2	3837.7	−2.8	0.3	124.8	536.9	257.0	808.1	143.4	537.8	184.6	593.1	244.1	664.7
		4039.7	−0.6	10.7										
96.7	494.5	4196.7	1.1	19.1	150.0	586.5	295.6	901.5			214.1	665.8	282.3	773.0
		4358.5	2.8	28.4										
125.0	546.6	4525.4	4.4	39.0			334.2	998.8			243.6	739.1	320.6	882.2
		4639.5	5.6	47.2										
153.3	601.3	4755.9	6.7	56.9			372.8	1100.0			273.1	813.8	358.8	993.0
		4874.8	7.8	69.6										
181.7	658.6	4935.2	8.3	78.5			411.4	1205.0			302.5	889.9	397.0	1105.6
		4996.1	8.9*	92.1										
210.0	718.6	4935.2	8.3	152.5			450.0	1313.5			332.0	967.8	435.3	1220.4
		4874.8	7.8	162.2										
238.3	781.0	4755.9	6.7	174.6							361.5	1047.4	473.5	1337.4
		4639.5	5.6	183.3										
266.7	845.9	4525.4	4.4	190.1							391.0	1128.9	511.8	1456.7
		4413.6	3.3	195.6										
295.0	913.3	4304.0	2.2	200.3							420.5	1212.2	550.0	1578.2
		4143.8	0.6	206.2										
323.3	983.0	3988.4	−1.1	211.1							450.0	1297.3		
		3837.7	−2.8	215.2										
351.7	1054.9	3691.6	−4.4	218.8										
		3503.5	−6.7	222.7										
380.0	1129.0	3322.9	−8.9	226.0										
		3107.3	−11.7	229.4										
408.3	1205.3	2862.8	−15.0	232.6										
		2596.2	−18.9	235.3										
436.7	1283.6	2281.0	−23.9	237.6										
		1964.8	−29.4	238.9										
465.0	1363.8	1655.0	−35.6	239.2										
		1247.7	−45.0	237.8										
493.3	1445.9	867.1	−56.1	233.9										
		507.3	−70.6	226.0										
521.7	1529.8	244.3	−87.2	214.0										
		71.3	−109.4	194.7										
550.0	1615.3	15.6	−130.0	175.2										

* Critical (approx.)

Enthalpy of benzene

P = kPa, T = °C, H = kJ/kg
Base: H = 0 at 0 K

Ideal gas		Saturation			P = 1000		P = 3000		P = 5000		P = 10000	
T	H	P	T	H	T	H	T	H	T	H	T	H
−130.0	73.1	1.1	−20.0	−316.4	178.3	362.9	250.4	439.5	280.0	315.2	280.0	296.5
		7.1	12.8	266.0								
−97.1	86.7	27.9	43.3	−214.8	212.3	421.3	284.5	518.2	298.5	479.5	307.8	374.9
		63.2	65.6	−175.3								
−64.2	105.6	127.4	87.8	−134.0	246.3	484.7	318.7	595.2	317.0	524.4	335.6	465.4
		234.1	110.0	−91.0								
−31.3	129.7	398.8	132.2	−46.3	280.2	550.4	352.8	671.7	335.6	593.2	363.3	561.7
		570.3	148.9	−11.5								
1.7	158.7	791.3	165.6	24.3	314.2	618.5	386.9	749.3	354.1	640.8	391.1	651.3
		1069.6	182.2	61.3								
34.6	192.3	1413.5	198.9	99.7	348.1	689.1	421.1	828.1	372.6	686.9	413.5	718.5
		1832.5	215.6	139.5								
67.5	230.3	2337.0	232.2	181.2	382.1	762.0	455.2	908.5	391.1	732.3	435.9	782.6
		2727.2	243.3	210.4								
100.4	272.4	3165.4	254.4	241.0	416.0	837.1	489.3	990.6	410.7	780.3	458.3	844.7
		3503.5	262.2	263.7								
133.3	318.5	3762.1	267.8	280.8	450.0	914.4	523.5	1074.2	425.5	816.3	480.7	905.7
		4007.9	272.8	297.2								
166.3	368.3	4237.1	277.2	313.0			557.6	1159.5	440.2	852.3	503.1	966.2
		4416.1	280.6	326.1								
199.2	421.6	4538.9	282.8	335.9			591.7	1246.2	450.0	876.4	525.6	1026.4
		4664.5	285.0	347.2								
232.1	478.1	4728.4	286.1	353.9			625.9	1334.4			548.0	1086.5
		4793.0	287.2	362.1								
265.0	537.8	4858.4	288.3*	374.0			660.0	1424.1			570.4	1146.7
		4825.6	287.8	407.6								
297.9	600.4	4728.4	286.1	430.2							592.8	1207.1
		4664.5	285.0	435.2								
330.8	665.8	4538.9	282.8	440.9							615.2	1267.7
		4416.1	280.6	444.4								
363.8	733.8	4237.1	277.2	447.3							637.6	1328.7
		4007.9	272.8	448.9								
396.7	804.2	3762.1	267.8	448.6							660.0	1389.9
		3503.5	262.2	446.7								
429.6	876.9	3165.4	254.4	442.4								
		2727.2	243.3	433.9								
462.5	951.7	2337.0	232.2	423.6								
		1832.5	215.6	406.2								
495.4	1028.6	1413.5	198.9	387.4								
		1069.6	182.2	367.6								
528.3	1107.3	791.3	165.6	347.3								
		570.3	148.9	326.8								
561.3	1187.8	398.8	132.2	306.1								
		234.1	110.0	278.8								
594.2	1269.9	149.5	93.3	258.6								
		76.0	71.1	232.3								
627.1	1353.7	34.7	48.9	207.1								
		13.8	26.7	183.3								
660.0	1438.8	4.6	4.4	161.2								
		1.1	−20.0	139.0								

* Critical (approx.)

Enthalpy-entropy data for methane

$T = °C$, $P =$ kPa, $H =$ kJ/kg, $S =$ kJ/(kg · K)

$P = 101.3$			$P = 500$			$P = 1000$		
T	H	S	T	H	S	T	H	S
82.2	756.6	10.631	82.2	753.9	9.797	82.2	750.4	9.431
100.7	801.4	10.754	100.7	799.0	9.921	100.7	795.9	9.555
119.3	847.3	10.874	119.3	845.1	10.041	119.3	842.3	9.676
137.8	894.3	10.991	137.8	892.3	10.159	137.8	889.7	9.795
156.3	942.5	11.105	156.3	940.6	10.274	156.3	938.3	9.910
174.8	991.8	11.218	174.8	990.1	10.387	174.8	988.0	10.023
193.3	1042.3	11.328	193.3	1040.8	10.498	193.3	1038.8	10.135
211.9	1094.1	11.437	211.9	1092.7	10.607	211.9	1090.9	10.244
230.4	1147.2	11.544	230.4	1145.8	10.714	230.4	1144.4	10.352
248.9	1201.5	11.650	248.9	1200.3	10.820	248.9	1198.8	10.458
267.4	1257.1	11.755	267.4	1256.0	10.925	267.4	1254.6	10.563
285.9	1314.0	11.858	285.9	1313.0	11.029	285.9	1311.7	10.667
304.4	1372.3	11.961	304.4	1371.3	11.132	304.4	1370.1	10.770

$P = 3000$			$P = 5000$		
T	H	S	T	H	S
82.2	736.7	8.831	82.2	723.1	8.536
100.7	783.5	8.959	100.7	771.3	8.668
119.3	831.1	9.084	119.3	820.2	9.796
137.8	879.6	9.204	137.8	869.7	8.919
156.3	929.1	9.322	156.3	920.1	9.039
174.8	979.6	9.437	174.8	971.4	9.156
193.3	1031.2	9.550	193.3	1023.7	9.270
211.9	1083.9	9.661	211.9	1077.1	9.383
230.4	1137.8	9.770	230.4	1131.6	9.493
248.9	1192.9	9.877	248.9	1187.2	9.601
267.4	1249.2	9.983	267.4	1244.0	9.708
285.9	1306.8	10.088	285.9	1302.1	9.814
304.4	1365.6	10.192	304.4	1361.3	9.918

$T = -50$			$T = 0$			$T = 50$			$T = 100$		
P	H	S	P	H	S	P	H	S	P	H	S
25.0	462.1	10.324	25.0	569.3	10.757	25.0	681.5	11.134	25.0	800.1	11.475
315.6	457.3	8.995	315.6	566.1	9.435	315.6	679.1	9.814	315.6	798.3	10.157
606.2	452.5	8.642	606.2	562.8	9.088	606.3	676.7	9.471	606.2	796.5	9.815
869.9	447.5	8.423	869.9	559.5	8.876	869.9	674.3	9.262	896.9	794.7	9.608
1187.5	442.4	8.262	1187.5	556.1	8.722	1187.5	671.9	9.111	1187.5	792.9	9.459
1478.1	437.3	8.132	1478.1	552.7	8.599	1478.1	669.5	8.992	1478.1	791.1	9.341
1768.8	432.0	8.023	1768.8	549.3	8.497	1768.8	667.1	8.893	1768.8	789.3	9.245
2059.4	426.5	7.927	2059.4	545.9	8.410	2059.4	664.7	8.809	2059.4	787.5	9.162
2350.0	420.9	7.840	2350.0	542.4	8.332	2350.0	662.2	8.735	2350.0	785.7	9.090
2640.6	415.2	7.761	2640.6	538.9	8.262	2640.6	659.8	8.669	2640.6	783.9	9.025
2931.3	409.3	7.688	2931.3	535.4	8.199	2931.3	657.4	8.609	2931.3	782.1	8.967
3221.9	403.2	7.619	3221.9	531.8	8.140	3221.9	654.9	8.554	3221.9	780.3	8.914
3512.5	397.0	7.554	3512.5	528.2	8.086	3512.5	652.5	8.503	3512.5	778.5	8.866
3803.1	390.5	7.491	3803.1	524.6	8.035	3803.1	650.0	8.456	3803.1	776.7	8.821

Enthalpy-entropy data for methane (Continued)

$T = -50$			$T = 0$			$T = 50$			$T = 100$		
P	H	S	P	H	S	P	H	S	P	H	S
4093.8	383.8	7.431	4093.8	521.0	7.987	4093.8	647.6	8.412	4093.8	774.9	8.779
4384.4	376.9	7.372	4384.4	517.3	7.914	4384.4	645.1	8.371	4384.4	773.1	8.739
4675.0	369.8	7.314	4675.0	513.6	7.898	4675.0	642.7	8.332	4675.0	771.4	8.702
4965.6	362.3	7.257	4965.6	509.8	7.856	4965.6	640.2	8.295	4965.6	769.6	8.667
5256.3	354.6	7.201	5256.3	506.1	7.817	5256.3	637.8	8.259	5256.3	767.8	8.634
5546.9	346.6	7.145	5546.9	502.3	7.778	5546.9	635.3	8.226	5546.9	766.1	8.602
5837.5	338.3	7.090	5837.5	498.5	7.741	5837.5	632.8	8.193	5837.5	764.3	8.572
6128.1	329.7	7.034	6128.1	494.7	7.076	6128.1	630.4	8.162	6128.1	762.6	8.543
6418.8	320.8	6.978	6418.8	490.8	7.671	6418.8	627.9	8.132	6418.8	760.8	8.515
6709.4	311.6	6.923	6709.4	487.0	7.637	6709.4	625.5	8.104	6709.4	759.1	8.488
7000.0	302.3	6.867	7000.0	483.1	7.605	7000.0	623.0	8.076	7000.0	757.3	8.462

$T = 150$			$T = 200$			$T = 250$		
P	H	S	P	H	S	P	H	S
25.0	926.4	11.792	25.0	1061.1	12.093	25.0	1205.0	12.382
315.6	924.9	10.475	315.6	1060.0	10.777	315.6	1204.1	11.066
606.2	923.5	10.134	606.2	1058.9	10.436	606.2	1203.3	10.726
896.9	922.1	9.928	896.9	1057.8	10.231	896.9	1202.4	10.522
1187.5	920.8	9.780	1187.5	1056.7	10.084	1187.5	1201.5	10.374
1478.1	919.4	9.664	1478.1	1055.6	9.968	1478.1	1200.6	10.259
1768.8	918.0	9.568	1768.8	1054.5	9.873	1768.8	1199.8	10.165
2059.4	916.6	9.486	2059.4	1053.5	9.792	2059.4	1198.9	10.084
2350.0	915.2	9.415	2350.0	1052.4	9.721	2350.0	1198.1	10.014
2640.6	913.8	9.352	2640.6	1051.3	9.659	2640.6	1197.2	9.952
2931.3	912.5	9.295	2931.3	1050.3	9.603	2931.3	1196.4	9.896
3221.9	911.1	9.243	3221.9	1049.2	9.552	3221.9	1195.6	9.846
3512.5	909.7	9.196	3512.5	1048.1	9.505	3512.5	1194.8	9.799
3803.1	908.4	9.152	3803.1	1047.1	9.461	3803.1	1193.9	9.756
4093.8	907.0	9.111	4093.8	1046.0	9.421	4093.8	1193.1	9.717
4384.4	905.7	9.073	4384.4	1045.0	9.384	4384.4	1192.3	9.679
4675.0	904.4	9.036	4675.0	1044.0	9.348	4675.0	1191.5	9.644
4965.6	903.0	9.002	4965.6	1043.0	9.315	4965.6	1190.7	9.612
5256.3	901.7	8.970	5256.3	1041.9	9.283	5256.3	1189.9	9.580
5546.9	900.4	8.940	5546.9	1040.9	9.253	5546.9	1189.1	9.551
5837.5	899.1	8.910	5837.5	1039.9	9.225	5837.5	1188.3	9.523
6128.1	897.8	8.882	6128.1	1038.9	9.198	6128.1	1187.6	9.496
6418.8	896.5	8.856	6418.8	1037.9	9.172	6418.8	1186.8	9.471
6709.4	895.2	8.830	6709.4	1036.9	9.147	6709.4	1186.0	9.446
7000.0	893.9	8.805	7000.0	1035.9	9.123	7000.0	1185.3	9.422

Enthalpy-entropy data for propane

$T = °C$, $P = kPa$, $H = kJ/kg$, $S = kJ/(kg \cdot K)$

P = 101.3			P = 500			P = 1000		
T	H	S	T	H	S	T	H	S
82.2	436.0	6.573	82.2	427.6	6.256	82.2	416.2	6.102
102.6	476.3	6.684	102.6	468.8	6.369	102.6	458.9	6.219
123.0	518.3	6.792	123.0	511.7	6.480	123.0	503.0	6.333
143.3	562.0	6.900	143.3	556.1	6.589	143.3	548.4	6.445
163.7	607.5	7.007	163.7	602.2	6.697	163.7	595.3	6.555
184.1	654.8	7.113	184.1	650.0	6.804	184.1	643.7	6.663
204.4	703.8	7.217	204.4	699.4	6.910	204.4	693.7	6.770
224.8	754.5	7.321	224.8	750.5	7.014	224.8	745.3	6.875
245.2	806.9	7.424	245.2	803.2	7.118	245.2	798.5	6.981
265.6	861.0	7.527	265.6	875.6	7.221	265.6	853.2	7.084
285.9	916.7	7.628	285.9	913.5	7.323	285.9	909.5	7.187
306.3	974.0	7.729	306.3	971.0	7.424	306.3	967.3	7.288
326.7	1032.9	7.829	326.7	1030.1	7.524	326.7	1026.6	7.389

Enthalpy-entropy data for propane

T = −50			T = 0			T = 50			T = 100		
P	H	S	P	H	S	P	H	S	P	H	S
25.0	219.8	6.084	25.0	293.8	6.382	25.0	377.8	6.664	25.0	472.4	6.936
26.9	219.7	6.069	44.1	293.1	6.273	95.9	376.0	6.407	315.6	467.0	6.448
28.9	219.6	6.056	63.2	292.4	6.204	166.8	374.2	6.299	606.2	461.4	6.314
30.8	219.5	6.043	82.3	291.7	6.152	237.7	372.4	6.228	896.9	455.6	6.229
32.8	219.5	6.032	101.4	291.0	6.111	308.6	370.6	6.175	1187.5	449.5	6.164
34.7	219.4	6.021	120.5	290.3	6.077	379.6	368.7	6.132	1478.1	443.0	6.110
36.7	219.3	6.010	139.5	289.6	6.047	450.5	366.8	6.095	1768.8	436.2	6.063
38.6	219.3	6.000	158.6	288.8	6.021	521.4	364.8	6.063	2059.4	429.0	6.020
40.6	219.2	5.991	177.7	288.1	5.998	592.3	362.8	6.035	2350.0	421.2	5.979
42.5	219.2	5.982	196.8	287.4	5.977	663.2	360.8	6.009	2640.6	412.8	5.940
44.4	219.1	5.973	215.9	286.6	5.957	734.1	358.7	5.985	2931.3	403.6	5.900
46.4	219.1	5.965	235.0	285.9	5.940	805.0	356.6	5.963	3221.9	393.2	5.860
48.3	219.0	5.957	254.1	285.2	5.923	875.9	354.4	5.942	3512.5	381.2	5.818
50.3	219.0	5.949	273.2	284.4	5.907	946.9	352.2	5.923	3803.1	366.6	5.769
52.2	218.9	5.942	292.3	283.6	5.893	1017.8	349.9	5.904	4093.8	347.0	5.709
54.2	218.9	5.935	311.4	282.9	5.879	1088.7	347.6	5.886	4384.4	310.9	5.607
56.1	218.8	5.928	330.5	282.1	5.865	1159.6	345.2	5.868	4675.0	238.2	5.408
58.0	218.7	5.922	349.6	281.7	5.854	1230.5	342.7	5.851	4965.6	220.1	5.357
60.0	218.7	5.915	368.6	281.3	5.843	1301.4	340.1	5.835	5256.3	212.0	5.333
61.9	218.6	5.909	387.7	280.9	5.832	1372.3	337.5	5.819	5546.9	206.8	5.317
63.9	218.6	5.903	406.8	280.5	5.822	1443.2	334.8	5.803	5837.5	202.9	5.304
65.8	218.5	5.897	425.9	280.1	5.812	1514.2	332.0	5.787	6128.1	199.9	5.294
67.8	218.5	5.892	445.0	279.7	5.803	1585.1	330.5	5.775	6418.8	197.4	5.285
69.7	218.4	5.886	464.1	279.3	5.794	1656.0	328.9	5.763	6709.4	195.2	5.277
71.7	218.4	5.881	483.2	278.8	5.786	1726.9	327.2	5.751	7000.0	193.4	5.270

$T = °C$, $P = kPa$, $H = kJ/kg$, $S = kJ/(kg \cdot K)$

P = 3000			P = 5000		
T	H	S	T	H	S
82.2	379.9	5.746	82.2	137.4	5.131
102.6	408.3	5.910	102.6	242.7	5.417
123.0	461.5	6.048	123.0	396.1	5.817
143.3	513.2	6.175	143.3	466.9	5.992
163.7	564.7	6.296	163.7	527.8	6.134
184.1	616.7	6.412	184.1	585.8	6.264
204.4	669.6	6.525	204.4	643.0	6.387
224.8	723.6	6.636	224.8	700.2	6.504
245.2	778.8	6.745	245.2	758.0	6.618
265.6	835.2	6.852	265.6	816.5	6.728
285.9	893.0	6.957	285.9	876.1	6.837
306.3	952.1	7.061	306.3	936.7	6.943
326.7	1012.6	7.163	326.7	998.5	7.048

T = 150			T = 200			T = 250			T = 300		
P	H	S	P	H	S	P	H	S	P	H	S
25.0	577.8	7.201	25.0	693.8	7.460	25.0	820.3	7.714	25.0	956.7	7.96
315.6	573.7	6.716	315.6	690.6	6.977	315.6	817.6	7.232	315.6	954.5	7.48
606.2	569.4	6.585	606.2	687.2	6.848	606.2	814.9	7.105	606.2	952.3	7.35
896.9	565.1	6.504	896.9	683.9	6.769	896.9	812.2	7.027	896.9	950.1	7.27
1187.5	560.7	6.444	1187.5	680.5	6.711	1187.5	809.5	6.970	1187.5	947.8	7.22
1478.1	556.1	6.395	1478.1	677.0	6.664	1478.1	806.7	6.925	1478.1	945.6	7.17
1768.8	551.4	6.353	1768.8	673.5	6.625	1768.8	804.0	6.887	1768.8	943.3	7.14
2059.4	546.6	6.316	2059.4	669.9	6.591	2059.4	801.2	6.854	2059.4	941.1	7.11
2350.0	541.7	6.282	2350.0	666.3	6.560	2350.0	798.4	6.825	2350.0	938.8	7.08
2640.6	536.6	6.251	2640.6	662.6	6.532	2640.6	795.5	6.799	2640.6	936.5	7.05
2931.3	531.3	6.222	2931.3	658.9	6.507	2931.3	792.7	6.775	2931.3	934.3	7.03
3221.9	525.8	6.194	3221.9	655.1	6.483	3221.9	789.8	6.753	3221.9	932.0	7.01
3512.5	520.2	6.168	3512.5	651.2	6.460	3512.5	786.9	6.733	3512.5	929.7	6.99
3803.1	514.3	6.142	3803.1	647.3	6.439	3803.1	784.0	6.714	3803.1	927.4	6.97
4093.8	508.2	6.117	4093.8	643.3	6.419	4093.8	781.0	6.695	4093.8	925.1	6.95
4384.4	501.8	6.092	4384.4	639.3	6.399	4384.4	778.1	6.678	4384.4	922.7	6.94
4675.0	495.2	6.068	4675.0	635.2	6.381	4675.0	775.1	6.662	4675.0	920.4	6.92
4965.6	488.2	6.043	4965.6	631.0	6.363	4965.6	772.1	6.646	4965.6	918.1	6.91
5256.3	480.9	6.019	5256.3	626.8	6.345	5256.3	769.1	6.631	5256.3	915.8	6.89
5546.9	473.4	5.994	5546.9	622.5	6.328	5546.9	766.0	6.616	5546.9	913.5	6.88
5837.5	465.5	5.969	5837.5	618.2	6.311	5837.5	763.0	6.602	5837.5	911.1	6.87
6128.1	457.3	5.944	6128.1	613.8	6.295	6128.1	759.9	6.588	6128.1	908.8	6.86
6418.8	449.0	5.919	6418.8	609.4	6.278	6418.8	756.9	6.575	6418.8	906.5	6.84
679.4	440.6	5.895	6709.4	605.0	6.263	6709.4	753.8	6.562	6709.4	904.2	6.83
7000.0	432.2	5.871	7000.0	600.5	6.247	7000.0	750.8	6.549	7000.0	901.8	6.82

Enthalpy-entropy data for ethylene

$T = °C$, $P = kPa$, $H = kJ/kg$, $S = kJ/(kg \cdot K)$

P = 101.3			P = 500			P = 1000		
T	H	S	T	H	S	T	H	S
−84.4	211.9	6.817	−84.4	212.6	6.310	−84.4	212.6	6.105
−65.0	246.2	6.940	−65.0	232.9	6.423	−65.0	224.3	6.188
−45.6	271.6	7.056	−45.6	259.7	6.547	−45.6	243.7	6.292
−26.1	298.0	7.168	−26.1	288.2	6.667	−26.1	274.6	6.423
−6.7	325.7	7.276	−6.7	317.4	6.781	−6.7	306.2	6.546
12.8	354.6	7.380	12.8	347.4	6.890	12.8	337.9	6.661
32.2	384.7	7.482	32.2	378.4	6.995	32.2	370.3	6.770
51.7	416.1	7.582	51.7	410.5	7.096	51.7	403.4	6.875
71.1	448.7	7.679	71.1	443.8	7.196	71.1	437.5	6.977
90.6	482.6	7.775	90.6	478.2	7.293	90.6	472.6	7.076
110.0	517.7	7.869	110.0	513.7	7.388	110.0	508.7	7.173
129.4	554.1	7.961	129.4	550.5	7.482	129.4	545.9	7.268
148.9	591.7	8.053	148.9	588.4	7.574	148.9	584.3	7.361

Enthalpy-entropy data for ethylene

T = −50			T = 0			T = 50			T = 100		
P	H	S	P	H	S	P	H	S	P	H	S
25.0	267.9	7.451	25.0	336.9	7.730	25.0	414.4	7.990	25.0	500.3	8.237
68.3	266.6	7.150	194.4	333.7	7.114	315.6	410.4	7.230	315.6	497.2	7.480
111.6	265.4	7.000	363.9	330.3	6.920	606.2	406.2	7.027	606.2	494.2	7.280
154.9	264.1	6.899	533.3	326.9	6.798	896.9	402.0	6.902	896.9	491.1	7.158
198.2	262.8	6.822	702.8	323.4	6.707	1187.5	397.8	6.810	1187.5	487.9	7.069
241.5	261.5	6.760	872.2	319.8	6.634	1478.1	393.4	6.735	1478.1	484.8	6.998
284.8	260.1	6.707	1041.6	316.1	6.572	1768.8	388.9	6.672	1768.8	481.6	6.938
328.1	258.8	6.660	1211.1	312.3	6.517	2059.4	384.3	6.617	2059.4	478.3	6.887
371.4	257.4	6.619	1380.5	308.4	6.468	2350.0	379.6	6.567	2350.0	475.1	6.841
414.7	256.0	6.582	1549.9	304.4	6.423	2640.6	374.8	6.521	2640.6	471.8	6.800
458.0	254.6	6.549	1719.4	300.2	6.381	2931.3	369.8	6.479	2931.3	468.4	6.763
501.3	253.2	6.517	1888.8	295.9	6.342	3221.9	364.7	6.439	3221.9	465.0	6.728
544.6	251.7	6.488	2058.3	291.4	6.304	3512.5	359.5	6.402	3512.5	461.6	6.696
587.9	250.3	6.461	2227.7	286.8	6.268	3803.1	354.1	6.366	3803.1	458.1	6.665
631.2	248.8	6.435	2397.1	281.9	6.233	4093.8	348.5	6.331	4093.8	454.6	6.636
674.4	247.3	6.411	2566.6	276.8	6.199	4384.4	342.7	6.297	4384.4	451.1	6.609
717.7	245.7	6.387	2736.0	271.5	6.165	4675.0	336.8	6.264	4675.0	447.5	6.583
761.0	244.3	6.365	2905.4	265.8	6.131	4965.6	330.6	6.231	4965.6	443.9	6.557
804.3	243.2	6.346	3074.9	259.7	6.096	5256.3	324.1	6.198	5256.3	440.2	6.533
847.6	242.1	6.327	3244.3	253.1	6.061	5546.9	317.4	6.166	5546.9	436.5	6.510
890.9	241.0	6.308	3413.8	245.9	6.025	5837.5	310.5	6.134	5837.5	432.8	6.487
934.2	239.8	6.291	3583.2	237.9	5.986	6128.1	303.2	6.102	6128.1	429.0	6.465
977.5	238.7	6.274	3752.6	228.7	5.944	6418.8	295.7	6.069	6418.8	425.2	6.443
1020.8	237.5	6.257	3922.1	217.7	5.896	6709.4	287.9	6.037	6709.4	421.3	6.422
1064.1	236.3	6.241	4091.5	207.9	5.852	7000.0	279.8	6.004	7000.0	417.4	6.401

$T = °C$, $P = kPa$, $H = kJ/kg$, $S = kJ/(kg \cdot K)$

P = 3000			P = 5000		
T	H	S	T	H	S
−84.4	212.6	5.779	−84.4	212.6	5.628
−65.0	224.3	5.862	−65.0	224.3	5.711
−45.6	231.2	5.926	−45.6	231.2	5.775
−26.1	230.9	5.965	−26.1	230.9	5.813
−6.7	245.2	6.048	−6.7	215.0	5.802
12.8	292.0	6.217	12.8	204.1	5.810
32.2	332.9	6.356	32.2	282.4	6.076
51.7	371.9	6.480	51.7	333.9	6.240
71.1	410.4	6.595	71.1	379.5	6.376
90.6	448.8	6.703	90.6	422.8	6.498
110.0	487.7	6.807	110.0	465.2	6.612
129.4	527.1	6.908	129.4	507.4	6.720
148.9	567.4	7.005	148.9	549.9	6.822

T = 150			T = 200			T = 250		
P	H	S	P	H	S	P	H	S
25.0	594.5	8.474	25.0	696.9	8.702	25.0	807.3	8.924
315.6	592.2	7.718	315.6	695.1	7.948	315.6	805.8	8.170
606.2	589.8	7.520	606.2	693.2	7.751	606.2	804.2	7.974
896.9	587.4	7.400	896.9	691.3	7.632	896.9	802.7	7.856
1187.5	585.0	7.313	1187.5	689.3	7.546	1187.5	801.1	7.770
1478.1	582.6	7.244	1478.1	687.4	7.478	1478.1	799.6	7.703
1768.8	580.1	7.186	1768.8	685.5	7.421	1768.8	798.0	7.647
2059.4	577.7	7.137	2059.4	683.6	7.373	2059.4	796.5	7.600
2350.0	575.2	7.093	2350.0	681.7	7.331	2350.0	795.0	7.558
2640.6	572.8	7.054	2640.6	679.8	7.293	2640.6	793.4	7.512
2931.3	570.3	7.019	2931.3	677.8	7.259	2931.3	791.9	7.488
3221.9	567.8	6.986	3221.9	675.9	7.228	3221.9	790.3	7.457
3512.5	565.3	6.956	3512.5	674.0	7.199	3512.5	788.8	7.429
3803.1	562.8	6.928	3803.1	672.0	7.172	3803.1	787.3	7.404
4093.8	560.3	6.902	4093.8	670.1	7.147	4093.0	785.7	7.379
4384.4	557.7	6.877	4384.4	668.2	7.124	4384.4	784.2	7.357
4675.0	555.2	6.853	4675.0	666.2	7.101	4675.0	782.7	7.335
4965.6	552.6	6.831	4965.6	664.3	7.080	4965.6	781.1	7.315
5256.3	550.1	6.809	5256.3	662.4	7.060	5256.3	779.6	7.296
5546.9	547.5	6.789	5546.9	660.4	7.041	5546.9	778.1	7.277
5837.5	544.9	6.769	5837.5	658.5	7.022	5837.5	776.6	7.260
6128.1	542.3	6.750	6128.1	656.5	7.005	6128.1	775.0	7.243
6418.8	539.7	6.731	6418.8	654.6	6.988	6418.8	773.5	7.227
6709.4	537.1	6.713	6709.4	652.7	6.971	6709.4	772.0	7.211
7000.0	534.5	6.696	7000.0	650.7	6.956	7000.0	770.5	7.196

APPENDIX

D

SHORT GUIDE TO SI

Quantity	Unit	SI symbol	
	Base units		
Length	meter	m	
Mass	kilogram	kg	
Time	second	s	
Electric current	ampere	A	
Thermodynamic temperature	kelvin	K	
Luminous intensity	candela	cd	
	Supplementary units		
Plane angle	radian	rad	
Solid angle	steradian	sr	
	Derived units		
Acceleration	meter per second squared	—	m/s^2
Activity (of a radioactive source)	disintegration per second	—	disintegration/s
Angular acceleration	radian per second squared	—	rad/s^2
Angular velocity	radian per second	—	rad/s
Area	square meter	—	m^2
Density	kilogram per cubic meter	—	kg/m^3
Electric capacitance	farad	F	$A \cdot s/V$
Electric field strength	volt per meter	—	V/m
Electric inductance	henry	H	$V \cdot s/A$

Quantity	Unit	SI symbol	
	Derived units		
Electric potential difference	volt	V	W/A
Electric resistance	ohm	Ω	V/A
Electromotive force	volt	V	W/A
Energy	joule	J	N · m
Entropy	joule per kelvin	—	J/K
Force	newton	N	kg · m/s^2
Frequency	hertz	Hz	(cycle)/s
Illumination	lux	lx	lm/m^2
Luminance	candela per square meter	—	cd/m^2
Luminous flux	lumen	lm	cd · sr
Magnetic field strength	ampere per meter	—	A/m
Magnetic flux	weber	Wb	V · s
Magnetic flux density	tesla	T	Wb/m^2
Magnetomotive force	ampere	A	—
Power	watt	W	J/s
Pressure	newton per square meter	—	N/m^2
Quantity of electricity	coulomb	C	A · s
Quantity of heat	joule	J	N · m
Radiant intensity	watt per steradian	—	W/sr
Specific heat	joule per kilogram-kelvin	—	J/(kg · K)
Stress	newton per square meter	—	N/m^2
Thermal conductivity	watt per meter-kelvin	—	W/(m · K)
Velocity	meter per second	—	m/s
Viscosity, dynamic	newton-second per square meter	—	N · s/m^2
Viscosity, kinematic	square meter per second	—	m^2/s
Voltage	volt	V	W/A
Volume	cubic meter	—	m^3
Wavenumber	reciprocal meter		(wave)/m
Work	joule	J	N · m

Decimal multiples

Multiplication factors	Prefix	SI symbol
1 000 000 000 000 = 10^{12}	tera	T
1 000 000 000 = 10^{9}	giga	G
1 000 000 = 10^{6}	mega	M
1 000 = 10^{3}	kilo	k
100 = 10^{2}	hecto	h
10 = 10^{1}	deka	da
0.1 = 10^{-1}	deci	d
0.01 = 10^{-1}	centi	c
0.001 = 10^{-3}	milli	m
0.000 001 = 10^{-6}	micro	μ
0.000 000 001 = 10^{-9}	nano	n
0.000 000 000 001 = 10^{-12}	pico	p
0.000 000 000 000 001 = 10^{-15}	femto	f
0.000 000 000 000 000 001 = 10^{-18}	atto	a

Source: Reprinted from *AIChE J.*, **17**:511(1971).

Avoid use of prefixes in denominators (except kg).

The use of hecto, deka, deci, and centi prefixes should be avoided except when used in areas and volumes.

SI symbols are not capitalized unless the unit is derived from a proper name, e.g., Hz for H. R. Hertz. Unabbreviated units are not capitalized, e.g., hertz, newton, kelvin. Only T, G, M prefixes are capitalized.

Except at the end of a sentence, SI units are not to be followed by periods.

Four or more digits in a group should be separated in groups of three with no comma, e.g., 1 983 212.322 7. not 1,983,212.3227.

With derived unit abbreviations, use center dot to denote multiplication and a slash for division, e.g., newton-second/meter^2 = N $\cdot$ s/m^2.

Conversion tables

To convert from	To	Multiply by
angstrom	meter (m)	1.000 000* E − 10
atmosphere (normal)	newton/meter^2 (N/m^2)	1.013 250* E + 05
barrel (for petroleum, 42 gal)	meter^3 (m^3)	1.589 873 E − 01
British thermal unit (International Table)	joule (J)	1.055 056 E + 03
Btu/(lbm · °F) (c, heat capacity)	joule/kilogram-kelvin [J/(kg · K)]	4.186 800* E + 03
Btu/hour	watt (W)	2.930 711 E − 01
Btu/second	watt (W)	1.054 350 E + 03
Btu/(ft^2 · h · °F) (heat transfer coefficient)	joule/meter^2-second-kelvin [J/(m^2 · s · K)]	5.678 264 E + 00
Btu/(ft^2 · h) (heat flux)	joule/meter^2-second [J/(m^2 · s)]	3.154 591 E + 00
Btu/(ft · hr · °F) (thermal conductivity)	joule/meter-second-kelvin [J/(m · s · K)]	1.730 735 E + 00
calorie (International Table)	joule (J)	4.186 800° E + 00
cal/(g · °C)	joule/kilogram-kelvin [J/(kg · K)]	4.186 800* E + 03
centimeter	meter (m)	1.000 000* E − 02
centimeter of mercury (0°C)	newton/meter^2 (N/m^2)	1.333 22 E + 03
centimeter of water (4°C)	newton/meter^2 (N/m^2)	9.806 38 E + 01
centipoise	newton-second/meter^2 (N · s/m^2)	1.000 000* E − 03
centistoke	meter^2/second (m^2/s)	1.000 000* E − 06
degree Celsius	kelvin (K)	$t_K = t_C + 273.15$
degree Fahrenheit	kelvin (K)	$t_K = (t_F + 459.67)/1.8$
degree Rankine	kelvin (K)	$t_K = t_R/1.8$
dyne	newton (N)	1.000 000* E − 05
erg	joule (J)	1.000 000* E − 07
farad (international of 1948)	farad (F)	9.995 05 E − 01
fluid ounce (U.S.)	meter^3 (m^3)	2.957 353 E − 05
foot	meter (m)	3.048 000* E − 01
foot (U.S. survey)	meter (m)	3.048 006 E − 01
foot of water (39.2°F)	newton/meter^2 (N/m^2)	2.988 98 E + 03
foot^2	meter^2 (m^2)	9.290 304* E − 02
foot/second^2	meter/second^2 (m/s^2)	3.048 000* E − 01
foot^2/hour	meter^2/second (m^2/s)	2.580 640* E − 05
foot-pound-force	joule (J)	1.355 818 E + 00
foot^2/second	meter^2/second (m^2/s)	9.290 304° E − 02
foot^3	meter^3 (m^3)	2.831 685 E − 02
gallon (U.S. liquid)	meter^3 (m^3)	3.785 412 E − 03
gram	kilogram (kg)	1.000 000* E − 03
horsepower (550 ft · lbf/s)	watt (W)	7.456 999 E + 02
hour (mean solar)	second (s)	3.600 000 E + 03
inch	meter (m)	2.540 000* E − 02

*An asterisk after the sixth decimal place indicates the conversion factor is exact and all subsequent digits are zero.

Conversion tables (Continued)

To convert from	To	Multiply by
inch of mercury (60°F)	newton/meter^2 (N/m^2)	3.376 85 E + 03
inch of water (60°F)	newton/meter^2 (N/m^2)	2.488 4 E + 02
inch^2	meter^2 (m^2)	6.451 600* E − 04
inch^3	meter^3 (m^3)	1.638 706 E − 05
kilocalorie	joule (J)	4.186 800* E + 03
kilogram-force (kgf)	newton (N)	9.806 650* E + 00
knot (international)	meter/second (m/s)	5.144 444 E − 01
liter	meter^3 (m^3)	1.000 000* E − 03
micron	meter (m)	1.000 000* E − 06
mil	meter (m)	2.540 000* E − 05
mile (U.S. statute)	meter (m)	1.609 344* E + 03
mile/hour	meter/second (m/s)	4.470 400* E − 01
millimeter of mercury (0 °C)	newton/meter^2 (N/m^2)	1.333 224 E + 02
minute (angle)	radian (rad)	2.908 882 E − 04
minute (mean solar)	second (s)	6.000 000 E + 01
ohm (international of 1948)	ohm (Ω)	1.000 495 E + 00
ounce-mass (avoirdupois)	kilogram (kg)	2.834 952 E − 02
ounce (U.S. fluid)	meter^3 (m^3)	2.957 353 E − 05
pint (U.S. liquid)	meter^3 (m^3)	4.731 765 E − 04
poise (absolute viscosity)	newton-second/meter^2 (N · s/m^2)	1.000 000* E − 01
poundal	newton (N)	1.382 550 E − 01
pound-force (lbf avoirdupois)	newton (N)	4.448 222 E + 00
pound-force-second/foot^2	newton-second/meter^2 (N · s/m^2)	4.788 026 E + 01
pound-mass (lbm avoirdupois)	kilogram (kg)	4.535 924 E − 01
pound-mass/foot^3	kilogram/meter^3 (kg/m^3)	1.601 846 E + 01
pound-mass/foot-second	newton-second/meter^2 (N · s/m^2)	1.488 164 E + 00
psi	newton/meter^2 (N/m^2)	6.894 757 E + 03
quart (U.S. liquid)	meter^3 (m^3)	9.463 529 E − 04
second (angle)	radian (rad)	4.848 137 E − 06
slug	kilogram (kg)	1.459 390 E + 01
stoke (kinematic viscosity)	meter^2/second (m^2/s)	1.000 000* E − 04
ton (long, 2240 lbm)	kilogram (kg)	1.016 047 E + 03
ton (short, 2000 lbm)	kilogram (kg)	9.071 847 E + 02
torr (mmHg, 0 °C)	newton/meter^2 (N/m^2)	1.333 22 E + 02
volt (international of 1948)	volt (absolute) (V)	1.000 330 E + 00
watt (international of 1948)	watt (W)	1.000 165 E + 00
watt-hour	joule (J)	3.600 000* E + 03
yard	meter (m)	9.144 000* E − 01

*An asterisk after the sixth decimal place indicates the conversion factor is exact and all subsequent digits are zero.

APPENDIX

E

STANDARD FORTRAN SUBROUTINES

This appendix includes standard Fortran subroutines for implementing (1) the Lee-Kesler method of prediction of compressibility factor, enthalpy, entropy, and isobaric heat capability departures, and fugacity coefficient and (2) the Soave equation of state for calculating equilibrium K values and bubble-point and dew-point pressures. Use of the methods is self-explanatory.

The Soave subprograms were written and tested by J. Richard Elliott, Research Assistant in Chemical Engineering, the Pennsylvania State University.

LEE-KESLER SUBROUTINES

THIS SERIES OF FORTRAN SUBROUTINES IS USED TO CALCULATE COMPRESSIBILITY FACTORS (CHAPTER 2), ENTHALPY DEPARTURES (CHAPTER 4), ENTROPY DEPARTURES (CHAPTER 4), AND FUGACITIES (CHAPTER 7) BY THE LEE KESLER (AICHE J 21 510 (1975)) REPRESENTATION OF THE PITZER CORRESPONDING STATES TECHNIQUE. THE EQUATIONS AND REFERENCES ARE GIVEN IN THE CHAPTERS NOTED. USERS MUST SUPPLY THEIR OWN MAIN PROGRAM TO SUPPLY CRITICAL PROPERTIES FOR THE DESIRED COMPOUNDS AS WELL AS THE TEMPERATURES AND PRESSURES FOR WHICH CALCULATIONS ARE TO BE MADE. ACCESS THE SUBPROGRAMS USING NORMAL CALL STATEMENTS. NECESSARY INPUT REQUIRED AND OUTPUT GENERATED ARE GIVEN BY THE COMMENT CARDS.

```
      SUBROUTINE LEEKES (PR,TR,W,NPHASE,Z,DELH,DELS,FUGC,DLCP,DLCV,LA)
C
C  THIS SUBROUTINE CALCULATES THE COMPRESSIBILITY FACTOR AND THE
C  DEPARTURES OF ENTHALPY,ENTROPY,ISOBARIC AND ISOCHORIC
C  HEAT CAPACITIES, AND THE FUGACITY COEFFICIENT
C  BY THE METHOD OF LEE AND KESLER
C
C  INPUT VARIABLES IN THE SUBROUTINE ARGUMENT
C  ----- --------- -- --- ---------- --------
C  PR    : REDUCED PRESSURE
C  TR    : REDUCED TEMPERATURE
C  W     : ACENTRIC FACTOR
C  NPHASE: PHASE CONDITION ; 0 = VAPOR PHASE
C                            1 = LIQUID AND DENSE PHASE
C  LA    : CODE FOR DESIRED ANSWER ; 0 = COMPRESSIBILITY FACTOR(Z) ONLY
C                                    1 = ENTHALPY AND Z
C                                    2 = ENTROPY AND Z
C                                    3 = ALL PROPERTIES
C
C  OUTPUT VARIABLES IN THE SUBROUTINE ARGUMENT
C  ------ --------- -- --- ---------- --------
C  Z     : COMPRESSIBILITY FACTOR
C  DELH  : ENTHALPY DEPARTURE IN REDUCED FORM
C  DELS  : ENTROPY  DEPARTURE IN REDUCED FORM
C  FUGC  : FUGACITY COEFFICIENT
C  DLCP  : ISOBARIC HEAT CAPACITY DEPARTURE IN REDUCED FORM
C  DLCV  : ISOCHORIC HEAT CAPACITY DEPARTURE IN REDUCED FORM
C
C  SUBROUTINE AND FUNCTION REQUIRED
C  ---------- --------
C  BILH  : TO SOLVE A MODIFIED BWR EQUATION FOR REDUCED DENSITY
C  GUESS : INITIAL GUESS FOR BILH
C
      DIMENSION B(13,2),VV(2),HH(2),SS(2),FC(2),CP(2),CV(2)
C
      DATA B/0.1181193,0.265728,0.15479,0.030323,0.0236744,0.0186984,0.0
     1,0.155488E-4,0.623689E-4,0.042724,0.060167,0.65392,0.0,0.2026579,0
     2.331511,0.027655,0.203488,0.0313385,0.0503618,0.016901,0.48736E-4,
     30.740336E-5,0.041577,0.037540,1.226,0.3942/
C
      PS(TR,WW)=EXP(5.92714+15.2518*WW-(6.09648+15.6875*WW)/TR-(1.28862+
     113.4721*WW)*ALOG(TR)+(0.169347+0.435765*WW)*TR**6)
C
      DO 104  M=1,2
```

```
      PP=PR
      IF (TR.GE.1.OR.PR.GE.1.0) GO TO 102
      IF (NPHASE.EQ.1) GO TO 100
      PX=(4.0-TR)/3.0*PS(TR,B(13,M))
      IF (PX.LT.PP) PP=PX
      CALL BILH (NPHASE,TR,PP,B(1,M),VV(M),HH(M),SS(M),FC(M),CP(M),CV(M)
     1,LA)
      VV(M)=VV(M)*PP/PR
      GO TO 104
  100 PX=(2.0+TR)/3.0*PS(TR,0.0)
      IF (PX.GT.PP) PP=PX
  102 CALL BILH (NPHASE,TR,PP,B(1,M),VV(M),HH(M),SS(M),FC(M),CP(M),CV(M)
     1,LA)
  104 CONTINUE
      Z=(W*(VV(2)-VV(1))/0.3942+VV(1))*PR/TR
      DELH=W*(HH(2)-HH(1))/0.3942+HH(1)
      DELS=W*(SS(2)-SS(1))/0.3942+SS(1)
      FUGC=W*(FC(2)-FC(1))/0.3942+FC(1)
      DLCP=W*(CP(2)-CP(1))/0.3942+CP(1)
      DLCV=W*(CV(2)-CV(1))/0.3942+CV(1)
      RETURN
      END

      SUBROUTINE BILH (LY,TR,PR,C,VR,DH,DS,FC,CP,CV,LA)
C
C     ALL VARIABLES GENERATED BY SUBROUTINE LEEKES
C
      DIMENSION C(13)
      DH=0.0
      DS=0.0
      FC=0.0
      CP=0.0
      CV=0.0
      LX=LY
      TR2=TR*TR
      IF (PR.GE.1.0) GO TO 100
      IF (TR.GE.1.0) LX=0
      GO TO 102
  100 LX=1
  102 IF (C(13)-0.001) 104,112,112
  104 IF (LX.EQ.0.AND.PR.LE.0.9) GO TO 106
      IF (LX.NE.0.AND.TR.LE.1.0) GO TO 108
      IF (PR.GT.2 .OR.TR.GT.1.2) GO TO 110
      TT=AMAX1(1.04,TR)
      D=1.0/GUESS(TT,1.0/PR,13)
      GO TO 120
  106 D=1.0/(GUESS(TR,PR,1)+TR/PR)
      GO TO 120
  108 PX=PR-EXP(1.6919-3.4469/TR+1.8968*TR2)
      IF (PX.LT.0.0) PX=0.0
      D=(17.065-(26.752-(32.810-17.821*TR)*TR)*TR)*(1.0+EXP(TR*(-16.454+
     1(21.499-4.0957*TR)*TR))*PX)**0.12
      GO TO 120
  110 D=1.0/GUESS(TR,1.0/PR,25)
      GO TO 120
  112 IF (LX.EQ.0.AND.PR.LE.0.9) GO TO 114
      IF (LX.NE.0.AND.TR.LE.1.0) GO TO 116
      IF (PR.GT.2 .OR.TR.GT.1.2) GO TO 118
```

```
      TT=AMAX1(1.04,TR)
      D=1.0/GUESS(TT,1.0/PR,49)
      GO TO 120
  114 D=1.0/(GUESS(TR,PR,37)+TR/PR)
      GO TO 120
  116 PX=PR-EXP(3.8455-5.7665/TR+2.2066*TR2)
      IF (PX.LT.0.0) PX=0.0
      D=(15.092-(7.7106-(4.7768-5.8333*TR)*TR)*TR)*(1.0+EXP(TR*(-20.037+
     1(28.425-8.1055*TR)*TR))*PX)**0.12
      GO TO 120
  118 D=1.0/GUESS(TR,1.0/PR,61)
  120 D2=D*1.005
      BB=C(1)-C(2)/TR-(C(3)+C(4)/TR)/TR2
      CC=C(5)-(C(6)-C(7)/TR2)/TR
      DD=C(8)+C(9)/TR
      AA=C(10)/(TR2*TR)
      GD2=C(11)*D2*D2
      Y2=D2*(1.0+(BB+(CC+AA*(C(12)+GD2)*EXP(-GD2))*D2)*D2+DD*D2**5)*TR
      DO 122  M=1,40
      D1=D
      DQ=D*D
      GD2=C(11)*DQ
      EXG=EXP(-GD2)
      D5=D**5
      Y=D*(1.0+BB*D+(CC+AA*(C(12)+GD2)*EXG)*DQ+DD*D5)*TR
      D=D1-(D1-D2)*(Y-PR)/((Y-Y2)+1.0E-10)
      ADELD=ABS(1.0-D/D1)
      IF (ADELD.LT.0.00005) GO TO 124
      IF (ADELD.GT.0.07) D=D1-0.07*(D1-D)/ADELD
      Y2=Y
  122 D2=D1
  124 VR=1.0/D
      IF (LA.EQ.0) GO TO 128
      BN=C(2)/TR+(2.0*C(3)+3.0*C(4)/TR)/TR2
      CN=(C(6)-3.0*C(7)/TR2)/TR
      DN=-C(9)/TR
      IF (LA.EQ.1) GO TO 126
      DS=ALOG(PR/(TR*D))-((BB+BN)*D+0.5*(CC+CN)*DQ+0.2*(DD+DN)*D5+AA/C(1
     11)*((C(12)+1.0+GD2)*EXG-C(12)-1.0))
      IF (LA.EQ.2) GO TO 128
      FC=(2.0*BB*D+1.5*CC*DQ+1.2*DD*D5+AA/C(11)*(0.5*(C(12)+1.0)-(0.5*(C
     1(12)+1.0)-GD2*(C(12)-0.5+GD2))*EXG)-ALOG(PR/(TR*D)))/2.302585
      CV=2.0*(C(3)+3.0*C(4)/TR)/TR2*D-3.0*C(7)/(TR*TR2)*DQ+3.0*AA/C(11)*
     1((C(12)+1.0+GD2)*EXG-C(12)-1.0)
      PV=1.0+2.0*BB*D+(3.0*CC+AA*(3.0*C(12)+GD2*(5.0-2.0*(C(12)+GD2)))*E
     1XG)*DQ+6.0*DD*D5
      IF (PV.EQ.0.0) PV=1.0E-20
      PT=1.0+(BB+BN)*D+(CC+CN-2.0*AA*(C(12)+GD2)*EXG)*DQ+(DD+DN)*D5
      CP=CV-1.0+PT*PT/PV
  126 DH=TR*((BB-BN)*D+(CC-0.5*CN)*DQ+(DD-0.2*DN)*D5+AA/C(11)*(1.5*(C(12
     1)+1.0)-(1.5*(C(12)+1.0+GD2)-GD2*(C(12)+GD2))*EXG))
  128 RETURN
      END
      FUNCTION GUESS (T,P,M)
C     ALL VARIABLES CONTROLLED BY SUBROUTINE BILH
      DIMENSION B(72)
      DATA B/-0.02055,0.05889,-0.31637,0.09909,0.08331,-0.73241,-0.17493
     1,-0.10797,1.2020,0.02604,0.30346,-0.95175,2.6061,-4.1747,2.1084,-0
```

```
     2.98547,-1.3852,1.2391,2.9973,7.2464,-9.9768,-7.1592,4.6867,3.2203,
     3-0.08166,0.31876,-0.12461,2.3172,-4.3137,2.2275,-2.9905,9.9960,-7.
     48360,1.7415,-5.8844,5.2288,-0.07506,0.29964,-0.54329,0.20400,-0.07
     5730,-0.75824,-0.17735,-0.49280,1.9649,-0.06316,0.83648,-1.6951,1.5
     6209,-3.5829,2.8271,2.7662,-0.91922,-4.4611,-4.4664,9.7687,-2.4752,
     7-2.5527,1.5071,0.49801,-0.03649,0.20198,-0.05564,2.0740,-3.2646,1.
     82733,-2.7232,8.6640,-6.3535,1.6822,-5.4447,4.6117/
      T2=T*T
      GUESS=B(M)*T+B(M+1)+B(M+2)/T2+(B(M+3)*T+B(M+4)+B(M+5)/T2+(B(M+6)*T
     1+B(M+7)+B(M+8)/T2+(B(M+9)*T+B(M+10)+B(M+11)/T2)*P)*P)*P
      RETURN
      END

      SUBROUTINE  LKMIX(NC,CPC,CTC,CW,X,PC,TC,W)

C  SUBROUTINE CALCULATES PSEUDOCRITICAL PROPERTIES OF MIXTURE
C  BY METHOD OF LEE AND KESLER

C  INPUT VARIABLES IN THE SUBROUTINE ARGUMENT
C  ----- --------- -- --- ---------- --------
C  NC  : NUMBER OF COMPONENTS
C  CPC : COMPONENT CRITICAL PRESSURE IN PSIA
C  CTC : COMPONENT CRITICAL TEMPERATURE IN DEGREES RANKIN
C  CW  : COMPONENT ACENTRIC FACTOR
C  X   : MOLE FRACTION
C
C  OUTPUT VARIABLES IN THE SUBROUTINE ARGUMENT
C  ------ --------- -- --- ---------- --------
C  PC  : PSEUDO-CRITICAL PRESSURE
C  TC  : PSEUDO-CRITICAL TEMPERATURE
C  W   : ACENTRIC FACTOR OF MIXTURE
C
      DIMENSION CPC(1) ,CTC(1) ,CW(1) ,X(1)
      BI1 = 0.
      BI2 = 0.
      BI3 = 0.
      BI4 = 0.
      BI5 = 0.
      BI6 = 0.
      BI7 = 0.
      W = 0.
      DO 101 I=1,NC
      VI = (.2905-.085*CW(I))*CTC(I)/CPC(I)
      V13 = VI**.3333333
      XT = X(I)*SQRT(CTC(I))
      BI1 = BI1+X(I)*V13
      BI2 = BI2+X(I)*V13*V13
      BI3 = BI3+X(I)*VI
      BI4 = BI4+XT
      BI5 = BI5+XT*V13
      BI6 = BI6+XT*V13*V13
      BI7 = BI7+XT*VI
      W = W+X(I)*CW(I)
  101 CONTINUE
      BI3 = BI3+3.*BI1*BI2
      TC = (BI4*BI7+3.*BI5*BI6)/BI3
      PC = TC*(.2905-.085*W)*4./BI3
      RETURN
```

SOAVE SUBPROGRAMS

THE FOLLOWING SUBPROGRAMS SOLVE FOR THE BUBBLE POINT PRESSURE, THE DEW POINT PRESSURE, AND EQUILIBRIUM K-RATIOS OF FLUID MIXTURES IN VAPOR-LIQUID EQUILIBRIUM. ALL THE ROUTINES EXCEPT FOR FUGI ARE GENERAL IN THE SENSE THAT THEY MAY BE USED WITH ANY EQUATION OF STATE. FUGI IS THE ROUTINE WHICH SOLVES FOR THE EQUILIBRIUM PROPERTIES BASED ON THE REDLICH-KWONG-SOAVE EQUATION OF STATE. THE REQUIRED INPUTS AND DESIRED OUTPUTS FOR CALLING THE ROUTINES ARE DISCUSSED IN THE BEGINNING COMMENTS OF EACH ROUTINE.

```
      SUBROUTINE BUBLP(TC,PC,ACE,RGAS,T,P,X,Y,NC,INIT,ITMAX,IER)
C
C  LATEST REVISION    - JUNE 13, 1984
C
C  PURPOSE            - CALCULATE BUBBLE POINT PRESSURE BASED ON
C                       TEMPERATURE AND LIQUID COMPOSITION(INPUTS).
C
C  ARGUMENTS  TC      - VECTOR CRITICAL TEMPERATURES OF THE COMPONENTS
C                       (INPUT)
C             PC      - VECTOR CRITICAL PRESSURES OF THE COMPONENTS
C                       (INPUT)
C             ACE     - VECTOR ACENTRIC FACTORS OF THE COMPONENTS(INPUT)
C             RGAS    - GAS CONSTANT (EG. 83.1434 CC-BAR/(GMOL-K) )
C             T       - TEMPERATURE (INPUT)
C             P       - PRESSURE (INPUT OPTIONAL/OUTPUT)
C             X       - MOLE FRACTIONS IN THE LIQUID PHASE (INPUT)
C             Y       - MOLE FRACTIONS IN THE VAPOR PHASE (OUTPUT)
C             NC      - NUMBER OF COMPONENTS (INPUT)
C             INIT    - 0 INITIAL GUESS FOR P CALCULATED BY PSTART
C                       1 INITIAL GUESS FOR P AND Y PASSED FROM CALLING
C                         ROUTINE
C             ITMAX   - MAXIMUM NUMBER OF ITERATIONS PERMITTED
C
C  IER = 1 -- 9   => #OF COMPONENTS ABOVE CRIT TEMP
C      = 10       => LIQUID ROOT PASSED FROM FUGI WAS NOT REAL
C      = 20       => VAPOR ROOT PASSED FROM FUGI WAS NOT REAL
C      = 100      => ERROR RETURNED FROM FUGI CALCULATION
C      = 1000     => CALCULATIONS DETERMINED VAPOR & LIQUID ROOTS EQUAL
C                    (TRIVIAL SOLUTION)
C      = 2000     => FAILED TO CONVERGE IN ITMAX LOOPS
C      = 3000     => CONVERGED TO A NEGATIVE PRESSURE
C
C  REQD. ROUTINES     - PSTART, FUGI(SRKNR)
C
      IMPLICIT DOUBLE PRECISION(A-H,O-Z)
      DIMENSION X(10),Y(10),PHI(10)
      DIMENSION PHIL(10),PHIV(10),F(10)
      DIMENSION TC(10),PC(10),ACE(10)
      KOUNT = 0
      IER = 0
      DO 1 I=1,NC
 1    IF(TC(I) .LE. T)IER = IER+1
      IER1 = IER
      IF(INIT.EQ.0)CALL PSTART(TC,PC,ACE,T,P,X,NC)
      DO  50 I=1,NC
 50   PHIV(I) = 1
```

```
57    LIQ = 1
      IER1 = IER
      CALL FUGI(TC,PC,ACE,RGAS,PHIL,T,P,X,NC,LIQ,ZL,IERF)
      IF(IERF.GT.9) IER = IER1 + 100
      IF(IERF.GT.9) RETURN
      DO 51 I=1,NC
51    F(I) = PHIL(I)*X(I)*P
59    SUMY=0
      SUMP=0
      DO 52 I=1,NC
      Y(I) = F(I)/(P*PHIV(I))
52    SUMY = SUMY + Y(I)
      G = 1-SUMY
      IF(KOUNT .GT. 25) GO TO 22
      CHNG = P*(1-SUMY)
      GO TO 23
22    IF(G .EQ. GOLD)RETURN
      CHNG = G*(P-POLD)/(G-GOLD)
23    CONTINUE
      POLD = P
      GOLD = G
      PNEW = P - CHNG
      P = PNEW
      IF(P .LE.O)IER = IER1 + 3000
      IF(KOUNT .GT. ITMAX) IER = IER1 + 200
      IF(KOUNT .GT. ITMAX) RETURN
      LIQ = 0
      IF(DABS((POLD-P)/P).LE.1.D-5)GO TO 54
55    DO 60 I=1,NC
60    Y(I) = Y(I)/SUMY
      CALL FUGI(TC,PC,ACE,RGAS,PHIV,T,P,Y,NC,LIQ,ZV,IERF)
      IF(DABS((ZV-ZL)/ZV).LT.1.D-4) IER = IER1 + 1000
      IF(IERF.GT.9) IER = IER1 + 100
      IF(IERF.GT.9) RETURN
      KOUNT = KOUNT + 1
      GO TO 57
54    IF(DABS(SUMY-1).LE.1.D-5) GO TO 56
58    DO 53 I=1,NC
53    Y(I) = Y(I)/SUMY
      CALL FUGI(TC,PC,ACE,RGAS,PHIV,T,P,Y,NC,LIQ,ZV,IERF)
      IF(DABS((ZV-ZL)/ZV).LT.1.D-4) IER = IER1 + 1000
      IF(IERF.GT.9) IER = IER1 + 100
      IF(IERF.GT.9) RETURN
      KOUNT = KOUNT + 1
      GO TO 59
56    IF(IERF.EQ.1)IER=IER+10
      IF(IERF.EQ.2)IER=IER+20
      RETURN
      END

      SUBROUTINE DEWP(TC,PC,ACE,RGAS,T,P,X,Y,NC,INIT,ITMAX,IER)
C
C  PURPOSE            - CALCULATE DEW POINT PRESSURE BASED ON
C                       TEMPERATURE AND VAPOR COMPOSITION(INPUTS).
C
C  ARGUMENTS  TC      - VECTOR CRITICAL TEMPERATURES OF THE COMPONENTS
C                       (INPUT)
```

```
C             PC     - VECTOR CRITICAL PRESSURES OF THE COMPONENTS
C                       (INPUT)
C             ACE    - VECTOR ACENTRIC FACTORS OF THE COMPONENTS(INPUT)
C             RGAS   - GAS CONSTANT (EG. 83.1434 CC-BAR/(GMOL-K) )
C             T      - TEMPERATURE (INPUT)
C             P      - PRESSURE (INPUT OPTIONAL/OUTPUT)
C             X      - MOLE FRACTIONS IN THE LIQUID PHASE (OUTPUT)
C             Y      - MOLE FRACTIONS IN THE VAPOR PHASE (INPUT)
C             NC     - NUMBER OF COMPONENTS (INPUT)
C             INIT   - 0 INITIAL GUESS FOR P CALCULATED BY PSTART
C                       1 INITIAL GUESS FOR P PASSED FROM CALLING
C                         ROUTINE
C             ITMAX  - MAXIMUM NUMBER OF ITERATIONS PERMITTED
C
C  IER = 1 -- 9  => #OF COMPONENTS ABOVE CRIT TEMP
C      = 10      => LIQUID ROOT PASSED FROM FUGI WAS NOT REAL
C      = 20      => VAPOR ROOT PASSED FROM FUGI WAS NOT REAL
C      = 100     => ERROR RETURNED FROM FUGI CALCULATION
C      = 1000    => CALCULATIONS DETERMINED VAPOR & LIQUID ROOTS EQUAL
C                   (TRIVIAL SOLUTION)
C      = 2000    => FAILED TO CONVERGE IN ITMAX LOOPS
C      = 3000    => CONVERGED TO A NEGATIVE PRESSURE
C
C  REQD. ROUTINES    - PSTART, FUGI(SRKNR)
      IMPLICIT DOUBLE PRECISION(A-H,O-Z)
      DIMENSION X(10),Y(10),PHI(10)
      DIMENSION PHIL(10),PHIV(10),F(10)
      DIMENSION TC(10),PC(10),ACE(10)
      KOUNT = 0
      IER = 0
      DO 1 I=1,NC
1     IF(T .GE. TC(I) )IER = IER + 1
      IER1 = IER
      IF(INIT.EQ.0)THEN
      CALL PSTART(TC,PC,ACE,T,P1,Y,NC)
      P=P1/3
      CALL FUGI(TC,PC,ACE,RGAS,PHIL,T,P,Y,NC,1,ZL,IERF)
      ELSE
      CALL FUGI(TC,PC,ACE,RGAS,PHIL,T,P,X,NC,1,ZL,IERF)
      END IF
C  THE SECANT METHOD IS USED TO ITERATE ON P.
C  FIRST, THE OLD POINT IS INITIALIZED.
      LIQ = 0
      CALL FUGI(TC,PC,ACE,RGAS,PHIV,T,P,Y,NC,LIQ,ZV,IERF)
      IF(IERF .GT. 9) IER = IER1 + 100
      IF(IERF .GT. 9) RETURN
      SUMX=0
      DO 152 I=1,NC
      X(I) = PHIV(I)/PHIL(I)*Y(I)
152   SUMX = SUMX + X(I)
      GOLD = SUMX - 1
      POLD=P
      P=P*1.001D0
C  BEGIN SECANT ITERATION.
57    KOUNT=KOUNT+1
      LIQ = 0
      CALL FUGI(TC,PC,ACE,RGAS,PHIV,T,P,Y,NC,LIQ,ZV,IERF)
      IF(IERF .GT. 9) IER = IER1 + 100
```

```
59    SUMX=0
      DO 52 I=1,NC
      X(I) = PHIV(I)*Y(I)/PHIL(I)
52    SUMX = SUMX + X(I)
      G = SUMX - 1
      CHNG = G*(P - POLD)/(G - GOLD)
C  SLIGHT STEP LIMITING IS NECESSARY
      IF(DABS(CHNG/P).GT.0.40)CHNG=DSIGN(0.40D0,CHNG)*P
      POLD = P
      PNEW = P - CHNG
      GOLD = G
      P = PNEW
      IF(P .LE.0)IER = IER1 + 3000
      IF(P .LE.0)RETURN
      IF(KOUNT .GT. ITMAX) IER = IER1 + 2000
      IF(KOUNT .GT. ITMAX) RETURN
      LIQ = 1
      IF(DABS((POLD-P)/P) .LE. 1.D-5) GO TO 54
55    DO 60 I=1,NC
60    X(I) = X(I)/SUMX
      CALL FUGI(TC,PC,ACE,RGAS,PHIL,T,P,X,NC,LIQ,ZL,IERF)
      IF(IERF .GT.9)IER = IER1 + 100
      GO TO 57
54    IF(DABS(SUMX-1) .LE. 1.D-5)  GO TO 56
      DO 53 I=1,NC
53    X(I) = X(I)/SUMX
56    IF(IERF.EQ.1)IER=IER+10
      IF(IERF.EQ.2)IER=IER+20
      IF(DABS((ZV-ZL)/ZV).LE. 1.D-4) IER = IER1 + 1000
      RETURN
      END

      SUBROUTINE KVALUS(TC,PC,ACE,RGAS,T,P,X,Y,Z,CK,NC,INIT,ITMAX,IER)
C
C
C  PURPOSE               - CALCULATE EQUILIBRIUM K-RATIOS BASED ON AN
C                          EQUATION OF STATE FOR BOTH THE VAPOR AND
C                          LIQUID PHASES.  FOR BINARIES, THEY CAN BE
C                          DETERMINED ANALYTICALLY WITHOUT A FLASH
C                          ITERATION.  FOR MULTICOMPONENT SYSTEMS, A
C                          FLASH CALCULATION MUST BE PERFORMED AND THE
C                          FEED COMPOSITIION MUST BE SPECIFIED.
C
C  ARGUMENTS  TC       - VECTOR CRITICAL TEMPERATURES OF THE COMPONENTS
C                          (INPUT)
C             PC       - VECTOR CRITICAL PRESSURES OF THE COMPONENTS
C                          (INPUT)
C             ACE      - VECTOR ACENTRIC FACTORS OF THE COMPONENTS(INPUT)
C             RGAS     - GAS CONSTANT (EG. 83.1434 CC-BAR/(GMOL-K) )
C             T        - TEMPERATURE (INPUT)
C             P        - PRESSURE (INPUT)
C             X        - MOLE FRACTIONS IN THE LIQUID PHASE (OUTPUT)
C             Y        - MOLE FRACTIONS IN THE VAPOR PHASE (OUTPUT)
C             Z        - MOLE FRACTIONS OF THE FEED.  NECESSARY FOR
C                          CALCULATIONS WITH MULTICOMPONENT SYSTEMS
C                          BUT NOT FOR BINARIES. (INPUT OPTIONAL)
C             CK       - CALCULATED K-RATIOS. INPUT WHEN INIT=1 FOR
C                          INITIAL GUESSES (INPUT(OPTIONAL)/OUTPUT)
```

```
C            NC      - NUMBER OF COMPONENTS (INPUT)
C            INIT    - 0 INITIAL GUESS FOR CK CALCULATED BY IDEAL GAS
C                        APPROXIMATION
C                      1 INITIAL GUESS FOR CK PASSED FROM CALLING
C                        ROUTINE
C            ITMAX   - MAXIMUM NUMBER OF ITERATIONS PERMITTED
C
C           IER      - 100  ERROR RETURNED FROM FUGI CALCULATION
C                    - 1000 CALCULATIONS DETERMINED VAPOR & LIQUID
C                             ROOTS EQUAL (TRIVIAL SOLUTION)
C                    - 2000 FAILED TO CONVERGE IN ITMAX LOOPS
C                    - 3000 FOR BINARIES INVALID VALUES OF X'S WERE
C                             CALCULATED.  FOR MULTICOMPONENT SYSTEMS,
C                             RACRIC FAILED.
C
C  SUBPROGRAMS CALLED:  RACRIC,FUGI(SRKNR)
C
      IMPLICIT DOUBLE PRECISION(A-H,O-Z)
      DIMENSION PHIL(10),PHIV(10),CK1(10)
      DIMENSION X(10),Y(10),Z(10),CK(10)
      DIMENSION TC(10),PC(10),ACE(10)
      KOUNT=0
      VF=.5D0
      IF(INIT.EQ.1)GO TO 1000
C  COMPUTE INITIAL GUESSES FROM IDEAL GAS LAW AND GENERALIZED
C  VAPOR PRESSURE EQUATIONS
      DO 2 I=1,NC
      ZC=.29-.065*ACE(I)
      TR = T/TC((I))
      IF(T .GT. TC((I)) ) GO TO 12
10    PSAT = DEXP(  (4.92*ACE((I))+5.81)*DLOG(TR)
     &        - 0.0838D0*(4.92*ACE((I)) + 2.06)
     &        * (36.D0/TR-35-TR**6+42*DLOG(TR))    ) * PC((I))
      GO TO 2
12    RLPSAT  =  (16.26D0-73.85D0*ZC+90*ZC**2)
     &        * (1-1/TR)
     &        - 10**(-8.68*( TR-1.8+6.2*ZC )**2)
      IF(RLPSAT .LT. -100.)RLPSAT = -100
      PSAT = DEXP(RLPSAT)*PC((I))
      CK(I)=PSAT/P
2     CONTINUE
C  BEGIN ITERATION
1000  KOUNT=KOUNT+1
      IF(KOUNT.GT.ITMAX)GO TO 861
      IER=0
      IF(NC.EQ.2)THEN
C   IF BINARY, SOLVE ANALYTICALLY FROM MOLE FRACTION CONSTRAINTS
      IF(DABS(CK(1)-CK(2)).LT.1.D-12)GO TO 862
      X(1) = (1-CK(2))/(CK(1)-CK(2))
      X(2) = 1-X(1)
      Y(1) = X(1)*CK(1)
      Y(2) = X(2)*CK(2)
      IF(X(1).LT.0 .OR. X(1).GT.1)GO TO 862
      ELSE
C   IF MULTICOMPONENT, SOLVE FOR THE VAPOR TO FEED RATIO AND THEN
C   X AND Y FROM Z.
      CALL RACRIC(X,Y,Z,CK,VF,NC,IERR)
      IF(IERR.GT.99)GO TO 862
```

```
      DO 51 I=1,NC
      X(I) = Z(I)/(1 + VF*(CK(I) - 1))
51    Y(I) = CK(I)*X(I)
      END IF
C  CALCULATE NEW K-RATIOIS BASED ON THE NEW X AND Y USING THE
C  EQUATION OF STATE
      SUMX=0
      SUMY=0
      DO 71 I=1,NC
      SUMX = SUMX + X(I)
71    SUMY = SUMY + Y(I)
      DO 59 I=1,NC
      X(I) = X(I)/SUMX
59    Y(I) = Y(I)/SUMY
      LIQ = 0
      CALL FUGI(TC,PC,ACE,RGAS,PHIV,T,P,Y,NC,LIQ,ZV,IERFU)
      IF(IERFU .GT. 9) IER = IER + 100
      LIQ = 1
      CALL FUGI(TC,PC,ACE,RGAS,PHIL,T,P,X,NC,LIQ,ZL,IERFU)
      IF(IERFU .GT. 9) IER = IER + 100
      DO 56 I=1,NC
      CK1(I) = PHIL(I)/PHIV(I)
56    CONTINUE
C  DETERMINE WHETHER THE K-RATIOS HAVE CONVERGED
      DO 55 I=1,NC
      RERROR=(CK1(I)-CK(I))/CK1(I)
      CK(I)=CK1(I)
55    IF(DABS(RERROR).GT.1.D-4) GO TO 1000
      IF(DABS((ZL - ZV)/ZV) .LT. 1.D-4 )IER = IER +1000
      RETURN
861   IER=IER+2000
      RETURN
862   IER=IER+3000
CCC   WRITE(6,601)CK(1),CK(2)
601   FORMAT(1X,2(E11.4,1X))
      RETURN
      END

      SUBROUTINE RACRIC(X,Y,Z,CK,VF,NC,IER)
C
C  PURPOSE          - CALCULATE THE VAPOR TO FEED RATIO (VF)
C                     BY SECANT ITERATION ON THE RACHFORD AND
C                     RICE (1952) OBJECTIVE FUNCTION.
C
C                     IER - 100 VF CALCULATED TO BE GT 1
C                           200 VF CALCULATED TO BE LT 0
C                           1000 DIDNT CONVERGE IN 25 LOOPS
C
      IMPLICIT DOUBLE PRECISION(A-H,O-Z)
      DIMENSION X(10),Y(10),Z(10),CK(10)
      IER=0
      FOLD=0
      DO 10 I=1,NC
      FOLD=FOLD+(CK(I)-1)*Z(I)/( (CK(I)-1)*VF+1 )
10    CONTINUE
      VFOLD=VF
      VF=VF*1.001D0
      KOUNT=0
```

```
100   KOUNT=KOUNT+1
      IF(KOUNT.GT.45)GO TO 861
      F=0
      DO 20 I=1,NC
      F=F+(CK(I)-1)*Z(I)/( (CK(I)-1)*VF+1 )
20    CONTINUE
      CHANGE=F*(VF-VFOLD)/(F-FOLD)
      IF(DABS(CHANGE/VF).GT.0.4)CHANGE=DSIGN(0.4D0,CHANGE)*VF
      VFOLD=VF
      FOLD=F
      VF=VF-CHANGE
      IF(DABS(CHANGE/VF).GT.1.D-5)GO TO 100
      IF(VF.GT.1)IER=100
      IF(VF.LT.0)IER=200
      RETURN
861   IER=1000
      RETURN
      END

      SUBROUTINE FUGI(TC,PC,ACE,RGAS,PHI,T,P,X,NC,LIQ,Z,IER)
C
C  PURPOSE -   CALCULATE THE FUGACITY COEFFICIENTS
C              OF EITHER A GAS OR LIQUID ACCORDING TO THE
C              SOAVE EQN OF STATE.  THE ALGORITHM IS BASED ON
C              T. GUNDERSEN "NUMERICAL ASPECTS OF THE IMPLEMENTATION
C              OF CUBIC EQUATIONS OF STATE IN FLASH CALCULATION ROUTINES"
C              COMPUTERS IN CHEMICAL ENGINEERING,6(3), 245,(1982).
C              BRIEFLY, THE ALGORITHM CALLS FOR THE CALCULATION
C              OF THE EXTREMUM(Z**3-Z**2+Q*Z-R) WHEN EITHER A LIQUID
C              OR VAPOR ROOT IS DETERMINED NOT TO BE REAL, EVEN THOUGH
C              THE CALCULATION CALLS FOR A ROOT.  IF SUCH ROOTS
C              ARE FOUND, IER SIGNIFIES THIS.
C
C
C  ARGUMENTS   TC      - VECTOR CRITICAL TEMPERATURES OF THE COMPONENTS
C                        (INPUT)
C              PC      - VECTOR CRITICAL PRESSURES OF THE COMPONENTS
C                        (INPUT)
C              ACE     - VECTOR ACENTRIC FACTORS OF THE COMPONENTS(INPUT)
C              RGAS    - GAS CONSTANT (EG. 83.1434 CC-BAR/(GMOL-K) )
C              PHI     - FUGACITY COEFFICIENT OF COMPONENTS
C                        IN PHASE LIQ (OUTPUT)
C              T       - TEMPERATURE (INPUT)
C              P       - ABSOLUTE PRESSURE (INPUT)
C              X       - VECTOR MOLE FRACTIONS OF COMPONENTS IN
C                        PHASE LIQ (INPUT)
C              NC      - NUMBER OF COMPONENTS (INPUT)
C              LIQ     - 0 FOR VAPOR PHASE  (INPUT)
C                        1 FOR LIQUID PHASE
C              Z       - COMPRESSIBILITY FACTOR OF MIXTURE (OUTPUT)
C              IER     - 1 IF LIQUID ROOT WAS FOUND BUT WAS NOT REAL
C                        2 IF VAPOR  ROOT WAS FOUND BUT WAS NOT REAL
C                       10 NEGATIVE LOG CALCULATED
C                      100 RLNPHI > 174
C                     1000 SRKNR DID NOT CONVERGE
C
       IMPLICIT DOUBLE PRECISION(A-H,O-Z)
```

```
      DIMENSION X(10),PHI(10),ALA(10,10),ACTION(10,10),A(10),B(10)
      DIMENSION TC(10),PC(10),ACE(10)
      DATA KALL/0/
      IF(KALL.EQ.1)GO TO 100
      KALL=1
      THIRD=1.D0/3
      OMA = 1.D0/(9*(2**THIRD-1))
      OMB = (2**THIRD-1)/3
C  COMPUTE THE MOLECULAR PARAMETERS A AND B AND THEIR CROSS COEFFS
100   DO 5 I = 1,NC
      A(I) = OMA*RGAS*RGAS*TC(I)**2/PC(I)
      S = 0.48508 + 1.55171*ACE(I) - 0.15613*ACE(I)**2
      ALPHA = (1 + S*(1 - DSQRT(T/TC(I))))**2
      ALA(I,I) = A(I)*ALPHA
      B(I) = OMB*RGAS*TC(I)/PC(I)
5     CONTINUE
      DO 8 I = 1,NC
      DO 9 J = 1,NC
      ACTION(I,J)=ESTACT(I,J,T)
      ALA(I,J) = (ALA(I,I)*ALA(J,J))**0.5D0*(1-ACTION(I,J))
9     CONTINUE
8     CONTINUE
      AM = 0
      BM = 0
      DO 10 I = 1,NC
      DO 20 J = 1,NC
      AM = AM + ALA(I,J)*X(I)*X(J)
20    CONTINUE
      BM = BM + B(I)*X(I)
10     CONTINUE
C  BEGIN GUNDERSEN ALGORITHM TO DETERMINE Z
      BIGA = AM*P/(RGAS*RGAS*T*T)
      BIGB = BM*P/(RGAS*T)
C
      IER = 0
      Q = BIGA - BIGB - BIGB*BIGB
      RR = BIGA*BIGB
      F3 = 1.D0/27 - 1.D0/9 + Q/3 - RR
      IF(Q .GT. 1.D0/3 .AND. F3 .GT. 0 ) GO TO 1200
      IF(Q .GT. 1.D0/3 .AND. F3 .LE. 0 ) GO TO 1300
      IF(RR .GT. 1.D0/27 .AND. LIQ .EQ. 0) GO TO 1300
      Z1 = (1 - DSQRT(1-3*Q))/3
      Z2 = (1 + DSQRT(1-3*Q))/3
      F1 = Z1**3 - Z1*Z1 + Q*Z1 - RR
      F2 = Z2**3 - Z2*Z2 + Q*Z2 - RR
      IF(LIQ .EQ.1 .AND. F1 .GT. 0) GO TO 1200
      IF(LIQ .EQ.1 .AND. F1 .LE. 0) GO TO 861
      IF(F2 .LT. 0) GO TO 1300
      GO TO 862
1300  Z = 3
      CALL  SRKNR(Z,Q ,RR,IERF)
      IF(IERF .GT. 9)IER=1000
      GO TO 9000
1200  Z = 0
      CALL  SRKNR(Z,Q ,RR,IERF)
      IF(IERF .GT. 9)IER=1000
      GO TO 9000
```

```
861   IER = 1
      Z = Z1
      BIGB = BIGB*(1 + F1/RR)
      GO TO 9000
862   IER = 2
      Z = Z2
      BIGB = BIGB*(1 + F2/RR)
      GO TO 9000
9000  CONTINUE
C  CALCULATE FUGACITY COEFFICIENTS OF INDIVIDUAL COMPONENTS
      DO 50 ICOMP = 1,NC
      SUMXA = 0
      IER1 = IER
      DO 55 I=1,NC
55    SUMXA = SUMXA + X(I)*ALA(ICOMP,I)
      IF( (Z-BIGB).GT. 0 .AND. (1+BIGB/Z).GT.0) GO TO 51
      WRITE(6,351)T,P
351   FORMAT(1X,'WARNING - NEGATIVE LOG CALCD IN FUGI. T,P',2(E11.4,1X)/
     &        1X,'PHI SET TO 1,IER ON RETURN WILL BE 10')
      RLNPHI = 0
      IER = IER1 + 10
      GO TO 50
51    RLNPHI = B(ICOMP)/BM*(Z-1) - DLOG(Z-BIGB)
     &       - BIGA/BIGB*(2*SUMXA/AM - B(ICOMP)/BM)*DLOG(1+BIGB/Z)
      IF(RLNPHI.LT.-174)RLNPHI=-170
      IF(RLNPHI .LT. 174) GO TO 50
      WRITE(6,353)T,P,Z,IER
353   FORMAT(1X,'LOG OF PHI.GT.174 IN FUGI. T,P,Z,IER',3(F8.3,1X),I4/
     &        1X,'PHI SET TO 1,IER ON RETURN WILL BE 100')
      IER = IER1 + 100
      RLNPHI = 0
50    PHI(ICOMP) = DEXP(RLNPHI)
      RETURN
      END


      SUBROUTINE SRKNR(Z,A1,A0,IER)
C
C  PURPOSE    - CALCULATE Z FROM NEWTON-RAPHSON ITERATION
C
C      IER    - 200 DIDNT CONVERGE IN 25 ITERATIONS
C
      IMPLICIT DOUBLE PRECISION(A-H,O-Z)
      IER = 0
      KOUNT = 0
100   CONTINUE
      KOUNT = KOUNT + 1
      F = Z**3 - Z**2 + A1*Z - A0
      DF = 3*Z*Z - 2*Z + A1
      ZN = Z - F/DF
      ERR = DABS((ZN - Z)/ZN)
      Z = ZN
      IF(KOUNT .GT. 25)IER = 200
      IF(KOUNT .GT. 25)RETURN
      IF(ERR.GT. 1.D-8 ) GO TO 100
      RETURN
      END
```

```
      FUNCTION ESTACT(I,J,TEMP)
      IMPLICIT DOUBLE PRECISION(A-H,O-Z)
C
C  THIS ROUTINE CALCULATES THE BINARY INTERACTION PARAMETER.
C  A DUMMY VALUE OF ZERO IS PROVIDED HERE.  THE USER
C  CAN MODIFY THIS ROUTINE TO FIT HIS OWN PURPOSES.
C
      ESTACT = 0
      RETURN
      END


      SUBROUTINE PSTART(TC,PC,ACE,T,P,X,NC)
CCCCCCCCCCCCCCCCCCCCCCCCCCCCCCCCCCCCCCCCCCCCCCCCCCC
C
C  PURPOSE      - CALCULATE AN INITIAL GUESS FOR ROUTINES BUBLP
C                 OR DEWP FROM A GENERALIZED VAPOR PRESSURE EQUATION.
C
CCCCCCCCCCCCCCCCCCCCCCCCCCCCCCCCCCCCCCCCCCCCCCCCCCC
      IMPLICIT DOUBLE PRECISION(A-H,O-Z)
      DIMENSION PSAT(10),X(10),ZC(10)
      DIMENSION TC(10),PC(10),ACE(10)
      DO 2 I=1,NC
      ZC(I)=.29-.065*ACE(I)
      TR = T/TC((I))
      IF(T .GT. TC((I)) ) GO TO 12
10    PSAT(I) = DEXP(  (4.92*ACE((I))+5.81)*DLOG(TR)
     &          - 0.0838DO*(4.92*ACE((I)) + 2.06)
     &          * (36.DO/TR-35-TR**6+42*DLOG(TR))   ) * PC((I))
      GO TO 3
12    RLPSAT  =  (16.26DO-73.85DO*ZC((I))+90*ZC((I))**2)
     &          * (1-1/TR)
     &          - 10**(-8.68*( TR-1.8+6.2*ZC((I)) )**2)
      IF(RLPSAT .LT. -100.)RLPSAT = -100
      PSAT(I) = DEXP(RLPSAT)*PC((I))
3     CONTINUE
2     CONTINUE
      P = 0
      DO 5 I=1,NC
5     P = P + X(I)*PSAT(I)
      RETURN
      END
```

NAME INDEX

SUBJECT INDEX